MEMBRANE MOLECULAR BIOLOGY

Membrane Molecular Biology

Edited by C. Fred Fox
UNIVERSITY OF CALIFORNIA, LOS ANGELES

and A. D. Keith
PENNSYLVANIA STATE UNIVERSITY

SINAUER ASSOCIATES INC.
PUBLISHERS • Stamford, Conn.

MEMBRANE MOLECULAR BIOLOGY

© 1972 by SINAUER ASSOCIATES, INC.
20 Second Street, Stamford, Conn. 06905

Printed in the United States of America

Library of Congress Catalog Card No. 73-181-989

ISBN: 0-87893–190-2

Contents

v

Chapter Outlines

I. MEMBRANE COMPOSITION AND ISOLATION

·

II. PHYSICAL PROPERTIES OF MEMBRANES
·

Preface

This book was conceived with the idea of providing a comprehensive introduction to molecular studies of membranes and is directed toward students, teachers, and researchers in the biological, physical, and medical sciences. It emphasizes correlations between the genetic, functional, chemical, and physical approaches to the study of membranes and their component parts. We have attempted to present a didactic volume rather than a series of exhaustively referenced review articles. Though the individual chapters do not, as a result, provide extensive literature reviews, the references to the key techniques and principles are cited for those interested in a point of departure for research in membranology.

Many of the chapters have been used in membrane courses taught by the authors at the University of California Los Angeles, the University of California Berkeley, and the University of Chicago. The order of presentation is that which in our experience has proved most adaptable to a teaching approach. Section I discusses the chemistry and isolation of membranes and membrane components, providing a background for the presentation of physical studies in Section II. Section III integrates the information from chemical and physical studies into a molecular biological approach to the study of membrane function and assembly. This final section also treats a number of biological topics of medical importance, such as the assembly of the membranes of animal cell viruses and the properties of membranes of animal cells with tumor-inducing capacity.

We are especially grateful to our students and colleagues, who have provided us with valuable criticisms, and to Miss Sylvia Stevens for translating our hen-scratchings into technical drawings.

C. Fred Fox
A. D. Keith

xv

Contributors

P. D. BOYER, Department of Chemistry and the Molecular Biology Institute, University of California, Los Angeles

C. FRED FOX, Department of Bacteriology and the Molecular Biology Institute, University of California, Los Angeles

GODFREY S. GETZ, Departments of Biochemistry and Pathology, University of Chicago

HARVEY R. HERSCHMAN, Department of Biochemistry and the Molecular Biology Institute, University of California, Los Angeles

GEORGE HOLZWARTH, Department of Biophysics, University of Chicago

ALAN F. HORWITZ, Laboratory of Chemical Biodynamics, University of California, Berkeley

A. D. KEITH, Department of Biophysics, Pennsylvania State University

FERENC J. KÉZDY, Department of Biochemistry, University of Chicago

W. L. KLEIN, Department of Chemistry and the Molecular Biology Institute, University of California, Los Angeles

JOHN H. LAW, Department of Biochemistry, University of Chicago

R. J. MEHLHORN, Department of Genetics, University of California, Berkeley

ROBERT D. SIMONI, Department of Biological Sciences, Stanford University

WILLIAM R. SNYDER, Department of Biochemistry, University of Chicago

THEODORE L. STECK, Department of Medicine, University of Chicago

GABOR SZABO, Department of Physiology, University of California, Los Angeles

MARK E. TOURTELLOTTE, Department of Pathobiology, University of Connecticut

I

MEMBRANE COMPOSITION AND ISOLATION

John H. Law
William R. Snyder

Membrane Lipids

Isolated membranes are composed of about 40% lipids and 60% proteins. Any study of membranes must therefore be concerned with the lipid components. This chapter considers the chemistry and biochemistry of those lipids commonly located in membrane structures. Not all lipids are associated with membranes. Triglycerides are usually reserve food materials and are sequestered in fat cells of adipose tissue. Waxes, hydrocarbons, and fatty alcohols are frequently located on the exterior surface of plants and animals, where they serve a protective function. In certain water beetles, physiologically active steroids are released from special glands for use in defense, and many insects use fatty acid derivatives for forms of chemical communication.

In true membrane structures one usually finds *amphipathic* lipids, i.e., compounds with both a polar and a nonpolar region. Such structures are ideally suited for the formation of interfaces between a polar solvent and a hydrophobic region. Common lipid constituents of membranes are phospholipids and glycolipids in prokaryotic cells, and phospholipids, sphingolipids, glycolipids, and sterols in eukaryotic cells.

Table I

Substituents of Glycerolipids

STRUCTURE	NAME OF SUBSTITUENT GROUP	TYPE OF INTACT MOLECULE
$R_1; R_2$: $CH_3(CH_2)_n\overset{O}{\overset{\|}{C}}-$	Fatty ester; often saturated in R_1, unsaturated in R_2; n usually = 10–20	If R_3 = H, a diglyceride
$CH_3(CH_2)_nCH_2-$	Fatty ether, n usually > 10; usually at R_1	Glycerol ether
$CH_3(CH_2)_nCH=CH-$	Vinyl ether; n usually > 9; usually at R_1	Plasmalogen
R_3: $CH_3(CH_2)_n\overset{O}{\overset{\|}{C}}-$	Fatty ester	If R_1 and R_2 are esters, triglyceride
[glucose structure, CH₂OH, OH, HO, OH]	Glucoside or other mono- or polysaccharide	A glycosyl glyceride (a glycolipid)
$\overset{\ominus}{O}-\overset{O}{\overset{\|}{P}}-$ $\overset{\|}{OH}$	Phosphate ester	Phosphatidic acid
$\overset{\oplus}{H_3N}CH_2CH_2O\overset{O}{\overset{\|}{\underset{O\ominus}{P}}}-$	Ethanolamine phosphate ester	Phosphatidylethanolamine (cephalin)
$\overset{\ominus}{O_2}CCHCH_2O\overset{O}{\overset{\|}{\underset{O\ominus}{P}}}-$ $\underset{\oplus}{NH_3}$	Serine phosphate ester	Phosphatidylserine (cephalin)
$(CH_3)_3\overset{\oplus}{N}CH_2CH_2O\overset{O}{\overset{\|}{\underset{O\ominus}{P}}}-$	Choline phosphate ester	Phosphatidylcholine (lecithin)
[inositol structure, OH OH, HO, OH] $-O\overset{O}{\overset{\|}{\underset{O\ominus}{P}}}-$	Inositol phosphate ester	Phosphatidylinositol
$HOCH_2CHCH_2O\overset{O}{\overset{\|}{\underset{O\ominus}{P}}}-$ $\underset{OH}{}$	Glycerol phosphate ester	Phosphatidylglycerol
$\overset{OH}{}$ $OCH_2CHCH_2O\overset{O}{\overset{\|}{\underset{O\ominus}{P}}}-$ $PO_3^{\ominus}$ $OCH_2-CH-CH_2OR_1$ OR_2	Phosphatidylglycerol phosphate ester	Diphosphatidylglycerol (cardiolipin)
$\overset{\oplus}{H_3N}CH_2CH_2\overset{O}{\overset{\|}{\underset{O\ominus}{P}}}-$	Aminoethylphosphonate ester	A phosphonolipid

LIPID STRUCTURES

It is first necessary to review the structural features of common lipids.

Glycerolipids (Including Phospholipids and Glycolipids)

$$CH_2OR_1$$
$$R_2O \blacktriangleright C \blacktriangleleft H$$
$$CH_2OR_3$$

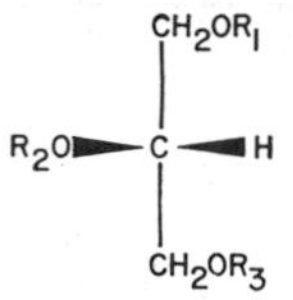

Glycerol has three hydroxyl groups which can be substituted to form a variety of compounds. It is conveniently pictured as shown in the accompanying illustration, for if the R_1 and R_3 groups are different, the natural glycerides have the stereochemistry shown. The usual substituents are shown in Table I. It is most important to realize that any natural phospholipid class, e.g., a phosphatidylethanolamine, consists of a complicated mixture of individual molecules differing in the complement of fatty acids at both the 1 and the 2 position. Only in relatively recent times has it been possible to separate out the individual components from such a mixture, for these often have very similar chemical and physical properties.

Sphingolipids (Including Certain Phospholipids and Glycolipids)

In a like manner, the structure of sphingolipids, which have as a structural backbone a long-chain amino alcohol rather than glycerol, can be generalized as shown in the accompanying illustration and in Table II.

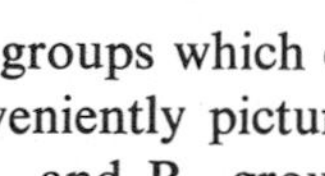

$$R_1 - CH - CH_2OR_3$$
$$NH$$
$$R_2$$

Sterols

A wide variety of sterol molecules occur in nature. The molecules have the same general ring structure, but differ in degree of unsaturation and substitution. Table III shows only the most abundant type of structure characteristic of a given class of organisms.

Minor Lipids

Several minor lipid components are often present in membranes either because they perform a vital functional role (e.g., vitamin A aldehyde in the retina) or possibly because they accumulate in the membrane as a result of their lipophilic character. Some examples of such compounds as an illustration of the varied structural features of lipids are given in the following list and in Figure 1.

Table II

Substituents of Sphingolipids

STRUCTURE	NAME OF SUBSTITUENT GROUP	TYPE OF INTACT MOLECULE
R_1: $CH_3(CH_2)_nCH-$ with OH		If n = 14, $R_2 = R_3 = H$ dihydrosphingosine
$CH_3(CH_2)_nCH=CH-CH$ with OH		If n = 12, $R_2 = R_3 = H$ sphingosine
$CH_3(CH_2)_nCH-CH$ with OH OH		If n = 13, $R_2 = R_3 = H$, phytosphingosine
R_2: $CH_3(CH_2)_nC-$ (C=O)	Fatty amide n = 14–24	
$CH_3(CH_2)_nCHC-$ (C=O) with OH	α-Hydroxy fatty amide	
R_3: $(CH_3)_3\overset{\oplus}{N}CH_2CH_2OP-$ (P=O, $O^{\ominus}$)	Choline phosphate ester	Sphingomyelin
$H_3\overset{\oplus}{N}CH_2CH_2OP-$ (P=O, $O^{\ominus}$)	Ethanolamine phosphate ester	
Oligosaccharide $-OP$ (P=O, $O^{\ominus}$)	Oligosaccharide phosphate ester	Phytoglycolipids
Glucose or galactose	Glucoside or galactoside	Cerebrosides
Oligosaccharide	Oligosaccharide	Hematosides, globosides, blood group substances, gangliosides
H		If R_2 is a fatty amide, a ceramide

1. Nonsteroid terpenoid compounds, including carotenoids (β-carotene), quinones (vitamin K_1), and prenols (polyisoprenoid alcohol)
2. Hydrocarbons, including the terpenoid squalene and odd-chain hydrocarbons
3. Wax esters
4. Sugar esters

Table III
Characteristics of Sterols

STRUCTURE	NAME	CHARACTERISTIC CELL
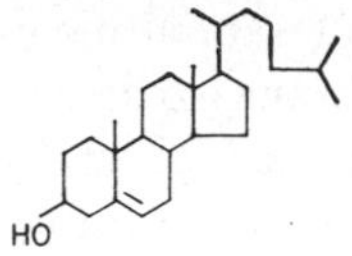	Cholesterol	Animal
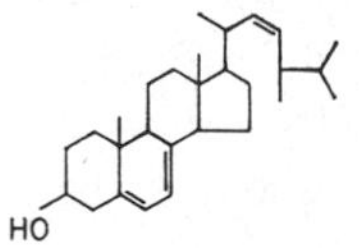	Ergosterol	Fungi
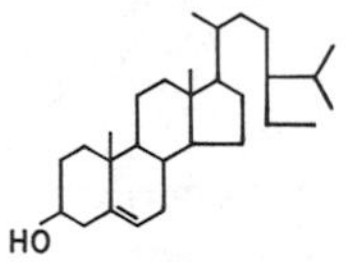	β-Sitosterol	Plants

5. Esters of polyols other than glycerol, such as ethylene glycol, butylene glycol lipids, and chlorinated derivatives (*Ochromonas danica*)
6. Lipopolysaccharides
7. Methyl esters of fatty acids

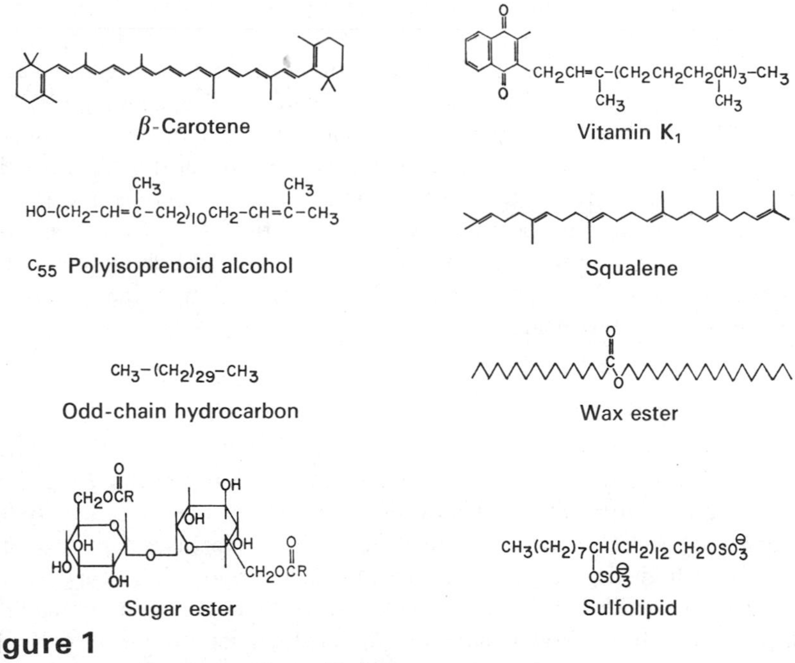
β-Carotene
Vitamin K₁
CH₂CH=C–(CH₂CH₂CH₂CH)₃–CH₃
HO–(CH₂–CH=C–CH₂)₁₀CH₂–CH=C–CH₃
C₅₅ Polyisoprenoid alcohol
Squalene
CH₃–(CH₂)₂₉–CH₃
Odd-chain hydrocarbon
Wax ester
Sugar ester
CH₃(CH₂)₇CH(CH₂)₁₂CH₂OSO₃⁻
Sulfolipid

Figure 1
Structures of some minor lipids.

LIPID DISTRIBUTION

Table IV summarizes what we know of the distribution of different lipid categories in nature. Again it should be noted that the occurrence of a given lipid in an organism is no assurance that that material is an essential membrane component or, indeed, that it is located in a membrane structure. Rather, the table is intended to point out which components are generally widespread in occurrence and which are limited in distribution. For example, phospholipids are ubiquitous, but sterols are not, to our knowledge, synthesized by prokaryotic cells. It is possible that they may yet be found in such cells, for until recent times it was believed bacterial cells also lacked sphingolipids, but now one group that contains these compounds has been identified.

LIPID BIOSYNTHESIS AND DEGRADATION

This section considers the sources of the building blocks and energy for lipid synthesis and the processes involved in assembly of the common membrane lipid complement. Many organisms can use preformed components; indeed, many have absolute requirements for certain lipid materials. For example, the lipid vitamins A, E, and K are essential to the growth of many animals. Mammals require polyunsaturated fatty acids, for they all lack the enzymatic machinery for converting oleic acid to linoleic acid. Certain animals, e.g., most insects, require dietary sterols for growth and reproduction. Common items of the diet of most animals are fatty acids or their esters, glycerol (in the form of triglycerides and phospholipids), and sterols. If these are lacking, most organisms are able to synthesize them from glucose. Pyruvate is converted to acetyl coenzyme A (acetyl CoA), which in turn can be converted either to fatty acids or to terpenoids and sterols. Glycerol phosphate can be produced by reduction of dihydroxyacetone phosphate and utilized for the synthesis of glycerolipids. Sphingolipid bases are available from the condensation of fatty acids with serine. Details of these procedures are considered below.

Fatty Acids

To a considerable degree our knowledge of fatty acid synthesis (1,28) has been obtained from studies with enzymes of the bacterium *Escherichia coli*. Unlike the fatty acid synthetase of eukaryotic cells, the *E. coli* system is a group of loosely associated enzymes which have been separated and purified to a high degree. In the bacterial cell, however, these enzymes are most likely associated with each other and with the cell membrane. For example, it has been shown that even the acyl carrier protein (ACP), one of the smallest macromolecular components of the *E. coli* synthetase, is

Table IV

Distribution of Various Lipid Categories in Nature

LIPID	EUBACTERIA	MYCOBACTERIA	FUNGI	PLANTS	PROTOZOA	INVERTEBRATES	VERTEBRATES
Saturated fatty acids	+[a]	+	+	+	+	+	+
Monounsaturated fatty acids	+[a]	+	+	+	+	+	+
Polyunsaturated fatty acids	−	+, rare[b]	+	+[e]	+[f]	+	+[c]
Cyclopropane fatty acids	+	+[d]	−	Rare	Rare	−	−
Branched-chain fatty acids	+	+[g]	Rare	Rare	Rare	Rare	Rare[h]
Phospholipids	+	+	+	+	+	+	+
Phosphonolipids	−	−	−	−	+	+	?[i]
Triglycerides	Traces	Rare	+	+	+[j]	+	+
Sterols	−	−	+	+	+	+	+
Sphingolipids	Rare[k]	−	+	+	+[l]	+[l]	+
Aliphatic hydrocarbons	−	?	?	+	?	+	+
Wax esters	?	?	?	?	?	+	+
Saturated ethers	Rare[m]	?	?	?	?	+	+
Plasmalogens	Rare[n]	?	?	?	?	+	+
Glycolipids	+	+	+	+	+	+	+
Sulfolipids	?	?	?	+	+	?	+
Quinones	+	+	?	+	+	+	+

[a] A few species lack unsaturated acids, which are replaced by branched-chain cyclopropane acids (see ref 1, p. 55).
[b] Some mycobacteria produce unconjugated polyene acids of a unique type.
[c] Vertebrates cannot convert oleic acid to linoleic acid and thus have a dietary requirement for polyunsaturated acids (see ref. 1, p. 1).
[d] Mycobacteria produce extremely long chain mono- or polycyclopropane acids (see ref. 2, p. 1).
[e] Malvaceous plants (e.g., cotton, okra, hibiscus) produce cyclopropene (double bond in ring) acids as well as saturated cyclopropane acids (see ref. 2, p. 1).
[f] Thus far found only in *Crithidia* (see ref. 2, p. 1).
[g] The major fatty acid of mycobacterial phospholipids is 10-methyl stearic acid (tuberculostearic acid).
[h] Ruminants have high contents derived from the bacterial flora of the rumen.
[i] A recent report indicates considerable amounts in human brain (see ref. 3).
[j] *Tetrahymena* lacks sterols, but contains the triterpene, tetrahymenol (see ref. 4).
[k] Limited so far to anaerobic *Bacteroides* (see ref. 5).
[l] Shorter chain bases are common in invertebrates (see ref. 6).
[m] Terpenoid ethers are found in strongly halophilic organisms (see ref. 7).
[n] Limited to certain anaerobes (see ref. 8).

localized in the cell envelope. The individual steps which have been demonstrated in the bacterial synthetase are:

1. Acetyl CoA carboxylase

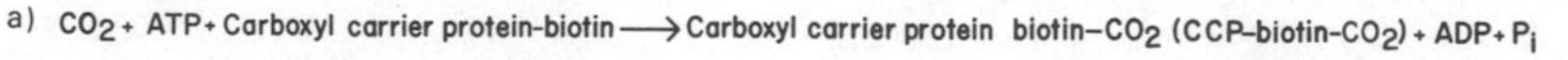

a) $CO_2 + ATP +$ Carboxyl carrier protein-biotin $\longrightarrow$ Carboxyl carrier protein biotin–CO_2 (CCP–biotin–CO_2) + ADP + P_i

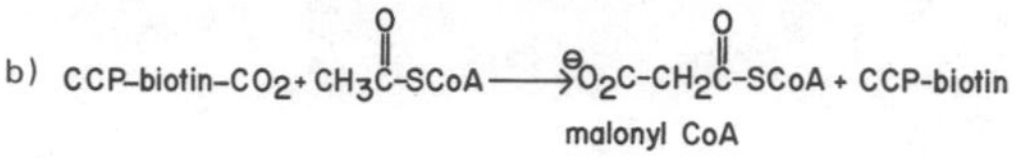

b) CCP–biotin–$CO_2 + CH_3\overset{O}{\underset{\|}{C}}$-SCoA $\longrightarrow \overset{\ominus}{O}_2C$-$CH_2\overset{O}{\underset{\|}{C}}$-SCoA + CCP–biotin

   malonyl CoA

2. Transacylases

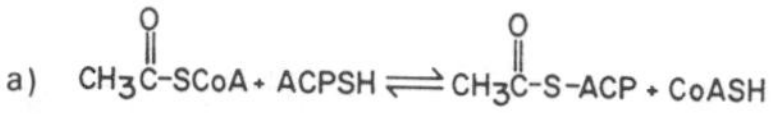

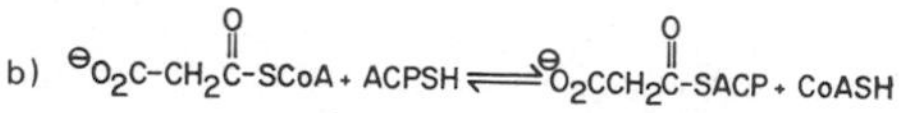

a) $CH_3\overset{O}{\underset{\|}{C}}$-SCoA + ACPSH $\rightleftharpoons CH_3\overset{O}{\underset{\|}{C}}$-S–ACP + CoASH

b) $\overset{\ominus}{O}_2C$-$CH_2\overset{O}{\underset{\|}{C}}$-SCoA + ACPSH $\rightleftharpoons \overset{\ominus}{O}_2CCH_2\overset{O}{\underset{\|}{C}}$-SACP + CoASH

3. Condensing enzymes

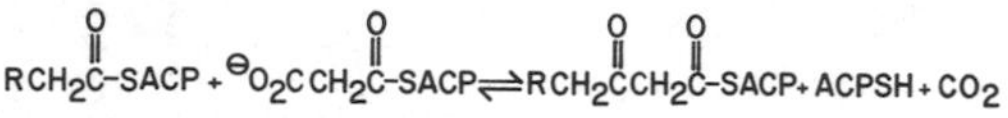

$RCH_2\overset{O}{\underset{\|}{C}}$-SACP + $\overset{\ominus}{O}_2C CH_2\overset{O}{\underset{\|}{C}}$-SACP$\rightleftharpoons RCH_2\overset{O}{\underset{\|}{C}}CH_2\overset{O}{\underset{\|}{C}}$-SACP + ACPSH + CO_2

4. β-Ketoacyl-ACP reductase

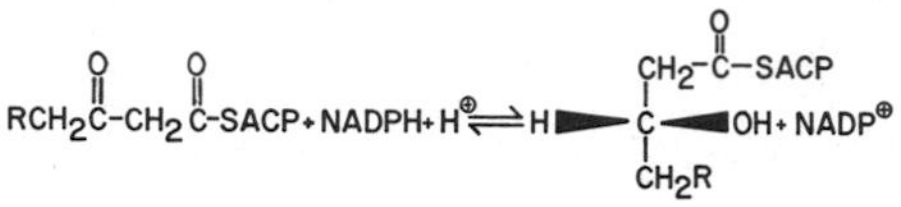

$RCH_2\overset{O}{\underset{\|}{C}}$-$CH_2\overset{O}{\underset{\|}{C}}$-SACP + NADPH + H$^\oplus \rightleftharpoons$ H $\blacktriangleright$ $\underset{CH_2R}{\overset{CH_2\overset{O}{\underset{\|}{C}}-SACP}{C}}\blacktriangleleft$ OH + NADP$^\oplus$

5. Enoyl ACP dehydratase

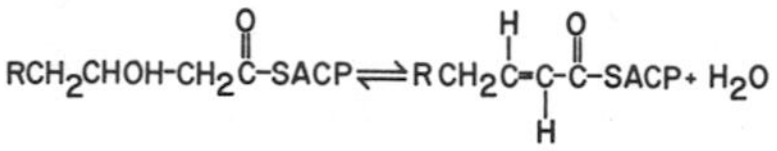

RCH_2CHOH-$CH_2\overset{O}{\underset{\|}{C}}$-SACP$\rightleftharpoons R CH_2\overset{O}{\underset{\|}{C}}$=$\overset{H}{\underset{H}{C}}$-$\overset{O}{\underset{\|}{C}}$-SACP + H_2O

6. Enoyl ACP reductase

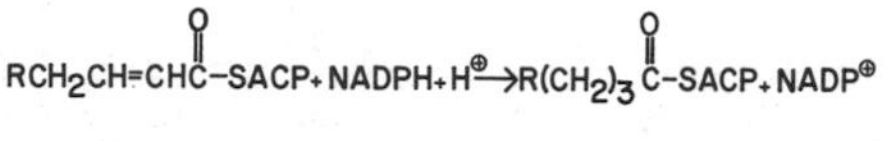

RCH_2CH=$CH\overset{O}{\underset{\|}{C}}$-SACP + NADPH + H$^\oplus \rightarrow R(CH_2)_3\overset{O}{\underset{\|}{C}}$-SACP + NADP$^\oplus$

The repetition of this group of reactions, commencing with R$=$H and terminating with R$=CH_3(CH_2)_{14}$, constitutes the biosynthesis of palmitoyl ACP by the over-all reaction:

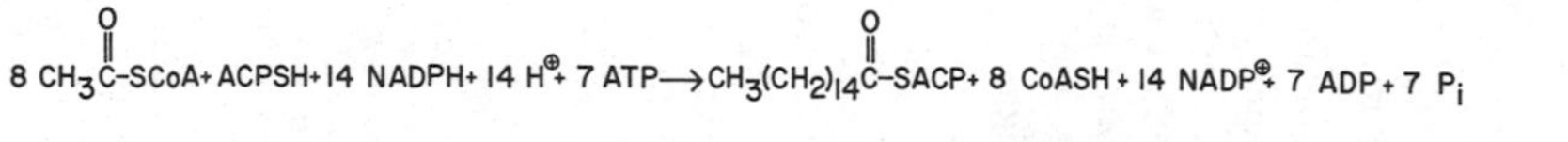

8 $CH_3\overset{O}{\underset{\|}{C}}$-SCoA + ACPSH + 14 NADPH + 14 H$^\oplus$ + 7 ATP $\longrightarrow CH_3(CH_2)_{14}\overset{O}{\underset{\|}{C}}$-SACP + 8 CoASH + 14 NADP$^\oplus$ + 7 ADP + 7 P_i

It can be seen that the process requires not only considerable energy in the form of adenosine triphosphate (ATP), but a supply of reduced triphosphopyridine nucleotide ($\sim$ 1 ATP + 1 NADPH/carbon atom). The latter can be produced in the cell cytoplasm primarily by operation of the pentose phosphate pathway. In order to carry out this sequence of reactions, the bacterial cell must have biotin and pantothenic acid in addition to ATP and NADPH. *E. coli* is capable of producing these materials, but auxotrophs for both biotin and pantothenic acid can be produced.

Palmitoyl ACP is used primarily for the synthesis of phospholipids, which are the major constituents of bacterial cell membranes. Under certain circumstances, the fatty acids may be released from ACP by a hydrolytic thiolase. The bacterial synthetase also produces monoolefinic fatty acids of 16 or 18 carbon chain lengths. The maximum size for the chains is controlled by the condensing enzymes, which have exceedingly low rates of reaction with palmitoyl ACP and octadecenoyl ACP.

It was noted earlier that glycerolipids often contain one saturated and one olefinic acid. In *E. coli*, the saturated acid is R_1, while the unsaturated acid is R_2, although in some bacteria, notably *Clostridium* and *Mycobacterium*, this situation is reversed. *E. coli* contains a membrane-bound enzyme which carries out the acylation of glycerol-3-phosphate in a specific manner so that saturated acyl ACP is used for 1-acylation and olefinic acyl ACP is used for the 2 position (9).

What is the source of olefinic fatty acids? Bloch and his colleagues have shown that an olefinic acid can be produced as a normal by-product of the *E. coli* synthetase by the simple expedient of introducing the olefinic linkage during chain synthesis and elongating the chain in front of the double bond (10). A special dehydrating enzyme is necessary, and this has been called β-hydroxydecanoyl thioester dehydrase. D-β-Hydroxydecanoyl ACP, a normal intermediate in fatty acid synthesis, can be acted upon by this enzyme to produce either the usual *trans*-2-decenoyl ACP or the *cis*-3-decenoyl ACP.

The latter compound, since it has no 2-double bond, is treated by the condensing enzyme as if it were a saturated acyl compound and is elongated to either the *cis*-9-hexadecenoyl ACP or the *cis*-11-octadecenoyl ACP (*cis*-vaccenoyl ACP).

The situation in eukaryotic cells is different and less well understood. In general, two components of fatty acid synthesis can be separated readily—the acetyl CoA carboxylase and the fatty acid synthetase. The carboxylase from the liver of birds exists in an inactive monomer and an active polymer form. The polymerization to the active form is effected by a variety of substances, especially citrate and isocitrate, which contribute to metabolic control of the carboxylase activity (11). The carboxylase is the first and rate-limiting enzyme in the synthetic process and is therefore a logical target for such regulation.

The synthetases of yeast and animal cells are large enzyme complexes (molecular weight 5×10^5 to 2.5×10^6) containing all the proteins necessary for carrying out individual steps analogous to those catalyzed by the bacterial enzymes. Acyl groups are transferred from acetyl and malonyl CoA to sulfhydryls of pantotheine units bound to the complex. The enzymatic machinery then cycles the chains until they reach an appropriate chain length, whereupon they are released as free acids. The cells contain CoA thiokinases for generating fatty acyl CoA derivatives, which are precursors of the fatty acid ester lipids:

$$RCOOH + ATP + CoASH \longrightarrow RC\overset{\overset{O}{\|}}{}-SCoA + AMP + PP_i$$

The production of olefinic fatty acids in eukaryotic cells is very different from that in most bacteria. The process consists of oxidative hydrogen removal from the CoA derivative of a preformed fatty acid chain:

$$CH_3(CH_2)_{16}\overset{\overset{O}{\|}}{C}-SCoA + NADPH + O_2 + H^{\oplus} \longrightarrow CH_3(CH_2)_7\overset{\overset{H\ H}{|\ |}}{C=C}(CH_2)_7\overset{\overset{O}{\|}}{C}-SCoA + NADP^{\oplus} + 2H_2O$$

The product has a *cis* double bond at the 9,10 position, regardless of the over-all chain length. Similar enzymes are involved in the production of polyunsaturated fatty acids. Certain bacteria also contain similar oxidative desaturation systems, and there may exist enzymes which introduce double bonds at positions other than 9,10. In some cases diene fatty acids can be produced.

Terpenoids and Sterols

In the case of fatty acid synthesis, cells have the problem of converting highly oxygenated products of glucose catabolism to the reduced and apolar aliphatic chain of methylene groups. The problem of terpenoid synthesis is similar, with the added necessity of introducing a regular branching at every fourth carbon atom. This is accomplished by the condensation of three acetate units in a branched rather than a linear configuration:

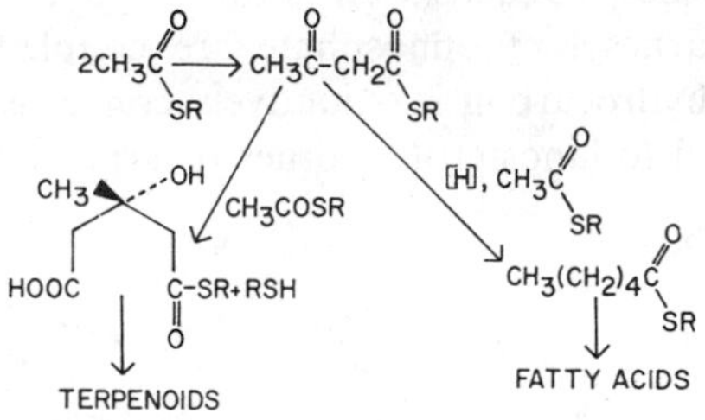

This branched intermediate, 3-hydroxy-3-methyl glutaryl CoA (HMG CoA), is then reduced to mevalonic acid:

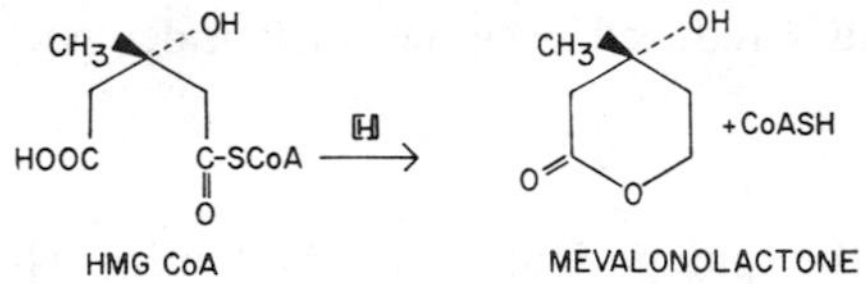

A series of ATP kinase reactions, accompanied by decarboxylation, yields isopentenyl pyrophosphate (IPP):

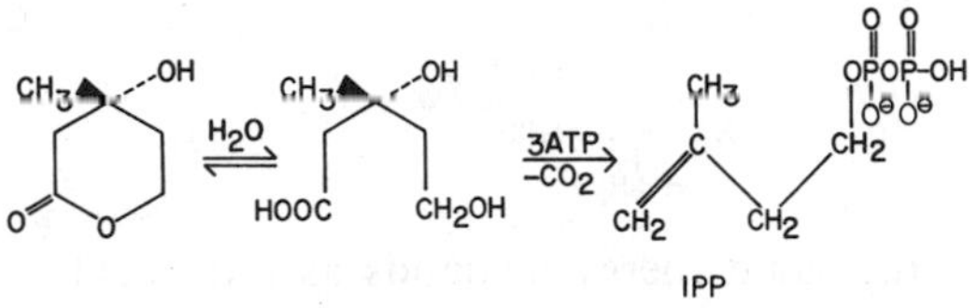

Double-bond isomerization yields the terpenoid chain initiator, dimethylallyl pyrophosphate (DMAP):

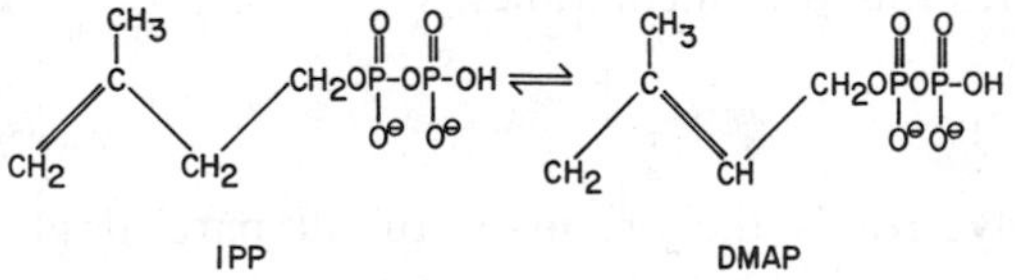

Chain propagation consists of the addition of allylic groups to the terminal methylene double bond of isopentenyl pyrophosphate:

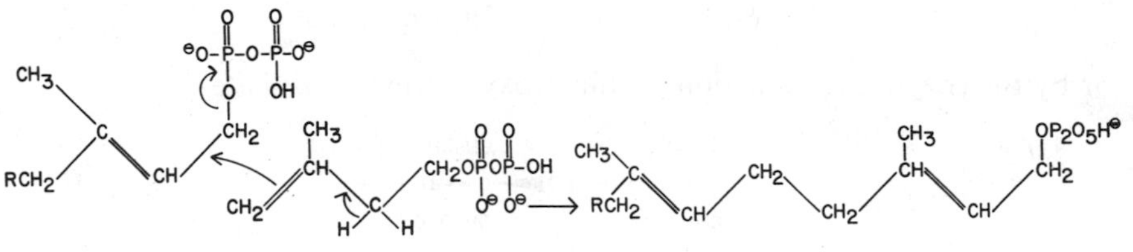

All natural terpenes are formed by such condensation reactions, although some involve the formation of *cis* rather than *trans* double bonds. In the case of polycyclic triterpenes, condensation is interrupted at the C_{15} stage, and two molecules of farnesyl pyrophosphate are coupled "head-to-head" to give squalene. This hydrocarbon is oxidatively converted to the 2,3-epoxide, which can be cyclized to lanosterol or other triterpenes:

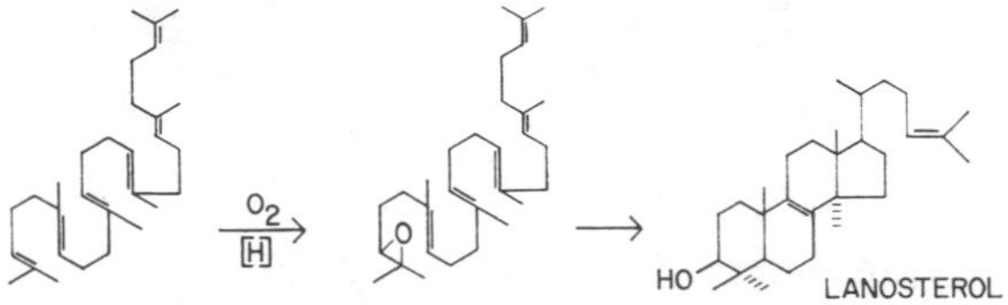

Note that the conversion of acetate to lanosterol requires more than 1 ATP per carbon atom and 2 reduced pyridine nucleotides per 5 carbon atoms.

Sphingolipids

The backbone of the sphingolipids is a polyolamine which originates from a fatty acid and the amino acid serine:

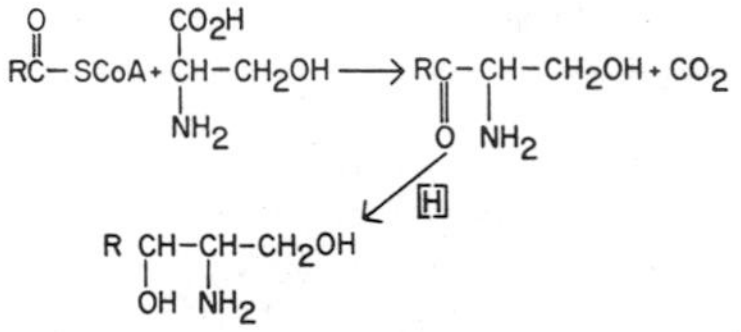

This process has the same energy demands as fatty acid synthesis, since a fatty acid forms the major portion of the molecule.

Having considered the processes leading to the fundamental parts of lipid molecules, we shall now consider methods for assembling these into the more complex structures found in membranes.

Phospholipids (12)

3-Phosphoglycerol is the precursor of all natural glycerolipids. This compound can arise either by a kinase reaction of glycerol and ATP:

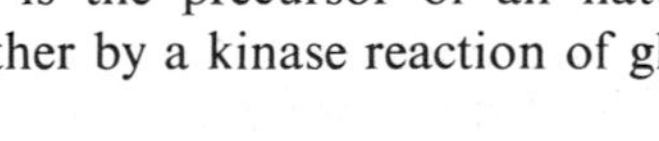

or by the enzymatic reduction of dihydroxyacetone phosphate:

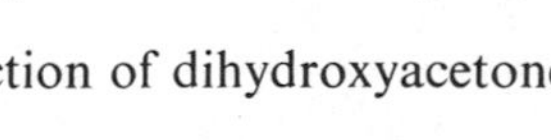

This compound can be acylated to form phosphatidic acid:

$$\underset{\text{HO}}{\blacktriangleright}\overset{\text{CH}_2\text{OH}}{\underset{\text{CH}_2\text{OPO}_3^\ominus}{\text{C}}}\blacktriangleleft\text{H} + \text{R}_1\overset{\text{O}}{\text{C}}\text{-SR}' \longrightarrow \text{R}_2\overset{\overset{\text{O}}{\parallel}}{\underset{\text{CH}_2\text{OPO}_3^\ominus}{\text{COCH}}}\overset{\text{CH}_2\text{OCR}_1}{}$$

$$\underset{\text{R}_2\overset{\text{O}}{\text{C}}\text{-SR}'}{+}$$

PHOSPHATIDIC ACID

R' = CoA or ACP

The enzymes that catalyze these acylation reactions are localized in cell membranes, and the specificity that accompanies these processes accounts for the unequal distribution of saturated and unsaturated fatty acids in natural glycerolipids.

The further reactions of phosphatidic acid are depicted in Figure 2. Examination of these schemes shows that the nitrogen-containing glycerolipids of eukaryotic cells are derived by dephosphorylation of phosphatidic acid to give a 1,2-diglyceride, to which the phosphorylated base is added from its cytidine nucleotide derivative. In bacterial cells, the nitrogen-containing lipids result from the reaction of CDP diglycerides with free bases. This would indicate that the phosphorus of the diester in the nitrogen-containing glycerolipids would be derived from 3-phosphoglycerol in bacteria, but from phosphoethanolamine or phosphocholine in animal cells. In all cells the anionic glycerolipids are derived from 3-phosphoglycerol by way of phosphatidic acid and CDP diglyceride. By either branch of the phosphatidic acid pathway, a molecule of CTP is required to provide the necessary energy for formation of the phosphodiester bond.

Glycolipids

Glyceroglycolipids and sphingoglycolipids consist of glycosidically linked sugars or oligosaccharides bound to a free primary hydroxyl at the 3 position of a diglyceride or the 1 position of a ceramide. The process of constructing these molecules consists simply in the transfer of the sugar units from nucleotides (e.g., UDP glucose) to the appropriate alcohols or growing glycolipid molecule.

Control of Lipid Synthesis

As in the case of other metabolic processes, the production of lipids is under various forms of control. As indicated earlier, acetyl CoA carboxylase, the first enzyme of fatty acid synthesis, is under metabolic control. HMG CoA reductase, which represents the first irreversible step in the synthesis of sterols, is controlled by products such as cholesterol and bile acids (13). Furthermore, the levels of synthetic enzymes in eukaryotic organisms are in

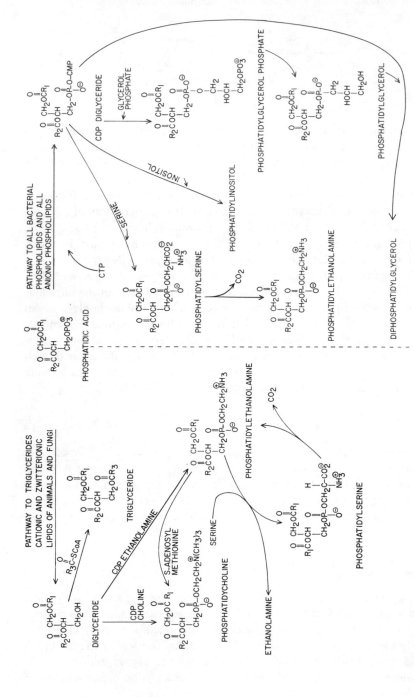

Figure 2
Reactions of phosphatidic acid.

a dynamic state in which enzyme synthesis is balanced by enzyme degradation. In starving rats, the level of acetyl CoA carboxylase is very low, but when such animals are fed, the enzyme levels rise rapidly to high values (14).

Temperature has a characteristic effect upon the fatty acid complement of cells. In organisms that live at the temperature of the environment, e.g., bacterial cells, the amount of unsaturated or branched-chain fatty acids is usually inversely proportional to the temperature (15). In mammals the temperature, and hence the fatty acid composition, is more uniform, although in certain animals, in which a portion of the body may be routinely exposed to low temperatures, a variant composition may exist (16). In the leg of the reindeer, for example, the proportion of unsaturated acids increases closer to the hoof. This phenomenon may be rationalized by the fact that unsaturated acids have lower melting points than saturated acids and they may impart a "liquid" character to the intact lipids at lower temperatures.

Lipid Catabolism

Since the lipid components of most membranes are in a dynamic state, catabolism must be considered as an important aspect of lipid metabolism. Lipids are first hydrolyzed to their component parts, which are either oxidized and discarded or reutilized. Lipolytic enzymes exist for all lipid classes. Triglycerides are hydrolyzed to yield di- and monoglyceride intermediates and finally free glycerol and fatty acids. Phospholipids may be deacylated, or the phosphodiester linkage may be split on either side of the phosphorus atom. Enzymes exist for the complete dismantling of sphingolipids and glycolipids. Even ethers are not immune to enzymatic attack.

Glyceride lipases (*17, 18*). Enzymes that hydrolyze triglycerides are widespread. In animals important examples are pancreatic lipase, the hormone-dependent lipase of adipose tissue, and lipoprotein lipase. Bacteria, fungi, and plants also produce glyceride lipases. Pancreatic lipase, the best studied of the lipolytic enzymes, is produced by mammalian pancreas and secreted into the intestine, where it attacks emulsions of triglycerides. The enzyme catalyzes the following reactions:

$$\text{triglyceride} \xrightarrow{\text{fast}} \begin{array}{c} \text{1,2-diglyceride} \\ \text{+ fatty acid} \end{array} \xrightarrow{\text{slow}} \begin{array}{c} \text{2-monoglyceride} \\ \text{+ fatty acid} \end{array}$$

The lipases of other sources catalyze similar reactions, but since triglycerides are not usually important components of membranes, they will not be discussed here.

Phospholipases. Of much more interest are the phospholipases which catalyze the hydrolysis of phospholipid molecules. The nomenclature of these enzymes has undergone several modifications; the most recent is indicated on the accompanying diagram. The A_2 group of phospholipases has been most extensively studied. The pancreatic enzyme is secreted as a zymogen which can be activated by a tryptic cleavage of a single small peptide. The enzyme is a single chain of molecular weight 14,000. The molecule has a compact form because of cross-linking by means of 6 disulfide bonds. It is rather resistant to denaturation and survives treatment at elevated temperatures. Several similar enzymes have been purified from snake venoms. These appear to have molecular properties similar to those of the porcine pancreas enzyme, although they sometimes exist in dimers of around 30,000 molecular weight. Definitive studies by van Deenen and de Haas have shown that phospholipase A_2 cleaves only glycerol lipids with a phosphate ester on the 3-hydroxyl and the L (or S) configuration at carbon atom 2.

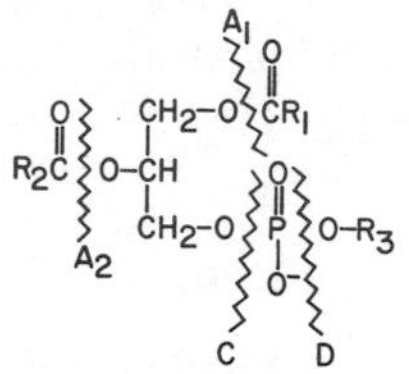

$$\begin{array}{c} A_1 \quad O \\ CH_2-O-CR_1 \\ O \\ R_2C-O-CH \qquad O \\ CH_2-O-P-O-R_3 \\ A_2 \qquad O^- \\ C \qquad D \end{array}$$

The properties of other phospholipases are summarized in Table V.

Glycosylases (19). A group of enzymes present in various tissues is responsible for removal of carbohydrate units from glycolipids. It has been postulated that certain metabolic diseases in which large amounts of sphingo-glycolipids accumulate in human tissue result from specific deficiencies of single glycosylases. The study of such conditions will doubtless lead to a better understanding of the role of lipids as structural components of membranes.

Sphingolipid Catabolism (19)

Ceramides can be cleaved enzymatically to fatty acids and free bases. Recently it has been found that dihydrosphingosine can be phosphorylated by a kinase and the phosphorylated base can be converted to the ethanol-amine phosphate and a fatty aldehyde.

Oxidative Degradation

The process of β-oxidation of fatty acids, known for many years, formally resembles fatty acid synthesis. The marked differences are that the oxidation takes place in mitochondria and utilizes different cofactors and even slightly different intermediates. By this process acetyl CoA and reduced pyridine nucleotides and flavoproteins are generated so that some of the energy expended in biosynthesis can be recovered in the form of ATP by oxidative phosphorylation.

Table V
Properties of Some Lipolytic Enzymes

ENZYME	SUBSTRATES	COFACTORS	SOURCES	PURITY	AVAILABILITY
Pancreatic lipase	Triglycerides, diglycerides	Ca^{++}, bile salts	Mammalian pancreas	Homogeneous; two forms in porcine pancreas	Commercially in partially purified form
Fungal lipases	Glycerides		Yeast, molds	Some in homogeneous form	Commercially in crystalline form
Phospholipase A_1	Phospholipids, glycerides	None	Widely distributed	Partially purified from brain	
Phospholipase A_2	Phospholipids	Ca^{++}	Snake venom, mammalian pancreas, bee venom	Crystalline, homogeneous	Crude snake venom
Phospholipase C	Phospholipids	Ca^{++}	Bacteria	Homogeneous	Commercially in partially purified form
Phospholipase D	Phospholipids	Ca^{++}	Plants	Partially purified	Commercially in partially purified form
Phosphatidate phosphohydrolase	Phosphatidic acid	None	Widely distributed, particulate	Partially purified from brain	

Data from Wakil, S. J. (ed.), *Lipid Metabolism*, Academic Press, New York, 1970, p. 185; Thompson, G., *Compr. Biochem* **18**,157 (1970); and Lennarz, W. J., *Annu. Rev. Biochem.* **39**, 359 (1970).

Under certain conditions fatty acids can be degraded from the ω-end by reactions initiated by hydroxylation of the terminal methyl group. ω-Oxidase is a microsomal enzyme which employs molecular oxygen and reduced pyridine nucleotide. In some microorganisms, aliphatic hydrocarbons can be similarly oxidized at one end. α-Oxidation of fatty acids leads to the α-hydroxy acids, which can be further oxidized to CO_2 and the n-1 acid. Glyceryl ethers can be cleaved by oxidative reactions which utilize atmospheric O_2 and produce fatty acids and glycerol.

Sterol Catabolism

Sterol fatty acid esters, which occur in some tissues, can be hydrolyzed to fatty acids and free sterols by cholesterol esterase or similar esterase enzymes. Free sterols can be oxidized to bile acids, which are excreted into the intestine. Generally animals excrete compounds with intact ring systems, although certain fungi are capable of oxidative ring-opening reactions.

ALTERATION OF LIPID COMPOSITION IN MEMBRANES

Two sorts of processes result in alterations in the lipid composition of membranes: *1*) normal processes which lead to a redistribution of lipid material within the cell or an in situ transformation of lipid molecules already deposited in membranes, and *2*) processes for altering membrane composition which may be manipulated at the will of the investigator. An example of the first process is the response to variations in temperature discussed above; examples of the second process are considered below.

Exchange of Phospholipids Between Organelles

It seems likely that newly synthesized lipid molecules are laid down in the cell membrane near the site of synthesis. The enzymes responsible for lipid biosynthesis seem to reside in membranes, a situation which adds logic to this hypothesis. Recently, however, investigators have become aware of exchange processes which result in translocation of lipid molecules. Wirtz and Zilbersmit (20) have shown that rat liver contains a soluble cytoplasmic protein which catalyzes exchange of phospholipids, especially phosphatidylcholine and phosphatidylethanolamine, between microsomes and mitochondria. If this situation proves a general one, earlier conclusions about the localization of biosynthetic enzymes may have to be revised.

Phospholipid Transformation

In addition to translocation process, the phospholipid composition of membranes can be altered by changes in the molecule in situ. Examples of lipid-transforming enzymes in the biosynthetic processes have already been presented. For example, phosphatidylserine can be enzymatically decarboxylated to phosphatidylethanolamine, and this compound can, in turn, be converted to phosphatidylcholine by transmethylation reactions. Lipolytic enzymes may catalyze the destruction of lipid molecules or the turnover and recycling of portions of the molecule. For example, certain secretory cells, when stimulated, show high turnover of certain phospholipids (21). A bacterial phospholipase can convert diphosphatidyglycerol to phosphatidylglycerol. Fatty acid chains of phospholipids can also be transformed by enzymatic processes. Double bonds in certain phospholipids can be alkylated to give cyclopropane rings or branched methyl groups, or polyunsaturated acid chains can be oxidized to give hydroperoxide groups. There is no definitive evidence that these processes take place within the cell membrane, but there are circumstantial indications that this is indeed the case.

Genetic and Chemical Techniques for Altering Membrane Composition

In order to gain information about the role of lipids in membrane structures, it has in some instances been possible to change the lipid composition and to investigate functional changes that accompany compositional variation. The methods employed include the use of drugs and specific inhibitors of parts of the metabolic pathways, use of mutant organisms, variation of environmental parameters, and removal of lipid components by enzymatic methods or by extraction with solvents.

Bacterial mutants lacking the β,γ-dehydrase enzyme for the production of unsaturated fatty acids have a requirement for olefinic acids or fatty acids with physical properties similar to those of olefinic acids. These mutants have been quite useful in probing membrane function, as will be discussed in Chapter 12. An *E. coli* mutant lacking glycerol phosphate dehydrogenase was unable to produce 3-glycerol phosphate from dihydroxyacetone phosphate and required glycerol for growth. Withholding glycerol from the medium permitted Hsu and Fox (22) to study the effect of phospholipid depletion on DNA and protein synthesis and on induction of the β-galactoside transport system. Recently, conditional lethal mutants which are blocked in phospholipid formation have been studied. These lack the ability to produce phosphatidic acid when they are cultured at elevated temperatures.

Biotin auxotrophs of bacteria and yeast can be used in much the same way as bacterial fatty acid auxotrophs, since the principal role of biotin is the carboxylation of acetyl CoA to give malonyl CoA for fatty acid synthesis. Indeed, *Lactobacillus* spp. have been used as a natural biotin auxotroph, since these organisms have a nutritional requirement for this vitamin (23). Unlike the β,γ-dehydrase mutants, a biotin auxotroph cannot make saturated fatty acids in the absence of biotin, and in most experiments both saturated and unsaturated fatty acids are supplied. As would be expected, saturated acids alone do not support growth in the absence of biotin.

A group of yeast mutants lacking the oxidative desaturase for the conversion of stearoyl CoA to oleoyl CoA and another group blocked in fatty acid formation are available (24). The fatty acid requirements of these cells seem to be more stringent than those of *E. coli* mutants, and they have been less extensively studied. When normal yeast cells are grown under strictly anaerobic conditions, they show a requirement for sterols and unsaturated fatty acids, since molecular oxygen is required for the production of these two classes of compounds. This provides the investigator with a means of controlling the lipid complement of the cell membrane. The control is less complete than might be desired, however, because the yeast cells have a limited tolerance for foreign structural components and may carry out further metabolic alterations on the administered compounds. For example, fatty acids may be shortened or lengthened, and sterol side chains may be methylated.

Mycoplasma have been extensively used to produce membranes of a desired composition, for they will utilize lipid components added to the growth medium. This subject is discussed in detail in Chapter 14. Many protozoa show requirements for specific types of lipids. Insects are one of the few groups of animals that generally lack the ability to make sterols and have dietary requirements for cholesterol or other sterols. Detailed studies of the use of cholesterol by insects have been carried out by Clayton (25), who has concluded that there is a general and a specific sterol requirement. The general requirement seems to be for a membrane structural component, which can be supplied by a number of different sterols, while a specific metabolic requirement for a very small amount of cholesterol is indispensable. It is likely that this specific requirement is necessary in part for the production of the important insect hormone, ecdysone.

Among the compounds which have proved valuable in the probing of membrane structure are polyene antibiotics and drugs which inhibit steps in sterol biosynthesis. Polyene antibiotics like filipin form complexes with sterols and disrupt sterol-containing membranes. Thus they are toxic to fungi, but innocuous to bacteria, which lack sterols.

FILIPIN

Two useful inhibitors for unsaturated fatty acid synthesis are the N-acetyl cysteamine derivative of 3-decynoic acid and sterculic acid (26). The former compound is a specific inhibitor of the bacteria β,γ-dehydrase, and the latter of the oxidative acyl CoA desaturase of eukaryotic cells. Use of the first of these inhibitors can block the production of unsaturated fatty acids in prokaryotic cells, while use of the latter can block production in eukaryotic cells.

$$CH_3(CH_2)_5C\equiv CCH_2\overset{O}{\overset{\|}{C}}SCH_2CH_2NH\overset{O}{\overset{\|}{C}}CH_3$$

S-(3–DECYNOYL)N–ACETYL CYSTEAMINE

$$CH_3(CH_2)_7 \triangle (CH_2)_7COOH$$

STERCULIC ACID

Methods have been developed for the extraction of membrane lipids without complete and irreversible inhibition of the function of the native membrane components. Fleischer and Fleischer (27) studied extensively the removal of phospholipids from mitochondria by aqueous acetone. In many cases, such extracted mitochondria maintain much of their structural features in spite of nearly complete removal of lipids. Functional properties of individual mitochondrial enzymes, lacking in the extracted organelles, are restored when lipid dispersions are added back to the organelle suspensions.

It is also possible to remove phospholipids from membranes by treatment with phospholipases. In many cases function lost in the process of lipid removal can be restored by lipid replacement.

METHODOLOGY

For the membranologist with limited experience in the lipid field, there are a number of useful references explaining the methodology in detail. The purpose here is to outline the strategies available.

The first problem is to achieve either complete extraction of the lipids or a partial extraction appropriate to the goal of the study. In general, lipids are soluble in organic solvents, especially alcohols, ether, chloroform, etc., but insoluble in water. Many exceptions exist, however, and caution is advised in both extraction and work-up procedures. Chloroform-methanol mixtures are often the solvents of choice, but careful washing of the solvent extract with aqueous solvent mixtures is necessary to remove nonlipid contaminants. Many amateurs have been led astray by the fact that sucrose and certain amino acids are remarkably soluble in organic solvents. Nonlipid contaminants can also be removed by chromatography on Sephadex using organic solvent mixtures.

Most lipid samples must be protected from oxidative degradation by storing them under nitrogen in the cold. It is usually advisable to exclude oxygen from solvents used for chromatography and to avoid high temperatures. Separation of crude mixtures of lipids can be achieved by several methods, including solvent partition, chromatography on silicic acid, chromatography on lipophilic Sephadex, and chromatography on ion-exchanging celluloses. Excellent thin-layer chromatographic (TLC) methods now exist for analytical separations of intact lipids of all classes. In some cases these methods can be adapted for preparative separation. Special modifications of the TLC techniques permit sensitive separations based upon particular structural features. For example, incorporation of silver nitrate into the silicic acid allows separation of lipids according to the number of double bonds in the aliphatic side chains. In one instance, natural phospholipids were first fractionated to yield pure classes, e.g., the phosphatidylcholine fraction, and these were then subfractionated by the degree of unsaturation. Finally specific enzymatic cleavage with phospholipase A_2, separation of the fatty acids and lyso compounds, and analysis of the fatty acid complement at the 1 position and the 2 position provided a rather complete analysis.

Degradative methods are often employed for analysis, and enzymatic and chemical hydrolyses have been used. Fatty acids can be removed by alkaline hydrolysis or by transesterification catalyzed by acids or bases. The method of choice should take into account any potential instability of the lipid components. For example, cyclopropane rings in fatty acids can be opened by strong acids.

Methyl esters of fatty acids are conveniently separated by gas chromatography. These are often identified by comparison of the chromatographic properties with known standards. This is a dangerous practice unless several different liquid phases of varying polarity are employed or unless the structures of the unknown acid esters are confirmed by an independent technique such as mass spectrometry.

Mild base hydrolysis of phospholipids has been used to remove fatty acids while preserving the phosphodiester structure. These water-soluble products can be separated by paper or ion-exchange chromatography. Detailed description of these and other analytical procedures can be found in the works listed under Recommended Reading.

REFERENCES

1. Wakil, S. J. (ed.). *Lipid Metabolism.* Academic Press, New York, 1970.
2. Gunstone, F. D. (ed.). *Topics in Lipid Chemistry.* Logos Press, London, 1970, vol. 1.
3. Alhadeff, J. A., and Daves, G. D. *Biochemistry* **9,** 4866 (1970).

4. Mallory, F. B., Gordon, J. T., and Conner, R. L. *J. Am. Chem. Soc.* **88**, 1362 (1963).
5. Labach, J. P., and White, D. C. *J. Lipid Res.* **10**, 528 (1969).
6. Gilliland, K. M., and Moscatelli, E. A. *Biochim. Biophys. Acta* **187**, 221 (1969).
7. Kates, M., Palameta, B., Joo, C. N., Kushner, D. J., and Gibbons, N. E. *Biochemistry* **5**, 4092 (1966).
8. Hagen, P.-O., and Goldfine, H. *J. Biol. Chem.* **242**, 5700 (1967).
9. Ray, T. K., Cronan, J. E., Mavis, R. D., and Vagelos, P. R. *J. Biol. Chem.* **245**, 6442 (1970).
10. Bloch, K. *Accounts Chem. Res.* **2**, 193 (1969).
11. Gregolin, C., Ryder, E., and Lane, M. D. *J. Biol. Chem.* **243**, 4227 (1968).
12. Hill, E. E., and Lands, W. E. M. *Lipid Metabolism.* Ed. S. J. Wakil. Academic Press, New York, 1970, p. 185.
13. Danielsson, H., and Tchen, T. T. *Metabolic Pathways,* 3rd ed. Ed. D. M. Greenberg. Academic Press, New York, 1968, vol. 2, p. 117.
14. Majerus, P. W., and Kilburn, E. *J. Biol. Chem.* **244**, 6254 (1969).
15. Marr, A. G., and Ingraham, J. L. *J. Bacteriol.* **84**, 1260 (1962).
16. Irving, L., Scmidt-Nielson, K., and Abrahamsen, N. *Physiol. Zool.* **30**, 93 (1956).
17. Hübscher, G. *Lipid Metabolism.* Ed. S. J. Wakil. Academic Press, New York, 1970, p. 279.
18. Thompson, G. A. *Compr. Biochem.* **18**, 157 (1970).
19. Gatt, S. *Chem. Phys. Lipids* **5**, 235 (1970).
20. Wirtz, K. W. A., and Zilbersmit, D. B. *J. Biol. Chem.* **243**, 3596 (1968).
21. Hokin, L. E., and Hokin, M. R. *Annu. Rev. Biochem.* **32**, 553 (1963).
22. Hsu, C. C., and Fox, C. F. *J. Bacteriol.* **103**, 410 (1970).
23. Hofmann, K. *Fatty Acid Metabolism in Microorganisms.* Wiley, New York, 1963.
24. Keith, A. D., Wisnieski, B. J., Henry, S., and Williams, J. C. In press.
25. Clayton, R. B. *Chemical Ecology.* Ed. E. Sondheimer and J. B. Simeone. Academic Press, New York, 1970, p. 235.
26. Raju, P. K., and Reiser, R., *J. Biol. Chem.* **242**, 379 (1967).
27. Fleischer, S., and Fleischer, B. *Methods in Enzymology.* Ed. R. W. Estabrook and M. E. Pullman. Academic Press, New York, 1967, vol. 10.
28. Vagelos, P. R. *Current Topics in Cellular Regulation.* Academic Press, New York, 1971, vol. 4, p. 119.

RECOMMENDED READING

Reviews

Kates, M., and Wassef, M. K. *Annu. Rev. Biochem.* **39**, 323 (1970).
Lennarz, W. J. *Annu. Rev. Biochem.* **39**, 359 (1970). These two reviews are typical of the up-to-date coverage of Annual Reviews. Similar reviews are available in older volumes.
van Deenen, L. L. M. *Progr. Chem. Fats Other Lipids* **8**, pt. 1 (1965). A comprehensive review of the chemistry of phospholipids and their role in membranes.
Florkin, M., and Stotz, E. H. (eds.). *Compr. Biochem.* **18**, (1970). A compendium of several review articles.

Greenberg, D. M. (ed.). *Metabolic Pathways*, 3rd ed. Academic Press, New York, 1968, vol. 2.

Wakil, S. J. *Lipid Metabolism*. Academic Press, New York, 1970. An excellent, up-to-date collection of review articles.

In addition, many review articles are available in the *Journal of Lipid Research* and in the following series:

Gunstone, F. D. (ed.). *Topics in Lipid Chemistry*. Logos Press, London.

Paoletti, R., and Kritchevsky, D. (eds.). *Advances in Lipid Research*. Academic Press, New York.

Methodology

Lowenstein, J. M. (ed.). *Methods in Enzymology*. Academic Press, New York, 1969, vol. 14. Good coverage of methods and preparations, including lipolytic enzymes.

Marinetti, G. V. *Lipid Chromatographic Methods*. Dekker, New York, 1967. A collection of useful analytical and chromatographic methods.

American Oil Chemists' Society. *Quantitative Methodology in Lipid Research*. The Society, Chicago. A collection of papers on various aspects of lipid analysis.

Folch, J., Lees, M., and Sloane Stanley, G. H. *J. Biol. Chem.* **226,** 497 (1957). A classic paper on lipid extraction, upon which many modifications are based.

Bligh, E. G., and Dyer, W. J. *Can. J. Biochem. Physiol.* **37,** 911 (1959). A popular procedure, also using chloroform-methanol mixtures.

Wells, M. A., and Dittmer, J. C. *Biochemistry* **5,** 3405 (1966). Useful methods for lipid analysis. See also the chapter by these authors in Lowenstein (listed above).

Fleischer, S., and Fleischer, B. *Methods in Enzymology*. Ed. R. W. Eastbrook and M. E. Pullman. Academic Press, New York, 1967, vol. 10.

Johnson, A. R., and Davenport, J. B. *Biochemistry and Methodology of Lipids*. Wiley, New York, 1971.

NOTE ADDED IN PROOF:

With reference to the biosynthesis of cardiolipin, recent work indicates that the pathway is

$$2 \text{ phosphatidylglycerol} \longrightarrow \text{cardiolipin} + \text{glycerol}$$

rather than the reaction shown (Hirschberg, C. B. and Kennedy, E. P. *Proc. Natl. Acad. Sci. U.S.A.* **69,** 648 (1972)).

2

Theodore L. Steck
C. Fred Fox

Membrane Proteins

It used to be that the lipids were the principal focus of the biochemical analysis of membranes. They can be extracted cleanly into organic solution, purified and fractionated by class, analyzed by well-established techniques, compared, and correlated back to their parent membranes (Chap. 1). Since lipids form regular arrays both in the isolated state and in natural membranes, physical techniques, particularly x-ray diffraction analysis, are also applicable.

The proteins of membranes have been less tractable. It has often been observed that membrane proteins are, by their very nature, insoluble under normal circumstances. In addition, upon extraction of the lipids, these proteins may form recalcitrant aggregates and lose their biological activity. They are thus difficult candidates for the hydrodynamic, optical, and other techniques developed for the study of soluble proteins. Furthermore, serious contamination of membrane preparations by adventitious nonmembrane proteins has little counterpart in the analysis of lipids.

The preparation of this manuscript was assisted by Grant P-578 from the American Cancer Society and by a Fellowship from the Schweppe Foundation to T.L.S. and by U.S.P.H.S. Grant GM-18233 and U.S.P.H.S. Research Career Development Award GM-42359 to C.F.F. We are grateful to Drs. J. A. Reynolds, S. J. Singer, and G. Guidotti for providing access to their manuscripts prior to publication.

Nevertheless, membrane proteins are currently under intensive, fruitful investigation. The serious methodological problems now being overcome are *1*) the adequate purification of the target membranes, *2*) the release of proteins from the membrane matrix, *3*) the isolation of the released proteins, *4*) the characterization of these proteins, *5*) the correlation of data obtained from the isolated proteins with the parent membrane, and *6*) the analysis of proteins while still membrane-bound.

Several central questions underlie much of the work in this field. Here are a few of them:

1. What is the physical and chemical character of membrane proteins? How are they adapted to their role in the membrane?

2. How many types of proteins are there in a membrane?

3. How do the proteins from various membranes resemble one another; how and to what purposes do they differ? Are there broad classes of similar (structural) proteins or a heterogeneous collection of specific proteins?

4. What associations do proteins make with other proteins, with lipids, and with other membrane components?

5. How and by what mechanisms are these proteins arranged and held in the membrane; are they fixed or mobile, symmetrical or asymmetrical in orientation; is there long-range order or a more random dispersion of components?

6. How do membrane proteins arise, how are they incorporated into the membrane; how do they turn over? Are they in equilibrium with a pool of cytoplasmic counterparts?

7. Does membrane assembly require a template to determine its specificity of organization, or can native membranes form de novo from their constituents?

8. What functions do membrane proteins perform? What is the physicochemical character of membrane enzymes, transport proteins, antigens, etc? What structural role do proteins play in stabilizing the membrane?

Many of these topics cannot be considered here. Membrane biogenesis, transport, and physical analysis are discussed in other chapters. Treatments of the enzyme systems of mitochondria, the endoplasmic reticulum, and other organelles fill entire volumes. Membrane proteins have been the subject of a recent symposium (74), and red blood cell membrane proteins in particular have been reviewed in detail (65). Our current understanding of membrane structure has also been well reviewed (121, 127).

PREPARATION OF MEMBRANES

The isolation of individual membrane proteins often follows from the purification of the parent membrane, which may in turn depend on the proper preparation of its organelle. From Chapter 3 the following seems clear.

1. Current practice permits fair or good purification of many cellular organelles, but very high purity (i.e., > 95%) may be difficult to achieve and to document. This fact may account for the popularity of bacteria (especially *Mycoplasma*), erythrocytes, and other cells possessing only a surface membrane and no other membranous organelles.

2. The membrane cannot be equated with its organelle. The preponderant fraction of the proteins in most organelles resides between and not within the membranes. The protein profile for the isolated membranes should be expected to differ greatly from that of the whole organelle, as has been shown for mitochondria (150).

3. The proteins trapped within sealed vesicles or organelles can be released by osmotic shock, analogous to the hypotonic lysis used in the preparation of erythrocyte ghosts and bacterial protoplasts. Sonication has been used for the same purpose; however, the small vesicles obtained may still bear entrapped protein. Surfactants also disrupt the membrane continuum and upset the permeability barrier. Digitonin may be of some utility here, since it seems circumscribed in its lytic action and need not dissolve the membrane. Other surfactants must be suspected of solubilizing membrane proteins and should therefore be used with care.

4. Adsorption of soluble proteins is favored in solutions of low ionic strength or mild acidity or in the presence of millimolar levels of divalent cations. Adsorption may be reversed by raising the ionic strength, eliminating divalent cations, and adjusting the pH to mild alkalinity. The amount of adsorbed protein can easily equal that of the membrane protein itself.

5. Degradation artifacts may occur if the preparations contain exogenous (namely, lysosomal) or endogenous proteases.

6. Genuine membrane proteins can be lost during membrane preparation by means other than proteolysis and detergent-mediated release. Alterations in ionic strength may elute some proteins. For example, preferential release of acetylcholinesterase from isolated erythrocyte membranes (but not from intact cells) occurs following incubation in 0.6–1.2 M NaCl; the ghosts remain grossly intact during the process. At the other extreme, many proteins are released by exposure to chelating agents or very low electrolyte levels.

7. We may summarize by observing that the question of which proteins in a membrane preparation are truly of the membrane and which are contaminants is a thorny one. The converse may also be problematical; namely, have any genuine membrane proteins been lost during isolation? We tend to assume that the boundaries of the living membrane are as sharply defined as they appear to be in thin-section electron micrographs. Functionally, however, certain proteins may adhere to the membrane, yet be readily dissociated. Hexokinase isozymes seem to be reversibly adsorbed to the outer mitochondrial membrane in this way. This is also the sense of studies which show that

all the presumably soluble glycolytic enzymes are represented on the erythrocyte ghost membrane, but are readily eluted by the alteration of ionic conditions (see Chap. 3). It has been proposed that the glycolytic chain is associated with the cell surface in vivo, perhaps providing a proximal energy source for membrane transport functions. Others have argued that true membrane proteins cannot be released without the dissolution of the membrane itself.

PREPARATION OF MEMBRANE PROTEINS: SOLUBILIZATION

Membrane proteins can be studied in situ or in solution. Investigators who take the membrane apart to analyze its proteins do so for various reasons. Some wish to study proteins chemically and must first remove the lipids; others seek to recover "native" lipid-protein complexes. Some demand total solubilization of all the proteins; for many others a selective extraction of individual proteins is desirable. Some require that native conformation and biological activity be preserved; others elect to thoroughly denature the proteins. For most, proteolysis would be an untoward hazard; but some rely on controlled proteolysis to release and solubilize special structures. Finally, there are those who take the membrane apart in order to elucidate how it is held together. For each of these, there are methods.

What Is Soluble?

When a membrane preparation is mixed with a detergent and the suspension immediately clarifies, should we assume that the membrane has dissolved? If no pellet forms when the sample is then centrifuged under conditions known to sediment the membranes, does a true solution exist? Although we all know better, optical clarity and failure to sediment are widely taken to signify solubilization. Of course, they only indicate a sharp reduction in the particle size. Whether the proteins and lipids have been reduced to their fundamental molecular state cannot be inferred. Terms such as *disaggregation*, *dispersal*, or *release* may therefore be more appropriate than *dissolution*. The crux of the issue seems to lie in the false analogy we make between the solubilization of membranes and the solubilization of crystalline solids. In the latter case, the molecules dissolve by passing homogeneously from one state to the other. However, membranes may disperse in as many ways as there are molecular combinations. For example, some proteins may be truly solubilized by elution without disruption of the membrane; on the other hand, membranes may be finely dispersed without solubilization of the proteins (e.g., by sonication).

In discussing solubilizing agents, we are forced to adopt the criteria employed by the various investigators. Rarely is this more exacting than a

10^7-g-min centrifugation, which will hardly pellet 70S ribosomes (containing dozens of proteins). More precise size estimates may be obtained by prolonged rate zonal centrifugation, analytical ultracentrifugation, or molecular sieving (gel filtration chromatography and gel electrophoresis).

Concepts of solubility vary with the investigator's assumptions about membrane structure. Those who believe that lipoprotein subunits are the basis of membrane structure expect soluble particles to contain both these species. Others feel that lipid-protein association denotes incomplete solubilization or reassociation. The same holds for analytical data suggesting multiprotein (e.g., > 20S) particles: Are they oligomers or aggregates?

Solubilization

Solubilization occurs when the interaction of a membrane component with the solvent is energetically favored over the interaction of the component with the membrane. This transition may be brought about by altering the solvent, the component, or the membrane matrix to which the component is bound. The chemical forces which stabilize membrane architecture seem to parallel those which govern protein and lipid structures: Coulombic, van der Waals, hydrophobic, and hydrogen bonding. Evidence for significant covalent lipid-protein association is scant, and interchain disulfide linkages do not seem to play a prominent role in stabilizing the membrane. The principal agents used to disperse the membrane components are ionic perturbations, denaturants, organic solvents, and surfactants. The components may be recovered as proteins and lipids in true solution, dispersed detergent complexes, discrete lipoprotein complexes, or higher aggregates of any size. The residual membrane matrix may be in evidence or may have disappeared.

Reaggregation of the dispersed components may occur. This process is favored by removal of the solubilizing reagent, by concentration of the sample, by freezing and thawing, by adjustment of the pH toward an apparent isoelectric point (usually around pH 4–5), and by the addition of salts, particularly divalent cations. The aggregates may not be readily redissolved. While not originally present, intermolecular disulfide bonds may be created by oxidation, forming tenaciously cross-linked aggregates. Prior alkylation of sulfhydryl groups may obviate the problem. Sometimes the reassociated components can be made to form structures with a morphology similar to that of membrane vesicles. These "membranes" testify to the propensity of proteins and lipids to associate, but give little evidence that the architectural specificity of the parent membrane has been restored (see under "Reassociation of Isolated Proteins and Lipids"). Likewise, the various lipid-protein complexes recovered from solubilized membranes may have formed after disruption and may not reflect native subunit structure.

Selective Extraction

We shall adopt the point of view that the membrane proteins are hetero-geneous and that each makes specific associations with its native locus in the membrane. The diversity of these associations is reflected in the many ways individual proteins can be selectively eluted from the membrane. The sig-nificance of selective elution, then, is not merely that certain components can be differentially solubilized and purified, but that specific physicochemical interactions between a protein and its particular site in the membrane are manifested in the elution.

Selective extraction is seen most often with mild treatments, such as nonionic detergents or manipulation of ionic conditions. Denaturants show less discrimination. Deoxycholate, for example, may reduce various mem-branes to stable lipid-protein particles (19, 62, 70), while sodium dodecyl sulfate (SDS) solubilizes membranes without evidence of lipoprotein sub-units (20, 59). However, even 6 M guanidine · HCl solubilizes only some of the red blood cell membrane proteins and leaves certain others associated with the lipids (66, 159).

Selective solubilization is particularly appropriate for the purification of membrane-associated enzymes. If the enzyme can be released without dis-rupting the remainder of the membrane, reconstitution experiments may be possible. If the factor being purified is measurable only in association with the rest of the membrane, such an approach is of obvious value. This situa-tion is well illustrated by studies on the functioning of the inner mitochon-drial membrane (92).

Alterations in Ionic Conditions

EFFECT OF REMOVING ELECTROLYTES

Several studies have provided evidence that the red blood cell mem-brane is stabilized by the presence of ions and is broken down upon their removal. It has long been noted that isolated ghost membranes are com-minuted in solutions of very low ionic strength. The initial fragmentation is actually an endocytic vesiculation which creates inside-out vesicles. Con-comitant with this breakdown is the simultaneous release of three major ghost components which account for about 25% of the ghost protein (23). The predominant high-molecular-weight species have been characterized by Marchesi and his associates as "spectrin" (69). Their release is facilitated by *1*) reduced ionic strength, *2*) elimination of divalent cations, and *3*) a mildly alkaline pH. Elevated ionic strength, millimolar levels of divalent cations, or acid pH retard the release of these proteins. The choice of buffer is also im-portant. Tris-HCl fosters the release of these polypeptides at ionic strengths at which phosphate or borate buffers do not (Yu and Steck, unpublished data). The effect of Tris may be related to the observation that organic cations such as tetramethyl ammonium salts potentiate the release of the proteins of this membrane (98). The release of spectrin is also temperature-dependent, taking several hours at 0–5°C but only a few minutes at 37°C (23). Since these polypeptides may contain a strong net negative charge, it is conceivable that the solutions of low ionic strength promote Coulombic repulsions which foster their dissociation from the membrane. It is possible that these proteins are bound through cation bridges; however, this hypothesis does not explain the stabilizing effects of acidic pH or inorganic buffers of high ionic strength (23, 98).

Other authors have employed extensive deionization as a means of releasing red blood cell membrane proteins. Harris (37, 38) used prolonged dialysis of human red blood cell membranes against distilled water to release a protein fraction of interesting electron microscopic appearance (see under "Associations Among Like Proteins"). Mazia and Ruby (71) not only dialyzed bovine erythrocyte ghosts extensively against distilled water (brought to pH 9 with ammonia), but introduced a deionizing resin to further the process. Unfortunately these authors included treatment with 0.1% Triton X-100 in their protocol, so that the effects of simple ion removal cannot be readily deduced. Furthermore, their data do not permit quantitation of the fraction of the membrane recovered. Hamaguchi and Cleve (35), investigating this protocol with human erythrocyte membranes, found that the release of proteins is incomplete, selective, and strongly dependent on the detergent treatment.

Reynolds and Trayer (98) pursued the same goals, using a prolonged incubation at room temperature in the presence of 5 mM ethylenediaminetetraacetic acid (EDTA) or 100 mM tetramethyl ammonium bromide to remove inorganic divalent and monovalent cations from the human erythrocyte membrane. By their criterion for solubility, namely, the 2.9×10^6 g-min supernatant, around 90% of the protein could be released after 4–5 days' incubation with these reagents. Phospholipid is associated with the released proteins, but at about 15–25% the level of the intact membranes; the lipid could be further reduced by prolonged dialysis. The released polypeptides represented those of the intact membrane proportionately. Physical studies suggested that they may have been in an aggregated state. In sharp contrast, the extraction of myelin and inner mitochondrial membranes by these methods caused little release of protein.

Selective protein release occurs when other membranes are subjected to divalent cation depletion. Neville (80) found that a major protein of rat liver plasma membrane was released by dialysis against 1 mM EDTA at neutral pH. Guidotti (33) found that this protein bears a strong resemblance to erythrocyte spectrin, besides being released in a similar fashion, and that both resemble myosin in certain compositional and physicochemical features.

Abrams (1) demonstrated that the release of an ATPase bound to *Strepto-coccus faecalis* protoplast membranes depended upon the removal of poly-valent cations, such as Ca^{++}, Mg^{++}, and spermidine. The binding was partially restored by mixing the membranes plus ATPase with 1 mM divalent cations. This elution behavior and other molecular characteristics (113) resemble those observed for the inner mitochondrial (F_1) ATPase (92). Chloroplast thylakoid membranes release a major protein, when washed with 1 mM EDTA (52), which also resembles the F_1 ATPase of mitochon-dria (92). It is interesting that all these proteins contain a large excess of acidic over basic side chains and a low content of hydrophobic residues compared with other membrane proteins which are not released by ionic manipulation.

Human erythrocyte ghosts may be washed repeatedly in hypotonic buffers (e.g., 5–10 mM Na phosphate, pH 7–8) without alteration. In con-trast, Burger *et al.* (10) observed that when *bovine* erythrocyte ghosts were washed three or more times they progressively shed a considerable fraction of their acetylcholinesterase into the supernatant solution, with a five-to six-fold increase in its specific activity. A roughly proportional amount of lipid was released concomitantly. Raising the ionic strength back to 0.15 M NaCl after hemolysis did not prevent this loss during subsequent washing. However, various divalent cations (1–5 mM) prevented the release.

HIGH IONIC STRENGTH

If a protein is bound electrostatically, raising the ionic strength might elute it without disrupting the membrane. A prototype of this reaction is the widely cited case of mitochondrial cytochrome c (14). Raising the level of NaCl also has a dramatic effect on erythrocyte membranes prepared at low ionic strength. Fairbanks *et al.* (23), for example, selectively eluted a single polypeptide from human erythrocyte ghosts by exposure to cold 0.1–0.5 M NaCl, but whether this component is actually an adsorbed cytoplasmic protein has not been determined. Mitchell and Hanahan (76) found that overnight incubation in cold 0.6–1.2 M NaCl released a fraction of the human erythrocyte membrane which was significantly enriched in acetyl-cholinesterase and membrane lipids. (The acetylcholinesterase appeared to be associated with the lipids, floating together on a 1.21 g/ml density barrier after ultracentrifugation.) The presence of small amounts of Tris buffer retarded the release. The residual ghosts were modified in appearance but grossly intact.

Concentrated solutions of certain neutral salts are known to be effective solubilizers and even denaturants. Their relative effectiveness agrees well with the Hofmeister (lyotropic) series for both cations and anions (131). For example, 2 M NaI appears to liberate a Na^+, K^+-dependent ATPase from pig kidney membranes which was not sedimented after a 2.5×10^7 g-min centrifugation (95). In one study, several such "chaotropic" salts at molar levels effected a partial release of membrane-bound proteins, but in addition, facilitated oxidation of the lipids (40). Recently, Marchesi *et al.* (46, 155) reported on the usefulness of lithium diiodosalicylate (0.3 M) in solubilizing erythrocyte membrane proteins.

PH

The influence of pH on the physicochemical properties of isolated membrane proteins has been studied in several contexts. Proteins isolated from red blood cell membranes by deionization (35), butanol extraction (94), or Triton X-100 treatment (Yu and Steck, unpublished data) were found to be most insoluble around pH 4 but soluble above and below that zone. Green and associates (32, 99) used insolubility in the neutral pH range as a principal criterion for the "structural proteins" of membranes, observing good solubility in the strongly acidic and basic ranges.

Extraction with mild alkali has been found to release certain inner mitochondrial membrane proteins (92). In other studies, Neville (79) incubated rat liver plasma membranes in 0.05 M K_2CO_3 at 25°C for 30 min. After a 6×10^5 g-min centrifugation, the clear supernatant was found to contain 70% of the total membrane protein. However, since these components could not be resolved in conventional polyacrylamide gel electrophoresis (PAGE) in the absence of concentrated urea plus acetic acid, they may not have been truly solubilized by the alkaline extraction. One of the eluted components, composing about 10% of the total membrane protein, has been further characterized (80). It seems to be specifically localized in the liver and in the plasma membranes of that organ. It has an unusual conformation, that of a highly helical two- or three-chain rod (81), similar to that found in tropomyosins and suggested for the structure of proteins saturated with SDS (96). As mentioned earlier, dialysis of the plasma membranes against 1 mM EDTA at neutral pH also released this protein.

Dilute acetic acid causes the selective release of certain proteins from red blood cell membranes (154).

Aqueous Denaturants

A variety of denaturants have been tested on membranes. Their mechanism of action cannot be specified, since it is not known in detail for soluble proteins, and we do not yet have a clear idea of how membrane structure is maintained. A most stimulating perspective on denaturation is offered by Tanford's treatise (131).

GUANIDINE · HCL

Concentrated solutions of this agent are widely used to denature globular proteins to linear random coils (131). Gwynne and Tanford (34) have shown that the polypeptides released from reduced and alkylated erythrocyte ghosts by 6 M guanidine · HCl also appear to be random coils free of intermolecular associations. Gel filtration in this solvent gave molecular weight estimates similar to those obtained with SDS-solubilized preparations.

Guanidine · HCl, however, does not lead to total release of the membrane proteins from their association with membrane lipids. For example, the procedure of Gwynne and Tanford (34) solubilized only about 60% of the membrane protein. The membrane residue has been found to be greatly enriched in certain major polypeptides, including the sialoglycoproteins, which are apparently not well extracted by guanidine · HCl (66, 159). If guanidine thiocyanate is used instead of the hydrochloride, no residue is found and total solubilization is achieved. Guanidine · HCl was also unable to solubilize red blood cell membrane protein aggregates following the extraction of lipids (66, 104), but fully dissolved membrane proteins could be dialyzed into concentrated guanidine · HCl without loss of solubility (104, 141).

UREA

Urea has been found useful as an adjunct in membrane solubilization (56, 67). However, unless used at acidic pH, it may extensively carbamylate proteins, thus modifying their electrophoretic mobilities and other properties (124). It has also been used to maintain the solubility of dissolved membrane proteins during electrophoresis at acidic pH.

Organic Solvents

Polar organic solvents extract and solubilize membrane lipids but denature and precipitate the proteins. Tanford (131) relates this phenomenon to the lowering of the dielectric constant of the medium with a concomitant increase in the electrostatic forces between the amphoteric proteins. In practice, the marked tendency toward aggregation may be suppressed by maximizing Coulombic repulsion between polypeptides through acidification and deionization of the extraction medium.

A variety of solvent systems have been examined, and their effects are quite distinctive. Those solvents which are immiscible with water may extract lipid into the organic phase while dispersing the proteins in the aqueous phase. Complexes of protein and lipid usually remain in the aqueous compartment; however, strong ionic complexes of proteins and phospholipids may partition into hydrocarbon solvents (14). In addition, certain hydrophobic proteins associated with lipids (e.g., the proteolipid of myelin) may prefer chloroform-methanol to water (119).

AQUEOUS 2-CHLOROETHANOL

This highly effective solvent and denaturant was used by Zahler and Wallach (148) to dissolve membrane proteins and dissociate them from lipids. Water-washed membranes were dissolved in 2-chloroethanol–water mixtures (9:1) at a pH of less than 2. These authors used gel filtration chromatography on Sephadex LH-20 (organophilic Sephadex beads) in this solvent system to separate dissolved proteins and lipids. The proteins remained soluble after dialysis into aqueous systems containing concentrated KI, guanidine · HCl, and urea.

Schnaitman (111) modified the above approach by substituting N,N-dimethylformamide for 2-chloroethanol.

AQUEOUS PYRIDINE

Blumenfeld (5) dissolved extensively deionized red blood cell membranes in cold 33% pyridine. When the pyridine was removed by repeated dialysis against cold water, turbidity developed. This material was centrifuged and the sediment was compared with the supernatant. Although the two fractions contained nearly equal amounts of protein, all the recoverable sialic acid partitioned with the soluble proteins. The two fractions had distinctly different amino acid and sugar profiles, suggesting that the method had fractionated the proteins into two classes.

ACIDIFIED PHENOL

Mixtures of phenol, organic acids, and water disperse membranes into their component proteins and lipids. Systems such as phenol–formic acid–water (14:3:3, w/v/v) (15) or phenol–acetic acid–water (130) have been explored. These mixtures can also solubilize lipid-free membrane protein powders (104).

ORGANIC ACIDS

Concentrated formic acid and acetic acid are capable of converting membranes into clear solutions. The nature of the dispersion has not been characterized.

BUTANOL AND PENTANOL

When erythrocyte membrane suspensions, washed free of salt, are shaken with cold *n*-butanol, almost all the lipid is extracted into the organic phase, while the proteins remain in the aqueous fraction; a small amount of insoluble protein collects at the interface (64, 94). Certain membrane enzymic activities survive the extraction. The protein becomes less soluble if the residual butanol is dialyzed away. Rapid addition of salt precipitates the protein; a gradual increase of the ionic strength does not (94).

The substitution of *n*-pentanol for *n*-butanol in this extraction protocol produced rather different results (151). The phospholipids and cholesterol were recovered along with the protein in the aqueous phase. The fact that the lipids appeared to travel with the proteins upon density gradient centrifugation suggests that the lipids remained associated with the dispersed proteins. This inference is strengthened by noting that pure lecithin is extracted into the pentanol phase in the absence of proteins (Getz, personal communication). Extraction of membranes with mixtures of butanol and pentanol of varying proportion caused the lipids to partition accordingly into both the aqueous and organic phases.

Kundig and Roseman (56) have recently described the liberation of active enzymes II, which catalyze the phosphorylation of sugars entering *E. coli*. Mixtures of butanol plus urea, but neither alone nor a variety of other agents, effected the release of these lipoproteins. Butanol extraction has also been useful in releasing transplantation antigens from membranes (49).

Tertiary amyl alcohol has been used in the preparation of microsomal electron-transfer complexes (63).

Detergents and Surfactants

Unlike the denaturants and solvents discussed above, rather low levels of detergents, often in a nearly stoichiometric relation to the amount of protein and/or lipid, are required for their full effect. This presumably is related to a high chemical affinity of the detergents for the proteins and/or lipids. Another reflection of this avid binding is the difficulty generally encountered in ridding proteins of the bound detergent. Being amphipathic, the detergents presumably act by binding hydrophobically to poorly soluble nonpolar moieties, creating water-soluble complexes. In doing so, they may denature the proteins and disperse the lipids or merely stabilize still-folded and functional proteins or lipoproteins in the aqueous phase. Salton (109) has considered the lytic action of different surfactants on membranes and cell surfaces.

NONIONIC DETERGENTS

Many commercially developed nonionic detergents have been applied to membrane solubilization; Triton X-100 is probably the most common, but Triton X-45 and X-114, Lubrol WX, Tergitol TMN, and Nonidet P40 are also useful. They are effective in the 0.1–10% concentration range, bringing about partial or total release of proteins from membrane systems. Their appeal derives from several features: *1*) their lack of charge permits characterization and purification of the proteins as polyanions by ion-exchange chromatography and electrophoresis; *2*) they do not generally denature proteins, so that active enzymes may be released, purified, and characterized in their presence; *3*) they may stimulate or activate solubilized membrane enzymes and may substitute for lipid molecules formerly associated with the enzyme; *4*) they may expose latent enzymic activity within membrane-bound structures by disrupting the barrier to the free access of substrate to enzyme; this mode of action is occasionally confused with true activation of membrane enzymes; and *5*) certain membrane systems may resist the action of these detergents, while others are quite susceptible; in this way the outer nuclear membrane can be selectively removed from nuclei with Triton X-100 (see Chap. 3).

These agents may disperse membranes into a form not sedimentable at 6×10^6 g-min (75). From the fact that certain membrane-bound enzymic activities migrated in 7% polyacrylamide gels when prepared in dilute Triton X-100, we may infer that these functions were not part of large complexes (17). Triton X-100 has been used under well-defined conditions to effect the selective solubilization of a few erythrocyte membrane proteins along with a proportional fraction of the lipids (Yu and Steck, unpublished data).

Some enzymes solubilized by nonionic detergents, such as those of the Triton series, are inactivated by storage in that detergent solution. Since removal of the detergent can lead to aggregation of the proteins, the situation can sometimes be remedied by exchanging the solubilizing detergent for a milder surfactant, such as those of the Tween series. Though the Tween detergents are usually poor solubilizing agents, they often have dispersive properties sufficient to prevent aggregation of the solubilized proteins. The detergent exchange can be accomplished simply by applying the solubilized membrane extract to a column of Sephadex G25 or G50 which has been equilibrated with new detergent (102).

CATIONIC DETERGENTS

Cetyltrimethyl ammonium bromide (CTAB) and other quaternary ammonium ion surfactants interact strongly with membranes (109). Little detailed information on the solubilization of membrane proteins is available. However, Heller (41) found that 2% digitonin, 2% deoxycholate, 8 M urea, and 6 M guanidine · HCl failed to truly solubilize bovine rhodopsin; that SDS decolorized it; but that 40 mM CTAB was effective in liberating the 28,000-dalton polypeptide with its original spectral properties intact. CTAB (1%) is apparently able to completely solubilize the red blood cell membrane, so that the polypeptides appear to be nonaggregated upon PAGE in this detergent (Fairbanks and Avruch, personal communication).

BILE ACIDS

The first types of anionic surfactants widely used in membrane extractions were the salts of bile acids, particularly sodium cholate and deoxycholate. These are mild enough in action to allow recovery of enzymic activity in many cases (17, 19, 56, 92). Bile salts may disrupt membranes into large, heterogeneous particles or more homogeneous lipoprotein dispersions (56, 70); they can also effect a major separation of the membrane proteins from the lipids; the phospholipids, at least, may enter into large micelles with the surfactant (88, 110). The proteins are sufficiently dispersed, in some cases, to penetrate 7% polyacrylamide gels electrophoretically (17).

SODIUM DODECYL SULFATE (SDS)

This widely used anionic detergent efficiently dissolves proteins, dissociates them from lipids, and brings both into solution as SDS complexes. The effect of SDS on proteins has recently been studied in detail. The interaction between the polypeptides and detergent was shown by Rosenberg *et al.* (103) to be primarily hydrophobic. Proteins of widely different compositions are all saturated at equilibrium with about 1.4 g SDS per gram of protein in the form of detergent monomers rather than micelles (90, 97). The proteins are denatured, but all assume a distinctive conformation as judged by hydrodynamic and optical rotatory dispersion measurements. A plausible model is that of a helical polypeptide chain folded near its middle, like a hairpin, to form a double helical rod which is stabilized by an envelope of bound SDS (96). This conformation is comparable to the helical rod structure of the tropomyosins; furthermore, it helps account for the regular relation between the electrophoretic mobility and molecular weight of proteins dissolved in SDS (24, 96; see under "SDS-PAGE Systems").

Other Methods

SONICATION

Irradiation with high-frequency sound waves shears membranes into small particles whose size reflects the intensity of the treatment. Since the technique relies on mechanical rather than chemical forces, a general separation of protein-protein or lipid-protein associations would not be expected. This mode of disruption has been used to release trapped proteins and to facilitate the action of chemical treatments. Mitochondrial subfractionation, in particular, has relied heavily on sonication in the preparation of functioning submitochondrial vesicles, fragments, particles, and proteins (21, 92). Sonic irradiation is also widely used in the preparation of transplantation antigens (49). Prolonged sonic irradiation may reduce membranes to small lipoprotein particles. There is the danger, however, that these may not reflect the native state of membrane constituents or their organization (135). There is also strong tendency toward aggregation in sonicates treated with inorganic salts (53).

SUCCINYLATION

Reaction with succinic anhydride has been used to solubilize various insoluble proteins, including membrane proteins. Amphoteric polypeptides are converted to polyanions as amino groups are replaced by carboxylate moieties. This modification should also minimize interactions with membrane lipids, which mainly have a net negative charge at neutral pH. Rosenberg and Guidotti (104) solubilized denatured, delipidated red cell membrane proteins with an excess of succinic anhydride. These authors point out, however, that incomplete reaction of the amino groups would create heterogeneity within originally homogeneous protein classes and thus complicate analysis.

PROTEOLYSIS

Digestion of membranes with proteases can release soluble polypeptide fragments of biochemical interest. Surface glycopeptides, some with blood group (144) and histocompatibility (117) antigen activities, have been liberated in this way. Trypsin digestion of submitochondrial particles has played a role in resolving the complex factors associated with oxidative phosphorylation (92). Recently, Forstner (27) found that mild papain digestion released heterogeneous intestinal surface membrane glycoprotein fragments in association with several species of surface hydrolase particles. In at least two instances, proteolysis seemed to cleave the exposed, hydrophilic portion of membrane proteins from a hydrophobic section buried in the membrane, thus liberating water-soluble, biologically active peptides from otherwise insoluble proteins (123, 144).

FRACTIONATION OF MEMBRANE PROTEINS

General Approaches

The fractionation of proteins from membranes is patterned after well-established techniques developed for soluble systems. Often these may be applied directly, but their use must often accommodate the water-insoluble nature of membrane constituents.

The first step in membrane protein purification is frequently the isolation of the membrane; if, however, the protein being sought is particularly abundant or readily isolated (e.g., cytochrome c), the entire cell or tissue may be extracted at once. Similarly, the membranes of organelles like mitochondria are not usually purified free of their contents prior to fractionation. However, the enclosed proteins should not be mistaken for membrane components.

The next step is the release of the target protein from the matrix. If this can be made selective, considerable purification may be possible. Mild techniques are more likely to be selective than denaturants or organic solvents (however, exceptions occur in both directions).

The centrifuge is one of the principal tools of fractionation. The dispersed membrane components may be heterogeneous in size and density and thus amenable to differential, rate zonal, or equilibrium density centrifugation. The analytical data obtained at the same time may help in characterizing the isolates.

The component may be released from the membrane in a truly soluble form or as part of a multimolecular complex. In either case, standard fractionation techniques may be applicable. Fractional precipitation with ammonium sulfate or other compounds used to "salt-out" proteins may be tried even on detergent-solubilized proteins. Many membrane proteins are insoluble in the neutral pH range. Even those polypeptides made soluble by nonionic detergents and the like may be reversibly precipitated as they are brought near their apparent isoelectric points (around pH 4–5 for erythrocyte and other membranes).

Proteins dispersed in a solvent, detergent, or denaturant may be dialyzed against water to remove these agents, but precipitation of the proteins usually occurs. If all the bound detergent is to be eliminated, extraction with polar solvents (e.g., 90% acetone or 80% ethanol) may be required. Such solvent extractions may also be used to remove lipids complexed to the extracted proteins; however, protein precipitation usually results. Membrane proteins may be dispersed free of lipids by strong solvents (e.g., 2-chloroethanol) or detergents (e.g., SDS) and separated from the lipids by gel filtration (59, 111, 148), gel electrophoresis (59, 110), or ion-exchange chromatography. If the proteins being investigated can adsorb to diethylaminoethyl (DEAE) cellulose, for example, lipid removal can be effected by extensive washing of the column with low-ionic-strength buffer solutions containing nonionic detergents. Subsequent elution at increased ionic strength may yield a protein fraction enriched in the protein/lipid ratio by as much as 1000-fold (102).

Gel filtration chromatography may be performed in the presence of denaturants, e.g., 6 M guanidine · HCl (34), by equilibrating the beads with the appropriate reagent prior to loading the sample. This technique provides analytical data on the molecular size of the solubilized components in addition to separating them. It appears that intact membranes, membrane fragments, and dissolved membrane components may bind tenaciously to anion exchangers such as DEAE-Sephadex or DEAE-cellulose. The tight binding may be electrostatic, owing to the polyanionic protein composition, but may also involve other short-range interactions. The samples may be processed in the presence of detergents or denaturants. For example, Spatz and Strittmatter (123) solubilized cytochrome b_5 and cytochrome b_5 reductase from rabbit liver microsomes in 1.5% Triton X-100 and applied the mixture to a

DEAE-cellulose column lacking detergent. The former protein was retained but the latter washed through with detergent-free buffer. The cytochrome b_5 could then be eluted with buffer containing 0.25 M thiocyanate and 0.25% deoxycholate.

This brief account can only hint at the variety of techniques available. Many rich details of their application are recounted in an entire volume devoted to methods used in analyzing two important membrane-centered processes, oxidation and phosphorylation (21). The remainder of this section will be devoted to a topic of particular interest, namely, the electrophoretic analysis of membrane proteins.

Electrophoresis

Zonal electrophoresis is widely used to characterize proteins by virtue of their charge and/or molecular weight. It plays a major role in purification procedures because it is a prime mode of ascertaining purity, and it may be adapted in several ways to the preparative scale.

A variety of liquid, semisolid, and solid media have been used to stabilize and support the electrophoretic column. Gels cast of cross-linked polyacrylamide have recently won increasing favor. The many attractive features of polyacrylamide gel electrophoresis (PAGE) have been made clear in a symposium devoted to this topic (143) as well as in recent discussions of its current application (13, 67). Among these virtues are simplicity, sensitivity, versatility, economy, speed, reliability, and reproducibility. The gels have excellent mechanical stability, chemical inertness, and purity; their lack of fixed charges obviates the troublesome electroendosmosis found in some other supporting media. High-resolution analysis of complex mixtures of proteins in microgram quantities may be obtained in the 10^3- to 10^6-dalton size range. The system is applicable over a wide pH range in the presence of a variety of additives such as detergents and denaturants. Gels may be stained, scanned spectrophotometrically, photographed, autoradiographed, and sliced for isotopic counting or chemical analysis.

It is commonly found that "solubilized" membrane proteins do not penetrate polyacrylamide gels. This is usually related to the fact that the proteins are not completely dispersed but are present as complexes or aggregates. In addition, some genuinely solubilized proteins may reaggregate upon exposure to the nearly neutral buffers of high ionic strength (0.1–0.3 M) usually used in electrophoresis. To maintain the proteins in true solution, a variety of denaturants and detergents have been introduced into standard PAGE protocols. Takayama *et al.* (130), for example, developed a procedure for analyzing insoluble, lipid-free mitochondrial proteins. The extracts were dissolved in phenol–acetic acid–water (2:1:1, w/v/v), made 2 M in urea, and layered onto 7.5% polyacrylamide gels which contained 35%

acetic acid and 5 M urea. The buffer reservoirs contained acetic acid but did not require urea, since this uncharged compound should not migrate electrophoretically from the gel. Neville (79, 80) described a similar urea–acetic acid system, while Tuppy *et al.* (137) modified the Takayama system by substituting propionic acid for acetic acid throughout. Panet and Selinger (84) added 0.1% of the nonionic detergent, Nonidet P40, to the Takayama system and reduced the sample with 0.5% mercaptoethanol. The presence of the acetic acid in this system serves several purposes: *1*) it facilitates solubilization of proteins in phenol, *2*) it enhances electrophoretic mobility by conferring a strong positive charge on the proteins, and *3*) it suppresses the accumulation of cyanate, which is formed by the decomposition of urea. Cyanate reacts with several protein side chains, including both amino and carboxyl groups (124), and would therefore modify the intrinsic charge characteristics of the proteins and muddle the electrophoretic analysis.

Demus and Mehl (15) have used gels into which a phenol–formic acid–water (14:3:3, w/v/v) mixture (without urea) was equilibrated following gel polymerization. Other denaturing systems have also been explored; by far the most widely used is electrophoresis in SDS, discussed next.

SDS-PAGE Systems

CONCEPTS

Several practical advantages accrue from the avid, extensive, and uniform binding of SDS to proteins (see above under "Detergents and Surfactants"). Total solubilization and dissociation of proteins and lipids are usually readily achieved by brief incubation of membranes with saturating levels of SDS. There is usually no need for prior lipid extraction, dialysis, and the like (23). The bound SDS confers a high electrophoretic mobility, hence shortening the time required for electrophoresis.

When the inter- and intramolecular associations are broken and disulfide bonds reduced, the SDS-saturated polypeptide chains appear to assume the shape of doubled-over helical rods. Their minor axes are uniformly about the dimensions of two adjacent helices ($\sim$ 18 Å), so that particle length is directly proportional to the polypeptide chain length (96). Furthermore, saturation with SDS minimizes the intrinsic charge variation among the proteins by conferring a similar charge per unit mass on all (97).

These two properties—uniform charge density and a direct relation between hydrodynamic behavior and molecular weight—lie behind the well-documented finding that the electrophoretic mobility of proteins in SDS-PAGE systems is inversely proportional to the logarithm of their molecular weights (116, 142). This simple, often linear, relation has been embraced by biochemists as a most expeditious adjunct to the estimation of the molecular weights of polypeptides, membrane and otherwise. (It should be noted that

many undenatured proteins [153] and single-stranded RNA molecules [86] obey a similar size-mobility relation on gels lacking detergent. They might be presumed to have uniform charge densities and molecular shapes, analogous to proteins in SDS-PAGE systems. Similar relations in gel filtration systems depend only on molecular shape and not on their charge.)

PRACTICE

Gels. An acrylamide concentration of around 5% (with 2.7–5% of that being N,N'-methylenebisacrylamide, the cross-linking compound) allows resolution and molecular weight estimation in the 20,000- to 200,000-dalton range. At 2–3% acrylamide, 10^6-dalton polypeptides enter the gel; however, the gels are so soft that agarose should be included to strengthen them mechanically (86). The acrylamide concentration may be extended to 20% for analysis of small polypeptides (13, 67).

Buffers. Various buffers, such as Tris acetate (23) or phosphate (67) have been used in the neutral pH range. Usually, SDS-PAGE is performed in a single continuous buffer and gel system. A discontinuous gel and buffer system may also be employed to augment resolution (57). Merely raising the ionic strength of the buffer system relative to that in the sample, however, tends to sharpen the protein zones as they enter the gel (67). This effect is probably not as important in zone sharpening as is the compacting of the proteins as they move from free solution into the highly retarding gel phase. (Since raising the ionic strength of the buffer prolongs the electrophoresis time, it is more convenient to minimize the electrolyte concentration in the applied sample solution.) EDTA is frequently added to the buffers and potassium omitted, because SDS forms insoluble salts with K^+ and heavy metals.

Samples. The sample should be saturated with SDS, both to effect total dissolution and to ensure uniform charge and hydrodynamic properties. About 1.4 g SDS per gram of protein is maximally bound at monomer equilibrium concentrations of greater than 0.5 mM. Maintaining a low ionic strength in the sample buffer raises the critical micelle concentration and enhances the binding of SDS monomers to the protein (97). In practice, making a sample of 1–2 mg protein per milliliter up to 1% SDS (i.e., 10 mg/ml) at $\mu < 0.1$ should be adequate. Dissolving the membranes in this high level of SDS, even if the run is made in a 0.1% SDS-PAGE system, also helps to inactivate contaminating proteases. These may remain transiently active in dilute (0.1%) SDS while the denaturation of the membrane proteins heightens their sensitivity to proteolysis (4, 23, 125). Recalcitrant samples have been dissolved by heating to boiling, lowering the pH, and adding 8 M urea (67).

Reduction of disulfide bonds must be carried out if intra- and inter-molecular associations are to be broken and if accurate molecular weights are to be estimated. 2-Mercaptoethanol (0.5–5%) or dithiothreitol (5–50 mM) is used for this purpose. Alkylation of the sulfhydryl groups may be performed (e.g., with 50–100 mM iodoacetamide) but is not usually necessary. (SDS may actually protect sulfhydryl groups from oxidation to disulfide links [24].) The presence of reducing agents in the gel during the run is not generally necessary.

It has been common practice to dialyze the sample solution against the gel electrophoresis buffer prior to the run, presumably to bring the level of SDS and other buffer system components to those of the gel system (67, 116). Recently, investigators have found that dialysis is unnecessary; furthermore, it may be undesirable since low-molecular-weight polypeptides may thereby be lost. Samples may therefore be dissolved in 1–2% SDS (for 15 min at 37°C or 1–2 min at 100°C) followed directly by electrophoresis on 0.1–1% SDS-containing gels (23, 57, 142).

Electrophoresis. The proteins are electrophoresed toward the anode at room temperature (since SDS is less soluble in the cold). A field of 3–12 V/cm and a current of 5–15 mA per gel tube are typical. Gels are usually run for a fixed time interval. However, running each gel to a constant migration distance of the tracking dye allows more facile intercomparison of gels, especially from run to run.

Fixation and staining. The electrophoretic pattern may be preserved indefinitely if the proteins in the gel are kept from diffusing. Overnight incubations in concentrated precipitants such as 20% sulfosalicylic acid (67) or 15–50% trichloroacetic acid (59) have been used for this purpose. Subsequent staining is often carried out in 0.25% Coomassie brilliant blue, a stain acclaimed for its high sensitivity. Staining, destaining, and storage of gels should be done in acidified media (e.g., 5–10% acetic acid), since keeping the proteins strongly positive in charge facilitates the binding of the anionic dye. Excess stain may be slowly eluted or removed electrophoretically (67).

Another approach to fixation and staining focuses on the rapid extraction of the SDS, which causes the denatured proteins to aggregate. Since proteins cannot be properly stained until the SDS is gone, removal of the detergent is an obvious first step. Soaking the gels in 25% isopropanol (23) or 45% methanol (142) for 10–15 hr accomplishes this goal. Actually, the Coomassie blue and acid can be included in the initial alcohol fixation step. Because of the efficient removal of SDS by alcohol, low levels of stain (e.g., 0.02–0.05%) are sufficient. In this way, the entire fixing, staining, and destaining cycle is completed within 36 hr. Alcohol also facilitates destaining

(23, 142), but its excessive use will elute the Coomassie blue from the protein bands.

Other stains are also useful. Amido black and Ponceau S, for example, are comparable to Coomassie blue, but stain less intensely (23). Certain glycoproteins do not adsorb any of these stains but react well with the periodic acid–Schiff (PAS) reagent for carbohydrates (23, 45). It is presumed that the heavy coat of oligosaccharides, composing 50% or more of the glyco-protein mass, blocks the binding of the protein stains.

The PAS procedure is a sensitive glycoprotein stain (23, 147). It may be difficult to quantitate, however, since the color intensity which develops is somewhat evanescent and sugar-dependent. For example, terminal sialic acids may show much more color than other moieties, presumably because of their exceptionally active oxidation by periodate.

PROBLEMS

Variations in mobility. Although, as Shapiro *et al.* (116) reported, variation of the SDS concentration between 0.1% and 1% does not substan-tially alter relative electrophoretic mobilities, Fairbanks *et al.* (23) found that material which yielded one sharp band in 1% SDS gels split into a doublet of two equally sharp, yet clearly distinct components at $\leq 0.2\%$ SDS. The mobility of individual members of complex mixtures of proteins is also somewhat dependent on the presence of the other proteins, particularly if a minor band migrates just ahead of a major one (23). While protein-protein and protein-lipid associations are generally believed to be abolished by this detergent, this is probably not always true (78).

Aggregation. In the ideal case, proteins are totally dissociated by reduc-tion in SDS prior to electrophoresis. In practice, multichain complexes may persist under these conditions. The complexes may appear as discrete bands or as diffuse zones; sometimes they are large and do not penetrate the gel. The appearance of protein adherent to the gel origin is a signal that an un-determined fraction of the sample has been excluded from the analysis; the material capping the top of the gel may only be a small part of that which could not penetrate.

Simple chemical treatment or the mere extraction of membranes with organic solvents may cause the irreversible aggregation of polypeptides, sometimes in a rather selective fashion (23). A single major polypeptide was found to be irreversibly aggregated by dispersing erythrocyte ghosts in con-centrated guanidine · HCl. The aggregation could be prevented by the prior reduction and alkylation of sulfhydryl groups, implying that denaturation of this protein promoted formation of intermolecular disulfide bonds which became inaccessible to conventional reducing agents even in the presence of 1% SDS (159).

Proteolysis. The analysis of membranes with the SDS-PAGE system is particularly susceptible to proteolysis artifacts. The problem begins with the potential contamination of membrane preparations by proteases. The membranes may bear these enzymes themselves or be contaminated by lysosomal proteases. In Chapter 3 (Fig. 2), it is observed that lysosomes may appear in every conventional subcellular fraction. This possibility is not often tested for or even anticipated. While red blood cells lack lysosomes, a trace of leukocytes remains associated with them no matter how well they are washed. Following lysis, white cell lysosomes may contaminate the ghosts by pelleting beneath them. The offending intruders are rich in proteases but can fortunate-ly be removed by aspiration of that firm tiny pellet (23).

When SDS is introduced into a supposedly purified membrane sample, the membrane proteins unfold and become many times more sensitive to the action of proteases, which may be released from latency in the lysosomes at the same time. At least some proteases appear to withstand the SDS denatura-tion long enough to cause extensive digestion of the vulnerable membrane proteins (125). Elevated electrolyte concentrations seem to potentiate the proteolysis (23; Fairbanks and Avruch, personal communication).

In nondenaturing systems partially digested proteins can retain their physical and even functional integrity and hence their identity through ex-tensive noncovalent associations. Since SDS denatures the proteins and separates the cleaved polypeptide chains from each other, however, the split products are dissociated from the parent molecules upon electrophoresis. SDS-PAGE is thus unsurpassed in its sensitivity to proteolysis; cleavage of one peptide bond per molecule may be manifested by a greatly altered gel pattern.

The controlled proteolysis of proteins has been useful in the analysis of membrane organization; however, uncontrolled proteolysis leads to confu-sion. If a proteolytic enzyme is deliberately added to membranes as a probe, it is imperative that its activity be entirely abolished before SDS is added. Appropriate inhibitors are available for some well-characterized proteases (4, 125). Washing the membrane is another way of eliminating the enzyme; however, traces of proteases could remain adsorbed to the membranes (4). In any event, it is important that a zero-time incubation control be per-formed to verify that proteolysis does not occur following the designated incubation period. If contamination with proteases is suspected, further membrane purification, rapid sample preparation in the cold, high concen-trations of boiling SDS during the solubilization step, minimization of ionic strength, and addition of EDTA may eliminate the problem (4, 23).

Molecular weight estimation artifacts. The SDS-PAGE system affords good estimates of polypeptide molecular weights with delightful ease. How-ever, a few complications should be borne in mind. First, while the logarithm

of molecular weight varies inversely with electrophoretic mobility, the relation is not always linear. Weber and Osborn (142) found that the use of a low proportion of cross-linking reagent in their 10% acrylamide gels (i.e., 1.35 g N,N′-methylenebisacrylamide per 100 g acrylamide) resulted in a concave log MW mobility curve. The use of two or three times this proportion of cross-linking agent obviated the concavity, but reduced the upper molecular weight limit for included polypeptides (142). Others have observed that reducing the total acrylamide but maintaining a high proportion of cross-linking agent yields linear curves for the high molecular weight range.

Often the log MW mobility curve is not linear for its entire length, but deviates at both its high and low ends toward the vertical. Therefore, calibration curves should be examined for the limits of linearity. Fairly accurate molecular weight estimates can still be obtained from smooth nonlinear curves by interpolation between standards of known molecular weight.

Another issue is that of the intramolecular cross-linking of sample proteins by disulfide bonds. These links might be expected to interfere with the folding of SDS-saturated proteins into the characteristic double helical rod conformation. Indeed, Fish *et al.* (24) have pointed out that unreduced proteins in SDS have an apparently smaller hydrodynamic size than if reduced, and hence have an anomalously enhanced mobility in SDS-PAGE. This tendency may be offset, however, because the diminished binding of SDS to disulfide-linked polypeptides (90) might lead to a diminished charge density and electrophoretic mobility. In any event, if disulfide-bonded proteins are used without reduction as molecular weight standards, the unknown proteins may be assigned artifactually high molecular weights, as pointed out by Fairbanks *et al.* (23) and noted by Bender *et al.* (4). Conversely, unreduced disulfide-linked polypeptides appear inappropriately small compared with reduced standards. Clearly, routine reduction of disulfide bonds should obviate this confusion.

Should the intrinsic charge on the polypeptides influence their mobility in SDS-PAGE? In general, the amino acid composition and conformation of a polypeptide can influence the amount of SDS bound, hence its charge density and conformation in an SDS solution (78). However, an empirical evaluation of 40 different soluble proteins suggests that natural variation in composition does not affect behavior in the SDS-PAGE system beyond about $\pm$ 10% of the mean (142). Recently, Tung and Knight (136) showed that the blocking of amino groups on proteins by reaction with maleic anhydride altered their mobility. But instead of the anodic mobilities increasing as the positive amino charges were eliminated, the mobilities were inexplicably decreased. It might be that the maleylated proteins bound less SDS than their native counterparts, as observed following succinylation (90), and thus migrated more slowly. Polylysylglutamic acid also bound much less SDS than natural proteins (90). Perhaps, then, the generalization being discussed should be limited to natural or unmodified proteins. The heavy glycosylation found in some glycoproteins may create still another exception to the general uniformity seen among "typical" proteins.

The major glycoprotein of the human erythrocyte membrane presents two interesting anomalies: *1*) it is not visualized with conventional protein stains (23), as discussed above, and *2*) its mobility does not vary with the concentration of acrylamide in the gels in the way that the mobilities of other proteins do (7, 114). This component appears to have a molecular weight of about 9×10^4 on SDS-PAGE (8, 23, 59), while analytical ultracentrifugation has suggested a value of 3×10^4 (144). A possible explanation for this discrepancy is that the high carbohydrate content (two-thirds of the total mass) of this glycoprotein confers unique properties with respect to its SDS-binding capacity, conformation in SDS, and the like. Thus its electrophoretic behavior cannot be interpreted by reference to simple polypeptide standards. The implications of these anomalies for other glycoproteins remain to be explored.

COMPOSITION OF MEMBRANE PROTEINS

We want to know how many different proteins there are in a single membrane, whether they are composed and ordered like other classes of proteins or adhere to unique principles, and how they vary according to their specialized roles and functions. Textbook answers to these questions cannot be formulated, since the data are just now coming in, but it is not too early to inspect the progress recently made.

Amino Acid Composition

Many quantitative amino acid determinations have been performed on a variety of membranes and protein fractions isolated from them. Table I presents illustrative data on the unfractionated proteins of the human erythrocyte membrane, two subfractions of the rat liver plasma membrane, and the plasma membrane and endoplasmic reticulum fractions of mouse Ehrlich ascites carcinoma cells. The amino acids are grouped as basic, acidic, neutral, and hydrophobic (see below).

Several features of these analyses are noteworthy:

1. There are no dominating amino acids or sets of amino acids, as in the collagens.

2. There are no significant levels of unusual amino acids.

3. The hydrophobic residues uniformly represent about 33–35 moles percent of the total (allowing 1% for tryptophan not determined). The acidic side chains consistently outnumber the basic side chains (about 20 moles percent to 12–14 moles percent).

Table I

Amino Acid Composition of Various Membrane Isolates

AMINO ACID	HUMAN ERYTHROCYTE MEMBRANE[a]	Rat Liver Plasma Membrane[b]		Ehrlich Ascites Cell[c]	
		LIGHT FRACTION	HEAVY FRACTION	PLASMA MEMBRANE	ENDOPLASMIC RETICULUM MEMBRANE
Lys	5.2	7.2	6.0	6.3	6.5
His	2.4	2.2	2.0	2.6	2.1
Arg	4.5	5.0	4.6	4.7	5.2
	12.1	*14.4*	*12.6*	*13.6*	*13.8*
Asp	8.5	8.6	9.7	8.8	8.7
Glu	12.2	11.1	10.9	10.1	10.6
	20.7	*19.7*	*20.6*	*18.9*	*19.3*
Thr	5.9	5.7	5.8	5.5	5.4
Ser	6.3	7.0	7.3	6.6	6.2
Pro	4.3	4.2	4.2	5.2	5.4
Gly	6.7	9.2	8.1	8.5	7.7
Ala	8.2	8.6	7.9	7.8	7.6
$\frac{1}{2}$Cys	1.1	0.5	0.9	Trace	Trace
	32.5	*35.2*	*34.2*	*33.6*	*32.3*
Val	7.1	6.9	7.1	6.6	6.7
Met	2.0	1.5	1.2	2.7	2.5
Ile	5.3	5.6	5.9	6.1	5.1
Leu	11.3	10.4	11.3	10.1	10.0
Tyr	2.4	2.7	2.6	3.1	3.4
Phe	4.2	4.7	4.6	4.8	4.8
Try	2.5	—	—	1.5	1.5
	34.8	*31.8*	*32.7*	*34.9*	*34.0*
NH_3	6.9	—	—	14.7	10.8

[a] Data from Rosenberg, S. A., and Guidotti, G. *J. Biol. Chem.* **243,** 1985 (1968).
[b] Data from Evans, W. H. *Biochem. J.* **166,** 833 (1970).
[c] Data from Wallach, D. F. H., and Zahler, P. H. *Biochim. Biophys. Acta* **150,** 186 (1968).
Data are grouped and summed as basic, acidic, neutral, and hydrophobic residues (see text). Values are expressed as moles percent (but sometimes do not total exactly 100%).

4. Where determined, significant ammonia was found; this suggests that a considerable fraction of the carboxylate side chains may be amidated; thus, the proteins really may not be strongly acidic, as it otherwise appears. In Wallach and Zahler's tabulation (141), this observation is extended to the other membranes on which NH_3 was determined.

5. Cysteine levels are rather low, consistent with the apparent paucity of disulfide bridges among membrane proteins.

No outstanding differences are evident among the profiles shown in Table I. Furthermore, the authors of these studies all make the point that their findings resemble data on still other membranes (not given here). While the two fractions of liver plasma membrane differ considerably in buoyant density, lipid content, enzyme profiles, and even protein patterns on PAGE, these distinctions are not manifested in their amino acid compositions. The plasma membrane and endoplasmic reticulum fractions of the ascites cells also show many physicochemical differences which are not reflected in their amino acid composition. The conclusion to be reached from these and other analyses is that a great similarity exists among the amino acid profiles of various membranes which does not reflect their biochemical differences.

Is this similarity in amino acid composition an indication that the proteins of all membranes may belong to a distinctive common class? It is more likely that the close resemblance derives from the tendency of mixtures of proteins, generally, to converge toward a common pattern which is unlike that of the individual constituents. Rosenberg and Guidotti (105) have illustrated this point by showing that the amino acid profiles of unfractionated proteins from various membrane sources resemble that of the total soluble cytoplasmic protein fraction of *E. coli*.

It can also be argued that many membrane preparations contain a large admixture of nonmembrane proteins and that the amino acid analyses obtained in their presence are therefore unrepresentative. Certainly this is true of intact mitochondria, where most of the protein is located between and not in the membranes. Adsorbed and entrapped soluble protein presents a similar problem in other membranes. However, in the case of the erythrocyte and Ehrlich ascites membranes cited in Table I, efforts were made to eliminate extrinsic proteins; yet these analyses still resemble the others.

Amino acid profiles have been examined for clues to how proteins are associated within the membrane. For example, if there is an excess of dicarboxylic acid residues, the proteins will carry a net negative charge and therefore repel most lipids at physiologic pH. Simple Coulombic bindings of protein to lipid, as proposed by the Davson-Danielli model, would be unfavorable. The interposition of divalent cations as bridges between carboxylate and phosphate anions has been suggested and is supported by certain selective extraction experiments with chelating agents (see under "Alterations in Ionic Conditions"). If, on the other hand, the carboxyl side chains are highly amidated, as Table I suggests, the proteins may actually carry a net positive charge at neutral pH, favoring ion-pairing between the basic amino acid side chains and the lipid phosphates (141). However, the rather low apparent isoelectric point (around pH 4–5) of at least red blood cell membrane proteins suggests that the carboxyl groups are not amidated (35, 94).

Since membrane proteins and lipids seem to interact hydrophobically (121), an increase in hydrophobic residues has been searched for in amino acid profiles. Evaluation of this issue is made difficult by *1)* the obscuring effects of nonmembrane proteins which contaminate some preparations, *2)* incomplete determination of amino acids such as tryptophan, *3)* our ignorance of how the individual proteins are disposed in the membranes, and *4*) variations in the classification of residues as nonpolar or hydrophobic.

The categorization of amino acids adopted here (see Table I) reflects the recent study of Nozaki and Tanford (83), who established a hydrophobicity scale based on the free energy of transfer of amino acid side chains from water to less polar solvents. Other authors might have placed alanine with the hydrophobes and taken tyrosine to be more hydrophilic. However, the free energy of transfer of alanine is about the same as that of histidine and threonine, while that of tyrosine falls between leucine and phenylalanine. This suggests not that the alanine methyl group is very polar, but only that it does not have a highly energetically unfavorable interaction with water, the prerequisite for hydrophobic bond formation. The quantitative hydrophobicity scale also allows each side chain to be weighted according to its free energy of transfer, since wide differences exist among them (83).

Although residues such as alanine and proline are not classified here as hydrophobic, they still could participate in van der Waals interactions with nonpolar lipid moieties within the membrane. These side chains do not flee from water, yet may make important nonpolar associations with the hydrocarbons.

Amino acid analyses, such as those in Table I, do not show an over-all increase in the hydrophobic content of membrane proteins. We believe this conclusion may be misleading, however, if it is not refined by looking beyond gross determinations on unfractionated polypeptides to values obtained on individual proteins and, importantly, on oriented fragments derived therefrom. Table II presents amino acid analyses on a series of proteins purified from various membrane sources. They are grouped into those readily eluted by manipulation of ionic conditions and those which cannot be eluted without disrupting the membrane matrix. Two trends are clear: *1*) the elutable proteins are enriched in ionic side chains and are low in hydrophobic residues compared with the nonelutable species, and *2*) there are greater compositional differences between the two proteins derived from the human erythrocyte membrane than between that membrane and the others listed in Table I. This is also true for the erythrocyte membrane glycoprotein depicted in Table III and discussed below. An even more hydrophobic protein than those in Table II, C_{55}-isoprenoid alcohol phosphokinase, was recently purified (158).

If a protein penetrates into the nonaqueous regions of the membrane, we might expect an increase in the hydrophobic content of *that part* of the molecule, while the surface-oriented portion would have a more hydrophilic composition. Data supporting this premise have been advanced through the study of the major red blood cell membrane sialoglycoproteins (144) and the cytochrome b_5 of rabbit liver microsomes (123). In each case, limited proteolysis of the membranes released the exposed, biologically active portions of the polypeptides into solution. In addition, the entire

Table II

Amino Acid Composition of Elutable and Nonelutable Membrane Proteins

	Elutable				Nonelutable		
AMINO ACID	HUMAN RBC SPECTRIN[a]	LIVER EIGEN PROTEIN[b]	S. FAECALIS ATPASE[c]	BOVINE F_1 ATPASE[d]	90,000 MW RBC PROTEIN[e]	BOVINE RETINAL RHODOPSIN[f]	BOVINE MYELIN PROTEOLIPID[g]
Lys *1.5*	6.7	8	6.1	6.2	4.7	4.3	4.5
His	2.6	1	1.7	1.7	1.9	1.7	2.3
Arg *.75*	5.8	7	4.5	5.8	4.8	2.6	2.6
	15.1	*16*	*12.3*	*13.7*	*11.4*	*8.6*	*9.4*
Asp	10.9	11	10.0	7.9	7.5	6.4	4.0
Glu	20.5	20	13.0	11.7	10.9	8.9	5.9
	31.4	*31*	*23.0*	*19.6*	*18.4*	*15.3*	*9.9*
Thr *.45*	3.6	5	6.7	5.8	6.2	7.2	8.4
Ser	4.1	5	6.3	6.2	5.8	5.1	5.2
Pro *2.6*	2.4	1	3.9	4.2	5.5	5.5	2.8
Gly	4.9	5	8.7	9.2	7.5	6.8	10.4
Ala *.75*	9.2	8	8.4	10.4	8.3	8.5	11.9
½Cys *1.0*	1.1	0	0.3	0.4	1.0	2.1	2.9
	25.3	*24*	*34.3*	*36.2*	*34.3*	*35.2*	*41.6*
Val *1.7*	4.7	6	6.8	7.5	7.3	8.5	7.4
Met *1.3*	1.7	2	2.3	2.1	1.9	3.4	1.3
Ile *2.95*	4.0	5	6.2	6.2	4.7	5.5	5.1
Leu *2.40*	12.4	12	9.3	8.7	14.1	8.5	11.5
Tyr *2.85*	2.0	2	3.3	2.9	2.5	4.7	4.4
Phe *2.65*	3.0	2	3.1	2.9	5.6	8.1	8.2
Try *3.0*	—	—	—	0	0.7	2.1	1.8
	27.8	*29*	*31.0*	*30.3*	*36.8*	*40.8*	*39.7*

[a] Data from Marchesi, S. L., *et al. Biochemistry* **9,** 50 (1970).
[b] Data from Neville, D. M., Jr. *Biochem. Biophys. Res. Commun.* **34,** 60 (1969).
[c] Data from Schnebli, H. P., *et al. J. Biol. Chem.* **245,** 1122 (1970).
[d] Data from Racker, E. (ed.). *Membranes of Mitochondria and Chloroplasts.* Van Nostrand-Reinhold, New York, 1970.
[e] Data from Guidotti, G. Personal communication.
[f] Data from Heller, J. *Biochemistry* **7,** 2906 (1968).
[g] Data from Wolfgram, F., and Kotorii, K. *J. Neurochem.* **15,** 1281 (1968).
Data are grouped and summed as basic, acidic, neutral, and hydrophobic residues (see text). Values are expressed as moles percent (but sometimes do not total exactly 100%).

957   975   1051   1060   1216   1204   1282

molecule could be independently solubilized and digested. Soluble peptides indistinguishable from those released from the intact membrane, plus small aggregating polypeptides, were generated from the solubilized parent proteins in both studies.

Amino acid analysis was performed on the intact molecules and their fragments (Table III). The various fragments were either more hydrophilic or more hydrophobic than the original polypeptides. The hydrophilic por-

Table III

Amino Acid Composition of Red Blood Cell Sialoglycoprotein, Rabbit Microsomal Cytochrome b₅, and Their Fragments

AMINO ACID	Sialoglycoprotein[a]			Cytochrome b₅[b]		
	INTACT	SOLUBLE PEPTIDE	INSOLUBLE PEPTIDE	INTACT	SOLUBLE PEPTIDE	AGGRE-GATING PEPTIDE
Lys	3.5	4.3	2.0	7.8	10.3	2.5
His	3.8	4.9	3.8	5.0	7.2	0
Arg	4.1	3.3	3.9	2.8	3.1	2.5
	11.4	*12.5*	*9.7*	*15.6*	*20.6*	*5.0*
Asp	6.0	7.6	2.4	11.3	10.3	10.0
Glu	10.0	4.7	7.0	10.6	14.4	2.5
	16.0	*12.3*	*9.4*	*21.9*	*24.7*	*12.5*
Thr[c]	13.8	23.8	6.2	7.1	7.2	7.5
Ser[c]	13.6	23.8	6.0	7.1	7.2	7.5
Pro	6.5	4.0	4.5	3.6	3.1	5.0
Gly	6.8	3.5	10.7	5.0	6.2	2.5
Ala	6.8	6.1	7.9	7.1	5.2	10.0
½Cys	0	0	0	0	0	0
	47.5	*61.2*	*35.3*	*29.9*	*28.9*	*32.5*
Val	7.7	4.9	8.1	5.0	4.1	7.5
Met	0	0	2.1	2.1	1.0	5.0
Ile	4.5	2.8	14.3	5.7	4.1	10.0
Leu	4.5	1.6	11.6	10.6	9.3	12.5
Tyr	3.6	1.5	2.4	3.6	3.1	5.0
Phe	3.5	0	5.1	2.8	3.1	2.5
Try	—	—	—	2.8	1.0	7.5
	23.8	*10.8*	*42.6*	*32.6*	*25.7*	*50.0*
MW	31,400	10,000	?	16,072	11,079	4,579

[a] Data from Winzler, R. J. *Red Cell Membrane Structure and Function.* Ed. G. A. Jamieson and T. J. Greenwalt. Lippincott, Philadelphia, 1969, p. 157.
[b] Values calculated from Spatz, L , and Strittmatter, P. *Proc. Natl. Acad. Sci. U.S.A.* **68,** 1042 (1971).
[c] The large fraction of serine plus threonine residues in the sialoglycoprotein and its soluble peptides reflects the linkage sites for the oligosaccharides.
Data are grouped and summed as basic, acid, neutral, and hydrophobic residues (see text). Values are expressed as moles percent (but sometimes do not total exactly 100%).

tions corresponded to those released from the intact membrane and bore the biological activity. The hydrophobic portions, which were not released, are presumed to represent "anchors" or "tails" which linked these proteins to the membrane matrix.

Our original questions concerning the uniqueness of membrane protein composition can now be seen in clearer perspective. As with other mixtures of proteins, the over-all amino acid analysis is a low-resolution indicator; the comparison of individual protein profiles is more edifying; and a glimpse of even the gross organization of amino acids within the polypeptides is quite illuminating.

The topic of blocked terminal amino acids deserves mention. While no systematic survey of the ends of the polypeptide chains of membranes has been performed, several studies revealed that the N-terminal amino groups and/or the C-terminal carboxyls were not free to react. The importance of reserving this type of analysis for well-characterized polypeptides rather than membrane protein mixtures is that *1*) adsorbed or trapped extraneous proteins must be eliminated and *2*) unsuspected proteolysis may have created spurious reactive termini in polypeptides which normally have none.

Heller (41) has found that the rhodopsin of the bovine retina lacks both N- and C-terminal reactivity. No N-terminus was found for either of two subunits of the ATPase of *S. faecalis* membranes (113). Likewise, Guidotti found that the two major high-molecular-weight (200,000- to 250,000-dalton) proteins of the erythrocyte membrane, spectrin, lacked free terminal amino group activity (33). Winzler (144) has found no terminal α-amino groups detectable in the major sialoglycoprotein of the human erythrocyte membrane. The report that "miniproteins" lack N-terminal reactivity (58) must be discounted since this material is evidently not actually peptide in origin (16). Since the polypeptides just enumerated are some of the few purified membrane proteins for which end-group analysis has been reported, we must eagerly await further data on other polypeptides.

Carbohydrate Composition

Cell membranes, particularly the plasma membrane, bear an array of oligosaccharides in the form of glycolipids and glycoproteins. The sugars of cell surface glycoproteins have been obtained both from isolated membrane proteins and from glycopeptide fragments released proteolytically from intact cells and membrane preparations.

Several generalizations can be forwarded at this early time, based on the apparent resemblance in composition and structure between nonmembrane (30) and membrane glycoproteins (9, 27, 29, 42, 55, 73, 77, 87, 118, 128, 144): *1*) the carbohydrate units are short heterosaccharides of specific sequences; *2*) the preponderant monosaccharides are galactose, glucose, N-acetyl-galactosamine, N-acetylglucosamine, mannose, fucose, and sialic acid, the last two being situated only at nonreducing termini; *3*) the oligosaccharides are linked either O-glycosidically to the hydroxyl of a serine or threonine residue or N-glycosidically to asparagine; *4*) there may be many hetero-saccharide species on a given membrane; the carbohydrate profile varies characteristically among glycoproteins and seemingly among organelles; *5*) many membrane proteins bear no detectable sugar.

Two major membrane glycoproteins have recently been purified and analyzed. Since they contrast sharply, their comparison is edifying. Bovine rhodopsin contains one oligosaccharide unit on each 28,000-dalton polypeptide chain (42). This heterosaccharide is composed of three mannose and three N-acetylglucosamine residues, and is joined to the protein by an N-acetylglucosaminyl–asparagine linkage.

A major sialoglycoprotein component of human erythrocyte membranes has been characterized by Winzler and his colleagues (144). It has not been proved to be a single homogeneous molecule. In fact, several biological activities are associated with this glycoprotein fraction (e.g., the MN blood group antigen, influenza receptor activity, and phytohemagglutinin binding), and three sialoglycopeptide species have been resolved by SDS-PAGE (23). Unlike the retinal pigment, which contains only around 4% sugar, this material is about 64% carbohydrate. It carries much of the membrane's sugar, including most of the sialic acid. Its other principal monosaccharide constituents are galactose, mannose, fucose, acetylglucosamine, and acetylgalactosamine (144). Some of these are assembled into tetrasaccharides containing two sialic acids, a galactose, and an N-acetylgalactosamine. There are many such units linked via the N-acetylgalactosamine to serine and threonine residues on each polypeptide chain (144). Some of these heterosaccharide units may be "incomplete," reflecting variation in their synthesis or degradation.

Other oligosaccharides are also present in this complex glycoprotein. They are presumably linked N-glycosidically to asparagine residues, have a different composition, and are larger than the O-glycosidic tetrasaccharides. Kornfeld and Kornfeld (55) have delineated the structure of one of these heterosaccharide units. There is a single branched chain; the two (nonreducing) branches are galactose → N-acetylglucosamine and sialic acid → galactose → N-acetylglucosamine. These are joined to an inner core composed of two mannose and one N-acetylglucosamine residues. These oligosaccharides are potent receptors for the kidney bean protein which agglutinates erythrocytes (55). Fucose-containing glycopeptide fragments have also been recently characterized (161).

Membrane Protein Multiplicity

How many different polypeptide species exist in a single membrane? Enumeration of known membrane functions attributable to proteins (e.g., enzymes, transport systems, certain antigens) might run to dozens of different polypeptides. The contribution of many of these proteins to the total membrane mass is small. For example, estimates of the number of Na^+, K^+-dependent ATPase sites per human erythrocyte range around 200 (18). The number of acetylcholinesterase active sites on the same membrane may be calculated from the data of Bellhorn *et al.* (3) to be around 6000. Since there are approximately five million polypeptide chains per red cell membrane (23), these functional proteins constitute a minute portion of the total. At the other end of the spectrum, some functional proteins such as rhodopsin constitute the predominant polypeptide species of their membrane (see below).

Various authors have suggested that membranes are constructed of fundamental polypeptide units which dominate their protein composition. One such theory is that of "structural protein" developed by Green and associates. The essential characteristics of this material, believed to contribute 30–50% of the protein of a variety of membranes and organelles (99), were insolubility at neutral pH, solubility at the extremes of pH and in detergents, stoichiometric binding of phospholipids and other physiologically important molecules, and an apparent molecular size homogeneity of around 23,000 daltons. Over the course of several years, however, additional findings have modified and broadened this concept so that the central unitary hypothesis has been weakened and membrane protein heterogeneity embraced (32).

Another unifying hypothesis was proposed by Mazia and Ruby (71). They saw a similarity of amino acid composition among a variety of "structural" proteins; this genus was dubbed the "tektins." No detailed discussion of the molecular character of these proteins was developed, and their suggestion has languished.

Recently, Laico et al. (58) detected a material of low molecular weight (apparently about 6000 daltons) in a variety of membrane preparations. They termed this component the "miniproteins" and suggested a central role for it in membrane architecture. Subsequent investigation, however, has indicated that this material is not actually polypeptide in origin and its significance is therefore uncertain (16).

In contrast to these unitary hypotheses, recent investigations have indicated a considerable heterogeneity in the molecular size and electrophoretic mobility of membrane proteins. Schnaitman (111), for example, found that the polypeptides of the inner and outer mitochondrial membranes of rat liver numbered 23 and 12, respectively, and were different except for one possibly common component. The rough and smooth microsomal membranes each exhibited about 15 bands, some of which were shared by both membranes while others were distinctly different. The outer mitochondrial membrane also bore a particular resemblance to the microsomes in three polypeptide species (111).

The erythrocyte membrane also shows considerable heterogeneity among the 15 or so polypeptides readily resolved by SDS-PAGE (8, 23, 35, 59). Furthermore, several major components have been shown to differ characteristically in their elution properties, composition, etc. Likewise, bacterial membrane proteins are quite polydisperse, as illustrated by *Mycoplasma* (Chap. 14) and *E. coli* membranes (Chap. 12). Briefly, the envelope of *E. coli* can be resolved into two membranous structures. The most peripheral of these is a complex assembly of uncertain biological function. The inner one is the more typical cytoplasmic membrane; it performs the functions attributed to the surface, inner mitochondrial, and nuclear membranes of higher cells. Six proteins are discernible by PAGE of the outer membranous layer and 27 are enriched in preparations of the cytoplasmic membrane (112). Though the outer and inner membranous layers have not yet been obtained as absolutely pure preparations, no protein appears to contribute equally to both membranous layers. The 27 polypeptides assigned to the cytoplasmic membrane do not constitute its total protein complement, as many membrane-associated enzymes are present in too low a quantity to be detected by PAGE.

The molecular weights of membrane polypeptides vary widely, ranging from below 20,000 to above 200,000 daltons per chain. A class of polypeptides may exist in various membranes whose molecular weights exceed the myosin heavy chains ($\sim$ 200,000 daltons), placing them among the largest polypeptides known (33, 34).

One recurrent issue—the purity of the membranes—should temper the general conclusion that the proteins of each membrane are heterogeneous. Several investigations intending to analyze membrane proteins actually studied an entire organelle or fragments which retained nonmembranous contents. These extrinsic proteins naturally created the impression that the membrane protein composition was extremely complex. For example, Kiehn and Holland (50) described the multitude of polypeptide components they obtained from various subcellular fractions as membranous, although the proteins entrapped within the organelles had not been removed. Zahler *et al.* (150), on the other hand, maintained a careful distinction between "membranes" and "organelles." They also clearly demonstrated that the electrophoretic pattern of whole bovine liver mitochondria mostly reflected the soluble proteins and not the membranes, in keeping with the observation that two-thirds of the mitochondrial protein is nonmembranous. In addition, they carefully considered the degree of cross-contamination of one organelle or membrane with another, a further source of artifactual membrane protein multiplicity.

It seems that some biological membranes are composed primarily of a few species of major polypeptides, plus a complement of other functional proteins at the trace level. The arboviruses, for example, have only one discernible polypeptide in their lipoprotein envelope (120, 129); it is a glycoprotein in at least one instance (9). It is of interest that these viral envelopes are derived from the plasma membrane of infected cells when the virions exit by budding through virus-modified regions of the surface membrane.

Another intriguing example is a bacterial strain, *Bacillus* PP, obtained from a culture of *B. megaterium* KM (85). Whereas the electrophoretic patterns of polypeptides from several bacterial species show a multitude of membrane components, this unusual strain revealed a single 32,000-dalton component which constituted at least 90% of the ghost membrane protein. The component had an electrophoretic mobility indistinguishable from a minor component of the presumed parent strain, and may represent an overproduction of that specific gene product (85).

The visual pigment, rhodopsin, has been estimated to be the only prominent polypeptide in the rod outer segments of bovine retinas (6, 16, 58). The polypeptides of rabbit sarcoplasmic reticulum are dominated by two or three major components, one of which is the Ca^{++}-dependent ATPase (62). Even in the complex human erythrocyte membranes, six major bands constitute over 70% of the membrane protein mass (23). Furthermore, four of these six bands may be eluted by simple ionic manipulation, leaving a single component composing over 40% of the residual stroma protein. Central nervous system myelin contains only three major polypeptides (119).

How are we to distinguish between extrinsic nonmembrane proteins and intrinsic membrane proteins? This issue may not be readily solved since the cell itself may not regard the membrane as a discrete structure but rather in organizational continuity with the cytoplasm. We have no ready rules to declare a "border" protein nonmembranous. This issue is well illustrated by a recent study on the proteins of hepatic microsomes by Hinman and Phillips (44). These investigators dismissed previous polydisperse electrophoretic patterns (111, 150) as reflecting contamination of the membranes by adsorbed and entrapped proteins. In contrast, they washed their microsomes until a single 52,000-dalton polypeptide band composing 30–40% of the protein, dominated the electrophoretic profile. In order to accept their conclusions, however, we must embrace their preparative procedures. These included washes with 0.14 M NaCl, 1.0 M NaCl, and 0.1 M $Na_2 CO_3$–0.1 M Na HCO_3, which appear to remove ribosomes but not phospholipid or microsomal enzymes. A wash with 0.075% deoxycholate was chosen to release trapped proteins. The effect of the latter on the membrane constituents was not described (44).

It is difficult to characterize membrane proteins as a class. They seem as diverse as any other group of proteins collected at a common locus in biological space (e.g., plasma proteins, cytoplasmic proteins). Their variations in size, composition, structure, and solubility seem to reflect a fundamental heterogeneity whose antecedents must reside in functional specificity.

Membrane Protein Associations

Proteins are the dominant component of membranes (54); yet their participation in membrane structure cannot, at present, be specified beyond conjecture. Optical studies—such as circular dichroism, optical rotatory dispersion, and infrared spectroscopy—have illuminated the secondary conformation of membrane proteins but not their disposition within the membrane. Physical studies based on differential scanning calorimetry, electron spin resonance, nuclear magnetic resonance, x-ray diffraction, etc., have represented the lipids as being predominantly in a bilayer array, but, with few exceptions (Chap. 4), have not revealed the intimate habitats of individual membrane proteins.

Protein-Protein Associations

There are serious methodological difficulties in studying the contacts made between membrane proteins in situ. If, on the other hand, the proteins are solubilized before analysis, their behavior may not reflect their original disposition so much as interactions secondary to denaturation and/or release from the membrane. For example, polypeptides have been recovered as simple oligomers after gentle solubilization (123), but it is unclear whether this reflects their native state or the aggregation of proteins released from (hydrophobic?) association with the membrane.

ASSOCIATIONS AMONG UNLIKE POLYPEPTIDES

Many extended supramolecular structures manifest specific interprotein interactions in their order, stability, and function; viral coats, ribosomes, microtubules, and myofibrils are a few. We might hope to find similar multiprotein arrays participating in membrane organization; however, we do not. Even Green and his associates (31), who have argued strongly for the formation of membranes from repeating lipoprotein subunits, reason that it is the lipid which directs their assembly.

Nevertheless, many workers feel that long-range order may exist within the plane of the membrane and that the proteins maintain it. One facet of this premise is the visualization of submembrane repeating units by electron microscopy (127). Another concept is that cooperative interactions among proteins may mediate membrane-wide function (11). Some workers conceive of a concerted response of wide tracts of proteins to focal perturbations. A dramatic example of the latter was recently offered by Sonenberg (122). He reported that the intrinsic fluorescence of human erythrocyte membrane proteins was significantly altered by minute amounts of human growth hormone—about 70 molecules of hormone per ghost (containing millions of polypeptides). The effect exhibited great specificity and was interpreted as being mediated by some far-reaching membrane alteration, such as cooperative changes in protein conformation or a change in viscosity within the membrane.

The most carefully studied associations of unlike membrane polypeptides are those which mediate oxidative phosphorylation in the inner mitochondrial membrane. Whether it functions as a solid-state array of integrated enzymes or as a vectorially organized barrier across which chemiosmotic potentials are generated, this membrane clearly demonstrates that a high degree of interprotein interaction is central to the respiration-driven synthesis of ATP (91, 92, 138). Analogous arrays may exist in chloroplast and endoplasmic reticulum membranes. Whether such functional integration should be taken as a general model of membrane protein organization or as a special type of membrane modification is an unanswered question.

ASSOCIATIONS AMONG LIKE PROTEINS

While the question of long-range order within the membrane cannot now be answered, there is evidence that identical or closely related polypeptides are sometimes associated as oligomers and polymers in membranes. Such complexes have been detected in both solubilized proteins and intact membranes.

Some of the proteins which have been released into aqueous solution by the removal of ions have been shown to manifest characteristic interchain associations. High-molecular-weight erythrocyte membrane proteins, called spectrin, were prepared by Marchesi and Steers (69). These proteins aggregated into coiled filaments following the introduction of ATP plus Ca^{++} or Mg^{++}. Visualized by negative staining, the filaments were 40–60 Å wide and of variable length. Similar structures can be seen on the surfaces of trypsin-treated, isolated erythrocyte membranes.

A rather similar preparative method was used by Harris (37, 38) to prepare an erythrocyte ghost protein fraction of strikingly different appearance. Rings or toruses, composed of 10 similar subunits, were the dominant image seen on negative staining. Frequently, four rings would be seen stacked into hollow cylinders.

The ATPase solubilized from the membranes of *S. faecalis* by the removal of divalent cations also has an oligomeric structure (113). Electron micrographs of negatively stained preparations reveal a homogeneous population of particles whose units are arranged in a planar hexagonal array. Each of the six units seems from ancillary physical studies to be composed of 33,000-dalton subunits, termed α and β. The entire particle is therefore $\alpha_6\beta_6$ and has a molecular weight of 385,000. It is somewhat similar to ATPase (F_1) particles of the inner mitochondrial membrane (92, 113).

Cytochrome b_5 solubilized from rabbit liver microsomes with Triton X-100 readily aggregated into oligomers unless denatured in urea or SDS (123). In the absence of detergents octamers formed, while in the presence of 0.4% deoxycholate it appeared to be a dimer. Its state in the membrane is not known.

Studies on the associations between polypeptides while still in the human erythrocyte ghost have been performed by applying mild cross-linking reagents to intact membranes and assessing their effect on the proteins by electrophoresis in SDS (160). Under optimal conditions, certain preferential cross-linking occurred. The pattern varied with the agent and experimental conditions. The two major proteins corresponding to spectrin (69) could be complexed to one another and apparently each to itself. This is consistent with their tendency to associate, discussed above. Another major polypeptide was specifically and reversibly linked into a dimeric form by disulfide-bond formation, but not by amino group–directed reagents. One polypeptide may have formed tetramers when the ghosts were treated with formaldehyde. Some components, notably the sialoproteins, were unreactive under the various conditions explored. It would appear that some of the polypeptides in this membrane are associated with like molecules, perhaps as oligomers or as close neighbors in a common domain, which can be cross-linked selectively.

Protein-Lipid Associations

Many membrane-associated enzymes are inactive when rendered free of lipid, but may be reactivated by its return. This lipid dependence should be a primary consideration in the development of rationales for membrane enzyme solubilization and purification. Since this topic has been reviewed recently (106, 134), only a few well-characterized systems are considered here.

A lipid requirement for enzymic activity was initially observed by Fleischer and his coworkers (25, 26) in their studies on the mitochondrial electron-transfer system. Mitochondria which had been depleted of phospholipids by aqueous acetone extraction were almost completely dependent on added phospholipid for respiratory activity. Respiratory activity could be restored by unfractionated mitochondrial phospholipid, lecithin, or cardiolipin; at low lipid concentrations, cardiolipin was approximately 10 times more effective than lecithin. When the extent of reactivation was calculated on the basis of phospholipid bound to the lipid-depleted mitochondrial particles, lecithin, cardiolipin, and total mitochondrial phospholipids were equally effective. Phospholipids dispersed by sonication (108) or by dialysis against butanol-cholate (26) were equally successful in restoring respiratory activity. Phospholipids prepared in a Potter-Elvehjem homogenizer were far less potent in restoring activity. Thus, the mode of preparation of the lipid suspension is a critical determinant of its efficacy in the activation of an isolated membrane enzyme.

The mitochondrial D(-)β-hydroxybutyric acid dehydrogenase has an absolute specificity for lecithin and is not activated by other phospholipids (48). Optimal activation requires an acyl unsaturated lecithin; totally saturated lecithins such as dimyristoyllecithin give only partial activation.

Activation reaches a maximum when the lipid/protein ratio is approximately 100:1, but mixed phospholipids containing approximately 30% lecithin are as effective as lecithin alone.

Isoprenoid alcohol phosphokinase is a membrane-associated enzyme of *Staphylococcus aureus* which catalyzes the phosphorylation of a C_{55}-isoprenoid alcohol intermediate in peptidoglycan synthesis. When freed of lipid, it exhibited an absolute requirement for either phosphatidylglycerol or cardiolipin (43). Certain other phospholipids, such as phosphatidylcholine and phosphatidylserine, produced a slight activation, but it was not clear whether this effect was an intrinsic property of these lipids or was caused by impurities in the lipid preparations. Activation by both phosphatidylglycerol and cardiolipin required the simultaneous presence of detergents such as deoxycholate, but these detergents were extremely poor activators when employed in the absence of phospholipid. It is conceivable that deoxycholate functions to solubilize the C_{55}-isoprenoid alcohol substrate.

The constitutive sugar-specific phosphotransferases of *E. coli* membranes exhibit an absolute requirement for phosphatidylglycerol (56; see Chap. 10). Other *E. coli* phospholipids and ionic and nonionic detergents have little or no activating potential. Activation is optimal only when the components are added in the sequence: enzyme IIB, divalent cation, phosphatidylglycerol, enzyme IIA. The reconstituted system has a particulate character and can be recovered by sedimentation. The optimal weight ratio of lipid to enzyme IIB is approximately 1:1. The physical structure of the reconstituted system has not yet been determined. Studies with ^{32}P-labeled phosphoryl donor substrate have indicated that the lipid activator does not function as an intermediate in the enzymic reaction.

An inducible β-glucoside–specific enzyme II (enzyme II^{bgl}) of the *E. coli* phosphotransferase system has been solubilized and purified free of lipid (102). Enzyme II^{bgl} is similar to the constitutive sugar-specific phosphotransferases in its lipid specificity, i.e., it is activated by an anionic lipid preparation consisting almost entirely of phosphatidylglycerol, but not by phosphatidylethanolamine. It differs from the constitutive phosphotransferases, however, in that it is activated by nonionic detergents as well as or better than it is activated by aqueous dispersions of phosphatidylglycerol. The detergent activation curves show an optimum at the critical micelle concentrations of Triton X-100, Triton X-114, Triton X-45, and Tween 40; detergent concentrations exceeding the critical micelle concentration are inhibitory. The detergent optima for activation are independent of the amount of the enzyme II^{bgl}–containing protein in the reaction mixture.

The *E. coli* membrane enzyme which transfers galactose from UDP-galactose to galactose-deficient lipopolysaccharide also exhibits an absolute dependence on phospholipids (107). Activation has been achieved with phosphatidylethanolamine, phosphatidic acid, phosphatidylglycerol, and cardiolipin. Phosphatidylcholine, which is not formed by *E. coli*, and phosphatidyl-serine are poor activators. An α,β-unsaturated (dioleyl) synthetic derivative of phosphatidylethanolamine is a potent activator, while totally saturated (dipalmitoyl) phosphatidylethanolamine is without effect. A preparation of *E. coli* phosphatidylethanolamine which contains cyclopropane fatty acids, but no double bonds, is also an excellent activator. This indicates that the physical structure imparted by the double bond, and not the double bond itself, is the determinant of the activation potential of phosphatidylethanola-mine. Phospholipids containing cyclopropane fatty acids are similar in their physical properties at air-water interfaces to the analogous phospholipids which contain *cis* double bonds in place of the cyclopropane rings (see Chaps. 1 and 12).

Reconstitution of the active glycosyl transferase system has been accomplished in films at the air-water interface (100, 101). A monomolecular film of phosphatidylethanolamine was first formed at the air-water interface. Lipopolysaccharide was then added to the subphase, and the formation of a composite film of these two components was detected by an increase in the surface pressure (see Chap. 5) and confirmed by a direct determination of the composition of the monolayer. From determinations of the molecular surface areas and the molar ratios of the two constituents of the binary mixture, the authors concluded that phosphatidylethanolamine and lipopolysaccharide were arranged in a side-by-side orientation with the fatty acid side chains pointed perpendicular to the plane of the surface. When the glycosyl trans-ferase was introduced beneath the binary mixture of phosphatidylethano-lamine and lipopolysaccharide, a ternary film was formed, as indicated by an increase in surface pressure. Reconstitution of enzymic function was demonstrated by injecting ^{3}H-galactose-labeled UDP-galactose into the subphase below the ternary film and measuring galactose transfer to lipopolysaccharide.

Reassociation of Isolated Proteins and Lipids

Membrane proteins and lipids can be dissolved and separated from one another. When the solubilizing agent is removed from the isolated proteins, an amorphous precipitate usually results. However, if the lipid fraction is re-introduced prior to removal of the dispersing agent, membrane-like struc-tures can form under certain conditions. They may take on the appearance of sheets or closed vesicles and may have a familiar trilaminar cross-section on electron microscopy (12, 126, 132, 149). Their buoyant density and com-position may also resemble those of the parent membrane. In addition, mem-brane components such as cytochrome oxidase (72) and the ATPase of sarcoplasmic reticulum (62) can be purified as lipoprotein complexes in deoxycholate; removal of the surfactant brings about their spontaneous coalescence into membrane vesicles.

Reaggregation has often been studied with model protein-phospholipid mixtures (14, 126); recently, however, purified glycolipids have also been re-associated with proteins derived from *Mycoplasma* membranes to form vesicles (12). The initial interaction between protein and phospholipid seems to be electrostatic in nature, while short-range and hydrophobic interactions may occur secondarily (14, 36, 47, 51, 152). Some water-soluble membrane proteins interact electrostatically with phospholipids to form complexes which are insoluble in water but are soluble in hydrocarbon solvents (14, 39, 60).

The clear implication of these provocative phenomena is that membranes may be recreated in vitro out of their constituents. An underlying hypothesis in these studies is that membrane assembly is best described by a thermodynamic model. In analogy with the refolding of denatured proteins, it is postulated that a mixture of soluble membrane components will, under the appropriate conditions, spontaneously come together to re-form a native membrane, guided only by the minimization of free energy. The molecules themselves carry all the organizational information required for faithful de novo membrane assembly. An explicit paradigm for such concepts was offered by Wallach and Gordon (140) in the form of a set of complex equilibrium equations for multiple interacting membrane components at various stages of assembly.

An opposing point of view holds that natural membrane assembly is guided by a template which is the preexisting membrane. Membranes do not *form* but *grow*, i.e., by sequential additions to preexisting structures. Studies on membrane biogenesis seem to support the template model over its thermodynamic alternative (see Chaps. 12 and 13).

Do reassociation phenomena demonstrate the reconstitution of membranes from their fundamental units? We must first note that nonmembrane proteins and lipids or lipid mixtures may also form these associations. Just as vesicles can be prepared from mixtures of dissolved membrane components (132, 149), the addition of poly-L-lysine to dispersed phosphatidyl-serine also produces vesicles (36); these artificial complexes, furthermore, exhibit a relative immobilization of the lipid fatty acyl chains and an augmentation of the α-helical content of the polypeptides, features regularly seen in natural biomembrane structures (121). While membrane-derived proteins have been observed to bind to and alter the permeability of synthetic phospholipid bilayers (93), so have a range of soluble proteins (51); in the former case, however, manifestations of biological specificity were observed.

Deoxycholate disperses mitochondrial cytochrome oxidase into lipoprotein complexes which reaggregate into vesicle-like structures when the detergent is removed; this coalescence is dependent on the presence of phospholipids (72). Green and his colleagues (31) have taken the view that the bile salts have solubilized or "depolymerized" the mitochondrial enzyme complex to its fundamental "repeating units," namely, 50- to 100-Å lipo-protein particles. The appearance of vesicles after a reduction in detergent concentration is interpreted as representing a de novo reassembly process. Since similar transitions were observed with other membranes dispersed in bile salts, Green *et al.* have generated a theory for the subunit assembly of membranes in general. Stoeckenius (126), however, has reexamined these experiments. His findings indicate that the bile salts dispersed the inner mitochondrial membrane into *fragments* bearing the 50-Å particles seen by Green *et al.* However, complete solubilization of the membrane to funda-mental subunits did not occur. The vesicle-like structures that formed subse-quently represented a coalescence of preexisting fragments and not de novo membrane assembly.

Likewise, it has been claimed for the sarcoplasmic reticulum that de-oxycholate-dispersed lipoprotein particles can reassociate into vesicles (70). Furthermore, MacLennan *et al.* (62) have purified and characterized the ATPase from these dispersions and found that the isolated enzyme lipoprotein can also form membranes when the surfactant is removed. However, as in the case of cytochrome oxidase, inspection of the electron micrographs indicates that the bulk of the preparation is not monodisperse (although 90-Å particles are seen) but is in the form of large fragments (62), so that the objections raised by Stoeckenius (126) seem applicable here. In this regard, it is inter-esting that Selinger *et al.* (115) also prepared particles from sarcoplasmic reticulum with deoxycholate. Their preparations did not form vesicles when the surfactant was removed and by electron microscopy were seen to be com-posed of a homogeneous dispersion of *nonaggregated* 100- to 200-Å particles.

To what extent then, do the reconstituted membranes recapture the biological properties of the parent? A major handicap in these studies is that the details of structure and assembly for the original membranes are not known. Certainly the reactivation of membrane enzymes by selected lipids indicates a specificity which may mirror native interactions. But for membrane re-formation, more stringent criteria have yet to be met. Racker (92) has developed this point in a discussion of the reconstitution of inner mitochon-drial membrane functions. Various complex activities have been recovered in reassembly experiments; however, topographical integrity has not been re-stored. Likewise, dissolved *Mycoplasma* membrane proteins and lipids have been reassociated into vesicles (132), but recent scrutiny suggests that their proteins are not normally disposed (156, 157). Freeze-etch electron micro-scopy has also indicated that the reassembled species lack the globular intramembrane particles prominent in native membranes (133). In contrast, freeze-fracture micrographs of the membranes formed from reassociated sarcoplasmic reticulum ATPase complexes did exhibit globular particle profiles identical with the parent membrane (62). This is consistent with the premise, discussed above, that the ATPase lipoprotein complexes are not true soluble subunits but membrane fragments with their native organizational information conserved (126).

It seems clear that the avid association of proteins and lipids plays a major role in the assembly and stability of membranes. Whether native membranes can arise in the absence of a template is much less certain. An ultimate challenge to this line of experimentation is to mix together the truly solubilized components of, for example, erythrocyte and mitochondrial inner membranes, remove the dispersing agent, and recover, in time, a set of red blood cell ghosts and mitochondrial inner membranes.

Membrane Protein Organization

Although the detailed molecular organization of no biological membrane is known, recent investigations have prompted the casting of several structural models (52, 121, 127, 139, 140). The diversity among these reflects the many facets recognized in membrane structure; however, enough overlap exists among current models that a common theme may be detected.

The lipid bilayer remains, despite careful scrutiny and criticism (54), a central concept of membrane structure. Much of the discussion of the organization of proteins has focused on how they might interact with the lipids. The early concepts of protein monolayers electrostatically bound to the polar heads of the lipid leaflet are not widely advocated today. Those proteins which seem most clearly associated by Coulombic interactions, such as binding cytochrome c to the inner mitochondrial membrane, are not believed to represent the essence of membrane proteins. In fact, they are frequently considered not to be intrinsic membrane proteins at all. A more tenacious, nonionic bonding is looked for in "integral" membrane proteins (121).

The last few years have brought the documentation and general acceptance of nonpolar associations between the membrane proteins and lipids. Since hydrophobic lipid groups are thought not to extend to the membrane surface, they must interact with proteins penetrating into the interior of the membrane. Obviously, the composition and conformation of the proteins will not only accommodate these deep associations, but determine them, presumably by minimization of the free energy of the protein–lipid–water system (121).

The models mentioned above differ mainly in emphasis: How much of the lipid leaflet is covered over with protein and how much is bare? How much of the protein penetrates the lipid? How regular an array does the composite structure assume? How asymmetrical is the membrane? None ignores the potential for both electrostatic and nonpolar associations between lipids and proteins; after all, protein-protein associations exploit both types of bonding for their stability.

There is scanty but engaging evidence that the proteins of membranes may be laterally mobile. Blasie and Worthington (146) have observed a temperature-dependent fluidity in the plane of the membrane for the proteins of retinal outer segments. In another study, the spreading of species-specific antigens in the hybrid surface of heterokaryons was monitored by immunofluorescence after cell fusion (28). The antigens comingled at a rapid rate which was highly temperature-dependent. This suggested that their translation in the membrane may depend on its state of fluidity. It is also possible that the apparent movement of the antigens reflected their reversible dissociation and equilibration with a cytoplasmic pool, a different mechanism by which membrane proteins might move about.

The distribution of functional membrane proteins in the plane of the membrane may be patchy or diffuse, generalized or focal. Enzymes detected by histochemical staining or membrane subfractionation commonly exhibit a homogeneous distribution over large areas. However, submembrane heterogeneity is also frequently observed, particularly in the specialized plasma membranes of tissue cells such as hepatocytes or intestinal mucosal cells (see Chap. 3). It is easier to account conceptually for a random dispersion of proteins throughout a membrane than for the complex submembrane organization in these cells. Membrane antigens may similarly be generally dispersed or show patchy foci of reactivity (2, 82).

Membranes, in general, distinguish between the compartments they separate and deal with them differentially. This implies a structural and functional asymmetry across the plane of membrane. The evidence to date supports this hypothesis.

A most dramatic formulation of such vectorial organization is the chemiosmotic mechanism for oxidative phosphorylation proposed by Mitchell. In essence, the components of the inner mitochondrial membrane are assembled in an asymmetrical array that sustains a proton gradient across the membrane as the energy source for ATP synthesis (92). Whether or not this hypothesis is correct, it has stimulated careful study of the topography of the inner mitochondrial membrane (138). Racker (91) has recently summarized these data and their implications. At this early time, it seems likely that the respiratory chain is asymmetrically arranged in native membranes. Furthermore, the functional assembly probably traverses the membrane thickness. The experimental approaches to the analysis of the two faces of this membrane have involved the use of differential substrate and inhibitor accessibility to membrane enzymes, electron microscopy and cytochemistry, antibodies against purified components, and selective elution and readdition techniques. Of particular value is the preparation of impermeable membrane vesicles of an "inside-out" orientation, so that each side of the inner mitochondrial membrane can be probed selectively (91, 138).

The enzymes of less highly specialized membranes show a similar "sidedness;" it would appear that they function at one face of the membrane or the other, but not at both. Transport-linked enzymes, such as the Na^+, K^+-dependent ATPase, seem to address both sides of the membrane, but anisotropically.

It has been pointed out that enzymes may constitute only a small fraction of the membrane proteins and may not accurately represent their general organization; for example, transport systems may span the thickness of the membrane, but the bulk of the other proteins might be confined to the polar surfaces (127). Recent studies on the major polypeptides of the red blood cell membrane, however, indicate that they may be *1*) oriented rather than mobile or randomly disposed; *2*) asymmetrical, in that they do not equally populate both faces of the membrane; and *3*) nonuniform in their reaction to probes (such as radioactive labels or proteolytic enzymes). Furthermore, at least some of these major proteins may penetrate through the thickness of the membrane (4, 7, 8, 89, 125). Evidence is also accumulating that the oligosaccharides of erythrocyte glycoproteins are oriented toward the extracellular space. Therefore, the major polypeptides show the same type of topographic specificity as do enzymes, and we may consider them to be functionally organized within the membrane.

CONCLUSIONS

The key to understanding membrane proteins lies in their specificity. Conclusions based on collective averages obscure their great heterogeneity. The over-all molecular size range, composition, and conformational content of membrane proteins are not distinctively different from those of nonmembrane proteins. Taken individually, however, they show considerable diversity in molecular and submolecular character which hopefully can be related to their role in the membrane.

Each membrane has a different protein profile. Some membranes contain several major proteins; some contain only one or a few principal species. The polypeptides of a given membrane are likewise diverse in molecular size, composition, orientation, reactivity, and mode of association with the membrane matrix.

Certain proteins are glycosylated by the covalent attachment of small, characteristic heterosaccharide units. Protein associations with lipids are noncovalent, but often show particular specificities. Some proteins may associate with one another in the membrane, forming oligomers or functional (multienzyme) arrays.

The proteins may be free to move laterally in the plane of the membrane, but appear to be organized and oriented with respect to the membrane's two faces. The detailed architecture of proteins in membranes is not known.

REFERENCES

1. Abrams, A. *J. Biol. Chem.* **240,** 3675 (1965).
2. Aoki, T., Hammerling, U., deHarven, E., Boyse, E. A., and Old, L. J. *J. Exp. Med.* **130,** 979 (1969).
3. Bellhorn, M. B., Blumenfeld, O. O., and Gallop, P. M. *Biochem. Biophys. Res. Commun.* **39,** 267 (1970).
4. Bender, W. W., Garan, H., and Berg, H. C. *J. Mol. Biol.* **58,** 783 (1971).
5. Blumenfeld, O. O. *Biochem. Biophys. Res. Commun.* **30,** 200 (1968).
6. Bownds, D., and Gaide-Huguenin, A. C. *Nature* **225,** 870 (1970).
7. Bretscher, M. S. *Nature New Biology* **231,** 229 (1971).
8. Bretscher, M. S. *J. Mol. Biol.* **58,** 775 (1971).
9. Burge, B. W., and Straus, J. H., Jr. *J. Mol. Biol.* **47,** 449 (1970).
10. Burger, S. P., Fujii, T., and Hanahan, D. J. *Biochemistry* **7,** 3682 (1968).
11. Changeux, J. -P., Thiéry, J., Tung, Y., and Kittel, C. *Proc. Natl. Acad. Sci. U.S.A.* **57,** 335 (1967).
12. Cole, R. M., Popkin, T. J., Prescott, B., Chanock, R. M., and Razin, S. *Biochim. Biophys. Acta* **233,** 76 (1971).
13. Crambach, A., and Rodbard, D. *Science* **172,** 440 (1971).
14. Dawson, R. M. C. *Biological Membranes: Physical Fact and Function.* Ed. D. Chapman. Academic Press, New York, 1968, p. 203.
15. Demus, H., and Mehl, E. *Biochim. Biophys. Acta* **203,** 291 (1970).
16. Dryer, W. J., Papermaster, D. S., and Kühn, H. *Ann. N.Y. Acad. Sci.* **195** (1972). In press.
17. Dulaney, J. T., and Touster, O. *Biochim. Biophys. Acta* **196,** 29 (1970).
18. Dunham, P. B., and Hoffman, J. F. *Proc. Natl. Acad. Sci. U.S.A.* **66,** 939 (1970).
19. Emmelot, P., Feltkamp, C. A., and Vaz Dias, H. *Biochim. Biophys. Acta* **211,** 43 (1970).
20. Engelman, D. M., Terry, T. M., and Morowitz, H. J. *Biochim. Biophys. Acta* **135,** 381 (1967).
21. Estabrook, R. W., and Pullman, M. E. (eds.). *Methods in Enzymology.* Academic Press, New York, 1967, vol. 10.
22. Evans, W. H. *Biochem. J.* **166,** 833 (1970).
23. Fairbanks, G., Steck, T. L., and Wallach, D. F. H. *Biochemistry* **10,** 2606 (1971).
24. Fish, W. W., Reynolds, J. A., and Tanford, C. *J. Biol. Chem.* **245,** 5166 (1970).
25. Fleischer, S., Brierley, G., Klouwen, H., and Slautterback, D. B. *J. Biol. Chem.* **237,** 3264 (1962).
26. Fleischer, S., and Klouwen, H. *Biochem, Biophys. Res. Commun.* **5,** 378 (1961).
27. Forstner, G. G. *Biochem. J.* **121,** 781 (1971).
28. Frye, L. D., and Edidin, M. *J. Cell Sci.* **7,** 319 (1970).
29. Glick, M. C., Comstock, C. A., Cohen, M. A., and Warren, L. *Biochim. Biophys. Acta* **233,** 247 (1971).
30. Gottschalk, A. (ed.). *Glycoproteins.* Elsevier, New York, 1966.
31. Green, D. E., Allmann, D. W., Bachmann, E., Baum, H., Kopaczyk, K., Korman, E. F., Lipton, S., MacLennan, D. H., McConnell, D. G., Perdue, J. F., Rieske, J. S., and Tzagoloff, A. *Arch. Biochem. Biophys.* **119,** 312 (1967).

32. Green, D. E., Haard, N. F., Lenaz, G., and Silman, H. I. *Proc. Natl. Acad. Sci. U.S.A.* **60,** 277 (1968).
33. Guidotti, G. Personal communication.
34. Gwynne, J. T., and Tanford, C. *J. Biol. Chem.* **245,** 3269 (1970).
35. Hamaguchi, H., and Cleve, H. *Biochim. Biophys. Acta* **233,** 320 (1971).
36. Hammes, G. G., and Schullery, S. E. *Biochemistry* **9,** 2555 (1970).
37. Harris, J. R. *Biochim. Biophys. Acta* **188,** 31 (1969).
38. Harris, J. R. *J. Mol. Biol.* **46,** 329 (1969).
39. Hart, C. J., Leslie, R. B., Davis, M. A. F., and Lawrence, G. A., *Biochim. Biophys. Acta* **193,** 308 (1969).
40. Hatefi, Y., and Hanstein, W. G. *Proc. Natl. Acad. Sci. U.S.A.* **62,** 1129 (1969).
41. Heller, J. *Biochemistry* **7,** 2906 (1968).
42. Heller, J., and Lawrence, M. A. *Biochemistry* **9,** 864 (1970).
43. Higashi, Y., and Strominger, J. L. *J. Biol. Chem.* **245,** 3691 (1970).
44. Hinman, N. D., and Phillips, A. H. *Science* **170,** 1222 (1970).
45. Holden, K. G., Yim, N. C. F., Griggs, L. J., and Weisbach, J. A. *Biochemistry* **10,** 3105 (1971).
46. Jackson, R. L., Segrest, J. P., and Marchesi, V. T. *Fed. Proc.* **30,** 1280 (1971).
47. Ji, T. H., and Benson, A. A. *Biochim. Biophys. Acta* **150,** 686 (1968).
48. Jurtshuk, P., Jr., Sekuzu, I., and Green, D. E. *J. Biol. Chem.* **238,** 3595 (1963).
49. Kahan, B. D., and Reisfeld, R. A. *Science* **164,** 514 (1969).
50. Kiehn, E. D., and Holland, J. J. *Biochemistry* **9,** 1716 (1970).
51. Kimelberg, H. K., and Papahadjopoulos, D. *J. Biol. Chem.* **246,** 1142 (1971).
52. Kirk, J. T. O. *Annu. Rev. Biochem.* **40,** 161 (1971).
53. Kirk, R. G. *Proc. Natl. Acad. Sci. U.S.A.* **60,** 614 (1968).
54. Korn, E. D. *Annu. Rev. Biochem.* **38,** 263 (1969).
55. Kornfeld, R., and Kornfeld, S. *J. Biol. Chem.* **245,** 2536 (1970).
56. Kundig, W., and Roseman, S. *J. Biol. Chem.* **246,** 1407 (1971).
57. Laemmli, U. K. *Nature* **227,** 680 (1970).
58. Laico, M. T., Ruoslahti, E. I., Papermaster, D. S., and Dreyer, W. J. *Proc. Natl. Acad. Sci. U.S.A.* **67,** 120 (1970).
59. Lenard, J. *Biochemistry* **9,** 1129 (1970).
60. Lesslauer, W., Wissler, F. C., and Parsons, D. F. *Biochim. Biophys. Acta* **203,** 199 (1970).
61. Lester, R. L., and Fleischer, S. *Biochim. Biophys. Acta* **47,** 358 (1961).
62. MacLennan, D. H., Seeman, P., Iles, G. H., and Yip, C. C. *J. Biol. Chem.* **246,** 2702 (1971).
63. MacLennan, D. H., Tzagoloff, A., and McConnell, D. G. *Biochim. Biophys. Acta* **131,** 59 (1967).
64. Maddy, A. H. *Biochim. Biophys. Acta* **117,** 193 (1966).
65. Maddy, A. H. *Semin. Hematol.* **7,** 275 (1970).
66. Maddy, A. H., and Kelly, P. G. *Biochim. Biophys. Acta* **241,** 114 (1971).
67. Maizel, J. V., Jr. *Fundamental Techniques in Virology.* Ed. K. Habel and N. P. Salzman. Academic Press, New York, 1969, p. 334.
68. Marchesi, S. L., Steers, E., Jr., Marchesi, V. T., and Tillack, T. W. *Biochemistry* **9,** 50 (1970).

69. Marchesi, V. T., and Steers, E., Jr. *Science* **159,** 203 (1968).
70. Martonosi, A. *J. Biol. Chem.* **243,** 71 (1968).
71. Mazia, D., and Ruby, A. *Proc. Natl. Acad. Sci. U.S.A.* **61,** 1005 (1968).
72. McConnell, D. G., Tzagoloff, A., MacLennan, D. H., and Green, D. E. *J. Biol. Chem.* **241,** 2373 (1966).
73. Meezan, E., Wu, H. C., Black, P. H., and Robbins, P. W. *Biochemistry* **8,** 2518 (1969).
74. *Membrane Proteins.* Little Brown, Boston, 1969; also *J. Gen. Physiol.* **54,** no. 1, part 2 (1969).
75. Miller, D. M. *Biochem. Biophys. Res. Commun.* **40,** 716 (1970).
76. Mitchell, C. D., and Hanahan, D. J. *Biochemistry* **5,** 51 (1966).
77. Muramatsu, T., and Nathenson, S. G. *Biochemistry* **9,** 4875 (1970).
78. Nelson, C. A. *J. Biol. Chem.* **246,** 3895 (1971).
79. Neville, D. M., Jr. *Biochim. Biophys. Acta* **133,** 168 (1967).
80. Neville, D. M., Jr. *Biochim. Biophys. Acta* **154,** 540 (1968).
81. Neville, D. M., Jr. *Biochem. Biophys. Res. Commun.* **34,** 60 (1969).
82. Nicolson, G. L., Masouredis, S. P., and Singer, S. J. *Proc. Natl. Acad. Sci. U.S.A.* **68,** 1416 (1971).
83. Nozaki, Y., and Tanford, C. *J. Biol. Chem.* **246,** 2211 (1971).
84. Panet, R., and Selinger, Z. *Eur. J. Biochem.* **14,** 440 (1970).
85. Patterson, P. H., and Lennarz, W. J. *Biochem. Biophys. Res. Commun.* **40,** 408 (1970).
86. Peacock, A. C., and Dingman, C. W. *Biochemistry* **7,** 668 (1968).
87. Pepper, D. S., and Jamieson, G. A. *Biochemistry* **8,** 3362 (1969).
88. Philippot, J. *Biochim. Biophys. Acta* **225,** 201 (1971).
89. Phillips, D. R., and Morrison, M. *Biochemistry* **10,** 1766 (1971).
90. Pitt-Rivers, R., and Impiombato, F. S. A. *Biochem. J.* **109,** 825 (1968).
91. Racker, E. *Essays Biochem.* **6,** 1 (1970).
92. Racker, E. (ed.). *Membranes of Mitochondria and Chloroplasts.* Van Nostrand Reinhold, New York, 1970, p. 127.
93. Redwood, W. R., Müldner, H., and Thompson, T. E. *Proc. Natl. Acad. Sci. U.S.A.* **64,** 989 (1969).
94. Rega, A. F., Weed, R. I., Reed, C. F., Berg, G. G., and Rothstein, A. *Biochim. Biophys. Acta* **147,** 297 (1967).
95. Rendi, R. *Biochim. Biophys. Acta* **198,** 113 (1970).
96. Reynolds, J. A., and Tanford, C. *J. Biol. Chem.* **245,** 5161 (1970).
97. Reynolds, J. A., and Tanford, C. *Proc. Natl. Acad. Sci. U.S.A.* **66,** 1002 (1970).
98. Reynolds, J. A., and Trayer, H. *J. Biol. Chem.* **246,** 7337 (1971).
99. Richardson, S. H., Hultin, H. O., and Green, D. E. *Proc. Natl. Acad. Sci. U.S.A.* **50,** 821 (1963).
100. Romeo, D., Girard, A., and Rothfield, L. *J. Mol. Biol.* **53,** 475 (1970).
101. Romeo, D., Hinckley, A., and Rothfield, L. *J. Mol. Biol.* **53,** 491 (1970).
102. Rose, S. Ph.D. Thesis. The University of Chicago, 1971.
103. Rosenberg, R. M., Crespi, H. L., and Katz, J. J. *Biochim. Biophys. Acta* **175,** 31 (1969).
104. Rosenberg, S. A., and Guidotti, G. *J. Biol. Chem.* **243,** 1985 (1968).

105. Rosenberg, S. A., and Guidotti, G. *Red Cell Membrane Structure and Function.* Ed. G. A. Jamieson and T. J. Greenwalt. Lippincott, Philadelphia, 1969, p. 93.
106. Rothfield, L., and Finkelstein, A. *Annu. Rev. Biochem.* **37,** 463 (1968).
107. Rothfield, L., and Pearlman, M. *J. Biol. Chem.* **241,** 1386 (1968).
108. Rouser, G. *Am. J. Clin. Nutr.* **6,** 681 (1958).
109. Salton, M. R. J. *J. Gen. Physiol.* **52,** 227s (1968).
110. Salton, M. R. J., and Schmitt, M. D. *Biochem. Biophys. Res. Commun.* **27,** 529 (1967).
111. Schnaitman, C. A. *Proc. Natl. Acad. Sci. U.S.A.* **63,** 412 (1969).
112. Schnaitman, C. A. *J. Bacteriol.* **104,** 890 (1970).
113. Schnebli, H. P., Vatter, A. E., and Abrams, A. *J. Biol. Chem.* **245,** 1122 (1970).
114. Segrest, J. P., Jackson, R. L., Andrews, E. P., and Marchesi, V. T. *Biochem. Biophys. Res. Commun.* **44,** 390 (1971).
115. Selinger, Z., Klein, M., and Amsterdam, A. *Biochim. Biophys. Acta* **183,** 19 (1969).
116. Shapiro, A. L., Vinuela, E., and Maizel, J. V. *Biochem. Biophys. Res. Commun.* **28,** 815 (1967).
117. Shimada, A., and Nathenson, S. G. *Biochemistry* **8,** 4048 (1969).
118. Shimizu, S., and Funakoshi, I. *Biochim. Biophys. Acta* **203,** 167 (1970).
119. Shooter, E. M., and Einstein, E. R. *Annu. Rev. Biochem.* **40,** 635 (1971).
120. Simons, K., and Kääriäinen, L. *Biochem. Biophys. Res. Commun.* **38,** 981 (1970).
121. Singer, S. J. *Structure and Function of Biological Membranes.* Ed. L. I. Rothfield. Academic Press, New York, 1971, p. 145.
122. Sonenberg, M. *Proc. Natl. Acad. Sci. U.S.A.* **68,** 1051 (1971).
123. Spatz, L., and Strittmatter, P. *Proc. Natl. Acad. Sci. U.S.A.* **68,** 1042 (1971).
124. Stark, G. R. *Methods Enzymol.* **11,** 590 (1967).
125. Steck, T. L., Fairbanks, G., and Wallach, D. F. H. *Biochemistry* **10,** 2617 (1971).
126. Stoeckenius, W. *Membranes of Mitochondria and Chloroplasts.* Ed. E. Racker. Van Nostrand Reinhold, New York, 1970, p. 53.
127. Stoeckenius, W., and Engelman, D. M. *J. Cell. Biol.* **42,** 613 (1969).
128. Straus, J. H., Jr., Burge, B. W., and Darnell, J. E. *J. Mol. Biol.* **47,** 437 (1970).
129. Straus, J. H., Jr., Burge, B. W., Pfefferkorn, E. R., and Darnell, J. E., Jr. *Proc. Natl. Acad. Sci. U.S.A.* **59,** 533 (1968).
130. Takayama, K., MacLennan, D. H., Tzagaloff, A., and Stoner, C. D. *Arch. Biochem. Biophys.* **114,** 223 (1966).
131. Tanford, C. *Adv. Protein Chem.* **23,** 121 (1968); **24,** 1 (1970).
132. Terry, T. M., Engelman, D. M., and Morowitz, H. J. *Biochim. Biophys. Acta* **135,** 391 (1967).
133. Tillack, T. W., Carter, R., and Razin, S. *Biochim. Biophys. Acta* **219,** 123 (1970).
134. Triggle, D. J. *Recent Progress in Surface Science.* Ed. J. F. Danielli, A. C. Riddiford, and M. D. Rosenberg. Academic Press, New York, 1970, vol. 3, p. 273.

135. Tsukagoshi, N., and Fox, C. F. *Biochemistry* **10**, 3309 (1971).
136. Tung, J.-S., and Knight, C. A. *Biochem. Biophys. Res. Commun.* **42**, 1117 (1971).
137. Tuppy, H., Swetly, P., and Wolff, I. *Eur. J. Biochem.* **5**, 339 (1968).
138. Van Dam, K., and Meyer, A. J. *Annu. Rev. Biochem.* **40**, 115 (1971).
139. Vanderkooi, G., and Green, D. E. *Proc. Natl. Acad. Sci. U.S.A.* **66**, 615 (1970).
140. Wallach, D. F. H., and Gordon, A. S. *Regulatory Functions of Biological Membranes.* Ed., J. Järnefelt. Elsevier, New York, 1968, p. 87.
141. Wallach, D. F. H., and Zahler, P. H. *Biochim. Biophys. Acta* **150**, 186 (1968).
142. Weber, K., and Osborn, M. *J. Biol. Chem.* **244**, 4406 (1969).
143. Whipple, H. E. (ed.). *Ann. N.Y. Acad. Sci.* **121**, 305 (1964).
144. Winzler, R. J. *Red Cell Membrane Structure and Function.* Ed. G. A. Jamieson and T. J. Greenwalt. Lippincott, Philadelphia, 1969, p. 157.
145. Wolfgram, F., and Kotorii, K. *J. Neurochem.* **15**, 1281 (1968).
146. Worthington, C. R. *Fed. Proc.* **30**, 57 (1971).
147. Zacharius, R. M., Zell, T. E., Morrison, J. H., and Woodlock, J. J. *Anal. Biochem.* **30**, 148 (1969).
148. Zahler, P. H., and Wallach, D. F. H. *Biochim. Biophys. Acta* **135**, 371 (1967).
149. Zahler, P., and Weibel, E. R. *Biochim. Biophys. Acta* **219**, 320 (1970).
150. Zahler, W. L., Fleischer, B., and Fleischer, S. *Biochim, Biophys. Acta* **203**, 283 (1970).
151. Zwaal, R. F. A., and Van Deenen, L. L. M. *Biochim. Biophys. Acta* **150**, 323 (1968).
152. Zwaal, R. F. A., and Van Deenen, L. L. M. *Chem. Phys. Lipids* **4**, 311 (1970).
153. Zwaan, J. *Anal. Biochem.* **21**, 155 (1967).
154. Maddy, A. H., and Kelly, P. G. *Biochim. Biophys. Acta* **241**, 290 (1971).
155. Marchesi, V. T., and Andrews, E. P. *Science* **174**, 1247 (1971).
156. Metcalfe, J. C., Metcalfe, S. M., and Engelman, D. M. *Biochim. Biophys. Acta* **241**, 412 (1971).
157. Metcalfe, S. M., Metcalfe, J. C., and Engelman, D. M. *Biochim. Biophys. Acta* **241**, 422 (1971).
158. Sandermann, H., Jr., and Strominger, J. L. *Proc. Natl. Acad. Sci. U.S.A.* **68**, 2441 (1971).
159. Steck, T. L. *Biochim. Biophys. Acta* **255**, 553 (1972).
160. Steck, T. L. *J. Mol. Biol.* **65** (1972). In press.
161. Thomas, D. B., and Winzler, R. J. *Biochem. J.* **124**, 55 (1971).

3

Theodore L. Steck

Membrane Isolation

Biological membranes are closed surfaces which circumscribe the cell and organize its contents into functional compartments, the organelles. The membranes are also specialized to mediate processes at the boundaries of these compartments and to participate in the activities of their organelles. An understanding of the molecular basis of these functions requires the purification of these membranes, which, in turn, demands the isolation of these organelles. Current methodology relies almost exclusively on centrifugation techniques. The principles involved have been analyzed (5) and a vast literature of practical experience summarized (1, 7). de Duve (3, 4, 4a) has set forth a valuable overview of the isolation of organelles. Progress in the isolation of membranes has been reviewed for plasma membranes (8, 9), mitochondria (6), and microsomes (2).

The general *modus operandi* of subcellular fractionation can be stated here. The isolation of a class of particles from a mixture entails a series of steps, each of which disperses the components according to some physico-chemical function (e.g., their size, density, charge, chemical affinity). Usually, no one technique can separate the desired component from all contaminants; rather, by passing through a series of steps, the unique set of physical co-ordinates for each particle is manifested and physical separation affected.

The preparation of this manuscript was aided by American Cancer Society Grant P-578.

For example, Anderson *et al.* (11) have shown that a variety of viruses have similar sedimentation rates and buoyant density characteristics which differ from those of the subcellular particles (Fig. 1). A virus "window" was thus defined by these physical indices which, it was hoped, might facilitate the search for elusive tumor viruses in tissue homogenates.

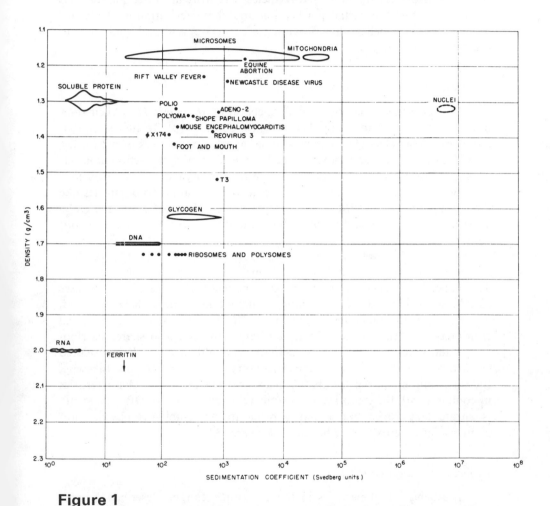

Figure 1

Sedimentation coefficients and equilibrium densities of cell components and viruses. Most of the viruses cluster in an area devoid of cellular particles (the "virus window"). Note that the combination of these two modes of fractionation brings about separation of the particles not achieved by the application of either alone. (From Anderson, N. G., *et al. Natl. Cancer Inst. Monogr. 21*, 253, 1966.)

Current methodology is imperfect and often inadequate. Some of the difficulties lie in *1*) the overlapping of the sizes and densities of the various organelles, *2*) the physical heterogeneity within some classes of organelles, *3*) the dispersal of some membranes into a broad range of fragment sizes and forms upon cell disruption, and *4*) difficulties in establishing the identity and purity of isolated fractions. Nevertheless, centrifugal fractionation has achieved at least the partial purification and characterization of the major membranous organelles, thus providing a cornerstone for cell biology over the past two decades.

Separation Techniques

The centrifuge offers two ways of separating particles: *1*) by their rates of sedimentation (*s*) and *2*) by their buoyant densities (*ρ*). To the extent that these two parameters characterize and set apart given classes of organelles, centrifugation serves as a powerful technique. Centrifugal fractionation has progressed on the assumption that a given type of organelle is homogeneous in its *s*-*ρ* values. While this is often true, it does not apply to organelles like the plasma membrane (PM), the endoplasmic reticulum (ER), and the Golgi apparatus which fragment during cell disruption to yield particles of various sizes.

A second assumption has been that a given biological function (such as an enzymic activity) is discretely apportioned to a single subcellular locus (3). This concept has generally been found to be valid. However, in some cases certain functional homologies among different membranes have emerged. For example, the outer membrane of the nucleus is not only in physical continuity with the rough endoplasmic reticulum, but also shares its ribosomes and glucose-6-phosphatase activity; it might therefore be considered to partake of both the endoplasmic reticulum and nuclear domains. Likewise, the outer mitochondrial membrane bears enzymes and other characteristics in common with the smooth endoplasmic reticulum. The NADP-dependent* isocitrate dehydrogenase is found to reside in the soluble cytoplasm, mitochondria, and peroxisomes, but not elsewhere (33).

Differential Rate Centrifugation

The behavior of particles in the centrifuge can be described by this familiar relation:

$$s = 2r^2(\rho_p - \rho_m)/9\eta \tag{1}$$

where *s* is the sedimentation coefficient of the particle (its velocity per unit

---

* NAD, nicotinamide adenine dinucleotide; NADH, nicotinamide adenine dinucleotide (reduced); NADP, nicotinamide adenine dinucleotide phosphate.

centrifugal field), r is its radius, and ρ_p is its density; ρ_m is the density and η the viscosity of the surrounding medium (all in cgs units). The sedimentation coefficient is given in seconds, but is more usually expressed in Svedberg units, $S = s \times 10^{13}$ sec. Equation 1 applies to spherical particles; non-spherical particles move somewhat more slowly in proportion to their larger frictional coefficient (5).

Equation 1 suggests that particles sediment as a function of their size and density. When cells are broken and the homogenate is centrifuged at 1000 × g for 10 min (i.e., a time integrated force of 10^4 g-min), the nuclei are pelleted preferentially, while the smaller and lighter particles tend to remain in suspension. The supernatant can be centrifuged again at progressively higher forces and/or for longer time periods to obtain pellets enriched in mitochondria, lysosomes, and microsomes. Table I indicates how the relative size of particles relates to their sedimentation coefficients and the forces usually employed to pellet them.

This procedure, called *differential centrifugation*, is the earliest, the simplest, and the most widely used fractionation technique. Its resolving power, however, is rather limited. Because the homogenate fills the entire centrifuge tube at the outset, slowly sedimenting particles at the bottom of the tube enter the pellet before the rapidly sedimenting particles traverse the length of the tube. The result invariably is an impure pellet, differentially enriched in the fastest sedimenting species. The closer the sedimentation coefficients, the worse the cross-contamination. In practice, reasonable separation of two species of particles requires at least an order of magnitude difference in their s values. The technique is therefore most useful as a preliminary enrichment step, and the terms "nuclear," "mitochondrial," and "microsomal" fractions should be viewed as operational rather than analytical designations (3). For example, the usual "mitochondrial" fraction actually is rich in lysosomes and peroxisomes and may contain considerable plasma membrane, Golgi vesicles, and endoplasmic reticulum. (A less misleading system of designations might be $\sim 10^4$, $\sim 10^5$, and $\sim 5 \times 10^6$ g-min pellets, etc.)

Figure 2 illustrates the distribution of rat liver organelle markers in a standard differential centrifugation protocol. Most species are broadly dispersed, with significant cross-contamination in each fraction. While the mitochondrial and lysosomal markers are roughly unimodal, the patterns for plasma membrane and endoplasmic reticulum are more complex. Their distribution reflects the range of fragment sizes created by the homogenization step. The bimodal appearance of the markers is also of note. The plasma membrane in the nuclear fraction probably represents the large bile canalicular regions of the cell surface, while the unspecialized membranes form vesicles which are distributed through the other fractions. The glucose-6-phosphatase in the nuclear pellet is probably derived from the outer nuclear membrane

Table I

Characteristics of Rat Liver Homogenate Particles

FRACTION	PROTEIN (%)	DIAMETER (μ)	PELLETING FORCE (g·min)	SEDIMENTATION COEFFICIENT (S)	EQUILIBRIUM DENSITY (g/ml)
Erythrocytes		6	10^4	10^6	~1.20
Liver cells	100^a	15–20	10^3	10^7	~1.20
Nuclei	15	5–11	$5-10 \times 10^3$	10^6-10^7	~1.32
Golgi apparatus	2	1–3	4×10^4	10^5	1.06–1.14
Mitochondria	25	0.7–1.1	10^5	10^4	1.18–1.21
Lysosomes	2	0.4–0.6	4×10^5	5×10^3	1.20–1.22
Peroxisomes	2.5	0.4–0.6	4×10^5	5×10^3	1.22–1.24
Rough microsomes	12	0.05–0.2	$3-10 \times 10^6$	$0.3-2 \times 10^3$	1.13–1.25
Smooth microsomes	8	0.05–0.3	$3-10 \times 10^6$	$0.3-2 \times 10^3$	1.10–1.20
Plasma membranes	2				
Bile fronts		2–10	10^4	10^6	1.16–1.18
Vesicles		0.1–0.7	10^5-10^7	$0.3-5 \times 10^3$	1.12–1.15
Soluble proteins	30	<.01	$>10^8$	2–20	~1.30

^{a} Corresponds to about 220 mg protein/g fresh liver.

These values, pooled from references 2, 4, 9, 11, 13, 24, 36, 38, 56, 61, and elsewhere, depict the behavior of subcellular components upon centrifugation. They represent either averages or ranges of peak values from the various reports considered. Some entries are estimates based on indirect information; for example, no sedimentation coefficients for the Golgi apparatus or bile front plasma membranes were found in the literature, but reasonable values could be derived from their size and density and the forces necessary to pellet them from suspension. Values for size and prevalence derived from centrifugation data agreed well with morphometric analysis (61). Size is given as the diameter of a sphere of volume equivalent to that of the particle. The g·min column gives the integrated forces typically used to pellet the particles. The sedimentation coefficients refer to solutions of 0.25 M sucrose at 5° C. Equilibrium densities are for solutions of 0.25 M sucrose at 5° C. Equilibrium densities refer to sucrose gradients.

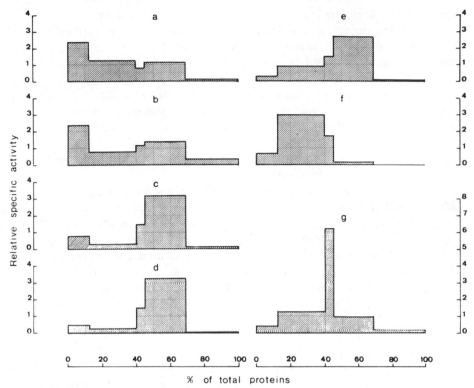

% of total proteins

Figure 2

Distribution of enzyme markers after differential centrifugation. Rat liver homogenate was centrifuged under the following integrated forces to prepare these major fractions: "nuclear," the 1×10^4 g-min pellet; "heavy mitochondrial," the 3.3×10^4 g-min pellet; "light mitochondrial," the 2.5×10^5 g-min pellet; "microsomal," the 3×10^6 g-min pellet; and "soluble," the 3×10^6 g-min supernatant. The relative protein content in the five fractions is indicated on the abscissa (cumulatively, in order of their isolation). The ordinates indicate the relative specific activity of the presumptive enzyme markers. Plasma membranes are represented by (*a*) adenosine diphosphatase and (*b*) adenosine monophosphatase. Endoplasmic reticulum markers are (*c*) inosine diphosphatase, (*d*) glucose-6-phosphatase, and (*e*) NADH–cytochrome c reductase. The mitochondria are followed in (*f*) by cytochrome oxidase. Lysosomes are represented by acid phosphatase (*g*). (From Wattiaux-deConinck, S., and Wattiaux, R. *Biochim. Biophys. Acta 183*, 121, 1969.)

as well as from the tags of rough endoplasmic reticulum which remain adherent to the nuclei. The low levels of NADH–cytochrome c reductase in the nuclear fraction may reflect the localization of this enzyme in the smooth and not the rough endoplasmic reticulum; the enrichment of this marker in the mitochondrial fraction, compared with the other endoplasmic reticulum markers, may be attributable to its presence in the outer mitochondrial membrane.

The utility of differential centrifugation can be enhanced. Recycling a fraction one or more times often effects considerable enrichment. For example, if centrifugation pellets 95% of the desired component and 20% of a slow-moving contaminant, then resuspension and repetition of the centrifugation twice more should, in theory, yield about 85% of the component and only about 1% of the contaminant.

Differential centrifugation can be used in an analytical mode to characterize subcellular particles. For instance, Anderson and his associates (20) have recently obtained precise information about the sedimentation constants and physical heterogeneity of various organelles by measuring the clearance of organelle markers from suspensions of homogenates centrifuged for a series of time periods and forces. Such data aid in devising optimal sedimentation conditions as well as in describing the properties of the particles directly.

Differential Density Gradient Sedimentation

Another means of enhancing the resolving power of differential centrifugation is to layer the homogenate onto a medium of a higher density. Particles of sufficient sedimentation constant and density are pelleted through the barrier without the obligatory contamination which accompanies conventional differential centrifugation. Nuclei have been extensively purified in this fashion (1, 15).

A more general approach to high-resolution differential centrifugation is to load the homogenate as a thin layer atop a shallow density gradient. The tube is centrifuged briefly to bring the particles into the gradient in the form of zones whose positions reflect their sedimentation rates. The technique is known as *rate zonal centrifugation*. The density gradient serves to stabilize the fluid column against convection and swirling during centrifugal accelerations and against mechanical perturbation during the collection of fractions.

The presence of a gradient complicates the estimation of sedimentation coefficients, since ρ_p, ρ_m, and η vary with the solute concentration and even r may be altered if the particles shrink in the highly osmotic gradients (5). In addition, the densities in the gradient should be less than those of the particles, otherwise the particles will not sediment. These considerations have not hindered the great success of this technique (7).

Equilibrium (Isopycnic) Density Gradient Centrifugation

When a particle sediments through a gradient to a point where its density corresponds to that of the surrounding medium, it stops its centrifugal motion since, according to Equation 1, $\rho_p = \rho_m$ and $s = 0$. Such particles are said to be in buoyant equilibrium. Each species of particles typically forms a band at a characteristic *isopycnic* density and can be fractionated on that basis. In these experiments, the particles can be loaded onto or below the gradient, or mixed homogeneously throughout, since (in simple gradients, see ref. 53) they should always come to the same equilibrium points. Centrifugation must be sufficient (frequently $\sim 10^8$ g-min) to allow a close approximation to equilibrium, since sedimentation rate approaches zero as equilibrium density is neared.

Continuous gradients can be constructed using any of a variety of mixing devices (1). Discontinuous density gradients are formed by layering a series of solutions of measured volume and density increments in a centrifuge tube. The advantages of the latter technique are that the gradient mixing chamber is obviated and fewer fractions may need to be collected, because the particles tend to accumulate as sharp zones at the prescribed density interfaces. However, the appearance of such well-separated bands may be misleading. Improperly chosen density barriers can artifactually split the desired organelles into two portions, each of which may be needlessly thrown together with contaminants which would band elsewhere on continuous gradients. At best, the results equal those of well-fractionated, continuous gradients, which are to be preferred, at least until the equilibrium density distributions of the desired and undesired particles are known with assurance.

With the exception of nuclei, the equilibrium density of most membranous organelles on sucrose gradients falls in a narrow range (1.06–1.25) with considerable overlap, as seen in Figures 1 and 3 and Table I. Most organelles exhibit a unimodal, but relatively broad, distribution; however, heterogeneity in the composition and physical state of fragments of plasma membrane is reflected in their equilibrium density dispersion (see below). Furthermore, the equilibrium densities of membranous particles are not fixed, but vary with several biological and experimental factors, as discussed later in this chapter, under "Strategies for Fractionation."

In concluding this brief discussion of centrifugal techniques, it is instructive to reconsider Figures 2 and 3. They show that frequently neither sedimentation rate nor equilibrium density alone clearly separates one type of organelle from all the rest. However, the combined effect of these two fractionation modes is more encouraging, as illustrated in Figure 1. Here not only viruses but various species of organelles are "windowed" by the sequential application of the two techniques. Even so, it is frequently necessary to harvest only a narrow slice of a broadly dispersed organelle in order to

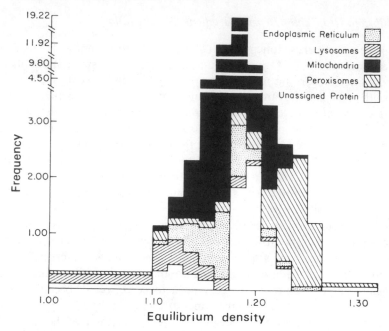

Figure 3

Equilibrium density distribution of some rat liver organelles. Particles pelleting between 24,400 and 342,000 g-min were suspended and centrifuged to equilibrium in sucrose gradients. A series of density steps were collected, and the composition of each was computed following assay of several characteristic enzyme markers. The ordinate represents the percent of the total protein in the histogram contributed by the components depicted. The unassigned protein may represent Golgi apparatus, plasma membranes, or other particles not assayed for. Note that the low density of the lysosomes in this experiment was induced by the administration of Triton WR-1339 to the animals prior to sacrifice, as discussed in the text. (From Leighton, F., *et al. J.Cell Biol. 37,* 482, 1968.)

achieve satisfactory purification. Thus, yield suffers at the expense of purity. These difficulties have called forth a host of methods by which organelles and their membranes can be moved to "open windows" in the s-ρ map and their centrifugal isolation thereby effected. These strategies are considered later in this chapter.

Other Approaches to Fractionation

Centrifugation techniques often fail to provide analysis and preparation of subcellular components at a level of resolution equal to the complexity and subtlety of biological systems. Therefore, complementary techniques are potentially of great value.

Various forms of chromatography have been explored in purifying subcellular particles and, particularly, viruses (1). Anion-exchange columns seem to bind plasma membranes and microsomes very tightly so that elution is rather difficult. Petersen and Kuff (44) have used this effect to separate ribosomal complexes from rough endoplasmic reticulum membrane.

Zonal density gradient electrophoresis has also been used preparatively in subcellular fractionation, and analytically to characterize various membranes, including the inner and outer mitochondrial membranes (49). Recently, lysosomes have been prepared and subfractionated using a carrier-free continuous electrophoresis system (52).

Cells, organelles, and viruses have been found to partition between two immiscible aqueous solutions of high polymers and can be thus fractionated by the countercurrent distribution technique. It appears that the surface charge composition is a major determinant of the partitioning of particles between the two phases (10).

Microsomal vesicles and other particles have been sorted and analyzed according to size by passage of the membranes through a graded series of porous membrane (e.g., Millipore) filters.

These techniques indicate that subcellular particles may be characterized and isolated by means other than centrifugation. However, none of them has yet demonstrated enough general advantage to be widely used in subcellular fractionation.

Other techniques which have been developed for separating macromolecules on the basis of specific biochemical affinities might be profitably applied to membrane fractionation. For example, antibodies prepared specifically against protein components of the outer surface of organelles could be used as agglutinins or could be coupled covalently to a solid matrix for column or batch adsorption of the target particles. Similarly, *affinity chromatography* couples antigens or competitive enzyme inhibitors covalently to macroscopic beads so that antibodies or enzymes with avid binding sites for the immobile moieties are selectively retained by the solid matrix. This technology has already been applied to sorting lymphocytes by their immunologic specificity for the solid-state haptens (57) and could be extended to isolating membranes on the basis of their specific enzyme content. The considerable difficulties inherent in these approaches should be apparent, and it is little wonder that the centrifuge continues to dominate the field of subcellular fractionation.

STRATEGIES FOR FRACTIONATION

In developing or adopting an isolation protocol the investigator must consider many experimental options, such as how to break the cells, what medium befits the homogenization protocol, what type of gradient is optimal, how to control the behavior of the particles, and how to assess the purified product. Some choices can be grounded in principle, others in accrued experience, still others in creative intuition. This section is addressed to these choices.

Cell Disruption

HOMOGENIZATION

Many techniques are employed to reduce cells and tissues to a suspension of subcellular particles (1, 7). The essential event is the breaking of the plasma membrane. This structure is differentially vulnerable to disruption by shearing forces, particularly if the cell is first swollen in mildly hypotonic media. Homogenization should be as gentle as possible so that the released organelles are not themselves disrupted. It is usually found that not all the cells can be broken, for the most fragile nuclei may break down before the toughest cells. Nuclear damage releases polybasic nucleoproteins which adsorb to the polyanionic membranes, thus promoting the aggregation of particles (a potential hazard to subsequent fractionation). The breakdown of lysosomes should likewise be avoided, since exposure to their autolytic enzymes may adversely affect the membranes. For these reasons, harsh techniques such as freezing and thawing or ultrasonic irradiation (sonication), useful in preparing homogenates for enzyme studies, are not as applicable to subcellular fractionation techniques. Sonication and similar high-shear techniques, however, are effective in reducing purified membranes to microfragments which can be subfractionated centrifugally (21, 47).

The most popular disruption devices are the coaxial pestle homogenizers, such as the Dounce or Potter-Elvehjem design (1). Homogenization should be monitored by phase-contrast microscopy and halted when a maximum of cells and minimum of nuclei have been disrupted. Wallach (9) has advocated plasma membrane disruption by equilibrating the cell suspension under 50–75 atm of nitrogen gas. Rapid decompression leads to the quantitative fragmentation of the plasma membrane and endoplasmic reticulum to small vesicles while preserving the integrity of the other membranous organelles.

Cells lacking internal organelles have been emptied of their soluble contents by osmotic stress. For example, the plasma membranes of mammalian red blood cells, bacterial protoplasts, and mycoplasma have been isolated as intact "ghosts" by lysis with hypotonic buffers (8).

How much physical disruption of organelles takes place during homogenization? The endoplasmic reticulum is the most sensitive structure in this regard. It seems invariably to become reduced to small vesicles which generally populate the microsomal fraction. The disruption of nuclei and lysosomes by hypotonic lysis or too vigorous homogenization is obviously undesirable, but occurs frequently. It is common, for example, to recover a portion of the lysosomal enzymes among the soluble proteins following cell disruption. The outer mitochondrial membrane may also be disrupted by routine homogenization, as indicated by the appearance of the marker enzyme, monoamine oxidase, among the microsomal vesicles (56). The Golgi apparatus, itself a composite of associated vesicles and tubules, may be dispersed by homogenization and recovered in the mitochondrial and microsomal, as well as in the nuclear, fractions.

The plasma membrane is obviously also disrupted by homogenization. Depending on the cell type and on the conditions applied, it may be recovered as an intact, sealed envelope (or ghost), as large sheet-like fragments, or as small, sealed vesicles, which may sediment with the nuclear, mitochondrial, or microsomal fractions (8). As with the endoplasmic reticulum, vesiculation appears to occur by a process of pinching off rather than by the release of open fragments which subsequently manage to seal their torn edges. The vesicles entrap and retain the adjacent cytoplasm in the process. The pinching off of nerve endings into "synaptosomes" encloses other organelles, such as mitochondria and synaptic vesicles, conferring novel sedimentation properties on the sequestered organelles.

When a membrane sheds vesicles into the compartment it encloses, the vesicles acquire an inside-out rather than a normal orientation. This may cause the appearance of otherwise hidden surface functions and the burying of otherwise exposed markers within the inverted vesicles. Useful preparations of inside-out inner mitochondrial (18, 36) and erythrocyte (54) membranes have been achieved. It can be assumed that pinocytic or phagocytic vesicles, such as those isolated from amebae following the ingestion of polystyrene beads (62), are bounded by an inverted plasma membrane, perhaps specialized for this function.

The tendency of membranes to fragment creates certain technical difficulties, but also offers some advantages. Since the purified product frequently represents only a portion of the target membrane, one must ask whether the fraction recovered reflects the whole or only a specialized component. For example, rat liver plasma membranes may be isolated as vesicles or sheet-like fragments which equilibrate at different densities and have different properties (8, 24); they may, in fact, represent different topographic areas of the complex liver plasma membrane system. An investigator may harvest one class of fragments without the other, but incorrectly consider his fraction to represent the entire plasma membrane.

On the other hand, fragmentation of complex membrane systems could aid in their analysis. In the case of the liver plasma membrane fragments just mentioned, controlled protocols can lead to selective isolation of both types of membrane for comparative study (24). Again subfractionation of the dispersion of endoplasmic reticulum vesicles recovered from the microsomes could lead to an appreciation of topographic heterogeneity and functional organization within that membrane system.

A final comment on tissue homogenization is offered. Current subfractionation techniques call for the disruption of highly ordered cell systems and the isolation of classes of particles from the randomized mixtures. The attendant loss of organizational information is enormous and should be of concern to the membranologist. Since most tissues contain a mixture of cell types, pooling of their homologous organelles during fractionation obscures their inherent biochemical specialization. Neuron-rich and glia-rich fractions from brain, for example, have been found to yield mitochondria of clearly different biochemical and physical properties (29). Yet these different species are regularly pooled and studied as "brain mitochondria." Heterogeneity in the properties of liver mitochondria has been associated with the organization of the liver lobules (see under "Mitochondrial Heterogeneity," later in this chapter). Unfortunately, sorting of cell types prior to cell disruption has usually proved difficult.

The loss of biological order (information) is no less great when the membrane systems within cells are fragmented during homogenization. Access to spatial or topographic integration of the whole may be lost. At both the cellular and subcellular level, it would seem that the correlation of high-resolution cytochemical analysis with subfractionation technology offers the best approach at present to these problems.

MEDIA

During homogenization, subcellular particles are diluted into an artificial medium, which is generally very unlike the cytosol. The ideal medium, of course, should not alter the properties of the particles, but should foster their isolation. In practice, however, considerable perturbation of the organelles may occur. For example, the loss of metabolites from, and metabolic alterations of, the nuclei during their isolation in aqueous media have stimulated development of the fractionation of rapidly freeze-dried tissues in nonaqueous solvents (1).

Early fractionation studies were often performed in distilled water or physiologic salt solutions. In the former case, the severe hypotonic exposure can lead to the swelling and rupture of nuclei, lysosomes, mitochondria, etc. In addition, certain membrane proteins may be selectively released at very low ionic strength, while some otherwise soluble cytoplasmic proteins may adsorb to the membranes.

Isotonic salt solutions resembling the intracellular milieu promote aggregation of the subcellular particles, especially when pelleted. Divalent cations (in the millimolar range) and acidic pH (especially between pH 4 and 6) do likewise (1). Freezing membrane particles in these media may greatly increase this tendency to aggregate. Subsequent fractionation steps cannot be expected to disperse the aggregates fully and effect a satisfactory separation. The adoption of isotonic (0.25 M) sucrose, usually lightly buffered to pH 7–8 (e.g., with 0.01 M Tris · HCl) tends to avoid these aggregation problems (see refs. 1 and 7 for examples).

Homogenization media must be suited to the particular problem and system at hand. For example, hypertonic sucrose solutions preserve mitochondrial morphology but inhibit oxidative phosphorylation. Alkaline media (pH 8–9) are suitable for work with isolated membranes, but nuclei homogenized in such buffers may be disrupted. Acidic media and divalent cations stabilize nuclei and some cell surfaces against disruption by homogenization, but foster particle aggregation. Detergents and surfactants (such as digitonin) have been used to selectively disrupt some membranes, while leaving other structures intact.

Centrifugal separations are also usually carried out in media of low ionic strength, lightly buffered at or above neutral pH, with little or no divalent cation present (1, 7). However, organelles and especially viruses are sometimes banded in salt gradients. At the molar concentrations employed, aggregation does not appear to be a problem, but strong salt solutions have been found to extract membrane proteins, lipids, and enzymes and may therefore alter the membrane considerably (37).

Adsorbed or Trapped Extraneous Proteins

Membrane preparations may be substantially freed of contaminating particles and yet still bear considerable artifactually associated soluble protein. Two general mechanisms can be named: adsorption and trapping.

In media of low ionic strength, adsorption of some normally soluble proteins to the membranes may occur. The contaminant may be removed by washes in isotonic salt solutions. In this way, various soluble enzymes and considerable protein have been removed from preparations of rat liver plasma membranes (8). Hexokinase isozymes are predominantly soluble cytoplasmic enzymes, but may be recovered adsorbed to isolated mitochondria. The outer mitochondrial membrane appears to have high-affinity binding sites for these enzymes. The physiological significance of the reversible association is presently under investigation.

Adsorption effects are exemplified by the red blood cell membrane (8). The association of residual hemoglobin with purified "ghost" membranes depends primarily on the pH and ionic strength of the lysis medium (23). Furthermore, the chain of glycolytic enzymes is bound to these membranes under certain ionic conditions, but may be eluted by suitable buffer washes. The locus of these enzymes in the intact red cell is debated.

Soluble protein may be trapped when extended membranes pinch off to form vesicles. The enclosed material can account for a large fraction of the total vesicle protein. The sequestered protein may be released via osmotic shock by transferring particles from isotonic sucrose solutions to hypotonic buffers (8, 9). Finally, many isolated membranes are in fact organelles or portions thereof with their original compartments enclosed. The membranes of the Golgi apparatus, lysosomes, peroxisomes, nuclei and certain cell surfaces, for example, have not yet been prepared entirely free of their original contents.

Altering Particle Size

We have seen that isolating particles in their s-ρ window is complicated by the overlapping physical properties of the organelles. One approach to this problem is to selectively change the sedimentation properties of the target particles at an advantageous point in the purification, thus moving them away from the contaminating species.

In general, the more forcefully a membrane is sheared by homogenization, the smaller are the fragments obtained (8, 9). Several authors have utilized differential homogenization in the purification of plasma membranes. For example, both Coleman *et al.* (19) working with rat liver and Boone *et al.* (16) working with HeLa cells used gentle homogenization to prepare large fragments of plasma membranes which were recovered in the low-speed pellet. Vigorous homogenization of that material selectively reduced the plasma membrane component to small vesicles, which were then readily separated from the rapidly sedimenting contaminants. In becoming vesicles the membranes also became less dense (perhaps by virtue of the enclosed aqueous space), which also can facilitate fractionation (see below).

Quigley and Gotterer (46) also isolated plasma membrane vesicles by inducing a change in their sedimentation properties. In this case, the mitochondrial fraction of cells from the intestinal mucosa was found to be rich in Na^+,K^+-ATPase,* a plasma membrane marker. After "aging" this material for 5 days at 0–$2°C$ or for 3–5 hr at $37°C$, the ATPase-bearing membranes could be separated from the mitochondria as small vesicles by brief rate zonal centrifugation. Though not specified, the aging may have caused large plasma membrane fragments to break down to small vesicles, thus fostering the separation observed.

It has already been pointed out that uncontrolled breakdown of membranous structures can undermine isolation protocols. This seems particularly true of some plasma membrane preparations from rat liver (8). There the large membrane sheets appear to shed small vesicles as they are washed, frozen and thawed, homogenized, etc. These vesicles sediment slowly and

---

* ATPase, adenosine triphosphatase.

may be lost, thereby contributing to the generally poor yield (5–15%) frequently obtained. A similar fate attends the isolation of plasma membranes from the intestinal brush borders and other surfaces. Even intact, viable cells can apparently shed significant amounts of their surface as small vesicles(8).

The effective particle size may be altered by aggregation, which increases sedimentation rate. A good example is the method of subfractionating microsomes developed by Dallner (2). When 15 mM CsCl is added to a microsomal mixture, the rough endoplasmic reticulum vesicles appear to selectively aggregate, presumably by the screening of their highly negative net charge. Brief centrifugation through a discontinuous sucrose gradient (also containing CsCl) brings the aggregated rough vesicles to the pellet while the smooth vesicles accumulate at the 0.25/1.3 M sucrose barrier. Dallner has also found that the smooth microsomes isolated in this way can be subfractionated. Adding Mg^{++} to these membranes induced the selective aggregation of a characteristic fraction which could be separated from the nonaggregated vesicle fraction on a discontinuous sucrose density gradient (2).

Altering Particle Densities

A membrane or organelle is a composite structure. Its buoyant density bears a simple relation to the densities of its components and their relative proportions within the particle. Thus, microsomes increase in density with their content of ribosomes, which are, in turn, made dense by their RNA content (see Fig. 1). Similarly, nuclei reflect the high density of their DNA and RNA, peroxisomes approach the equilibrium density of their concentrated protein contents, and the buoyancy of Golgi vesicles may relate to their enrichment in triglycerides. Water (or solvent) also contributes to the density of organelles and membranes, but in a more complex fashion; this topic is considered separately in the next section.

Parsons (41) has found that the relative proportions of protein and phospholipid (or phospholipid plus galactolipid) correlate well with the equilibrium density of a series of membranes (Fig. 4). His linear curve extrapolates to reasonable density values for pure hydrated protein and liquid phospholipid. Why other elements such as neutral lipid, carbohydrate, and water are not reflected in the observed density is not stated; nor is it clear that equally linear correlations could not be obtained by choosing another set of parameters. Furthermore, Parson's own data (41) indicate that microsomes and outer mitochondrial membranes have the same equilibrium density but differ considerably in their protein/phospholipid ratio. The explanation for this observation may lie in the fact that the microsomal vesicles enclose a sucrose-impermeable water space which contributes to their buoyancy, while the mitochondrial membrane fragments lack this water compartment (see below). The concept of correlating density with chemical composition clearly is sound; however, it would seem that the contribution of all the components should be considered.

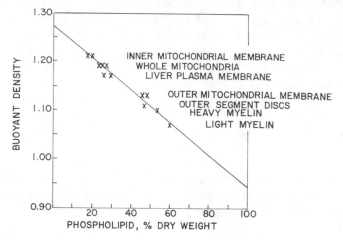

Figure 4

Correlation of equilibrium density with composition for a series of membranes. The abscissa expresses the fractional content of phospholipid (or phospholipid plus galactolipid) as percent dry weight. (Modified from Parsons, D. F. *Can. Cancer Conf. 7,* 193, 1967.)

The composition of particles, and hence their densities, can be manipulated experimentally. Perhaps the simplest example of density alteration is the disruption of a composite structure to liberate components of different densities which can be separated on that basis. The mitochondrion can be broken into fragments of outer mitochondrial membrane plus inner membrane enclosing the proteinaceous matrix. Because the latter particles are more dense than the former, the two mitochondrial subfractions can be separated from each other and from intact mitochondria on sucrose gradients (see below, under "Isolation of Membrane Systems"). Similarly, plasma membranes isolated from rat liver have been homogenized to yield open fragments of density 1.21 g/ml and closed vesicles (ρ = 1.13) which have distinctly different biochemical profiles (24).

Another straightforward way to alter membrane density selectively is by the specific adsorption of suitable substances. Boone *et al.* (16) found that the binding of specific antibodies to isolated HeLa cell plasma membrane fragments increased the fragment density significantly. In a like manner, high-density viruses which bind tightly and specifically to plasma membranes might be expected to alter the membrane buoyancy (45). The treatment of microsomal mixtures with digitonin increases the density of the plasma membrane fragments selectively, thus aiding in the identification of the component (56). Such techniques, however, have not found general application in membrane isolation protocols.

Membranous vesicles can be fractionated at times on the basis of the contents of their enclosed compartments. For example, the membranes surrounding polystyrene-latex beads phagocytosed by amebae have been readily isolated by virtue of the very low density of the ingested particles (62). Another technique of this type employs the uptake of low-density detergent into lysosomes to effect their separation from mitochondria and peroxisomes (4, 33). Triton WR-1339 is administered intravenously to rats 2–4 days prior to sacrifice. The lysosomes in the liver homogenate take up large amounts of the detergent so that their equilibrium densities are shifted from around ρ 1.21 to ρ 1.12. This removes them from the density range of the mitochondria and peroxisomes (Fig. 3). Alternatively, dextran has been administered to the rats to make the lysosomes more dense.

The sequestration of insoluble lead phosphates within vesicles has been applied in subfractionating microsomal vesicle mixtures (34). The microsomes were allowed to hydrolyze glucose-6-phosphate (an activity characteristic of liver endoplasmic reticulum) and the released phosphate was precipitated in situ by the presence of millimolar Pb^{++}. The enclosed precipitate brought those vesicles rich in glucose-6-phosphatase to a higher equilibrium density during gradient centrifugation. By this technique it was shown that glucose-6-phosphatase appears evenly distributed throughout the hepatocyte rough endoplasmic reticulum, confirming the histochemical data (34). Similarly, the density of rat liver mitochondria has been shown to increase in proportion to the accumulation of insoluble calcium phosphate salts during respiration-driven ion transport (28).

Density labeling can be used in the study of membrane biogenesis. Luck (35) found that the mitochondria of a mutant strain of *Neurospora crassa* decreased in density when the choline in their growth medium was increased from 1 to 10 μg/ml. By following the change in the mitochondrial density profile following this transition, Luck concluded that these organelles replicated by the random accretion of new molecules into existing mitochondria which were undergoing simple division. The assimilation of dense isotopes or analogues of normal constituents would be a useful tool in the study of membrane biosynthesis and turnover (see Chap. 12).

Effect of Gradients on Particle Densities

The buoyant density of sedimenting particles is not indifferent to the composition of the medium they traverse. The solutes in the gradient affect the association of solvent with the particles, hence influence their mass and density (5, 9, 13). There are three possible water compartments to be considered: *1*) water of hydration, *2*) a solvent space freely permeable to the gradient solutes, and *3*) an osmotically active aqueous space, impermeable to the gradient solutes.

The hydration of membranes and organelles derives from the great affinity of their proteins, carbohydrates, and ionized groups for water. The bound water lowers the density of the particles. (This effect is somewhat mitigated by the fact that bound water has a density considerably above unity.) Concentrated solutions of sucrose, salts, and other gradient solutes affect the activity of water, and hence alter the degree of hydration of a particle. This can lead to small differences in the equilibrium density of a particle in various gradient media.

Beaufay and Berthet (13) have estimated the extent of hydration of mitochondria, lysosomes, and peroxisomes by measuring the difference between their equilibrium densities in sucrose-H_2O and sucrose-D_2O gradients. (Heavy water has a density of about 1.10 g/ml.) The degree of hydration differs among these organelles, leading to unequal shifts in density and some improvement in the separation of peroxisomes from mitochondria in sucrose-D_2O gradients. All in all, hydration effects represent a minor factor in centrifugal fractionations and are not reckoned with in most investigations.

Most of the water in cells and their membranous organelles is confined within closed, semipermeable boundary membranes; these bodies are therefore, in fact, *vesicles*. The behavior of this aqueous compartment in density gradient centrifugation depends upon whether the vesicle membrane is permeable to the gradient solute. If the membrane is permeable to the solute, the particle exhibits no osmotic response to the gradient. As the vesicle moves through the gradient, its solvent space assumes the composition of the surrounding medium and thus the sedimenting particle continuously increases in density. An equilibrium point eventually is reached where the density of the gradient, the fluid space, and the rest of the particle are the same. Thus, the permeable aqueous compartment does not influence, but rather assumes, the equilibrium density of the particle as a whole.

Gradients formed of glycerol, D_2O, and probably molar salt solutions penetrate many membranes and therefore fall in the above category. Sucrose permeates the outer (but not the inner) mitochondrial membrane and the lysosomal and peroxisomal boundary membranes as well (4).

In contrast, solutes which cannot penetrate the vesicle membrane create an osmotic force which acts to shrink the aqueous compartment. The fluid compartment contracts until the intravesicular osmotic activity (which arises from solutes sequestered within the vesicle) equals that of the external medium. Osmotic equilibrium then obtains, and further volume changes are resisted by the osmotic balance.

Any density gradient can thus be an osmotic gradient as well. Sucrose, mannitol, and related compounds are impermeable to many membranes, but not to others (for example, compare the inner and outer mitochondrial membranes). Membranes which are permeated by simple sugars seem to exclude large polymers of sucrose (Ficoll) and glucose (glycogen and dextran) as well as serum albumin and other large polymers used as gradient solutes.

As a vesicle moves through an osmotically active gradient, its aqueous space continuously shrinks. Thus, during its sedimentation, the particle as well as the surrounding medium smoothly increase in density. At some point, the particle and gradient have the same density; then buoyant equilibrium and osmotic equilibrium obtain.

The density at which buoyant equilibrium is achieved depends on the relation of the osmotic activity of the gradient solute to its density. For example, smooth microsomes have an equilibrium density of around 1.15 in sucrose gradients but equilibrate at $\rho \sim 1.03$ in polysucrose (Ficoll) gradients (9). Similar dramatic differences are seen for isopycnic densities of mitochondria and lysosomes on sucrose and glycogen gradients (13). The explanation lies in the relatively high osmotic activity per unit of density in sucrose solutions. The osmotic activity of sucrose solutions in the equilibrium density range of most particles is above the molar level, while the large polymers of sucrose and glucose probably do not exceed 0.1 osmolar. The osmotic space of the vesicles in sucrose is almost entirely depleted so that the density of the particle approaches that found in permeant solute gradients (where the fluid compartment does not contribute to the equilibrium density of all). In contrast, vesicles centrifuged to equilibrium in large-polymer gradients retain a significant water space and are therefore much less dense.

The isopycnic density of fluid-filled vesicles may also be influenced by factors more subtle than the osmotic activity in the gradient. For example, the equilibrium density of vesicles in Ficoll or dextran gradients depends on the pH, ionic strength, and divalent cation concentration of the gradient medium (9). An explanation for these ionic effects has been offered in the form of a theoretical model (53). It seems that the charged groups fixed within the vesicle lumen (i.e., on the inner surface of the membrane and on the enclosed proteins) cause the accumulation of diffusible electrolytes within the vesicle water space, according to the Donnan equilibrium. Although freely permeable, the excess electrolyte ions constrained inside the vesicle exert an osmotic force which tends to expand that water space, just as osmotic solutes outside the vesicle have a shrinking effect. The volume of the vesicle fluid space must therefore satisfy a Donnan-osmotic equilibrium across its membrane which is reflected in the equilibrium density of the vesicle (53).

A useful prediction of the theory is that closely related membranous vesicles of low internal charge density may be best fractionated on gradients of low osmotic activity and ionic strength, since under these conditions small differences in their charge composition will lead to the greatest difference in buoyant density. An example of this prediction is the separation of plasma membrane and endoplasmic reticulum vesicles from microsomal mixtures in Ficoll gradients, where careful titration of the membranes with Mg^{++} was essential for optimal resolution (see later in this chapter under "Microsomal Subfractionation," and ref. 9).

Another case in point is the separation of inside-out from right-side-out erythrocyte membranous vesicles (54). It appears that the charge density on the outside of this membrane is high compared with that on its cytoplasmic face, so that the two types of vesicles have different charge densities within their aqueous spaces and can be separated on that basis. As predicted, gradients of dextran but not of sucrose or glycerol foster this separation (53).

Finally, Donnan-osmotic equilibrium effects have been exploited in the isolation of erythroid cells in albumin gradients as a function of the pH-dependence of their equilibrium densities (50).

Assessing the Product

The means most frequently employed to characterize subcellular fractions are electron microscopy and enzyme assay. The former technique, while of obvious importance in qualitative assessment, is difficult to apply in the quantitative mode. For example, does the visualization of 1 mitochondrion per 100 microsomes, even if accurate, signify a small or large contamination? The dry weight of the mitochondrion is around 10^{-13} g, which is equivalent to the mass of about 100 microsomal membrane vesicles of 0.2-μ diameter!

The identification of enzyme functions characteristic of the various organelles has provided a firm basis for monitoring subcellular fractionation (cf. 3, 7, 8). Several parameters can be evaluated using simple enzyme assays. By measuring the total and specific activity of an appropriate marker enzyme in the initial homogenate and in the isolated fractions, the recovery and relative purification of the organelle may be ascertained. The reciprocal of the relative purification (i.e., specific activity in the homogenate divided by specific activity in the fraction) also estimates an upper limit on the percentage of total cell protein contributed by that organelle species. For example, if a Golgi apparatus marker, galactosyl transferase, shows a 50-fold enrichment in specific activity in the isolate compared with the homogenate, then those Golgi vesicles must constitute one-fiftieth (2%) or less of the cell protein. Of course, if the same marker is present on other particles or if the isolated product is seriously contaminated, this upper limit can be quite inaccurate.

Contamination can also be quantified. This is done by measuring the specific activity of enzymes associated with each contaminating organelle in both the target fraction and the purified contaminant. The ratio of the specific activity in the sample to that in the pure contaminant is an estimate of the fraction of the sample protein contributed by that contaminant. For example, if a plasma membrane isolate contains a mitochondrial marker enzyme, succinic dehydrogenase, at a specific activity one-tenth that of a pure mitochondrial preparation, then 10% of the plasma membrane fraction protein is mitochondrial contaminant. A general approach to the analysis of subcellular fractions is beautifully illustrated in the evaluation of liver particle preparations by Leighton *et al.* (33).

The use of enzyme markers as representatives of their organelles has some serious drawbacks:

1. Uncontrolled enzyme activation, inactivation, or release may occur during fractionation, thus complicating quantitative analysis.

2. A given marker may not represent the entire class of organelle under study. For example, glucose-6-phosphatase and other endoplasmic reticulum markers are not homogeneously distributed throughout the microsomal population (see under "Subfractionation of Microsomes"). In general, at least some organelles are expected to be heterogeneous in structure, composition and function, including enzyme content. Discerning these subtle specializations does not render the marker enzymes useless, but rather refines the resolution of the spectrum of subcellular particles from bands to fine lines.

3. The assumption that each bound enzyme is localized in only one organelle has greatly facilitated subcellular fractionation studies (3), but it is now known to be incorrect in several instances. In addition, some substrates can be attacked by a variety of enzymes borne on separate organelles. They may be distinguishable on other grounds, however, such as their structure-bound latency (in the case of lysosomes), their activation by specific effectors (e.g., Na^+,K^+-stimulated ATPase), pH optima, etc.

4. A membrane isolate can be heavily contaminated with another species of particle which, for want of a marker, remains undetected and unsuspected. Electron microscopic or other ancillary techniques might be useful in these instances.

5. Misleading conclusions may be reached because of the large differences in the relative abundance of the various organelles. For example, mitochondria constitute approximately 25% of the total liver cell protein (Table I). Therefore, the specific activity of a mitochondrial marker enzyme can be increased only about four times over that in the initial homogenate. The entire lysosomal or plasma membrane population could contaminate the mitochondrial fraction with only a 10% reduction in the specific activity of the mitochondrial markers.

On the other hand, Table I also suggests that a 25-fold purification of the plasma membrane can still leave it contaminated with an equal weight of extraneous protein. Since most liver plasma membrane preparations show less purification than that (8), the liver plasma membrane is either more extensive than is estimated here, more contaminated than is commonly reckoned, or inaccurately represented by the enzyme markers under study.

ISOLATION OF MEMBRANE SYSTEMS: EXAMPLES

Nuclear Membranes

Although the nucleus is the organelle most readily purified, its membrane has only recently been prepared for biochemical evaluation. The nuclear envelope is composed of two separate membranes, the outer of which is covered with ribosomes and seems to be in structural and functional continuity with the rough endoplasmic reticulum. The two nuclear membranes are traversed by numerous "pore" complexes whose fenestrations are not empty, but have a characteristic granular ultrastructure. The possibility of functionally specific inner membrane–DNA complexes has received some experimental support and adds to the challenge of preparing meaningful membrane isolates.

Nuclei are the largest and most dense of the organelles, and almost all purification methods draw upon this happy circumstance (1). While a variety of homogenizing media have been employed, current practice favors isotonic or slightly hypertonic sucrose solutions of low ionic strength, buffered near neutrality. The presence of millimolar Ca^{++}, Mg^{++}, or other polyvalent cations stabilizes the membrane; higher divalent cation levels promote aggregation, while the absence of these ions or alkaline media invites nuclear disruption. Nuclei are initially pelleted from dilute homogenates by low gravitational forces (10^4 g-min). Such pellets may be contaminated by whole cells, mitochondria, plasma membrane, Golgi complexes, and endoplasmic reticulum. The last mentioned is found either as fragments which adhere to the nuclei or as tags in continuity with the outer nuclear membrane.

The nuclei may be pelleted through sucrose density barriers (1.28–1.29 g/ml) into which other membranous organelles do not pass. Even the tags of endoplasmic reticulum are stripped from the nuclei by this procedure. Based on these considerations, Blobel and Potter (15) have presented a single-step discontinuous density gradient purification procedure of high yield and purity.

Five types of treatment have been employed in separating nuclear membranes from the nucleoplasm: exposure to detergents, hypotonic media, concentrated salt solutions, ultrasound, and deoxyribonuclease. Neutral detergents, such as 0.5% Triton X-100, have been found to disrupt various cell membranes, including the outer nuclear membrane, while the inner nuclear membrane and nucleoplasm apparently remain intact (15). While the full effect of such detergent treatment on the nuclear envelope has not been assessed, this technique offers the only current method for separating inner and outer membranes.

Hypotonic media have long been known to disrupt nuclei and release their nucleoprotein. Zbarsky *et al.* (63) applied this technique (following detergent treatment) in isolating inner nuclear membranes.

Recently, three groups have published extensive studies on the purification of nuclear membranes, all relying on media of high ionic strength to disperse adherent nucleoprotein from the envelope fragments. These solutions were 0.5 M $MgCl_2$ (14), 1.5 M KCl (26), and 0.3 M K citrate (30). In two cases the nuclei were previously sonicated to facilitate extraction of the nucleoplasm (26, 30), while deoxyribonuclease digestion preceded the salt extraction in the other study (14). Final purification was accomplished by differential and sucrose density gradient centrifugation.

The membrane fragments recovered by these various authors had equilibrium densities of ~ 1.18–1.22. In general, they retained their ribosomes, pore complexes, and double membrane structure, although vigorous sonication seemed to disrupt the double membrane configuration in favor of small vesicles and sheets (26). Chemical and biochemical analysis revealed considerable similarity among these preparations. Discrepancies may relate to the disparate disruption and fractionation protocols. For example, the various salt treatments might have caused unequal elution of RNA, DNA, or protein from the membranes. Similarly, the failure to recover glucose-6-phosphatase activity (26) could relate to the thermal lability of the nuclear but not microsomal enzymes (30).

Mitochondrial Membranes

Interest in the isolation of mitochondrial membranes is heightened by the recognition that their topographical organization is intimately associated with their metabolic function (6, 51). The mitochondrial enzymes are compartmented either by sequestration within, or by physical association with, the membranes. Between the simple, smooth outer membrane and the inner membrane (made complex by its multiple infoldings, the cristae) lies the intermembrane space. The compartment within the inner membrane is the matrix space.

PREPARATION AND PURIFICATION OF MITOCHONDRIA

Following removal of nuclei, mitochondria are pelleted at 5–20×10^4 g-min. They are often further purified by equilibrium banding ($\rho \sim 1.18$–1.21) in sucrose gradients (3, 5, 13, 33). This step, however, may not entirely remove the iso-dense contaminants, which often include lysosomes, peroxisomes, plasma membrane, and endoplasmic reticulum (22). If respiratory studies are the aim, repeated differential centrifugation is usually elected so that the adverse effects of hypertonic sucrose gradients are avoided.

An example will indicate how even small amounts of contamination can vex mitochondrial membrane studies. Vignais and Nachbaur (58) showed that lysosomes which travel with mitochondria are lysed by the same hypotonic treatment that releases the outer mitochondrial membranes, yielding lysosomal membranes which band with the outer membrane on discontinuous sucrose gradients. Such lysosomal contamination may lead to an incorrect assessment of the enzymic and chemical composition of the outer membranes. While the lysosomal mass is only a small fraction of that of the mitochondria, the lysosomal membranes could well constitute a major contamination of the outer mitochondrial membrane isolate. Furthermore, damage may be caused by lysosomal hydrolases. Prior administration of Triton WR-1339 shifts lysosomes to a lower equilibrium density so that they can be separated from the mitochondria prior to the lysis step.

MITOCHONDRIAL HETEROGENEITY

The mitochondria of tissues such as rat liver usually exhibit a reproducible, narrow, symmetrical distribution of equilibrium densities on sucrose gradients. The equilibrium density of mitochondria, however, can be experimentally altered. When the growth medium of *Neurospora crassa* cells is varied with respect to limiting choline or leucine levels, the mitochondrial phospholipid/protein ratios and hence their densities are characteristically changed (35). Similarly, when rat liver mitochondria are allowed to accumulate calcium phosphate during respiration, granules of this salt accumulate within the matrix and the mitochondria become more dense in proportion to the Ca^{++} accumulated (28). In each of these systems the density distribution of the altered state is narrow, unimodal, and symmetrical. These results provide evidence that the mitochondria are homogeneous in their response to these perturbations; if this were not the case, a broad range of new densities would be observed.

The distribution of rat liver mitochondrial sedimentation constants is rather broad but appears to be unimodal and symmetrical. de Duve attributes the broad dispersion to random biological and experimental factors rather than to heterogeneity within the mitochondrial population (22). Recent studies, however, indicate otherwise. The sedimentation rates of mitochondria-bearing ornithine aminotransferase and malate dehydrogenase differ significantly, while their equilibrium densities do not (55). Their size distributions must presumably be different. These differences are most striking in mouse liver mitochondria, where the mean particle diameters for the two enzymes differ by more than 15%, and the relative specific activity of the two enzymes varies over a 10-fold range across the sedimentation profile. This heterogeneity was demonstrated in the mitochondria of isolated hepatocytes, so that contribution from other cell types within the liver was ruled out. Other tissues show no marked heterogeneity of the distribution of this pair of enzymes. These authors suggest that ornithine aminotransferase might be enriched in the small mitochondria of the cells surrounding the central vein of liver lobules (55).

Two peaks of cytochrome oxidase activity within rat liver mitochondria have been demonstrated by rate zonal centrifugation (48). The morphology of the particles in the two fractions differ and suggest that the heterogeneity may reflect differences in the metabolic state of the particles.

Various investigators have found that "heavy" and "light" mitochondria (i.e., those pelleted at integrated forces of $\sim 0.5 \times 10^5$ g-min and 1×10^5 g-min, respectively) differ in the specific content of certain enzymes. Katyare *et al.* (31) found differences in respiratory control, response to triiodothyronine, and turnover of mitochondrial proteins, from which they suggested that the light and heavy fractions represented young and old mitochondria; i.e., that the sedimentation differences reflected heterogeneity in maturation.

Another source of mitochondrial heterogeneity derives from mixtures of cell types. Hamberger *et al.* (29) found that the mitochondria derived from neuronal and glial cells of brain showed different equilibrium density distributions as well as distinctive biochemical features, suggesting that the mitochondrial mixtures in whole brain homogenates can be subfractionated on density gradients.

SEPARATION OF OUTER AND INNER MITOCHONDRIAL MEMBRANES

In the last few years techniques devised to fractionate the mitochondrial membranes have greatly stimulated analysis of the topographical organization of this organelle (6, 51). Two means are currently used to selectively disrupt the outer membrane: osmotic swelling and digitonin. The effect of hypotonic media on the mitochondrial membranes is complex. The outer membrane is apparently freely permeable to a variety of solutes with a molecular weight less than $\sim 10,000$ daltons, while the inner membrane is impermeable to even small solutes. In hypotonic media, the matrix space swells, causing the cristae to unfold. Eventually, the outer membrane becomes passively stretched and is presumably disrupted by the swollen inner compartment.

Digitonin is known to disrupt many membranes, presumably through its specific interaction with cholesterol. The outer membrane, rich in cholesterol, thus succumbs to dilute digitonin solutions while the cholesterol-poor inner membrane remains intact.

These disruption techniques yield various quantities of inner and outer membrane plus unbroken mitochondria. The retention of the contents of the intermembrane and matrix spaces also varies among the procedures employed. It is, of course, important to distinguish between the purified inner membrane and the inner membrane containing residual matrix components usually obtained in these procedures.

The separation of inner and outer membrane from undisrupted particles and from each other has been performed by both differential and equilibrium density gradient centrifugation. In the former case, the inner membrane structures are both large and dense and sediment at low forces (10^5 g-min) normally used to pellet intact mitochondria. The outer membrane is recovered as small fragments which require more than 10^6 g-min for pelleting. When centrifuged to equilibrium on sucrose density gradients, the inner membrane plus matrix bands at ρ 1.21, the intact mitochondria at ρ 1.19, and the outer-membrane at ρ 1.13, thus permitting a ready separation (6, 41).

The outer and inner membranes differ considerably in their chemical composition and enzymes profile; the intermembrane and matrix spaces also have distinctive compositions (51). The outer membrane is more akin to microsomal (presumably smooth endoplasmic reticulum) membranes than to the inner membrane. It possesses a heterogeneous spectrum of enzymes rather than the integrated pathways of the inner membrane plus matrix. Like the outer membrane of the nucleus, its function may ultimately be shown to address the cytoplasmic compartment rather than the interior of the organelle.

INSIDE-OUT AND RIGHT-SIDE-OUT INNER MITOCHONDRIAL MEMBRANES

A principal technique in the study of oxidative phosphorylation and related functions has been the preparation of "submitochondrial particles" by the disruption of mitochondria. Various studies have revealed characteristic differences in the functions of vesicles produced by sonication and digitonin treatment. The hypothesis was advanced that sonication produced vesicular fragments of the inner membrane that were "inside-out," i.e., that had their original inner surface (which faced the matrix) on the outside of the vesicle. Digitonin vesicles were supposedly normally oriented. Studies on Ca^{++} transport and the accessibility of enzymes to exogenous reagents, as well as electron microscopy, supported this view (cf. 36). In a recent study digitonin fragments were prepared that were also inside-out (18). This area of mitochondrial membranology has not yet reached maturity, but one can anticipate much further exploration of these systems, since differential studies of sidedness probe the essence of membrane function.

Golgi Apparatus

The Golgi apparatus has recently attracted great interest among membranologists. Morphologically, it is a composite of closed cisternae, tubules, and vesicles organized in characteristic arrays. Furthermore, it seems to be a repository of multienzyme glycosylation systems which build specific oligosaccharide units onto secreted and membrane-bound proteins. Golgi vesicles also appear to participate in the secretion of complex carbohydrates, proteins, and lipoproteins.

Early studies on the isolation of the Golgi apparatus established that *1)* this structure is mechanically fragile, *2)* the vesicles are of a relatively light equilibrium density ($\rho < 1.127$) on sucrose gradients, and *3)* they can be identified by their characteristic morphology and their intense staining for carbohydrate. Glutaraldehyde fixation, Ca^{++}, and 0.1 M phosphate buffer have been employed to stabilize the membranes. Despite precautions, this organelle is still disrupted by homogenization and is frequently recovered as small vesicles in the microsomal fraction (9, 17, 25).

These considerations led to the isolation method of Fleischer *et al.* (25). Bovine livers were gently homogenized and then centrifuged at 2.6×10^5 g-min. The supernatant was collected and centrifuged at 1×10^6 g-min. The pelleted membranes were then further purified on discontinuous sucrose gradients. The cleanest Golgi fraction was harvested from the ρ 1.05–1.09 region. It was uniquely enriched in a galactosyl transferase activity and showed characteristic heavy staining following incubation in osmium tetroxide. A dispersion of sacs, tubules, and vesicles in configurations reminiscent of Golgi membranes was seen by electron microscopy. Their preparation was estimated to be 80% pure Golgi vesicles and to constitute 2% of the starting Golgi material.

A different fractionation protocol has been developed by Morré *et al.* (38). Rat liver minces were gently homogenized and then centrifuged for 4×10^4 g-min. The supernatant was removed and the upper portion of the low-speed pellet further purified on a discontinuous sucrose gradient. The Golgi-rich fraction (ρ 1.12–1.14) was collected and washed three or four times by pelleting at 6×10^4 g-min. The morphology of the isolated membranes was that of the intact Golgi apparatus. Purity was estimated to exceed 80%.

A point of comparison between the methods of Fleischer *et al.* (25) and Morré *et al.* (38) is pertinent. Both groups recognized that fragmentation of the Golgi complex threatened its recovery and both employed gentle homogenization to avoid loss of small vesicles to the microsomal fraction. Morré *et al.* selected the 4×10^4 g-min pellet as a source of crude Golgi. This integrated force corresponds to our expectations for the sedimentation of the intact Golgi apparatus. Fleischer *et al.*, however, discarded the pellet obtained from a 2.6×10^5 g-min centrifugation. That supernatant contained only 24% of the galactosyl transferase marker which, furthermore, had a lower specific activity than the starting homogenate. It seems likely that most of the Golgi complexes remained relatively intact and were eliminated by this step. Their purified product was composed of the fragments which pelleted at higher speeds. Such considerations are of importance not only in obtaining a good yield but in obtaining a representative isolate, since the Golgi apparatus seems to be a heterogeneous, composite structure.

Plasma Membranes

The plasma membrane defines the boundary between the cytoplasm and the extracellular space; it is constituted and organized to mediate between these two compartments. The rapid advances in our understanding of its composition, structure, and function have been sustained by the availability of purified plasma membrane preparations. A review of methods for isolating plasma membranes has recently been published (8); a summary will suffice here.

DISRUPTION

The cell surface must be disrupted to be isolated. This step, no matter how gentle, can have a major impact on the subsequent isolation and evaluation of this membrane. Some examples are presented below.

The integrity and stability of the membrane may be altered upon isolation. The red blood cell membrane is a case in point. It is often purified by diluting erythrocytes into hypotonic salt solutions (23). This causes the cells to swell until holes appear in the membrane which allow the cytoplasmic constituents to equilibrate with the extracellular space. If, after the initial lysis, the medium is again made isotonic in saline and the preparation is incubated at 37°C, the membranous ghosts regain impermeability to Na^+, K^+, etc. After two or three hypotonic washes, however, the ghosts can no longer be resealed.

The isolated ghosts are altered in other ways. The membranes of intact erythrocytes are rather resistant to digestion by proteolytic enzymes; almost every major protein of purified ghosts, on the other hand, is highly sensitive to proteolysis. The same is true for the differential action of phospholipases on the intact and isolated red cell membranes. Furthermore, repeated washing in hypertonic salt removes no detectable protein or enzyme activity from intact erythrocytes; however, when isolated ghosts are treated in the same way, considerable membrane protein, lipid, and acetylcholinesterase activity is eluted (37). In another realm, it has long been noted that various membrane-bound enzymes exhibit great lability following isolation of the membranes but appear quite stable in the intact cell.

Some recent studies have indicated that the structure of the red cell membrane may depend in some respects on energy metabolism. By infrared spectroscopy, purified ghost membranes contain no protein of the β-conformation. However, the addition of ATP plus Mg^{++}, but neither by itself, induces a prominent anti-parallel β-signal (Graham and Wallach, personal communication). ATP plus Mg^{++} also causes endocytic vesiculation of purified ghosts, attributed to an energized conformational change in the membrane protein (42). Changes in the mechanical rigidity of isolated ghost membranes also occur when ATP plus Mg^{++} are added, again suggesting an energy-dependent membrane structure (32). These studies open a great issue for the membranologist; namely, is the membrane structure in living cells directly maintained by metabolic energy so that isolated membranes are at once no longer "native?"

All in all, these findings indicate that the purified membrane is an altered membrane. On the other hand, it may well be that we do not yet know how to protect membranes against the many perturbations associated with purification procedures and that direct study of this problem will indicate its remedy.

The physical form of the membrane may be altered upon isolation. The erythrocyte ghost is unusual in that the intact membrane is purified by gentle osmotic stress. Most tissue cell surfaces must be broken by homogenization, yielding open envelopes, large and small open fragments, and closed vesicles. The form the fragments assume depends on the mechanical properties of the membrane itself, the mode and intensity of homogenization, medium composition, etc. This variability can complicate plasma membrane isolation. Fragments of a plasma membrane may not all be of similar size; they will therefore sediment at unequal rates and may be recovered in the nuclear, mitochondrial, or microsomal fractions. Not all fragments are of the same density; vesicles may be significantly less dense than open sheets. Heterogeneity in the surface composition can lead to heterogeneity in the s-ρ profile of the membrane fragments. Isolation of one s-ρ species may lead to large losses, and the fraction recovered may be unrepresentative of the whole. All these possibilities do, in fact, occur (8).

Several strategies have evolved to treat this situation.

1. An attempt can be made to isolate intact membrane envelopes. This is desirable (since it yields a homogeneous population of membranes which represents the entire cell surface), but it is often difficult to achieve. Cultured HeLa cell membranes, however, are strong enough to yield large envelope fragments after conventional homogenization (16, 17). Furthermore, Warren and his colleagues (59) have advocated "toughening" the membrane with any of a variety of chemical agents.

2. The plasma membrane can be reduced to small vesicles which enter the microsomal fraction. Since considerable vesiculation unavoidably occurs during many homogenization procedures, this approach attempts to cooperate with the inevitable and convert the entire plasma membrane to vesicles. The difficult aspect of this approach lies in executing the separation of the plasma membrane from the other microsomal elements, which is more complex and arduous than other methods (see "Microsomal Subfractionation," below).

3. A readily prepared subfraction of the surface is isolated. Two of the earliest plasma membrane preparations were of this type: bile canalicular membranes from liver and brush border membranes from intestinal mucosa. In both cases, the specialized surface structures were more resistant to homogenization in hypotonic buffer than the rest of the plasma membrane; they retained their large size and characteristic morphology while the remainder

of the cell surface fragmented. Both structures could be sedimented under low gravitational force and then further purified. Neither preparation was totally immune to fragmentation; in both cases vesicles were continuously shed from these structures in response to mild shearing stresses (8). Since these are specialized structures, their biochemical comparison with the remainder of the cell surface is of great interest and is currently under investigation.

ISOLATION

From the foregoing, it should be evident that isolation of plasma membrane is more complex than that of the other organelles. Because it is difficult to generalize about plasma membrane isolation, several specific cases are considered briefly.

Red blood cell membranes. Most recent methods for the preparation of clean erythrocyte membranes employ hemolysis and washing in hypotonic buffers, as detailed in the careful study of Dodge *et al.* (23). The ghosts can be further reduced to vesicles by various treatments (8).

Liver plasma membranes. The first and most extensively studied plasma membrane preparation from solid tissue was that of Neville (39) for the rat liver. His procedure took advantage of the stability of the bile canaliculi to homogenization in hypotonic buffer. The bile front membrane complexes were pelleted with the nuclei and purified by repeated washes and isopycnic banding in sucrose gradients.

A variation on this procedure (19) pelleted the canalicular membrane complexes with the nuclei (at 10^4 g-min), thus eliminating the slowly sedimenting contaminants. The nuclear pellet was then more vigorously homogenized, which differentially fragmented the canalicular membranes into small vesicles and sheets. These were then freed of nuclei and mitochondria by a second 10^4 g-min centrifugation and further purified on sucrose gradients. Their buoyant density (~ 1.13) was significantly lower than that of the parent bile fronts (~ 1.17). This is attributable to the acquisition of a water compartment upon vesiculation, as well as to an increased lipid/protein ratio. Evans (24) has provided evidence that the vesicles and sheets represent different fractions of the canalicular membrane complex.

A different approach to liver cell surface isolation is to recover the plasma membrane by subfractionating the microsomal vesicle population. The results with this approach are comparable to those obtained by the Neville procedure. Other studies have found that liver plasma membrane fragments can be recovered in the mitochondrial fraction as well (8).

Cultured cells. The plasma membranes from L cells and HeLa cells have been recovered from homogenates as envelopes or large surface fragments which pellet with the nuclei and mitochondria (8). It would appear that some cells cultured in vitro have a tough membrane which withstands homogenization better than liver or other tissue cells (16, 17).

Warren (59) has advocated various chemical agents to further stabilize plasma membranes against fragmentation. Among these are acetic acid, sulfhydryl reagents, salts of heavy metals, divalent cations, Tween-20, and dimethyl sulfoxide. The advisability of such treatments is difficult to assess. Plasma membrane isolates obtained from cultured chick embryo fibroblasts with and without Warren's Zn^{++}–Tween-20 treatment were comparable except that the untreated membranes tended to vesiculate (43). Furthermore, these toughening agents seem to promote the adherence of cytoplasmic material to the inner surface of the plasma membrane, which may relate to their role as stabilizers, but which may also promote contamination of the product. Some of these treatments may also inactivate enzymes and extract membrane components.

Warren's methods seem to stabilize the cytoplasm as well. After treatment with 1 mM $ZnCl_2$, the cell contents appear aggregated and clearly separated from the distended cell surface, thereby facilitating the selective disruption and isolation of the surface. These effects were exploited in the procedure of Barland and Schroeder (12). Cells growing as monolayers in culture flasks were overlayed with 1 mM $ZnCl_2$ and fluorescein mercuric acetate, which caused swelling of the cells. Moderate rotary shaking of the flasks effected the release of large surface membrane fragments; the rest of the cell body remained intact and fixed to the vessel wall. The membranes were collected and washed at 6000 g-min to complete the purification. This simple technique dispensed with conventional cell homogenization and fractionation protocols. It further avoided freeing the cells from the flask walls by trypsin digestion, which may alter the cell surface considerably.

The isolation of *Amoeba proteus* surface membrane has also been facilitated by promoting the coalescence of the cytoplasmic contents (40). In this case, the cells were contracted by immersion in 50% glycerol. Their surface membranes subsequently expanded spontaneously, while the cell contents remained as a small congealed mass.

On the other hand, the surfaces of several types of cultured cells have been reduced to vesicles and then isolated by Wallach's method (8, 9).

Membranes from nervous tissue. There is considerable interest in various types of nerve membrane preparations, since neuronal behavior is intimately associated with membrane activity. The axolemma of giant squid axons, for example, can be freed of axoplasm by mechanical extrusion, leaving a functioning membrane complex suitable for electrophysiological investigations. The contents of both inner and outer compartments can be manipulated in such systems. However, satellite cells remain adherent to the membrane so that biochemical and biophysical studies are somewhat ambiguous.

Myelin has been prepared by a variety of methods and has been favored as a model in the biochemical and structural analysis of membranes. As is frequently pointed out, however, myelin is probably not representative of cell surfaces, but is rather a specialized derivative therefrom (8).

The presynaptic nerve terminals can be detached by homogenization and recovered as large, sealed vesicles (synaptosomes). These typically contain cytoplasm and organelles such as mitochondria and acetylcholine-rich synaptic vesicles. The membranes of retinal receptor discs have also been isolated and subjected to morphological and biochemical evaluation (8).

Microsomal subfractionation. Wallach (9) has advocated reducing the plasma membrane entirely to small vesicles and then purifying them by subfractionating the microsomal membranes with which they sediment. He has utilized the rapid decompression of a high-pressure nitrogen atmosphere to effect this disruption; more conventional homogenization methods also suffice (9).

To separate Ehrlich ascites carcinoma plasma membrane from endoplasmic reticulum vesicles, these steps seem important: *1*) the microsomal vesicles are freed of trapped soluble protein by hypotonic lysis; *2*) aggregated vesicles are dispersed by careful homogenization; *3*) the membranes are equilibrated with 1 mM Mg^{++} by dialysis; *4*) the mixture is loaded atop a ρ 1.09 Ficoll barrier and centrifuged for 3×10^7 g-min. The plasma membranes are recovered on top of the barrier, while the endoplasmic reticulum pellets. Further equilibrium centrifugation on continuous Ficoll gradients achieves greater purification of the surface membranes, including the separation of a low-density (Golgi) membrane fraction. The rationale underlying this procedure was discussed above, under "Effect of Gradients on Particle Densities."

The Wallach approach has also been successfully applied to rat liver, HeLa cells, mouse lymphoma cells, the thyroid, rat adipose tissue, and other cell types. However, in general, the optimum gradient pH and Mg^{++} concentration could well vary from tissue to tissue, depending on the charge characteristics of the membranes involved. Parenthetically, Dextran 110 (Pharmacia) seems to be of a higher purity, greater homogeneity, and lower viscosity than Ficoll, and has been substituted for it in these procedures.

Other tissues. Methods exist for preparing plasma membrane from intestinal brush borders, skeletal and smooth muscle, the thyroid, leukocytes, adipose tissue, the kidney, and microorganisms (bacteria, mycoplasma, yeast, and protozoans) (8).

MARKERS

Because the disrupted cell surface may not be identifiable by a clear-cut homogeneous morphology, a unique functional system, or a distinctive chemical component, biochemical markers are essential to any preparative method. Some distinguishing characteristics of the plasma membranes thus far studied are: *1*) a high cholesterol/phospholipid ratio (fortunately not influenced by the presence of extraneous soluble proteins), *2*) a high level of certain sphingolipids, and *3*) a high sialic acid specific content. These attributes are quantitative, not qualitative; relative, not absolute (8).

The enzyme activities most often associated with cell surface isolates are 5′-nucleotidase, Na^+,K^+-dependent ATPase, and a variety of phosphatase activities. Other enzymes have also been reported (8). Surface-specific heteroantibodies have been prepared. Furthermore, transplantation antigens seem to be concentrated on the plasma membrane. Viral receptor sites also seem to be confined to the cell surface (8, 45).

Microsomal Membranes

The microsomes are not an organelle but a heterogeneous mixture of vesicles and fragments operationally grouped as the slowest sedimenting particles of the subcellular homogenate. The rough and smooth endoplasmic reticula dominate this fraction, but equating microsomes with endoplasmic reticulum is misleading, since not all the endoplasmic reticulum travels with the microsomes, and not all the microsomes are derived from endoplasmic reticulum. Why the endoplasmic reticulum is so sensitive to mechanical disruption is not known, but even mild homogenization reduces most of it to small vesicles. However, some may remain either adherent to or in structural continuity with the outer nuclear membrane. Large reticulum fragments may also contaminate the mitochondrial fraction. Conversely, fragments of the plasma membrane, the Golgi apparatus, lysosomal membranes, and the outer nuclear and mitochondrial membranes have been recovered in the microsomal fraction. The subfractionation of the microsomes is obviously important since studies on microsomal mixtures often cannot be related unambiguously to the parent subcellular structures.

SUBFRACTIONATION OF MICROSOMES

The microsomal membranes either are smooth or bear arrays of ribosomes and polysomes morphologically analogous to the rough endoplasmic reticulum. Three major methods for separating smooth and rough microsomes have been elaborated: *1*) differential centrifugation of the microsomes in 0.88 M sucrose, an approach that effects some separation, but does not offer high resolution, owing to overlap in size and density distributions of the rough and smooth membranes; *2*) equilibrium density gradient centrifugation, based on the premise that vesicles bearing ribosomes are significantly more dense than smooth vesicles; and *3*) selective aggregation of rough microsomes in 15 mM CsCl solutions and separation by rate zonal sedimentation through a discontinuous sucrose gradient (2). The rough and smooth microsomes isolated in this manner differ in their chemical composition, enzyme content, and rate of turnover, as well as in their response to induction by phenobarbital (2).

Thines-Sempoux *et al.* (56) examined the heterogeneity of the microsomal population by determining the equilibrium density distribution of a number of microsomal constituents on sucrose gradients. They found that all the components evaluated showed a broad density dispersion so that the markers overlapped considerably. The median densities of the constituents varied in a manner suggesting that at least four classes of particles contributed to the mixture (Table II). Additional experiments showed that monoamine oxidase could be dissociated from the other markers in group A. For example, altering the gradient composition affected the density of these functions differentially. Further, the sedimentation rate of the 5'-nucleotidase was greater than that of the bulk of the microsomes, while the monoamine oxidase sedimented more slowly. Finally, when digitonin was added to the microsomal mixture, all the group A markers except monoamine oxidase were shifted to a higher equilibrium density; no other functions were affected in this manner (Table II). From these and other data, a provisional hypothesis can be formulated which identifies the enrichment in monoamine oxidase with outer mitochondrial membrane fragments; 5'-nucleotidase, alkaline phosphodiesterase I, and cholesterol with plasma membrane vesicles; the oxidoreductases with smooth endoplasmic reticulum; and the dense esterases with rough endoplasmic reticulum. The high median density of the RNA may reflect the contributions of both bound and free ribosomes (the latter presumably not at density equilibrium).

Two salutary features of this study should be noted. First, the authors did not attempt to isolate discrete components by chopping the density dispersion of the microsomes. Instead, they adhered to an analytical rather than preparative mode to obtain high-resolution data. Preparative methodology can flow naturally from such analytical studies. Second, these workers employed equilibrium densities rather than sedimentation rates as the indicator of heterogeneity. Since it is not clear at present that heterogeneity in composition is clearly reflected in the size of the vesicle fragments, this approach seems a discerning one.

SUBFRACTIONATION OF ROUGH MICROSOMES

The removal of ribosomes from the microsomal vesicles by gentle deoxycholate or ethylenediamine tetraacetate (EDTA) treatment leaves the bulk of the enzymes associated with the membrane fraction, indicating that the ribonucleoprotein particles are not the locus of most endoplasmic reticulum enzymes.

Dallner and his associates have evaluated heterogeneity within the rough

Table II

Median Densities of Microsomal Components in Sucrose-H_2O Gradients

	NORMAL MICROSOMAL FRACTION	MICROSOMAL FRACTION TREATED WITH 0.25% DIGITONIN
A. 5'-Nucleotidase	1.138	1.172
Alkaline phosphodiesterase I	1.145	1.171
Cholesterol	1.141	1.167
Monoamine oxidase	1.134	1.137
B. NADH: cytochrome c reductase	1.149	1.148
NADPH: cytochrome c reductase	1.157	1.154
Aminopyrine demethylase	1.152	1.152
Cytochrome b_5	1.145	1.151
Phospholipids	1.150	1.158
C. Glucose-6-phosphatase	1.169	1.168
Esterase	1.166	1.164
Nucleoside diphosphatase	1.172	1.167
Proteins	1.164	1.166
D. RNA	1.201	1.195

Rat liver microsomal mixtures, with and without pretreatment with 0.25% digitonin, were centrifuged to equilibrium and the gradients fractionated by collecting density increments of about 0.01 g/ml. A series of constituents was assayed in each fraction and a median density estimated for each marker. Four classes of particles (A–D) were suggested by the density distributions. Furthermore, group A was resolved into two components by the digitonin treatment.
Adapted from Thines-Sempoux, D., *et al. J. Cell Biol.* **43**, 189 (1969).

microsomes by rate zonal sedimentation on sucrose gradients. The dispersion of vesicles was found to vary in specific content of various markers and in response to the prior administration of phenobarbital (2). In a subsequent study (21), this group sonicated rough microsomes to create tiny fragments which were then centrifuged through sucrose gradients. The vesicles were dispersed roughly according to size. The smaller vesicles were associated with NADPH-linked functions while the larger fragments were enriched in NADH-dependent enzymes. The authors concluded that these enzyme systems are distributed through the rough endoplasmic reticulum in a nonrandom fashion.

On the other hand, Leskes and Siekevitz (34) found histochemically that glucose-6-phosphatase was homogeneously distributed within the rough endoplasmic reticulum. Furthermore, the entire population of rough microsomes appeared uniformly active in its in vitro glucose-6-phosphatase activity. However, the presence of the enzyme did vary from cell to cell in the *embryonic* liver, so that heterogeneity was demonstrated at the cellular but not the subcellular level at that early stage of development.

SUBFRACTIONATION OF SMOOTH MICROSOMES

Dallner has found that the addition of 5 mM Mg^{++} to smooth rat liver microsomes causes the bulk of the vesicles to aggregate. The aggregated and unaggregated fractions can be separated by rate centrifugation through a discontinuous sucrose gradient. The two fractions differ considerably with respect to lipid composition, turnover rate, and stimulation by phenobarbital (2). This fractionation has certain similarities to Wallach's separation of plasma membrane and endoplasmic reticulum vesicles from microsomal mixtures, suggesting that the nonaggregated microsomes of Dallner might bear surface fragments and/or Golgi vesicles (8). Dallner's group has recently offered another approach to the fractionation of smooth microsomes which further documents this heterogeneity (27).

Conclusions

The ultracentrifuge has fostered a powerful technology by which subcellular organelles and their membranes can be prepared and analyzed. Difficulties in purifying organelles and membranes arise from the overlap in their physical properties. Membrane isolation is further complicated by fragmentation (and the concomitant heterogeneity of the product) and by the tendency of membranes to adsorb or enclose extrinsic macromolecules. However, an understanding of the physicochemical behavior of the organelles and membranes has added depth to basically simple fractionation concepts. Thus, particles can selectively be made more dense or less, larger (by aggregation) or smaller (by fragmentation), inside-out or right-side-out, etc., and purified on that basis.

The analysis of isolated membranes has revealed that they are physically and chemically diverse with respect to species, cell type, organelle, and even within the organelle itself. (Some membrane components also vary characteristically during the lifetime of the cell or organism.) The various membrane systems of the cell appear to partake of certain common features while at the same time expressing significant biochemical specialization. The differences in composition and structure are presumably related to membrane function in ways we do not now clearly perceive. Fortunately, these differences are often manifested in the physical properties which influence the fractionation of the particles. In this way, high-resolution fractionation techniques can be expected to sustain the exploration of the role of the membranes in organizing the economy of the cell.

References

Reviews

1. Allfrey, V. G. *The Cell.* Ed. J. Brachet and A. E. Mirsky. Academic Press, New York, 1959, vol. 1, p. 193.
2. Dallner, G., and Ernster, L. *J. Histochem. Cytochem.* **16,** 611 (1968).

3. de Duve, C. *J. Theor. Biol.* **6**, 33 (1964).

4. de Duve, C. *Harvey Lect.* **59**, 49 (1965).

4a. de Duve, C. *J. Cell Biol.* **50**, 20D (1971).

5. de Duve, C., Berthet, J., and Beaufay, H. *Progr. Biophys. Biophys. Chem.* **9**, 325 (1959).

6. Ernster, L., and Kuylenstierna, B. *Membranes of Mitochondria and Chloroplasts.* Ed. E. Racker. Van Nostrand Reinhold, New York, 1969, p. 172.

7. Murray, R. K., Suss, R., and Pitot, H. C. *Methods Cancer Res.* **2**, 239 (1967).

8. Steck, T. L., and Wallach, D. F. H. *Methods Cancer Res.* **5**, 93 (1970).

9. Wallach, D. F. H. *The Specificity of Cell Surfaces.* Ed. B. D. Davis and L. Warren. Prentice-Hall, Englewood Cliffs, N.J., 1967, p. 129.

Experimental Research References

10. Albertsson, P.-Å. *Adv. Protein Chem.* **24**, 309 (1970).

11. Anderson, N. G., Harris, W. W., Barber, A. A., Rankin, C. T., Jr., and Candler, E. L. *Natl. Cancer Inst. Monogr.* **21**, 253 (1966).

12. Barland, P., and Schroeder, E. A. *J. Cell Biol.* **45**, 662 (1970).

13. Beaufay, H., and Berthet, J. *Methods of Separation of Subcellular Structural Components, Proc. 23rd Biochem. Soc. Meeting.* Ed. J. K. Grant. Cambridge University Press, Cambridge, 1963, p. 66.

14. Berezney, R., Funk, L. K., and Crane, F. L. *Biochim. Biophys. Acta* **203**, 531 (1970).

15. Blobel, G., and Potter, V. R. *Science* **154**, 1662 (1966).

16. Boone, C. W., Ford, L. E., Bond, H. E., Stuart, D. C., and Lorenz, D., *J. Cell Biol.* **41**, 378 (1969).

17. Bosmann, H. B., Hagopian, A., and Eylar, E. H. *Arch. Biochem. Biophys.* **128**, 51 (1968).

18. Christiansen, R. O., Loyter, A., Steensland, H., Saltzgaber, J., and Racker, E. *J. Biol. Chem.* **244**, 4428 (1969).

19. Coleman, R., Michell, R. H., Finean, J. B., and Hawthorne, J. N. *Biochim. Biophys. Acta* **135**, 573 (1967).

20. Cotman, C., Brown, D. H., Harrell, B. W., and Anderson, N. G. *Arch. Biochem. Biophys.* **136**, 436 (1970).

21. Dallman, P. R., Dallner, G., Bergstrand, A., and Ernster, L. *J. Cell. Biol.* **41**, 357 (1969).

22. de Duve, C. *Methods Enzymol.* **10**, 7 (1967).

23. Dodge, J. T., Mitchell, C., and Hanahan, D. J. *Arch. Biochem. Biophys.* **100**, 119 (1963).

24. Evans, W. H. *Biochem. J.* **166**, 833 (1970).

25. Fleischer, B., Fleischer, S., and Ozawa, H. *J. Cell Biol.* **43**, 59 (1969).

26. Franke, W. W., Deumling, B., Ermen, B., Jarasch, E.-D., and Kleinig, H. *J. Cell Biol.* **46**, 379 (1970).

27. Glaumann, H., and Dallner, G. *J. Cell Biol.* **47**, 34 (1970).

28. Greenawalt, J. W., Rossi, C. S., and Lehninger, A. L. *J. Cell Biol.* **23**, 21 (1964).

29. Hamberger, A., Blomstrand, C., and Lehninger, A. L. *J. Cell Biol.* **45**, 221 (1970).

30. Kashnig, D. M., and Kasper, C. B. *J. Biol. Chem.* **244**, 3786 (1969).

31. Katyare, S. S., Fatterpaker, P., and Sreenivasan, A. *Biochem. J.* **118,** 111 (1970).
32. La Celle, P. L., and Weed, R. I. *J. Clin. Invest.* **48,** 795 (1969).
33. Leighton, F., Poole, B., Beaufay, H., Baudhuin, P., Coffey, J. W., Fowler, S., and de Duve, C. *J. Cell Biol.* **37,** 482 (1968).
34. Leskes, A., Siekevitz, P., and Palade, G. E. *J. Cell Biol.* **49,** 264, 288 (1971).
35. Luck, D. J. L. *J. Cell Biol.* **24,** 445 (1964).
36. Malviya, A. N., Parsa, B., Yodaiken, R. E., and Elliott, W. B. *Biochim. Biophys. Acta* **162,** 195 (1968).
37. Mitchell, C. D., and Hanahan, D. J. *Biochemistry* **5,** 51 (1966).
38. Morré, D. J., Hamilton, R. L., Mollenhauer, H. H., Mahley, R. W., Cunningham, W. P., Cheetham, R. D., and Leguire, V. S. *J. Cell Biol.* **44,** 484 (1970).
39. Neville, D. M., Jr. *J. Biophys. Biochem. Cytol.* **8,** 413 (1960).
40. O'Neill, C. H. *Exp. Cell Res.* **35,** 477 (1964).
41. Parsons, D. F. *Can. Cancer Conf.* **7,** 193 (1967).
42. Penniston, J. T., and Green, D. E. *Arch. Biochem. Biophys.* **128,** 339 (1968).
43. Perdue, J. F., and Sneider, J. *Biochim. Biophys. Acta* **196,** 125 (1970).
44. Petersen, E. A., and Kuff, E. L. *Biochemistry* **8,** 2916 (1969).
45. Philipson, L., Lonberg-Holm, K., and Pettersson, U. *J. Virol.* **2,** 1064 (1968).
46. Quigley, J. P., and Gotterer, G. S. *Biochim. Biophys. Acta* **173,** 456 (1969).
47. Schrier, S. L., Giberman, E., Danon, D., and Katchalski, E. *Biochim. Biophys. Acta* **196,** 263 (1970).
48. Schuel, H., Berger, E. R., Wilson, J. R., and Schuel, R. *J. Cell Biol.* **43,** 125a (1969).
49. Sellinger, O. Z., and Borens, R. N. *Biochim. Biophys. Acta* **173,** 176 (1969).
50. Shortman, K., and Seligman, K. *J. Cell Biol.* **42,** 783 (1969).
51. Smoley, J. M., Kuylenstierna, B., and Ernster, L. *Proc. Natl. Acad. Sci. U.S.A.* **66,** 125 (1970).
52. Stahn, R., Maier, K.-P., and Hannig, K. *J. Cell Biol.* **46,** 576 (1970).
53. Steck, T. L., Straus, J. H., and Wallach, D. F. H. *Biochim. Biophys. Acta* **203,** 385 (1970).
54. Steck, T. L., Weinstein, R. S., Straus, J. H., and Wallach, D. F. H. *Science* **168,** 255 (1970).
55. Swick, R. W., Tollaksen, S. L., Nance, S. L., and Thomason, J. F. *Arch. Biochem. Biophys.* **136,** 212 (1970).
56. Thines-Sempoux, D., Amar-Costesec, A., Beaufay, H., and Berthet, J. *J. Cell Biol.* **43,** 189 (1969).
57. Truffa-Bachi, P., and Wofsy, L. *Proc. Natl. Acad. Sci. U.S.A.* **66,** 685 (1970).
58. Vignais, P. M., and Nachbaur, A. P. *Biochem. Biophys. Res. Commun.* **33,** 307 (1968).
59. Warren, L., Glick, M. C., and Nass, M. K. *J. Cell Physiol.* **68,** 269 (1966).
60. Wattiaux-deConinck, S., and Wattiaux, R. *Biochim. Biophys. Acta* **183,** 118 (1969).
61. Weibel, E. R., Stäubli, W., Gnägi, H. R., and Hess, F. A. *J. Cell Biol.* **42,** 68 (1969).
62. Wetzel, M. G., and Korn, E. D. *J. Cell Biol.* **43,** 90 (1969).
63. Zbarsky, J. B., Perevoschikova, K. A., Delektorskaya, L. N., and Delektorsky, V. V. *Nature* **221,** 257 (1969).

II

PHYSICAL PROPERTIES OF MEMBRANES AND MEMBRANE COMPONENTS

4

A. D. Keith
R. J. Mehlhorn

Membrane Lipid Structure as Revealed by X-Ray Diffraction

In the decade before the turn of the century, Overton observed that the swelling of plant cells and osmotic pressure occurred at different rates, suggesting a selective diffusion barrier at cell surfaces (1). Several years later, the astounding observation was made by Gorter and Grendel that the lipids in a variety of red blood cells could be spread into a monolayer having an area almost precisely twice that of the red blood cell surfaces (2). After another decade, evidence was presented, in the now well-known work of Danielli and Davson, that membranes are composed of a lipid bilayer coated with protein on each surface (3). This enduring membrane model owes much of its credibility to the evidence from x-ray studies during the next two and one-half decades.

Theory

Constructive interference between scattered x-rays occurs when the Bragg condition is met, i.e., $2D \sin \theta = n\lambda$, where D is the distance between electron-dense regions of the sample, λ is the wavelength of the x-ray source (usually 1.54 Å for ^{2}Cu radiation), and 2θ is the angle between the incident beam and the nth maximum of the scattered beam.

To a first approximation the sharpness of the x-ray diffraction peaks is a measure of the integral number of wavelengths that can be accommodated between scattering centers. Consequently, the sharpness of the diffraction peaks produced by a regular array of molecules is a measure of the extent of the array. When lipid chains are arranged in hexagonal arrays (Fig. 1), the size of the ordered regions can be estimated by an analysis of the widths of the inter-hydrocarbon-chain diffraction bands (4).

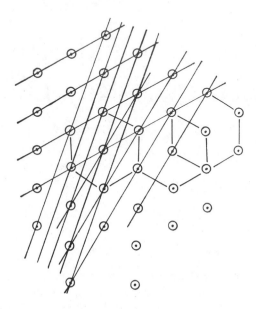

Figure 1

The average interchain distance in phospholipids is about 4.8 Å for hexagonal packing. This gives rise to a sharp peak at $D = 4.2$ Å and other weaker maxima. As the temperature is increased, the hydrocarbon packing becomes more irregular, and the sharp 4.2 Å peak gives way to a diffuse scattering maximum at about 4.6 Å. It is difficult to estimate the mean distance between hydrocarbon chains for this state from the x-ray data. Water produces a diffuse diffraction ring at $D = 3.3$ Å, which overlaps somewhat with the 4.6 Å hydrocarbon maximum. This imposes a lower concentration limit of about 5% on aqueous membrane dispersions for the 4.6 Å peak to be observed.

Large spacings, i.e., $D \gtrsim 30$ Å, can be observed using the methods of low-angle x-ray diffraction, where special care is exercised to reduce random scattering in the vicinity of the incident beam. This is of particular importance for membrane dispersions, where the scattering peaks are quite broad and hence difficult to observe.

A random dispersion of lipid bilayers contains a preponderance of orientations tending to diffract x-rays along the incident direction. Also, the scattering occurs over greater solid angles as the scattering angle increases. These considerations imply that the Fourier transform of the electron-density profile of the bilayer is proportional to $I \sin^2 \theta$, where I is the intensity corresponding to the scattering angle θ (5). Thus, there is a drastic intensity decrease with scattering angle, making it difficult to observe more than a few orders of diffraction.

Multilayers

X-ray studies of the 4.6 Å diffraction by oriented multilayers of egg lecithin have shown that such multilayers contain large groups of parallel hydrocarbon chains oriented normal to the lamellar planes (6). This orientation within the multilayers is enhanced by the addition of cholesterol. Scattering was observed in other directions as well, and some inferences about the distribution of hydrocarbon chain orientations could be made. The results are consistent with orientation studies of spin-labeled fatty acids in lecithin multilayers (7, 8).

Vesicles

Low-angle scattering by phospholipid vesicles produces an intensity maximum corresponding to a head-group separation of about 36 Å (5) in egg lecithin and 43 Å in Asolectin. This spacing decreases as the temperature is raised. About three diffraction bands characteristic of randomly dispersed bilayers can be seen in densitometer tracings of the photographs.

Packed Membranes

Membrane fragments packed by high-speed centrifugation give scattering curves analogous to the vesicle patterns. Extensive domains of bilayer structure have been observed in the membranes of erythrocyte ghosts, nerve ending from rat cerebral cortex, isolated sarcoplasmic reticulum, *Escherichia coli*, *Halobacterium*, and *Mycoplasma* (5). *Mycoplasma laidlawii* membranes enriched in selected fatty acids gave the expected long spacings. In these studies the first x-ray diffraction patterns from membranes of intact organisms were observed (9, 10). The low-temperature scattering peak was quite sharp, indicating the existence of continuous bilayer domains ~ 400 Å across and providing strong evidence for the bilayer as the predominant structural element in *Mycoplasma* membranes.

In a recent review (11), Segerman summarized the information gleaned from his x-ray studies of human erythrocyte ghosts. The 50% hydrated ghosts produced a scattering maximum at 4.2 Å at all temperatures between 0° and

48°C. This finding implies that the lipids are packed in hexagonal arrays throughout this temperature range. The scattering maximum is shifted, however, when the ghosts are dehydrated. Segerman attributes this shift to a rearrangement of the cholesterol molecules upon removal of the water molecules.

In their experiments, Wilkins *et al.* (5) observed the high-angle maximum in wet ghosts at a spacing of 4.6 Å at 4°C. The source of the discrepancy between these erythrocyte studies is not known.

Phase Changes

A lipid phase change observed by differential scanning calorimetry is seen by x-ray studies as a loss of long-range order as evidenced by a change of the high-angle peak from the sharp 4.2 Å hexagonal packing diffraction to a broad 4.6 Å peak. It is not clear from the data whether this phase change is accompanied by an appreciable change in the spacing between hydrocarbon chains. However, the low-angle data indicate that the bilayer width decreases as the temperature is increased. For example, the head-group separation for *Mycoplasma* membranes enriched with erucic acid changes from about 52 Å below the phase transition (10°C) to about 42 Å above the transition (37°C) (10).

We have studied the high-angle scattering in Asolectin vesicles between 6°C and 42°C and found a broad scattering peak corresponding to $D = 4.6$ Å throughout this temperature range (12). On the other hand, motion parameters obtained by spin-labeling techniques exhibit a phase transition at 15°C in the same vesicle preparations (13). Thus, a lipid phase change in this temperature range does not necessarily lead to hexagonal packing of the hydrocarbon chains.

Myelin

When, in the mid-1930s, x-ray diffraction was carried out on a variety of nerve fibers (14), it had already been known for several years that frog sciatic nerve fibers have anisotropic characteristics where the optical axis of the lipid components lies in an array normal to the long axis of the fiber (15). These studies set down the general range of spacings for myelin. Schmitt *et al.* observed that heating frog sciatic nerve at 42°, 44°, and 46°C caused a coincident loss of the action potential and a loss of sharpness in the x-ray diffraction patterns, thereby allowing a correlation between membrane order and neural function (14).

Myelin is composed of approximately 70–80% lipid and 20–30% protein. Most metabolically active cells have membranes which are 60–75% protein, the remainder of the composition being primarily polar lipids. Consequently, three important points should be stated with respect to myelin: *1*) based on its gross chemical composition, myelin does not a priori serve as an adequate membrane on which to constitute a general membrane structure; *2*) the lipid elements in myelin are considerably more complicated than those of a membrane such as that of yeast, *E. coli*, or a liver cell, and substantial chemical differences exist even among these different cell types; *3*) many other membranes have less uniformity in molecular composition over molecular dimensions than does myelin.

The familiar histology textbook illustration of nerve myelination showing a Schwann cell wrapped around a nerve fiber suggests a basic membrane structural unit of myelin. Two cell membranes are pressed against each other on their outer surfaces, while their cytoplasmic surfaces face away from this double-unit membrane.

Schmitt *et al.* presented four possible packing arrangements for the lipid-protein elements in myelin (16). These all utilize some form of lipid bilayer arrangement with protein interfaced at the bilayer boundaries. Two of these models considered lipid-protein hydrophobic associations, and the other two allowed for only ionic interactions between the lipid elements and protein elements. Subsequent work by Finean and others led to a refinement of the Schmitt treatment and the selection of a probable structure (17). This model contained a basic repeating unit of about 170 Å (one unit membrane thickness therefore would be 170/2 Å). Polar lipids and cholesterol were arranged in a bilayer with the polar surfaces interfaced with proteins. Recently, Blaurock carried out detailed low-angle x-ray experiments on myelinated frog sciatic nerve fibers (18). This work generally confirmed the model of Finean. All the experiments so far cited have a maximum resolution of a structural domain of about 30 Å. Therefore, the details of membrane structure still remain unknown even in myelin, the (so far) best characterized membrane system.

These studies do allow such statements as: *1*) the close-order packing of hydrocarbon chains is about 4.6 Å, *2*) there is a lipid bilayer of about 50 Å width, and *3*) a protein coat is on the surface of each bilayer. From chemical studies the lipid composition of some myelin sources is known; however, the locale of such molecules as cholesterol in the lipid zones is unresolved by these studies.

Recently, Casper and Kirschner carried out x-ray diffraction studies on rabbit optic, rabbit sciatic, and frog sciatic myelinated nerve fibers and claimed a resolving power of about 10 Å (19). These authors concluded that the electron-density profile of the unit membrane was asymmetrical although the over-all lipid dimensions were symmetrical about a central point. This symmetry of dimensions and asymmetry of electron density was explained on the basis that cholesterol has a higher concentration on the membrane outer bilayer-half surface than on the cytoplasmic side of the bilayer.

Since most membranes are composed of both heterogeneous lipid and protein, it is difficult to make statements about protein localization in the membrane. Blasie (20) has recently reported the localization of the photo-pigment protein in the frog retinal disc membrane. Since the photopigment protein is the principal protein, the protein heterogeneity problem is reduced. He reports that the 42-Å photopigment molecule is embedded in hydro-carbon-rich zones, 14 Å in the dark-adapted membranes and 21 Å in the bleached preparations.

REFERENCES

1. Overton, E. *Vierteljahrschrift der Naturforschende Gesellschaft* **44,** 88 (1899).
2. Gorter, E., and Grendel, F. *J. Exp. Med.* **41,** 439 (1925).
3. Danielli, J. F., and Davson, H. *J. Cell. Physiol.* **5,** 495 (1935).
4. Guinier, A. *X-ray Diffraction.* Freeman, San Francisco, 1963.
5. Wilkins, M. H. F., Blaurock, A. E., and Engelman, D. M. *Nature New Biol.* **230,** 72 (1971).
6. Levine, Y. K., and Wilkins, M. H. F. *Nature New Biol.* **230,** 69 (1971).
7. Hubbell, W. L., and McConnell, H. M. *Proc. Natl. Acad. Sci. U.S.A.* **64,** 20 (1969).
8. Libertini, L. J., Waggoner, A. S., Jost, P. C., and Griffith, O. H. *Proc. Natl. Acad. Sci. U.S.A.* **64,** 13 (1969).
9. Engelman, D. M. *J. Mol. Biol.* **47,** 115 (1970).
10. Engelman, D. M. *J. Mol. Biol.* **58,** 153 (1971).
11. Bolis, B., and Pethica, B. A. (eds.). *Membrane Models and the Formation of Biological Membranes.* North-Holland, Amsterdam, 1968, p. 52.
12. This work was carried out in collaboration with Dr. K. Palmer.
13. Eletr, S. Private communication.
14. Schmitt, F. O., Bear, R. S., and Clara, G. L. *Radiology* **25,** 131 (1935).
15. Schmidt, W. J. *Z. Zellforsch. Mikrosk. Anat.* **23,** 657 (1936).
16. Schmitt, F. O., Bear, R. S., and Palmer, K. J. *J. Cell. Comp. Physiol.* **18,** 31 (1941).
17. Finean, J. B., and Barge, R. E., *J. Mol. Biol.* **7,** 672 (1963).
18. Blaurock, A. E. *J. Mol. Biol.* **56,** 35 (1971).
19. Casper, P. L. D., and Kirschner, D. A. *Nature New Biol.* **231,** 46 (1971).
20. Blasie, J. K. *Biophys. J.* **12,** 191 (1972).

5

Ferenc J. Kézdy

Lipid Monolayers

Biological membranes are located at the boundary separating two distinct phases, such as air and liquid, liquid and gel, or gel and gel phases. The region of contact between the two phases is defined in thermodynamics as an interface, and a wealth of information is available concerning the laws governing the physical and chemical phenomena occurring at the interface. Because of their well-defined structure and their catalytic properties, biological membranes are more than simple thermodynamic interfaces. Nevertheless, by virtue of their location at an interface, many of their properties can be derived from thermodynamic principles, and all aspects of membrane chemistry can be fruitfully discussed in terms of thermodynamic concepts. For this reason the first part of this chapter is devoted to the basic notions of the physical chemistry of interfaces.

The most constant and well-defined structural element of biological membranes is a lipid bilayer. It is composed of lipid molecules possessing a hydrophobic and a hydrophilic end, with well-oriented and closely packed side chains at the separation of the two phases. Conceptually at least, it consists of two insoluble lipid monolayers, since insoluble monolayers at the air-water interface are also composed of lipid molecules oriented through the interaction of their hydrophilic groups with the solvent and through the hydrophobic interactions between their nonpolar side chains. All our present knowledge suggests that the structural similarities between membranes and monolayers are also reflected in the physical and chemical properties of the lipid molecules in the two organized structures. Therefore, the second part of our discussion is concerned with the physical and chemical properties of insoluble monolayers at the air-water interface. The study of monolayers rather than bilayers is justified not only by the theoretical simplicity of monolayer systems but also, most importantly, by the great ease with which insoluble lipid monolayers can be formed over large surfaces and in a reproducible manner. Finally, a variety of simple experimental techniques is available for their study, and for the time being our knowledge of monolayers is far more advanced than our understanding of bilayers.

SURFACE TENSION

On a macroscopic scale the region of contact between two homogeneous phases can be conceived of as a geometric surface of zero thickness. This simple idea of a surface soon vanishes when one considers the surface region at the level of molecular dimensions. The violent thermal motion of the molecules, the rapid exchange of molecules between the two phases, and the interpenetration of the two phases resulting in ever-changing local concentrations give a notion of a rather ill-defined surface region which gradually blends into the two homogeneous phases. Without trying to define the exact boundaries of this transition layer, we will accept its existence as a separate entity—the surface phase—characterized by the fact that its molecules are in an environment different from that of either of the bulk phases. To this surface phase we can then apply the laws of thermodynamics and derive relations describing its physical properties. The following thermodynamic functions will be used:

The enthalpy of a system, H, is defined by

$$H = E + PV \tag{1}$$

where E is the internal energy content, P the pressure, and V the volume of the system. At constant pressure, the free energy function, F, is given by

$$F = H - TS = E + PV - TS \tag{2}$$

where S is the entropy and T is the temperature of the system. The great usefulness of F as a function describing the state of a system lies in the fact that F has a minimum value when the system is at equilibrium.

Finally, the chemical potential or standard molar free energy, μ, is defined by

$$\mu = \mu_0 + RT \ln a = \frac{\partial F}{\partial n} \tag{3}$$

where n is the number of moles of a chemical component of the system, a its activity, and μ_o its standard chemical potential.

The following relations will also be used:

$$dF = -S\,dT + V\,dP + \sum \mu\,dn \qquad (4)$$

$$dE = T\,dS - P\,dV + \sum \mu\,dn \qquad (5)$$

Consider now a system composed of two phases at equilibrium such as a water droplet in air saturated with water vapor. Both phases possess a given value of F and the free energy of the system is at a minimum. What will happen if we enlarge the surface separating the two phases, e.g., by producing wrinkles on the surface? The free energy of the two bulk phases will not change since they do not undergo any change. Therefore all free energy change, if any, will be associated with the change in surface, ds, and be proportional to it:

$$dF = \gamma\,ds \qquad (6)$$

with γ as the proportionality factor. Thus γ is defined as the change in free energy associated with a change of surface.

$$\gamma = \left(\frac{\partial F}{\partial s}\right)_{T,p,n} \qquad (7)$$

If free energy is associated with the surface, then the total surface free energy is measured by the product of the surface area, s, and the free energy per unit surface, F_s; $F = F_s s$.

A change in surface will then produce $dF = d(F_s s)$ and therefore

$$\frac{dF}{ds} = s\frac{\partial F_s}{\partial s} + F_s\frac{\partial s}{\partial s} \qquad (8)$$

Since, at least for the pure liquid, F_s is independent of the total area $[(\partial F_s)/(\partial_s) = 0]$, from Equations 7 and 8 we obtain $\gamma = F_s$, i.e., γ also measures the free energy associated with a unit area. γ is called the surface tension and it has the dimensions of work per unit area, expressed in units of erg per square centimeter. Since work equals force times length, it also can be expressed as force per unit length, in units of dyne per centimeter. The equivalence of the dimensions of work per area and force per length results in a duality of concepts by which surface tension can be visualized. On the level of molecular dimensions, an increase in surface requires the insertion of new molecules between the strongly interacting surface molecules. The transfer of molecules from the bulk phase to the surface thus requires work, and this work is stored in the molecules at the newly created surface. On the other hand, surface tension can also be imagined as an "elastic skin" kept under tension by the pressure of the bulk phase. The surface tension is then a tangential force acting at every point of the surface.

Thermodynamic theory does not specify whether γ should be positive or negative. Experiments indicate, however, that for air-liquid interfaces γ is always greater than zero: At equilibrium (F minimum) and in the absence of other forces, all liquid droplets are spherical, i.e., they have a minimum surface for a given volume. A great variety of experimental methods is available for the measurement of the surface tension. The choice of the method often depends on the physical state of the phases surrounding the interface (gas-liquid or liquid-liquid), the amount of material available, and the desired accuracy. For the determination of the surface tension of liquids in contact with air, the du Noüy ring method is the most suitable. It is based on the measurement of the force necessary to detach from the surface a small wire ring initially submerged in the liquid. When the ring is lifted from the surface, the surface tension exerts a downward pull proportional to twice the circumference of the ring. The vertical component of this force is readily measured on a balance (Fig. 1A). There is one position of the ring with respect to the surface where the surface tension exerts a purely vertical pull (Fig. 1B), and in this position the balance registers the maximum force. If the ring has a radius r, the maximal pull is measured by the force $f = 2\gamma 2\pi r$, and thus the surface tension can be calculated from $\gamma = (f/4\pi r)$.

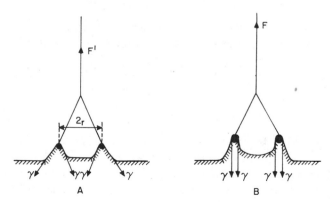

Figure 1
The du Noüy ring method.

Owing to the finite thickness of the wire and the fact that the contact angle is not always identical on the outer and inner circumferences of the ring, several corrections to this simple equation proved necessary. In practice it is more convenient to determine an empirical "ring constant," c, by calibrating the ring with liquids of known surface tension. With a recording electrobalance, time-dependent surface tension changes are also readily measured.

The surface tension of most pure organic liquids is in the range of 20 to 40 dyn/cm. Water has an exceptionally large standard surface tension: 72 dyn/cm at 25° C. This is a consequence of the strongly hydrogen-bonded structure of water in the liquid state, since this interaction has to be partially broken when molecules are transferred from the symmetrical environment of the bulk phase to the air-water interface. Liquid metals have surface tensions of the order of several hundred dynes per centimeter, again reflecting strong atomic interactions. Finally, the surface tension at the interface between two immiscible liquids is equal to or less than the difference of the surface tensions of the two liquids.

THE KELVIN EQUATION

The two concepts of surface tension, surface energy or a force acting parallel to the surface, are equivalent and can be used interchangeably in deriving theorems concerning the behavior of physical surfaces. In the following, we will give as an example the derivation of the Kelvin equation by using the concept of surface energy. The Kelvin equation establishes the important point that the curvature of the interface changes the physical and chemical properties of the bulk phase which it separates.

When a small droplet of liquid is in equilibrium with its vapor phase, the chemical potential of the molecules in the liquid phase is the same as the chemical potential of the molecules in the vapor phase, μ_a. Likewise, when a large volume of liquid of the same compound is separated by a plane interface from its vapor, at equilibrium the compound has the same chemical potential, μ, in both phases. The transfer of a small number of moles, dn, from the interior of the large liquid phase into the droplet requires expenditure of work, since the transfer results in a significant increase of the surface of the droplet. The surface of the droplet is $A = 4\pi r^2$ and the increase in surface area is

$$dA = 8\pi r \, dr \tag{9}$$

If the molar volume of the liquid is v, then the transfer of dn moles will result in an increase in volume of $dV = v \, dn$. The volume of the droplet is $V = \frac{4}{3}\pi r^3$ and thus the increase in volume is

$$dV = 4\pi r^2 \, dr = v \, dn \tag{10}$$

The energy associated with the newly created surface, $\gamma \, dA$, is equal to the change in free energy of the transferred material:

$$-dF = (\mu_a - \mu) \, dn = \gamma \, dA \tag{11}$$

Substituting into this equation the value of dn from Equation 10 and the value of dA from Equation 9 yields the Kelvin equation:

$$\mu_a - \mu = \frac{2\gamma v}{r} \qquad (12)$$

For an ideal gas, $\mu = RT \ln p$, where p is the partial pressure of the gas, and therefore Equation 12 can be written for a liquid in equilibrium with an ideal gas:

$$RT \ln \frac{p_a}{p} = 2\gamma \frac{v}{r}$$

In Table I are shown values of p_a/p calculated for water droplets at room temperature from Equation 12, as a function of the radius.

Table I

r (cm)	p_a/p	Δmp (°C)
10^{-4}	1.001	0.5
10^{-5}	1.01	2.0
10^{-6}	1.1	20.0
10^{-7}	2.8	

The results are surprising: Droplets with diameters of the order of 100 Å possess a much higher vapor pressure than larger liquid particles. Since the chemical potential also measures the solubility of solids, Equation 12 also predicts that finely divided solids will have a much higher solubility than larger particles of the same material. In the same way, the melting point of solids is also a function of particle size and γ, and Table I also shows the experimentally measured decrease of the melting temperature of small mercury particles. In general, on the scale of the average cell and even more on the scale of subcellular particles, the physical properties *and the chemical reactivity* of the constituting molecules are strongly influenced by the curvature of the surface, i.e., the morphology of the particles.

The Gibbs Equation

The previous sections considered the surface tension of pure liquids. When a second component is dissolved in a liquid, some of the solute molecules also migrate to the surface where they produce a change in the molecular interactions, thereby modifying the surface energy, i.e., the surface tension. The Gibbs equation defines the restrictions imposed by thermodynamics on such a change in surface tension.

As pointed out earlier, the surface of a liquid is hardly a geometrical plane of zero thickness; thermal motion and the perpetual condensation and evaporation of solvent molecules result in a surface region where the solvent concentration changes in a continuous manner from that of the gas phase to that of the liquid phase. The idealized surface of the liquid is defined by a plane located in such a way that if all solvent molecules were compressed into the liquid phase, the solvent concentration would be constant up to the plane of the ideal surface. The concentration would then change there in a discontinuous manner to the negligibly small value of the gas phase (Fig. 2). In other words, the surface is defined by a plane for which the areas a and a' of Figure 2 are equal.

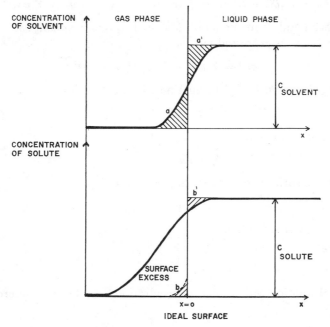

Figure 2
Surface excess and ideal surface.

In general the solvent-solvent and solute-solute molecular interactions are stronger than the solvent-solute interactions. This results in a tendency of the solvent and solute molecules to segregate from each other, and the solute molecules concentrate in the surface region of lower solvent concentration (Fig. 2). When in the ideal system the actual solute concentration is described as constant throughout the solution up to the ideal surface plane and zero beyond that plane, a certain amount of solute molecules is left unaccounted for. In our idealized treatment we will propose therefore that the solution consists of a liquid phase of constant solute concentration and an excess of solute molecules constituting a "surface phase." The surface excess, Γ, is defined as the number of moles of solute per unit surface area in excess with respect to the ideal solution, $\Gamma = n^s/A$. From Equations 2 and 3 we can then write for the liquid phase:

$$E = TS - PV + \sum \mu n$$

and for the surface phase

$$E^s = TS^s + \sum \mu n^s + \gamma A \tag{13}$$

Differentiation of Equation 13 yields

$$dE^s = T\ dS^s + S^s\ dT + \sum n^s\ d\mu + \gamma\ dA + A\ d\gamma + \sum \mu\ dn^s \tag{14}$$

On the other hand, from Equation 5 a small reversible change in energy of the surface phase is expressed by

$$dE^s = T\ dS^s + \sum \mu\ dn^s + \gamma\ dA \tag{15}$$

Subtracting Equation 15 from Equation 14 yields

$$0 = S^s\ dT + \sum n^s\ d\mu + A\ d\gamma$$

or at constant temperature

$$\sum n^s\ d\mu = -A\ d\gamma$$

For one single surface component this reduces to

$$n^s\ d\mu = -A\ d\gamma$$

and dividing by A we obtain

$$\frac{n^s}{A} = \Gamma = \frac{-\delta\gamma}{\delta\mu}$$

For an ideal solution with a solute concentration c, we have by differentiating Equation 3

$$\partial\mu = RT\ \partial\ln c = \frac{RT\ \partial c}{c}$$

and therefore

$$\Gamma = -\frac{c}{RT}\frac{\partial\gamma}{\partial c} \tag{16}$$

the usual form of the Gibbs equation. This equation, derived from thermodynamic considerations only, states that accumulation of solute molecules on the surface of a solution produces a decrease in surface tension. Conversely, if a solute does decrease the surface tension of the solvent, then it must accumulate on the surface of the solution.

The Gibbs equation does not specify the sign of Γ. In aqueous solutions most organic compounds show a positive value for Γ, i.e., they decrease the surface tension of water. Inorganic salts, on the other hand, usually have a negative Γ and they increase the surface tension of water upon dissolution. A "negative surface excess" actually expresses a depletion of the solute on the surface with respect to the liquid phase. In our simplified treatment of the surface region this would be equivalent to saying that the solution of uniform concentration is covered by a layer of pure solvent of finite thickness.

The general validity of the Gibbs equation has been verified by several experimental techniques. An accumulation of the solute in the surface region was demonstrated by collecting and analyzing the foam generated from the solution, or by measuring the surface emission of solutes labeled with a weak emitter, such as ^3H, ^{14}C, or ^{35}S. The most spectacular method made use of a surface microtome, in which a high-speed blade sliced off thin layers from the solutions and collected them for analysis. In a typical experiment it has been found that a 5×10^{-2} M phenol solution has a surface excess of 1.4×10^{14} molecules of phenol/cm^2, in excellent agreement with the value predicted from the surface tension of the solution by the Gibbs equation: 1.35×10^{14} molecules/cm^2. If we assume a molecular area of 20 A^2 for phenol, this would mean that 30% of the surface was covered by phenol molecules. By measurement of the optical anisotropy of the surface, the surface region has been found to be two to three molecules thick. This would mean then that the phenol was about 30 times more concentrated in the surface region than in the bulk solution. Such large concentration effects obviously have important consequences concerning the transport of material and the chemical reactions and equilibria in biological surfaces. They also lead to important new experimental methods, such as the isolation of proteins from dilute solutions by foaming, or the application of surfactants to the emulsification of lipids.

INSOLUBLE MONOLAYERS

Ideal Monolayers

At low solute concentrations the partitioning of solute molecules between the bulk phase and the surface phase is independent of the solute concentration, i.e., $\Gamma/c = K$. Then Equation 16 can be rearranged into $-(1/RT) \, d\gamma = K \, dc$, which yields by integration

$$-\frac{1}{RT} \gamma = Kc + C' \qquad (17)$$

For the pure solvent ($c = 0$), $\gamma = \gamma_o$ and thus $C' = -(\gamma_o/RT)$. Substituting this value into Equation 17 yields $(\gamma_o - \gamma)/RT = Kc = \Gamma$, or

$$\frac{\gamma_0 - \gamma}{\Gamma} = RT \qquad (18)$$

Γ has been defined as the number of moles of solute per unit area in the surface phase. Its inverse is then the area occupied by 1 mole of solute and it is called the *molar area*, $\sigma = 1/\Gamma$. The difference of surface tension of the pure solvent and that of the solution is called the *surface pressure*, $\pi = \gamma_o - \gamma$. In the "elastic skin" concept of surface tension, the surface pressure can be visualized as a spreading force tending to enlarge the surface and acting against the contracting action of the surface tension.

With these two new variables at hand, Equation 18 takes the final form

$$\pi\sigma = RT \tag{19}$$

Equation 19 is sometimes called the "two-dimensional gas law" because of its exact analogy with the ideal gas law. It is valid for the surface phase of any solute which forms a surface excess, however soluble the compound might be in the solvent. The most important application of Equation 19 is nevertheless to the particular class of substances which are completely insoluble in the solvent, usually water. With water-insoluble substances, all the "solute" is at all times on the surface of the liquid phase; but in order to obey Equation 19, being located on the surface is only one of the conditions. The substance also has to cover the whole surface in a uniform manner and all its molecules have to be in contact with the solvent. Not all water-insoluble compounds form such surface layers. When a droplet of mineral oil is deposited on the water surface, it remains there as a droplet. It spreads out somewhat and takes the shape of a flattened lens, as illustrated in Figure 3. At equilibrium, the forces acting on the circumference of the lens cancel each other, i.e.,

$$\gamma_{a/w} = \gamma_{a/o} \cos \beta + \gamma_{o/w} \cos \alpha \tag{20}$$

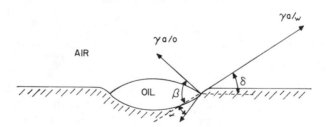

Figure 3
Oil lens on water.

The shape of the droplet is defined by the relative values of the three surface tensions, $\gamma_{a/o}$, $\gamma_{a/w}$, and $\gamma_{o/w}$; and the initially spherical oil droplet spreads until the angles α and β satisfy Equation 20. If $\gamma_{a/w} \geq \gamma_{a/o} + \gamma_{o/w}$, Equation 20 has no solution when the angles are small: The spreading force $\gamma_{a/w}$ is always larger than the sum of the contractive forces $\gamma_{a/o} \cos (\beta - \delta) +$

$\gamma_{o/w} \cos (\alpha + \delta)$. The lens then spreads spontaneously until the whole available surface is covered by a monomolecular layer of the water-insoluble material. The surface monolayer decreases the value of $\gamma_{a/w}$ (Equation 19) proportionally to the surface concentration. As a result, with a large amount of insoluble substance and a limited area, the spreading stops at a given surface concentration, establishing an equilibrium between a lens and a monolayer-covered aqueous phase.

The ability of a compound to form an insoluble monolayer at the air-water interface is determined by its chemical structure. Since the surface tension measures the energy of interaction of surface molecules, one of the requirements for spreading, namely a low value of $\gamma_{w/o}$, is fulfilled by a strong interaction between the compound and water. In other words, a hydrophilic group must be present on the molecule. On the other hand, the compound has to be insoluble in water and this requires a large hydrophobic structure. It is the delicate balance between hydrophilicity and hydrophobicity which produces a stable monolayer: In the absence of a soluble group, the layer collapses into a lens and if the solvation is too strong, then the compound becomes water-soluble. A typical monolayer-forming substance is then composed of a long carbon skeleton with a moderately solvated end group. With a given chain length, e.g., C = 15, weakly solvated end groups, such as halides, esters, or ethers, do not form stable monolayers, while the strongly solvated sulfonate or quaternary amino groups bring on the dissolution of the layer. With the same chain length, stable monolayers are formed with amide, alcohol, acid, or methyl ketone end groups. Conversely, with a given hydrophilic group the stability of the monolayer is a function of the chain length: compounds with short chains are water-soluble, while too long a chain might result in the inability to spread.

Finally, the stability of the monolayer also depends on the volatility of the compound; if the atmosphere in contact with the monolayer is not saturated with respect to the compound, the monolayer will evaporate. Benzene, for example, spreads readily to form a monolayer, but the film evaporates in air within a few seconds. In the laboratory this fact is used profitably to prepare monolayers of slowly spreading substances. The compound is dissolved and spread in a volatile solvent of high spreading coefficient, such as benzene or petroleum ether.

The most readily measurable property of monolayers is their surface pressure. In the simplest and most convenient experimental technique, the monolayer is spread out on the surface of the aqueous subphase in a Teflon-coated trough with a surface of 50–500 cm^2 and a depth of 0.5–2 cm. The surface tension is monitored with a du Noüy ring and the surface pressure is calculated from $\pi = \gamma_o - \gamma$. A direct measurement of the surface pressure is possible with the help of the Langmuir trough, where a mobile barrier floats on the surface and the monolayer is spread on one side of the barrier.

The other side of the barrier is in contact with a clean water surface and thus the horizontal force exerted on a barrier of length l is equal to $l \times (\gamma_o - \gamma)$. This force is measured by a balance after transduction into a vertical plane. The great advantage of the Langmuir trough is the possibility of changing the surface area and thus the compression of the monolayer by simply moving the barrier and compressing the film mechanically. It is, however, more difficult to use the trough for routine measurements because of the experimental problems of leakage of the monolayer around the edges of the barrier and because of the rather complicated mechanical devices used in the instrument.

The measurement of the surface pressure of monolayers yields valuable information concerning the behavior and state of the molecules composing the layer. For most substances at very low surface concentrations, Equation 19 has been shown to be valid. When n moles of substance is spread out on the surface, Equation 19 takes the form $\pi n \sigma = nRT$. Since $n\sigma$ is the total area of the trough, A, we can write

$$\pi A = nRT \tag{21}$$

or

$$\pi = nRT \frac{1}{A} \tag{21a}$$

Thus for a given number of moles of substrate in the monolayer, the surface pressure varies linearly with the inverse of the total area covered by the monolayer, and a plot of π versus $1/A$ yields a straight line passing through the origin and with a slope of nRT. By knowing the amount of material spread, we can thus determine n and from it, the molecular weight.

Equation 21 is valid only for very low surface concentrations with several thousand square Angstroms per molecule. At higher concentrations the area available for a molecule is decreased considerably by the area occupied by the molecule itself, as measured by the collision cross-section. The area thus excluded by 1 mole of substance is called the limiting molar area, A_o. For n moles of substrate, Equation 21 is then modified into

$$\pi(A - A_0 n) = nRT \tag{22}$$

or

$$\pi A = nRT + nA_0 \pi \tag{22a}$$

Equation 22 is again the close analogue of the three-dimensional gas law for a real gas. By plotting πA versus π, a straight line is again obtained with an intercept of nRT on the πA axis and a slope of nA_o. Thus both n and A_o can be obtained from the same experiment.

Experimental surface pressure data for insoluble monolayers are usually presented as a π versus A/Nn plot, i.e., surface pressure versus molecular area, A_m, as in Figure 4. According to Equation 22 this will yield an equilateral hyperbole with $\pi = 0$ and $A_m = A_o/N$ as asymptotes.

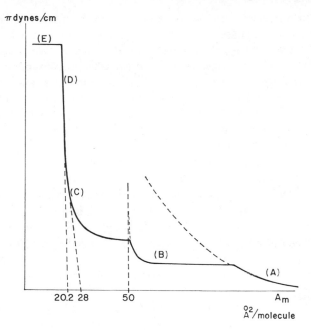

Figure 4
Force-area diagram.

By carefully spreading over ammonium sulfate solutions, even mono-layers of proteins can be prepared and their molecular weight determined with very small amounts of material. For some proteins, such as ovalbumin and pepsin, excellent agreement has been obtained for the molecular weights determined by the monolayer technique and by more conventional methods. With other proteins, such as β-lactoglobulin and insulin, the monolayer technique yielded a molecular weight twice the value of that obtained by other methods, indicating a strong dimerization of the protein in the monolayer state. For lipid molecules, the πA_m versus π plots usually yield very accurate molecular weights.

Monolayer States

In three dimensions, substances can be gaseous with independently moving molecules separated by large distances; they can be liquid with loosely packed molecules interacting moderately and possessing a reduced freedom of motion; and finally they can be solid with closely packed, strongly inter-acting molecules, whose motion is reduced to molecular vibrations. In a surface monolayer the different degrees of molecular interactions—between surface molecules and with the solvent—can result in several more thermodynamically distinct phases which have no analogue in the three-dimensional system. These states have their stability limited to a certain range of surface pressure and are characterized by a distinct mathematical function describing the force-area curves. These states reflect the different possible organizations of the molecules on the surface, organizations depending on the molecular interactions allowed by the conformations assumed by the molecules. The transition between these states is characterized by a more or less sharp break in the surface pressure-area curves.

Not all substances are able to assume all states, and the literature abounds in the description of states of dubious existence, observed with a questionable technique for one peculiar substance. The situation is further confused by the fact that at least four different nomenclatures have been proposed for surface states. In the following we will limit ourselves to a simplified description of surface phases as exemplified by isooleic acid monolayers. (The nomenclature used is that of Adam, who first studied systematically the monolayers of organic molecules.) The states are summarized in Table II; Figure 4 shows the characteristic surface pressure-area curves. From these curves and from several other experimental methods including optical methods, electron microscopy, and the measurement of surface potentials, several conclusions can be reached concerning the conformation and the interactions of surface molecules.

Table II

STATE	CONFORMATION	A_o/N (Å^2/molecule)	EQUATION
1. Gaseous (A)	Flat on surface	> 50	$\pi (A_m - A_o/N) = kT$
2. Liquid-expanded (B)	Coherent, randomly oriented chains	50	$(\pi - \pi_o)(A_m - A'_o/N) = kT$
3. Closely packed heads (C)	Vertical chains, liquid	28	$\pi = cst(A''_o/N - A_m)$
4. Closely packed chains (D)	Vertical chains, solid	20.2	$\pi = cst'(A'''_o/N - A_m)$
5. Collapsed (E)	Molecules in multilayers	20.2	$\pi = cst''$

1. The gaseous state is characterized by the absence of any interactions between the molecules in the monolayer. This state is adequately described by the two-dimensional gas law, indicating a random motion of individual molecules. It is thought that in this state even the hydrophobic side chain of the molecule is in contact with the water, at least part of the time, and the molecule is approximately horizontal on the surface, occupying the maximal molecular area.

2. At higher compressions a transition occurs into the liquid-expanded state, in which the side chains are lifted progressively from the surface as the monolayer is more and more compressed. The chains are still in the random coil conformation and the molecule occupies an area of about 50 Å^2 at the highest compression. All molecules are in contact with each other and they form a coherent layer with a relatively high compressibility. From surface viscosity measurements it can be concluded that the layer is liquid and the structure is loose. At the lower end of the surface pressure range of the liquid-expanded state, at pressures only slightly higher than the transition pressure, there are too few molecules per unit area to cover all the surface and the monolayer is composed of two distinct phases: Patches of liquid-expanded regions are in true equilibrium with gaseous molecules. In this pressure range the surface pressure is virtually independent of A_m since further compression results only in the condensation of more gaseous molecules until the whole surface is covered with loosely packed chains.

3. At still higher compressions the side chains are progressively oriented vertically as the pressure increases. The hydrophilic end groups are closely packed but still completely hydrated. Since they have a larger cross-section than that of the side chains, the limiting molecular area for this state is determined by the area occupied by these hydrated "heads." The state of closely packed heads is characterized by a low compressibility: The surface pressure rises sharply and almost linearly upon decrease of the molecular area once the heads are packed. This state is still "liquid" in the sense that the molecules possess some degree of freedom of translational motion in the plane of the surface.

4. Further increase of pressure squeezes out the solvent molecules from the layer, allowing the organization of the side chains parallel to each other and with the maximum packing. The compressibility is extremely low, the pressure-area curve is almost parallel to the vertical axis. The molecules have no translational freedom; the state of closely packed chains is a solid state in two dimensions.

5. Once a limiting pressure is attained, further compression of the layer results in a progressive collapse of the monomolecular structure. Large fragments of bilayers start protruding out from the surface and produce trilayers by folding over the existing solid monolayer, as illustrated in Figure 5. Since

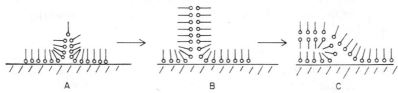

Figure 5
Collapsing monolayer.

the collapse is a time- and pressure-dependent process, slow decrease of the surface pressure is most often observed after a supercompressed state is obtained, and several hours may be required before an equilibrium is reached.

Properties of Monolayers

STABILITY

Monolayers are not indefinitely stable and are slowly destroyed either by collapse or by dissolution or evaporation. Collapse is time-dependent, but at the end a true equilibrium is established within the regions of monolayers and trilayers. The final equilibrium pressure, called the collapse pressure, is quasi-independent of the number of molecules occupying the surface over a large range of molar areas. The collapse phenomenon is reversible, and upon expansion the monolayer is slowly regenerated.

Evaporation and dissolution are also equilibrium phenomena; the monolayer is stable only as long as it is surrounded by a gas and a liquid phase saturated with respect to the compound in the monolayer. Immediately upon formation of the monolayer, a thin layer of water under the surface—called the δ-layer—also becomes saturated with the compound in the monolayer. The δ-layer is in dynamic equilibrium with the monolayer through rapid exchange of material. It is characterized by sharp gradient of solute concentration rather than by a constant solute concentration with a discontinuity at its boundaries. The δ-layer blends progressively into the bulk solution. Once the δ-layer is saturated with respect to the solute, further depletion of the monolayer by dissolution is governed by the laws of diffusion from the δ-layer into the aqueous solution. The rate of dissolution is thus dependent on the solubility of the compound in water, its diffusion coefficient, and its surface concentration. In other words, an increase in surface pressure and the stirring of the solution accelerate the dissolution. Qualitatively, evaporation shows a behavior similar to that of dissolution; high surface pressure and air currents close to the surface accelerate the evaporation of the monolayer. Finally, the phenomenon of dissolution can be reversed if the solution contains initially the monolayer-forming substance. The transfer of the molecules occurs again through the δ-layer, which forms a gradient between the solution of homogeneous solute concentration and the monolayer. Once the gradient is established, diffusion is again the rate-limiting process, and in the steady state transfer the surface concentration, n_t, at time t is given by the following equation

$$n_t = 2n_0(Dt/\pi)^{1/2} \tag{22b}$$

for an unstirred solution and low surface concentrations. In this equation n_o is the solute concentration in the solution and D the diffusion coefficient (5×10^{-6} cm^2 sec^{-1} for small molecules). Stirring of the solution decreases the thickness of the δ-layer, and accelerates the transfer process.

CONFORMATION OF MOLECULES

In the monolayer, the conformation of the molecules and their orientation with respect to the solvent are determined by the surface pressure, by hydrophobic and hydrophilic interactions, and by the randomizing effects of intramolecular rotations. At low compressions, maximum contact of the molecules with water is achieved and even weakly hydrophilic groups, such as double bonds in aliphatic chains, are fully hydrated. At higher surface pressures the optimal utilization of the available space is the dominant factor, and tight packing of the surface molecules into an almost crystalline arrangement is the final result. In the process of packing the molecules, the energy of hydration of the double bonds and that of translational and rotational motions have to be overcome. This loss of energy is only partially compensated for by hydrophobic interactions, and most of the work has to be provided by the compressive force. It is important to realize that although surface pressures are measured in very small units, of the order of a few dynes per centimeter, when these weak forces are translated into energies per molar area, they appear quite large. For example, the energy associated with 1 mole of triglyceride in the gaseous state at room temperature is $2 \times 300 = 600$ cal/mole. At the compression of 20 dyn/cm the molecules occupy approximately 80 $Å^2$, and thus the surface energy is $20 \times 2.4 \times 10^{-11} \times 80 \times 10^{-16} \times 6 \times 10^{23} = 2300$ cal/mole. The energy to be supplied by compression is thus 1.7 kcal/mole.

The most pronounced conformational changes are observed with protein monolayers. Most proteins completely unfold upon spreading into a monolayer at the air-water interface, as witnessed by the molar areas of approximately 15 $Å^2$/amino acid residue, in protein films at low compressions (less than 0.5 dyn/cm). Some proteins are able to maintain their native conformation in monolayers, while others are partially unfolded, and the degree of unfolding is a function of the surface pressure.

MIXED MONOLAYERS

When a mixture of two compounds is spread out as a monolayer, at very low surface pressures the system behaves as an ideal mixture. Each compound has its own partial surface pressure which it would have if it were alone on the surface, π_a and π_b, and the surface pressure is the sum of the partial pressures. If the two compounds form miscible monolayers, this additivity of partial pressures is maintained also at higher pressures, although the partial pressure of each compound is also influenced by the limiting molar area of the other:

$$\pi_a = n_a RT/(A - n_a A_{oa} - n_b A_{ob})$$

Total miscibility is usually observed for compounds of similar chemical structure, e.g., the triglycerides of two fatty acids of different chain length. For some pairs of compounds, such as cetyl sulfate and cetyl alcohol, or chlorophyll and vitamin K_1, molecular complexes have been observed in their mixed monolayers. The existence of these complexes is indicated by discontinuities at equimolar concentrations, when at constant pressure and at constant total number of moles the molar area is plotted against the mole fraction.

Immiscibility of two compounds in the monolayer results in surface heterogeneity. The two "surface phases" are distributed in separate homogeneous islands the size of which depends largely upon the method of spreading. When these individual regions are large, the heterogeneity is readily evidenced by large fluctuations in the surface potential (see below). When the surface is composed of small patches of the two compounds, it is not always easy to demonstrate surface immiscibility. The only reliable criterion for the existence of two surface phases comes from the application of the Gibbs phase rule, which predicts that the collapse pressure of a surface phase is independent of the presence of the other phase. Thus a monolayer composed of two surface phases shows a constant collapse pressure independent from the composition of the layer and identical to the collapse pressure of one of the compounds. Monolayers generated with two miscible compounds, on the other hand, have a variable collapse pressure depending on the composition of the mixture.

SURFACE POTENTIAL AND IONIZATION OF MONOLAYERS

When a metal plate coated with a weak α-emitter (^{241}Am or ^{210}Po) is placed over the surface of an aqueous solution, the ionizing radiation renders the air conductive and the electric potential of the plate is identical to that of a point immediately outside the aqueous phase. With the use of an electrometer one can then compare the potential of the plate with that of a saturated calomel electrode immersed into the solution and thus measure a potential difference between the air and the water phases called the *contact potential*, V_c. The formation of an insoluble monolayer on the surface modifies the value of the contact potential, and if the new potential is V_m, the *surface potential of the monolayer*, ΔV, is defined as

$$\Delta V = V_c - V_m \qquad (23)$$

Typical experimentally observed values of ΔV are of the order of a few hundred millivolts for most lipid monolayers. In the monolayer the permanent electric dipole moment of the surface molecules is oriented with respect to the surface, thus producing an electric bilayer. The surface potential of the monolayer, therefore, reflects the strength and the direction of these electric dipole moments attenuated somewhat by the redistribution of the ions in the liquid phase. If there is no compensating effect due to the ions in solution, the surface potential yields directly the molecular surface dipole moment, μ

$$\mu = (\tfrac{1}{4}\pi)\, \Delta V/n \tag{24}$$

where n is the number of molecules per square centimeter. Although the exact theoretical interpretation of surface potentials is beset by many difficulties, the experimental determination of the surface potential of lipid monolayers yields some important conclusions concerning the structure of monolayers.

Large fluctuation of the surface potential is the best evidence for surface heterogeneity, arising from the simultaneous presence of more than one surface phase. Also time-dependent change of surface potential always indicates a phase transition or a chemical reaction in the monolayer. In the same vein, the constancy of $\Delta V/n$ upon compression of the monolayer indicates the absence of any phase transition. Since the intrinsic dipole moment of an organic molecule, μ_i, is readily measured or calculated from group dipole moments, the measurement of the surface dipole moment allows one to determine the orientation of the molecules in the monolayer. If θ is the angle between the axis of the intrinsic dipole moment and a line perpendicular to the surface, then $\cos\theta = \mu/\mu_i$. With the help of this equation it is possible to show, for example, that the ethyl group in ethyl esters of long-chain fatty acids lies parallel to the surface at low surface pressures, but is completely surrounded by water and perpendicular to the surface at high compressions.

The largest changes in surface potential occur upon ionization of acidic or basic functions in the monolayer. The introduction of a negative or positive charge into the molecule results in the formation of an ionic double layer, since the charge in the layer induces the formation of a statistical layer of counterions in the solution. The surface potential is then described by

$$\Delta V = 4\pi n\mu + \psi$$

where ψ is the potential due to the ionic double layer. No satisfactory theoretical equation is available to describe ψ as a function of ionic strength, molecular structure, and dielectric constant. The value of ψ is nevertheless proportional to the degree of ionization in the monolayer, and by measuring ΔV as a function of the pH of the solution, the surface ionization constant K_a^s can be determined. It has been found experimentally that the surface ionization constant is in general quite different from the ionization constant in solution, K_a^b. At low ionic strength, fatty acids have a pK_a^s of 5.6 and a pK_a^b of 4.9, and long-chain amines have a pK_a^s of 10.1 compared with a pK_a^b of 10.6. This apparent shift of pK toward neutrality occurs because the pH of the solution close to the surface is not identical to the pH of the solution further away from the surface. The formation of carboxylate ions in the monolayer results in an increase of the local concentration of protons which act as counterions. This decrease of the surface pH with respect to the pH of the bulk phase then reduces any further ionization of the acid and thereby increases the apparent pK_a. In the same manner, the positively charged protonated amino group has hydroxide ions as counterions, thus increasing the surface pH with respect to that of the solution. Deprotonation of the amino group occurs then at a lower bulk pH and the apparent pK_a^s is lower than pK_a^b. As the ionic strength increases, less and less protons or hydroxide ions act as counterions and they are replaced in the bilayer by other ions. Therefore the surface pH and the pK_a^s are strongly ionic-strength-dependent.

CHEMICAL REACTIONS IN MONOLAYERS

The reactivity of molecules composing a monolayer should a priori differ from that of molecules in the aqueous solution. Not only are the conformation and the environment of the surface molecules different, but the frequency of their collision with reagent molecules in the solution is reduced because of the pronounced steric effects of the neighboring surface molecules. When we add to these factors the compression-dependency of the conformation, the possibly slow dissolution of the reaction product, and the variability of the concentration of the soluble reagent close to the surface beause of the surface potential and the Gibbs surface excess, we realize that surface reactions are much more complicated than their counterparts in homogeneous solution. Finally, the lack of easy and general experimental techniques for the determination of the time course of the reaction have discouraged many interested investigators from carrying out kinetic and mechanistic analyses of biologically important surface reactions. In the past few years many new experimental methods have finally been developed and interesting results have become available at an ever-increasing rate.

As in the study of the kinetics of any chemical reaction, the stoichiometry of the reaction has to be established, i.e., the chemical nature of the reagents and of the reaction products has to be determined. The reaction products of a monolayer reaction can be collected by collapsing the layer through compression, flooding the surface with a water-insoluble solvent, and separating the reaction product from the solvent. In the case of a water-soluble reaction product evaporation of the solution yields the reaction product. The chemical nature of the product can then be established by thin layer chromatography, vapor phase chromatography, or mass spectrometry. With lipid monolayers of moderate surface concentrations, 10^3 cm^2 surface usually yields enough material for this analysis. The sensitivity of the method can further be increased by the use of radioactively labeled reagents and the addition of unlabeled carrier molecules after the reaction is completed.

At constant total area, the time course of the reaction is readily followed by monitoring changes in the surface pressure and/or the surface potential. The largest changes in surface pressure are of course observed when the reaction product is water-soluble. In this case, from the surface pressure and the force-area curve of the starting material, the surface concentration at any moment can be calculated. Care should be taken to ascertain that the dissolution of the product is not the rate-limiting process, but is much faster than the chemical reaction. Dissolution usually obeys Equation 22b, while chemical reactions follow usually an exponential equation. If the reaction product remains on the surface, then the determination of the degree of conversion of the starting material from the surface pressure changes requires the measurement of the surface pressure of mixtures of reagent and product of known composition.

With the help of the Langmuir trough and an automated movement of the barrier, reactions can also be carried out at constant surface pressure. For a reaction yielding water-soluble products, the area of the surface film directly measures the amount of unreacted material. In the case of water-insoluble products, the changes in area during the reaction have to be calibrated again with known mixtures of reagent and product.

If the reaction product is water-soluble, the use of reactive material labeled with ^{14}C or ^3H and the monitoring of the surface radioactivity allow one to determine the rate of disappearance of the starting material.

When the reaction yields an ionized product remaining on the surface, the measurement of the surface potential is the most appropriate method to follow the time course of the reaction.

Once the number of molecules reacted as a function of the time has been determined, conventional kinetic analysis of the data yields the rate constant determining the kinetic behavior of the monolayer.

Detailed kinetic analysis of reactions in monolayers has been performed only for a few hydrolytic reactions and one enzymatic reaction. The alkaline hydrolysis of γ-stearolactone has been measured, and it has been found that the rate of the reaction is proportional to the number of molecules on the surface and the hydroxide ion concentration in the solution. At higher surface pressures the rate of the reaction decreases with increasing compression, indicating that the accessibility of the ester function to hydroxide ions is decreased with the close packing of the molecules. In the reverse reaction, the acid-catalyzed lactonization of γ-hydroxystearic acid, the rate constant of the reaction is again independent of the compression up to 10 dyn/cm. At higher compressions the rate constant falls off rapidly, indicating the unfavorable orientation of the acid and alcohol functions in a highly packed state. The reaction is first order with respect to the hydroxy acid and to hydrogen ions. Thus the reaction appears in every respect identical to a reaction of lactonization in the homogeneous solution, with the exception of the steric hindrance introduced by close packing of the reactive molecule.

The kinetic study of the pancreatic lipase-catalyzed hydrolysis of glycerol esters has been carried out by using di- and tri-esters of octanoic acid, which yield water-soluble reaction products. The enzyme is dissolved in the aqueous phase and the reaction monitored by measuring the surface tension. The reaction is first order with respect to both the substrate and the enzyme. The rate constant is independent of the surface pressure over a wide range of compressions. This then indicates that the enzymatic reaction occurs without penetration of the enzyme into the monolayer and that the enzyme-substrate interaction is limited to the hydrated portions of the reagents. The enzymatic specificity and the kinetic characteristics of the monolayer reaction are identical to those of the reactions carried out with water-soluble or emulsified substrates. Thus it appears that, at least for pancreatic lipase, the heterogeneity of the reaction medium and the organization of the substrate into a monolayer do not contribute anything intrinsically different to the efficiency of the catalytic process. The only unusual feature of the reaction lies in the ability of the lipase to interact with the surface phase without any noticeable unfolding, a property which is shared by only a few other proteins.

SUMMARY

The chemistry and physical chemistry of surfaces and membranes are situated between gas phase chemistry, where the behavior and the evolution of the system are determined by the sole properties of the individual molecules, and solution chemistry, where theories and experimentation are concerned with the bulk properties of large, statistically homogeneous systems. Our primary goal has been to demonstrate that chemical theories both on the molecular level and on the level of large systems can be fruitfully applied to surfaces, but that in every case allowance has to be made for the quasi-molecular dimensions of surface systems. Just as one single molecule cannot be treated from the thermodynamic point of view and the use of quantized energy levels is awkward in treating solutions, the elaboration of theories concerning membranes requires new concepts and a new methodology best adapted to the particular dimensions of membranes. In the past decade considerable progress has been made in this direction, and hopefully the time is not far off when membranology will be a scientific discipline of its own.

RECOMMENDED READINGS

1. Adamson, A. W. *Physical Chemistry of Surfaces*, 2d ed. Interscience, New York, 1967.
2. Davies, J. T., and Rideal, E. K. *Interfacial Phenomena*, 2d ed. Academic Press, New York, 1963.
3. Defay, R., and Prigogine, I. *Surface Tension and Adsorption*. Longmans, London, 1966.
4. Gaines, G. L., Jr. *Insoluble Monolayers at Liquid-Gas Interfaces*. Interscience, New York, 1966.

5. Gold, R. F. (ed.). *Molecular Association in Biological and Related Systems.* Advances in Chemistry Series No. 84. ACS Publications, Washington, D.C., 1968.
6. Gushee, D. E. (ed.). *Chemistry and Physics of Interfaces.* ACS Publications, Washington, D.C., 1965.
7. Osipow, L. I., *Surface Chemistry.* Reinhold, New York, 1962.
8. Rothen, A. Surface Film Techniques. *Physical Techniques in Biological Research,* 2d ed. Ed. D. H. Moore. Academic Press, New York, 1968, vol. IIA, p. 217.

Gabor Szabo

Lipid Bilayer Membranes

The most striking common structural feature of both artificial lipid bilayer and natural membranes is the presence of a thin ($\simeq 50$ Å) hydrophobic interior. In lipid bilayers the membrane interior is composed of the lipid hydrocarbon tails, while in natural membranes it is likely to contain hydrophobic protein moieties as well. The membrane surfaces of artificial bilayers are covered with the lipid polar head groups, while other hydrophilic moieties (e.g., protein, sialic acid) are also present in natural membranes. (For a timely discussion of membrane structure, see ref. 1.) Thus, lipid bilayers are particularly appropriate model systems for the study of those properties of natural membranes which result directly from the presence of the nonpolar membrane interior or the polar surface groups. For example, the electrical capacitance and lipophilic solute permeability in lipid bilayers are similar to those of natural membranes, and both these properties can be inferred as a simple consequence of the thin hydrocarbon-like membrane interior. However, lipid bilayers lack most of the functional characteristics of natural

Supported by NSF Grants GB16194 and GB30835, and USPHS Grant NS09931. The author thanks Dr. George Eisenman and Dr. Sally Krasne for their valuable comments and suggestions.

membranes, presumably because normal components of cell membranes, such as proteins (e.g., "permeases" and other enzymes) are absent.

The well-defined chemical composition of lipid bilayers makes possible the quantitative study of membrane structure and molecular interactions at the membrane surface as well as in the membrane interior. Study of bilayers combined with that of monolayers, as discussed in Chapter 5, should provide a powerful experimental system capable of guiding the development and verification of a useful formulation of membrane physical chemistry. Furthermore, the bilayer is an excellent structural substrate of well-known composition into which one may attempt to incorporate not only model molecules but also functionally active isolated membrane components for quantitative assay and study. Thus, the lipid bilayer is likely to become a useful tool in membrane molecular biology, both as a prototype for the physical and chemical properties of membrane systems and as a substrate for reconstitution of the functions of natural membranes.

This chapter presents recent methods of bilayer preparation and well-established physical, chemical, and permeability properties of bilayer membranes in order to provide the reader interested in the study of bilayer membranes with a general conceptual framework, a list of useful references, and some acquaintance with the present "state of the art" in this rapidly advancing field.

PREPARATION OF BILAYERS

A number of lipids (the most frequently studied is lecithin) spontaneously form bilayer-like multilamellar suspensions which are directly useful for studies of bilayer structure by such physical techniques as x-ray diffraction (2–4), differential thermal analysis (5), optical measurements (6), or electron spin (7) and nuclear magnetic resonance (8), as described elsewhere in this volume. However, owing to the absence of well-defined aqueous compartments, such lipid suspensions are usually not suitable for the measurement of membrane electrical (such as capacitance and conductance) or permeability properties. The planar or spherical single bilayer preparations, to be described below, have been developed particularly for the study of those membrane properties.

Planar Bilayers

Following the initial discovery of Mueller *et al.* (9) that stable bilayer membranes can be formed from phospholipid solutions, a number of techniques have been devised that permit the formation of planar bilayer membranes from a variety of lipids in a relatively simple manner (for a detailed review of these, as well as for specialized techniques, see refs. 10–13). The recent tendency toward the use of purified lipids of well-defined composition (14–16) is made increasingly practical by their commercial availability (for example, by such firms as Supelco and Sigma).

The formation of bilayers is relatively simple and resembles that of "black" soap films (or bubbles) (17), both involving basically the spontaneous drainage of a thin film to a thermodynamically more stable form of a bimolecular layer.* The lipid from which the bilayer membrane is to be formed is dissolved in a nonpolar solvent (e.g., hexane, decane, hexadecane, or a mixture of chloroform and methanol) at concentrations which are typically about 10–40 mg of lipid in 1 ml of solvent. This lipid solution is spread over a small hole (approximately 1 mm in diameter) punched in a thin Teflon or Lucite septum that separates two aqueous solutions. The lipid adheres to the septum and covers the hole, thus providing a thick membrane of lipid solution with oriented monolayers of lipid at each surface. As the lipid solution drains gradually from the membrane, interference colors appear, indicating that the membrane is several thousand angstroms thick. Usually the membrane spontaneously thins further to a state in which two surface monolayers come into close contact, as indicated by the appearance of a nonreflecting, optically black region, the bilayer membrane, which adheres to the supporting septum through a torus of bulk lipid solution. In addition to the pleasant display of interference colors, membrane thinning is worthy of visual observation not only to ensure that the membrane is planar and that it has thinned properly but also to estimate the bilayer area and assess such membrane characteristics as surface mobility (judged by the movement of inclusions).

The amount of solvent that remains after membrane thinning is not precisely known. Variation of membrane capacitance as a function of the chain length of hydrocarbon lipid solvent (e.g., hexadecane as compared with decane) indicates that the membrane thickness, and therefore the amount of residual solvent in the bilayer, varies with the nature of the solvent (18).

Although it is not clear to what extent membrane properties and structure are affected by any residual solvent, the presence of solvent is undesirable since it introduces an additional variable not present either in natural membranes or in spherical bilayer membranes. It is possible, usually at the expense of membrane stability, to form solvent-free planar bilayers. For example, glyceryl dioleate or oleylamine, both of which are liquid at room temperature, are capable of forming bilayer membranes without any solvent (19). The use of those long-chain hydrocarbon solvents (e.g., hexadecane) which were inferred to be almost totally excluded from bilayers (18) should also result in nearly solvent-free membranes.

The experimentally important problem of bilayer membrane stability is not well understood, although useful practical guidelines emerge as the result

* Indeed, bilayers can be pictured as "inside-out" soap bubbles, since hydrocarbon tails in the former, and polar groups in the latter, form the interior of the bimolecular leaflet.

of accumulating experimental observations and theoretical arguments (20). The nature of the lipid polar head groups and hydrocarbon tails, the temperature, and the composition of the aqueous solutions are all found to influence membrane stability.

Stable bilayer membranes can be formed from lipids having strongly polar neutral (glyceryl mono- or diesters), amphoteric (phosphatidylcholine, phosphatidylethanolamine), negatively charged (phosphatidylglycerol, phosphatidylserine at pH 7), or positively charged (primary amine) head groups provided a delicate balance is maintained between the size of the head group and of the hydrocarbon tail.* Thus, glyceryl mono- or dioleate both form stable bilayer membranes, while glyceryl trioleate or glyceryl monodecanate do not; dipalmitoyl lecithin forms stable membranes (at 50°C), but lysopalmitoyl lecithin does not (19).

Lipids having straight- or branched-chain saturated hydrocarbon tails as well as lipids that contain single or multiple double bonds or cyclopropyl groups in their hydrocarbon tails are all capable of forming stable bilayer membranes in the presence of an appropriate (e.g., glycerophosphoryl choline) polar head group.† The chemical composition of the lipid hydrocarbon tails has a marked effect, presumably owing to differences in tail-to-tail interactions, on how temperature alters the packing and fluidity of the hydrocarbon tails in the bilayer interior (5, 7). A striking example of this is seen when glyceryl monostearate and glyceryl monooleate membranes are compared. At room temperature membranes formed from decane solutions of the saturated lipid are rigid and show no surface mobility (judged by the lack of solvent inclusion movements), while membranes formed from unsaturated lipid solutions are highly deformable and show obvious surface mobility by the same criteria. When the temperature is increased to near 50°C from room temperature, a sharp transition occurs with the saturated lipid bilayers, which "melt" by the criteria of being deformable and surface-mobile. This type of transition also has striking effects on membrane permeability properties, as will be discussed later.

Planar bilayers are particularly well suited for the measurement of membrane electrical properties, since the membrane separates large aqueous volumes into which electrodes can be inserted without difficulty. On the other hand, the small membrane surface/aqueous volume ratio and the presence of a torus make isotopic flux measurements relatively difficult. An additional

---

* The aqueous salt concentration is expected to alter the interactions between the polar head groups in bilayers. This should alter somewhat not only the packing of lipids in the membrane but also the membrane stability. For example, charged lipids are expected to be more stable at high salt concentrations since head-to-head repulsions are minimized by the presence of coions.

† Oxidation of the membrane-forming lipids, which is most noticeable for hydrocarbon tails having conjugated double bonds, always results in greatly reduced membrane stability (21).

drawback of the technique is the uncertainty about the solvent that may remain in the bilayer. These difficulties may be overcome by forming bilayer vesicles, but at the expense of difficulties of a different nature.

Bilayer Vesicles (*Liposomes*)

It is also possible to prepare structures where a bilayer encloses an internal volume, thus forming a vesicle (10, 13, 22).

Small (typically 200–1000 Å in diameter) bilayer vesicles are usually prepared by ultrasonic dispersion of a lipid in an aqueous solution. The formation of undesirable multilayered vesicles (i.e., vesicles within vesicles) can be minimized by selecting optimal conditions of dispersion. Although the size distribution of the dispersed liposomes is rather heterogeneous, vesicles of nearly uniform size can be selected by such techniques as gel filtration (23). The internal composition of the liposomes is set by the dispersion medium, and the external medium can then be altered by passing the vesicle suspension through a gel filtration column preequilibrated with the solution into which the liposomes are to be resuspended.

Liposomes are solvent-free and therefore have a better defined composition than solvent-containing planar bilayers, although when vesicle size is very small (500 Å or less in diameter), the sharp curvature of the bilayer is likely to influence the membrane structure, not only for steric reasons (fewer molecules can be packed into the inner molecular layer than into the outer molecular layer) but also because of interactions between the diffuse ionic layers usually present at the interfaces (see, for example, Chap. 3 for the effective use of this phenomenon in red blood cell vesicle preparation). Bilayer vesicles are convenient for the study of isotopic tracer fluxes of relatively impermeant solutes because of the large membrane surface area available for a given volume of a suspension. On the other hand, the internal composition is difficult to alter and, more importantly, membrane electrical properties are practically impossible to measure.

Difficulties inherent in planar and vesicular bilayer preparations may be overcome by their combined use. In particular, liposomes may be used to examine the effects of added solvent on bilayer permeability.

An intermediate preparation which is expected to combine the advantages of planar bilayers and small liposomes has been described by Mueller and Rudin (10), who found that large bilayer vesicles (1–50 μ in diameter) can be formed from a number of lipids by repeated cycles of hydration and dehydration. Such large vesicles should be useful in bridging the gap between the usual small liposomes and the planar bilayers, since introduction of microelectrodes into the large vesicles allows their electrical properties to be measured simultaneously with their tracer permeabilities.

PHYSICAL PROPERTIES OF BILAYER MEMBRANES

Bilayer membranes are easily deformable elastic films that have rather low surface tension, in the range of 0–5 dyn/cm, depending somewhat on membrane composition (13).

Remarkably low ionic conductances, in the range of 10^{-9}–10^{-7} ohm^{-1} cm^{-2}, are typical of lipid bilayer membranes formed in aqueous solutions of hydrophilic electrolytes such as NaCl or KCl. Such an insulator-like behavior indicates the presence of a large energy barrier to the movement of ions across the membrane. The most likely origin of this barrier is the large energy required to bring the ion from the high-dielectric-constant aqueous phase into the low-dielectric-constant membrane interior formed by the nonpolar lipid hydrocarbon tails and residual lipid solvent. Indeed, the magnitude of observed bilayer conductances is consistent with values calculated for a 50-Å thin membrane having the properties of a bulk hydrocarbon phase such as tetradecane (24).

The large electrical capacitance (0.4–1.2 $\mu F/cm^2$) typical of bilayer membranes is also consistent with the presence of a thin, insulating membrane interior which separates ion-rich aqueous solutions. Such a structure closely resembles a parallel plate condenser, whose specific capacitance C (in microfarads per square centimeter) is related to the thickness d (in angstrom units) and dielectric constant ε of the membrane interior by the formula

$$C = 8.854 \frac{\varepsilon}{d} \tag{1}$$

Equation 1 can be used as a simple method of estimating the thickness of bilayers from the measurement of their capacitance, provided the dielectric constant of the bilayer interior is assumed to be invariant and, for example, equal to that of a pure hydrocarbon phase (typically, $\varepsilon = 2.1$).

While membrane capacitance can be used to estimate the thickness of the insulating membrane interior alone, optical reflectance measurements also yield estimates of total membrane thickness (presumably including polar head groups, ref. 12), while electron microscopy permits visualization of fixed and dehydrated bilayer membranes (10). Capacitance and optical reflectance measurements as well as electron microscopy of bilayer membranes indicate a thickness of the hydrocarbon layer in the range of 50 Å (13), depending somewhat on the chain length of the lipid hydrocarbon tails and of the hydrocarbon solvent (14, 18). This value agrees well with those inferred from x-ray diffraction measurements made on aqueous lipid bilayer dispersions (see Chap. 4 and ref. 3).

CHEMICAL PROPERTIES OF BILAYER MEMBRANES

Bilayer membranes present a large, macroscopically significant amount of surface area whose properties are readily available for study. An example of the unusual local environment at the membrane surface is provided by bilayer membranes formed from lipids having dissociable polar head groups. The resulting charged membrane surface attracts a cloud of oppositely charged ions in the aqueous phase near the membrane, creating a so-called "diffuse double layer" (25). Thus, if the membrane has a net negative charge, which results in a negative surface potential, the surface concentration of cations is increased while that of anions is lowered relative to the concentrations in the bulk aqueous phase (the opposite holds true for positively charged membranes).*

Surface ionic concentrations can be inferred from measurements of bilayer membrane conductance, which are theoretically expected to be proportional to the interfacial concentration of an ion when appropriate carriers of that ion are present in the membrane (see later in this chapter and ref. 16). Such bilayer measurements, in accord with monolayer surface potential measurements, show that an increasing membrane surface charge density, a low salt concentration, and a large electrical charge on the aqueous ion are factors that enhance the deviation between surface and bulk concentrations. Thus, a charged membrane surface accumulates molecules selectively on the basis of their charge; for example, multiply charged positive species such as cytochrome C (26) or Ca^{++} (27) are attracted to a highly negatively charged membrane surface preferentially, relative to monovalent ions. Note, however, that the binding of species to membrane surface groups is possible through specific interactions even in the absence of charged surfaces, e.g., Th^{4+} or UO_2^{2+} binds to surface phosphate groups of amphoteric lipids (27).

The presence of a surface charge must also alter the surface H^+ concentration, and thus the surface pH. This is most important for the understanding and interpretation of the titration behavior of acidic and basic membrane surface groups (16, 28, 29). Lastly, the strong electrical potential gradient near a charged membrane surface is expected to orient large, multiply charged macromolecules. Thus, for a negatively charged membrane, positively charged regions of molecules should face the membrane, while negatively charged groups should be as far as possible from the surface.

Although partitioning of a lipophilic species well into the interior of bilayers can be rationalized in terms of the variables that govern partitioning into nonpolar bulk phases, the behavior of the species at membrane-solution interfaces may be quite different owing to the possible existence of energy maxima or minima which alter the surface concentration. Thus, strongly amphiphilic molecules are expected to adsorb to the membrane surface with their lipophilic moiety buried in the membrane interior and their hydrophilic moiety in the aqueous phase; for example, chlorophyll molecules have been inferred to adsorb to lecithin bilayers with the phytyl chain penetrating the

---

* The same effects are seen in the aqueous subphase below monolayers of lipids having dissociable polar head groups, as discussed in Chap. 5.

membrane interior and the porphyrin ring facing the aqueous phase (30). Charged lipophilic molecules are also expected to be preferentially adsorbed to the membrane surface as a result of the surface energy minimum created by the balance of "hydrophobic interactions" tending to partition the molecule into the membrane interior and electrostatic forces tending to exclude the molecule from the membrane interior (31). The same reasoning should hold true for macromolecules (or parts of macromolecules) which are hydrophobic but charged.

PERMEABILITY OF BILAYER MEMBRANES

A model which is helpful in rationalizing the basic features of bilayer membrane permeability in terms of simple chemical concepts is shown in Figure 1. The upper part depicts schematically the arrangement of lipid molecules in the bilayer, using open circles to indicate lipid polar head groups and broken lines to indicate lipid hydrocarbon tails and solvent molecules. The lower part of Figure 1 shows a model which abstracts those features which are essential for understanding the permeability properties of the bilayer, approximating it as a homogeneous, thin, hydrocarbon-like, low-dielectric-constant membrane (represented by hatched lines) with the polar groups distributed uniformly at the surface. Such a model is consistent with

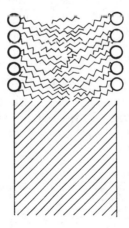

Figure 1

Relation between bilayer membrane structure (upper part) and the homogeneous model (lower part) taken to represent the bilayer. The lipid polar head groups (open circles) are uniformly distributed at the model membrane surface (heavy lines), with the lipid hydrocarbon tails and solvent molecules (broken lines) forming the homogeneous interior (hatched lines) of the membrane model.

previously discussed bilayer conductance and capacitance properties and is conceptually useful for the understanding of bilayer solute permeation to be discussed in this section.

Intrinsic Permeability of Bilayer Membranes

The simplest way in which a solute can permeate a bilayer membrane is by penetrating the membrane-solution interface (i.e., "dissolving" into the membrane) and diffusing across the membrane interior. The upper part of Figure 2 identifies those basic processes of solute transport which are involved in the permeation of bilayers by such a "solubility-diffusion" mechanism.

In general the magnitude of bilayer permeability is governed by the diffusion rate of solute molecules across the membrane interior and by the

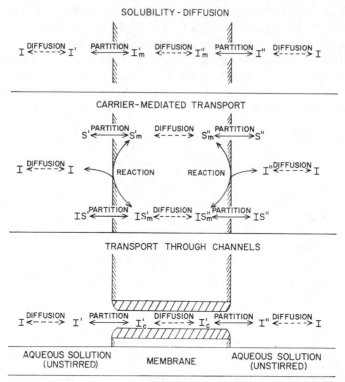

Figure 2

Permeation mechanisms for bilayer membranes. The steps involved in the transfer of a solute, I, between the left (') and right (") side of the membrane are identified on the figure for each of the three permeation mechanisms. See text for further explanation.

rate of solute partition across the membrane surfaces. Either one of these processes may limit the solute flow across the membrane depending on whether partition or diffusion is faster. In the particularly simple but practically important case where partitioning of solutes between the membrane and the salt solution is much faster than diffusion across the membrane interior, the partitioning of solute is expected to be at equilibrium at both membrane surfaces so that membrane permeability becomes simply proportional to the equilibrium partition coefficient and the diffusion coefficient of the solute in the membrane. This case appears to hold true for molecules whose true transmembrane permeability has been measured experimentally (i.e., water, urea, glycerol, erythritol; see ref. 13). In particular, water permeability of bilayers can be predicted rather closely from the known bulk equilibrium partition coefficient and diffusion coefficient of water in tetradecane, which is taken as a model of the membrane interior (24).

Solutes which actually flow across the membrane must be supplied and carried away by diffusion across relatively thick (typically 10^{-2} cm) unstirred aqueous layers invariably present at both membrane surfaces. The presence of these layers puts an upper limit on the experimentally measurable solute permeability, the value of which is determined by the rate of solute diffusion across unstirred layers alone, i.e., in the absence of the membrane. Thus for those molecules to which the membrane is highly permeable, solute movement is limited by the aqueous unstirred layers. These effects of practical importance are well documented for the diffusion of tritiated water across bilayer membranes (13).

Permeation of bilayers by the solubility-diffusion mechanism is not restricted to neutral molecules, although the partition coefficients of ions are considerably lower than those of neutral molecules of the same nature (because of the additional energy needed to bringing a charged molecule from the high-dielectric-constant aqueous phase to the low-dielectric-constant membrane interior), resulting in small membrane permeabilities of all but those ions which are unusually lipophilic or large. The experimentally observed membrane permeabilities of lipophilic ions, a well-documented example of which is the tetraphenyl boron anion (31, 32), are usually more complicated than expected for simple solubility diffusion. This is because diffusion through the unstirred aqueous layers, surface adsorption, and Coulombic repulsion within the membrane interior of the charged molecules are usually not negligible for these species (31).

We have seen that membrane diffusion-limited permeability is proportional both to the diffusion coefficient and to the partition coefficient of a solute. These parameters are both expected to depend on the packing as well as the chemical nature of the lipid hydrocarbon tails. A tighter packing of the lipid hydrocarbon tails is expected to reduce the mobility of solute molecules in the membrane and also to reduce the membrane-water partition coefficient by making it more difficult to insert a solute molecule into the closely packed membrane (28).

The solubility-diffusion type of permeation requires for its function only the presence of a liquid-like, nonpolar membrane interior and is therefore a mechanism which is intrinsic to all bilayer membranes. The permeability of biological membranes to large lipophilic molecules is also consistent with simple solubility diffusion (for a recent review, see ref. 33), indicating that such molecules see the biological membrane essentially as a bilayer. However, compared with the ionic permeability of cell membranes (10^{-4}–10^{-1} ohm^{-1} cm^{-2}), the electrical conductance of bilayer membranes in the presence of small hydrophilic ions such as Na$^+$, K$^+$, or Cl$^-$ is extremely low (10^{-7} ohm^{-1} cm^{-2}), indicating that these ions do not penetrate the membrane to any appreciable extent per se. Moreover, in the absence of specific additives, lipid bilayer membranes show much smaller ionic specificity (34) than natural membranes. It seems likely, therefore, that biological membranes contain components that are absent in bilayer membranes, and these are likely to mediate membrane permeability by the mechanisms outlined below.

Carrier-Mediated Permeability

The middle part of Figure 2 shows schematically a permeation mechanism capable of transporting solutes, such as small cations, which cannot penetrate the membrane significantly by simple solubility diffusion. The process postulates the existence of a lipophilic carrier molecule, S, which can solubilize the solute, I, in the form of a mobile carrier-solute complex in the membrane, $IS_{(m)}$, that in turn diffuses across the membrane and unloads the solute from the membrane by dissociation of the complex near the membrane surface. Note that the carrier, S, as well as the transported molecule, I, may be charged, so that the transport of both charged and neutral complexes is feasible by this mechanism.*

Carrier-mediated transport has two remarkable features: *1*) it permits a selective translocation which is governed not by intrinsic lipid solubility but by the rate (or equilibrium) constants of the reaction between the carrier and the transported solute; and *2*) it permits a coupling between the flow of the transported molecule and the electrical potential across the membrane whenever the carrier or the complex (or both) are charged.

NEUTRAL CARRIERS OF IONS

The best documented example of carrier-mediated permeation is the selective cationic permeability induced in bilayer membranes by neutral

---

* Theoretical expectations for the carrier-mediated transport have been worked out in detail for neutral carriers of neutral molecules (35), for neutral carriers of ions (36, 37), and for charged carriers of ions having a neutral ion–carrier complex (38, 39).

cation–complexing antibiotics, such as nonactin or valinomycin (see for example ref. 40). For these neutral carriers of monovalent ions, the membrane-bound neutral carrier, $S_{(m)}$, combines with an aqueous monovalent cation, I^+, to form the charged lipophilic ion–carrier complex, $IS_{(m)}^+$, in the membrane according to the heterogeneous surface reaction

$$I^+ + S_{(m)} \rightleftharpoons IS_{(m)}^+ \tag{2}$$

As a result of Reaction 2 the membrane is made selectively permeable to ions by the solubilizing action of the carrier. This is reflected not only in enhanced transmembrane fluxes of the carried ion but also in a greatly increased membrane conductance and in the appearance of transmembrane potentials in salt gradients.

The quantitative details of bilayer permeation in the presence of neutral ion carriers depend on whether the rate of ion transfer is limited by diffusion of the complexed ion through the membrane, heterogeneous surface reactions, or aqueous unstirred layer diffusion.

A range of particularly simple "linear" permeability properties is expected when the diffusion rate of the complexed ion across the membrane is slow compared with the rates of surface Reaction 2, allowing it to approach equilibrium. Under such conditions the membrane conductance is expected to be proportional both to the permeant ion concentration at the membrane surface and to the membrane carrier concentration (which is proportional to the aqueous concentration of free carriers).*

For membranes in their "linear" range, ionic selectivity, as indicated by conductance ratios for different cations, is expected to equal the permeability ratios deduced from membrane potential measurements (36). In turn, these selectivity parameters are expected to be predictable from the equilibrium chemistry of the ion-carrier interaction as measured by salt extraction into a nonpolar solvent approximating the membrane interior (41).

Regions of "linear" membrane properties, expected when the diffusion of the complex across the membrane is rate-limiting, have been found experimentally for neutral carriers of cations such as nonactin or trinactin (42). However, in the case of trinactin, which binds ions more strongly than nonactin, regular deviations from "linear" membrane behavior were also detected, indicating that the surface reactions can be displaced from equilibrium. Indeed, the rate constants of surface reactions can be determined

---

* The affinity of an ion to a carrier may be so high that a substantial amount of ion-carrier complex is formed even in the aqueous phases, resulting in a reduced concentration of the free carrier and therefore a reduced membrane conductance. High affinity of the carrier for the solute also tends to reduce the surface reaction rates, resulting in a diminished range of linear membrane conductance behavior. When binding of the ion to the carrier is essentially irreversible, almost all the carrier is complexed, and the complex should act as a lipophilic ion.

from conductance relaxation measurements, as done by Stark and Benz (43) for valinomycin, which acts in a manner similar to trinactin.

CHARGED CARRIERS OF IONS

A number of lipid-soluble weak acids* that induce proton permeability across bilayer membranes appear to do so as carriers of hydrogen ions (39). In this case the carrier, S^-, is the dissociated form of the acid which is negatively charged, while the "complex" is the undissociated acid that forms according to the heterogeneous surface reaction

$$H^+ + S_{(m)}^- \rightleftharpoons HS_{(m)}$$

Membrane potentials and conductances observed for this system are complicated functions of the weak acid and hydrogen ion concentration. However, for at least one of the weak acids, the observed membrane permeability properties can be rationalized on the basis of the charged carrier mechanism provided diffusion in the unstirred aqueous layers is taken into account (39).

A less well-characterized class of negatively charged ion-complexing antibiotics (44), such as monensin, are also believed to act as carriers of cations in bilayer membranes. These molecules form a neutral membrane-bound complex, $IS_{(m)}$, as the negative carrier $S_{(m)}^-$ combines with the cation I^+ following the surface reaction

$$I^+ + S_{(m)}^- \rightleftharpoons IS_{(m)}$$

For this system the only charged species in the membrane is the uncomplexed carrier whose concentration $[S_{(m)}^-]$ determines the conductive properties of the membrane. If $[S_{(m)}^-]$ is low, the membrane conductance is also proportionately low. However, a rapid *exchange* of cations across the membrane, which is related to the membrane concentration of the neutral complex $[IS_{(m)}]$, is still possible even in an apparently nonconducting membrane. Such an "electrically silent" Na^+–K^+ exchange can be detected in planar bilayer membranes by following the monensin-induced depletion of a transmembrane Na^+–K^+ gradient using an indicator such as the electrical potential developed by trace amounts of nonactin (19).

EFFECTS OF LIPID COMPOSITION ON CARRIER-INDUCED
BILAYER PERMEABILITY

Although the chemistry of ion-carrier interactions decides the extent to which the carrier selects a molecule for translocation across the membrane, the translocation of the complex itself depends on the nature of the bilayer

---

* These compounds are uncouplers of mitochondrial oxidative phosphorylation, apparently because of their ability to induce proton transport across the mitochondrial membrane. Neutral carriers of ions (such as valinomycin) may also act as uncouplers because of their ability to induce K^+ leaks in mitochondria (44).

membrane (28). Variables of membrane composition that affect the concentration and mobility of the complex in the membrane, such as the packing of the lipid hydrocarbon tails, are expected to alter greatly the carrier-induced bilayer permeability of an ion. For example, the nonactin-induced K^+ conductance is suppressed upon a phase transition ("freezing") which presumably packs the lipid hydrocarbon chains more tightly (45). The nature of lipid polar head groups may also influence the rate of the carrier-mediated translocation of ions by altering the ionic concentration at the membrane surface. Thus, the presence of negatively charged lipid head groups increases the nonactin-induced K^+ conductance by several orders of magnitude as a result of increasing the surface K^+ concentration (16).

Transport Through Channels

Up to this point we have taken the bilayer to be homogeneous in the plane of the membrane and have considered carriers to provide a localized, mobile, and energetically favorable environment in the membrane for the carried species. We will now examine the case in which there is a continuous pathway, or channel, of sites across the membrane interior, providing a fixed but energetically favorable environment for the translocation of a solute. Such a structure embedded in the membrane is shown schematically in the lower part of Figure 2 by hatched lines. The transfer of solutes can be visualized as a partition into the local environment of the channel, diffusion of the solute within the channel, and partition from the channel into the opposite aqueous phase.

Channels should provide distinct pathways of relatively large ionic conductance through the membrane. The appearance and dislocation of such conductance pathways have been detected on bilayer membranes in the presence of the channel-forming pentadecapeptide gramicidin A (46). Quantal increases of membrane conductance (1.7×10^{-11} ohm^{-1} in 0.2 M NaCl solutions) having a mean lifetime of 0.4 sec in glyceryl monooleate-decane bilayers (47) indicate that a channel can conduct many solute molecules during its existence, before it is dislocated by thermal motion.* On the basis of presently existing data, Urry (48) proposed a structure for the gramicidin A–induced channel. In this model two gramicidin A molecules folded in a $\pi_{(LD)}$ helix dimerize to form a rod-like structure having a lipophilic exterior and an internal channel lined with carbonyl oxygen residues. The oxygen residues provide a polar pathway favorable to cations, in accord with the experimentally observed gramicidin A–induced cationic membrane selectivity.

---

* In the presence of carriers such as nonactin, discrete conductance steps are neither expected nor experimentally observed (46).

Relatively large transmembrane channels should provide essentially aqueous pathways for ionic as well as neutral solute permeation. Such channels appear to be formed by complexing several molecules of polyene antibiotics, such as amphotericin B, with cholesterol-containing bilayer membranes. (For a lucid review of this system, see ref. 49.) As the number of channels is increased (by increasing the aqueous polyene concentration), a direct proportionality is observed between the flux of small neutral molecules and the conductance due to small ions, indicating that both neutral and charged molecules penetrate the membrane through the channels. The radius of the amphotericin B–induced channels has been estimated to be near 4 Å, since only those neutral solutes or ions whose diameter is less than about 7 Å can penetrate the channel. The amphotericin B–treated bilayer membrane is an excellent model, which mimics almost quantitatively the permeability properties of biological membranes to small molecules which are known not to penetrate by a simple solubility-diffusion mechanism.

An interesting compound (50) which, judged by the criteria of quantal conductance charges, also forms channels across the bilayer (51, 52) is the so-called excitability-inducing material (EIM) of Mueller and Rudin (50). This proteinaceous bacterial product of unknown composition has the remarkable property of inducing strongly voltage-dependent ionic conductances in bilayer membranes (i.e., the membrane conductance is practically suppressed when an electrical potential exists across the bilayer), and under favorable conditions the bilayer properties can mimic those of excitable membranes. Unfortunately, the molecular basis of the EIM function has not yet been elucidated.

CONCLUSIONS

Bilayer membranes are models having the dimensions and constituents of cell membranes. The bilayer composition is well defined and can be varied in order to test the validity of hypotheses concerning the function of natural membranes (e.g., the effects of fluidity of the membrane interior or membrane surface charges). Furthermore, studies of bilayer permeability model systems, such as those described in this chapter, not only provide a variety of possible permeation mechanisms but also reveal some basic parameters that govern the interaction of solute molecules with the membrane.

PERSPECTIVE

The prime role of membranes has long been recognized as one of providing compartments of distinct composition and of regulating the exchange of solutes between the interior and exterior of such compartments. But membranes are also involved in other vital cell processes such as the detection of extracellular signals (e.g., synaptic receptors or hormone receptors), the propagation of signals (e.g., nerve and muscle action potentials), and the translation of signals (e.g., excitation-contraction coupling in muscle or setting of the cyclase activity in the cyclic adenosine monophosphate system). Recently, a role of membranes has been suggested in such basic cell processes as DNA replication (53) and protein synthesis (54) as well as in a large number of enzyme systems that function only in association with membranes (e.g., $Na^+,-K^+$-activated ATPase or Ca^{++} sequestering by sarcoplasmic reticulum).

Despite the fundamental involvement of membranes in cell function, virtually nothing is known about the molecular mechanisms of membrane-associated functions. At present, the prime (and yet unattained) goal of model studies is the reconstitution of functional membrane-bound systems in order to elucidate the molecular basis of their function.

REFERENCES

1. Stoeckenius, W., and Engelman, D. M. *J. Cell. Biol.* **42,** 613 (1969).
2. Luzzati, V. *Biological Membranes.* Ed. D. Chapman. Academic Press, New York, 1968.
3. Levine, Y. K., and Wilkins, M. H. F. *Nature New Biol.* **230,** 69 (1971).
4. Chapman, D., Byrne, P., and Shipley, G. G. *Proc. R. Soc.* **290A,** 115 (1966).
5. Chapman, D., Ladbrooke, B. D., and Williams, R. M. *Chem. Phys. Lipids* **1,** 445 (1967).
6. Chapman, D., and Wallach, D. F. H. *Biological Membranes.* Ed. D. Chapman. Academic Press, New York, 1968.
7. McConnell, H. M. *The Neurosciences: Second Study Program.* Ed. F. O. Schmitt. Rockefeller University Press, New York, 1970.
8. Chan, S. I., Feigelson, G. W., and Seiter, C. H. A. *Nature* **231,** 110 (1971).
9. Mueller, P., Rudin, D. O., Tien, H. T., and Wescott, W. C. *Circulation* **26,** 1167 (1962).
10. Mueller, P., and Rudin, D. O. *Curr. Top. Bioenergetics* **3,** 157 (1969).
11. Mueller, P., and Rudin, D. O. *Laboratory Techniques in Membrane Biophysics.* Ed. H. Passow and R. Stämpfli. Springer, New York, 1969.
12. Tien, H. T., and Diana, A. L. *Chem. Phys. Lipids* **2,** 55 (1968).
13. Thompson, T. E., and Henn, F. A. *Membranes of Mitochondria and Chloroplasts.* Ed. E. Racker. Van Nostrand Reinhold, New York, 1970.
14. Taylor, J., and Haydon, D. A. *Discuss. Faraday Soc.* **42,** 51 (1966).
15. Hopfer, U., Lehninger, A. L., and Lennarz, W. J. *J. Membrane Biol.* **2,** 41 (1970).
16. McLaughlin, S. G. A., Szabo, G., Eisenman, G., and Ciani, S. M. *Proc. Natl. Acad. Sci. U.S.A.* **67,** 1268 (1970).

17. Adam, N. K. *The Physics and Chemistry of Surfaces.* Dover, New York, 1968.
18. Fettiplace, R., Andrews, D. M., and Haydon, D. A. *J. Membrane Biol.* **5**, 277 (1971).
19. Szabo, G. Unpublished observations.
20. Ohki, S., and Fukuda, M. *J. Theoret. Biol.* **15**, 362 (1967).
21. Huang, C., Wheeldon, L., and Thompson, T. E. *J. Mol. Biol.* **8**, 148 (1964).
22. Bangham, A. D. *Progress in Biophysics and Molecular Biology.* Ed. J. A. V. Butler and D. Noble. Pergamon Press, New York, 1968.
23. Huang, C. *Biochemistry* **8**, 344 (1969).
24. Finkelstein, A., and Cass, A. *J. Gen. Physiol.* **52**, 145s (1968).
25. Davies, J. T., and Rideal, E. K. *Interfacial Phenomena.* Academic Press, New York, 1963.
26. Steinemann, A., and Läuger, P. *J. Membrane Biol.* **4**, 74 (1971).
27. McLaughlin, S. G. A., Szabo, G., and Eisenman, G. *J. Gen. Physiol.* **58**, 667 (1971).
28. Szabo, G., Eisenman, G., McLaughlin, S. G. A., and Krasne, S. *Ann. N.Y. Acad. Sci.* In press.
29. Papahadjopoulos, D. *Biochim. Biophys. Acta* **163**, 240 (1968).
30. Steinemann, A., Alamuti, N., Brodmann, W., Marschall, O., and Läuger, P. *J. Membrane Biol.* **4**, 284 (1971).
31. Läuger, P., and Neumcke, B. *Membranes: A Series of Advances.* Ed. G. Eisenman. Dekker, New York, vol 2. In press.
32. LeBlanc, O. H. *Biochim. Biophys. Acta* **193**, 350 (1969).
33. Diamond, J. M., and Wright, E. M. *Annu. Rev. Physiol.* **31**, 581 (1969).
34. Papahadjopoulos, D. *Biochim. Biophys. Acta* **241**, 254 (1971).
35. Rosenberg, T., and Wilbrandt, W. *Exp. Cell Res.* **9**, 49 (1955).
36. Ciani, S., Eisenman, G., and Szabo, G. *J. Membrane Biol.* **1**, 1 (1969).
37. Läuger, P., and Stark, G. *Biochim. Biophys. Acta* **211**, 458 (1970).
38. Markin, V. S., Krishtalik, L. I., Liberman, E. A., and Topaly, V. P. *Biofizika* **14**, 256 (1969).
39. LeBlanc, O. H. *J. Membrane Biol.* **4**, 227 (1971).
40. Eisenman, G., Szabo, G., Ciani, S., McLaughlin, S. G. A., and Krasne, S. *Progress in Surface and Membrane Science.* Eds. J. F. Danielli, M. D. Rosenberg, and D. A. Codenhead. Academic Press, New York. In press.
41. Eisenman, G., Ciani, S., and Szabo, G. *J. Membrane Biol.* **1**, 294 (1969).
42. Szabo, G., Eisenman, G., and Ciani, S. *J. Membrane Biol.* **1**, 346 (1969).
43. Stark, G., and Benz, R. *J. Membrane Biol.* **5**, 133 (1971).
44. Pressman, B. C. *Membranes of Mitochondria and Chloroplasts.* Ed. E. Racker. Van Nostrand Reinhold, New York, 1970.
45. Krasne, S., Eisenman, G., and Szabo, G. *Science* **174**, 412 (1971).
46. Hladky, S. B., and Haydon, D. A. *Nature* **225**, 451 (1970).
47. Hladky, S. B., and Haydon, D. A. Studies of the unit conductance channel of gramicidin A. Manuscript, 1972.
48. Urry, D. W. *Proc. Natl. Acad. Sci. U.S.A.* **68**, 672 (1971).
49. Finkelstein, A., and Holz, R. *Membranes: A Series of Advances.* Ed. G. Eisenman. Dekker, New York, vol. 2. In press.
50. Mueller, P., and Rudin, D. O. *J. Theor. Biol.* **18**, 222 (1968).

51. Bean, R. C., Shepherd, W. C., Chan, H., and Eichner, J. *J. Gen. Physiol.* **53,** 741 (1969).
52. Ehrenstein, G., Lecar, H., and Nossal, R. *J. Gen. Physiol.* **55,** 119 (1970).
53. Watson, J. D. *Molecular Biology of the Gene.* Benjamin, New York, 1970.
54. Hendler, R. W. *Protein Biochemistry and Membrane Biochemistry.* Wiley, New York, 1968.

7

Alan F. Horwitz

Nuclear Magnetic Resonance Studies on Phospholipids and Membranes

The use of nuclear resonance in the studies of biological systems has grown rapidly in the last few years. Part of this growth has been in the use of nuclear magnetic resonance (NMR) to study membranes and their components. This review discusses almost exclusively the studies on phospholipids, membranes, and membrane proteins; work on solid lipids (1–5), on serum lipoproteins (6–10), or on cell water (11–19) is not considered. The review does not cover the complete literature, but rather selected important and representative studies concerning the applications of NMR or the interpretation of NMR spectra and the subsequent conclusions.

SELECTED PRINCIPLES OF NMR

Chemical Shifts (20–23)

A nucleus with a non-zero magnetic moment will interact with electro-

The original work discussed in this review was supported in part by a postdoctoral fellowship from the National Institutes of Health and in part by funds from the U.S. Atomic Energy Commission. The discussions with and comments of Dr. M. P. Klein, Dr W. J. Horsley, and Mr. Nelson Teng were extremely helpful.

magnetic radiation in the presence of an external magnetic field and absorb energy of frequency v according to Equation 1:

$$v = \frac{\mu H_0}{hI} = \frac{\gamma}{2\pi} H_0 \tag{1}$$

where μ is the nuclear magnetic moment, H_o is the external magnetic field strength, h is Planck's constant, I is the nuclear spin, and γ is the gyromagnetic ratio. Since each nucleus has its own characteristic gyromagnetic ratio, γ, for a given value of magnetic field different nuclei will absorb at different frequencies. The biologically important nuclei with nonvanishing magnetic moments are listed along with their nuclear spin, magnetic moment, resonant frequency, and relative sensitivity in Table I.

Table I

Characteristics of Some Biologically Important Nuclei with Nonvanishing Magnetic Moments

ISOTOPE	SPIN	NMR FREQUENCY (MHz FOR 10 = kG)	NATURAL ABUNDANCE	RELATIVE SENSITIVITY AT CONSTANT FIELD	MAGNETIC MOMENT (μ, IN UNITS OF eh/2Mc)
^{1}H	1/2	42.5759	99.985	1.00	2.79268
^{2}H	1	6.53566	1.5×10^{-2}	9.65×10^{-3}	0.857387
^{13}C	1/2	10.7054	1.108	1.59×10^{-2}	0.702199
^{14}N	1	3.0756	99.63	1.01×10^{-3}	0.40347
^{15}N	1/2	4.3142	0.37	1.04×10^{-3}	-0.28298
^{19}F	1/2	40.0541	100	0.833	2.62727
^{31}P	1/2	17.235	100	6.63×10^{-2}	1.1305

Data from Becker, E. *High-Resolution NMR.* Academic Press, New York, 1969.

The local magnetic field seen by a given nucleus is determined by the external magnetic field which is modified by local molecular shielding effects leading to chemical shifts, by the dipole interactions discussed in the next section, and by a through-the-bond coupling with neighboring nonequivalent nuclei (which is not of interest for the work discussed in this review). The chemical shift for a given nucleus is generally expressed with respect to an internal or external reference compound. For protons, tetramethylsilane (TMS) and 3-(trimethylsilyl)-1-propane sulfonic acid sodium salt (TSS) are the most commonly used standards. The chemical shift is generally represented by the symbol δ, expressed in parts per million (ppm), defined by Equation 2:

$$\delta = \frac{v_S - v_R}{v_R} \times 10^6 \tag{2}$$

where v_R is the value of the resonant frequency for the reference and v_S is that for the sample. Some authors express the chemical shift on a τ scale where τ is defined by Equation 3:

$$\tau = 10.000 - \delta \tag{3}$$

It should be readily apparent from Equations 1 and 2 that the chemical shift is field-dependent; therefore, the use of higher field strengths gives better chemical shift resolution.

Nuclear Spin Relaxation (20–23)

The shape and width of an NMR absorption are generally determined by *1)* the lifetime of the spin states, *2)* the dipole interactions with neighboring nuclei, *3)* magnetic field inhomogeneities (due to the sample as well as to the external magnetic field), *4)* interaction with paramagnetic substances, *5)* electric quadrupole coupling, *6)* nuclear spin-rotational coupling, and *7)* chemical shift anisotropy. Since most of the NMR studies of membranes have been with protons, and since only the first three contributions to the line width are of major importance, the following discussion considers the first three contributions using protons as an example.

The lifetime of the spin states is described by the spin-lattice relaxation time, T_1. In the absence of a magnetic field, at equilibrium, the two (for protons) spin states are equally populated. When the external magnetic field is turned on, the time necessary for the populations of the two spin states to come to $1-1/e$ of their new equilibrium value is equal to T_1.

The major mechanism contributing to the process of spin-lattice relaxation is transitions between spin states induced by nearby fluctuating magnetic dipoles with a frequency at or very near the resonant frequency of the nucleus. Since most proton magnetic resonance (PMR) experiments are performed at frequencies between 60 and 220 MHz (1 Hz = 1 cps), the random relative motion of the molecules provides fluctuating magnetic dipoles (from neighboring nuclei) at these frequencies.

A collection of randomly tumbling molecules at any given time has a distribution of components of rotational and relative translational motion from zero frequency out to a maximum frequency that is related to the inverse of the correlation time, $1/\tau_c$. The value of the rotational correlation time for a spherical molecule depends on the absolute temperature, T, the viscosity in poise, η, and the particle radius in centimeters, a, and can be estimated by the Stokes relation shown in Equation 4 (20, 22):

$$\tau_c = \frac{8\pi\eta a^3}{6KT} \tag{4}$$

At 20°C water has a correlation time of the order of 10^{-12} sec. A decrease in the rate of tumbling motion from $1/\tau_c$ to $1/\tau_c'$ (by lowering the temperature, for example) serves to increase the intensity of frequency components for all the frequencies below $1/\tau_c'$ and to decrease the intensity of the frequency components above $1/\tau_c'$. As a consequence the nuclei in the molecule are more efficiently relaxed by the decrease in tumbling rate if the resonant frequency, ν_0, is below $1/\tau_c'$ and are less efficiently relaxed if ν_0 is above $1/\tau_c'$. The most effective relaxation is when $2\pi\nu_0 = 1/\tau_c$.

The source of the fluctuating magnetic dipoles at these frequencies is generally neighboring nuclei that possess a magnetic moment. This dipole interaction has an r^{-6} distance dependence, and therefore intramolecular contributions are often larger than intermolecular contributions to the relaxation time. In addition to the dependence on distance, there is a dependence on the strength of the magnetic dipole moment; therefore, molecules with larger magnetic dipole moments are more efficient in relaxing nuclei. Paramagnetic molecules, which have a magnetic dipole moment several orders of magnitude larger than that of protons, can be extremely effective in relaxing nuclei.

The second major determinant of the line width is low-frequency dipole interactions between neighboring nuclei possessing a non-zero magnetic moment. The main effect of this interaction is to alter the value of the field seen by a given nucleus from that of the external field, e.g., $H_{1oc} = H_o + H_d$, where H_{1oc} is the local field, H_o is the external field, and H_d is the dipole field from neighboring nuclei. A proton at a distance of 1 Å from another proton broadens its absorption line by a value varying from zero to about 10^5 Hz depending on the angle, θ, between the external magnetic field vector and the vector joining the two nuclei. As in the case of spin-lattice relaxation, there is a dependence on the value of the nuclear magnetic dipole moment such that nuclei with larger magnetic dipole moments are more effective in broadening a given resonance line. The dipole interaction between nuclei is described by a relaxation time, T_{2d}, that can be estimated from the line width using the uncertainty principle.

Solids at very low temperatures have very broad line widths owing to the effects mentioned above. The line widths in solids are often studied by the method of second moments (20, 22, 24). The effect of molecular motion is to time-average the angular dependence of the low-frequency dipole interactions, thus greatly reducing the effect on the line width. When the rate of rotational motion, i.e., the tumbling time, is much greater than the maximum dipole-dipole contribution to the line width, these interactions are effectively averaged, and therefore the line widths are relatively narrow (and T_{2d} is relatively long). Since there are always components of motion near zero frequency, the T_{2d} relaxation time continues to increase with increased tumbling motion, $1/\tau_c$. In fact, for rapid motion such that $1/\tau_c \gg 2\pi\nu_0$, the intensity of the frequency components at low frequency are such that

$T_{2d} = 2T_1$. In many instances the extent of motional narrowing can be estimated by using Equation 5:

$$\Delta v \cong 2\pi \, \Delta v_0^2 \, \tau_c \tag{5}$$

where Δv is the line width (in Hz), Δv_0^2 is the rigid lattice second moment (Hz^2), and τ_c is the correlation time of the motion (22).

The third major contribution to the line width is magnetic field inhomogeneity. The local field seen by a given nucleus in the sample can be modified by external magnetic field gradients, magnetic susceptibility gradients within the sample, and anisotropies (orientation dependency) in the chemical shifts. The first of these contributions causes nuclei in different parts of the sample tube to experience different external magnetic field strengths. Generally this contribution is less than 0.5 Hz. The second contribution is due to gradients in the diamagnetic (or paramagnetic) susceptibility of the sample. This is a field-dependent parameter that can also result from a geometrically dependent diamagnetic susceptibility. The final contribution is due to geometrically dependent values of the chemical shift. This last contribution is generally seen only in solid, solid-like, or liquid crystalline samples, and may affect the line shape as well as the line width (22). When the tumbling rate of the molecules is much greater than the maximum value of the anisotropy in the diamagnetic susceptibility or the maximum value of the anisotropy in the chemical shift, the nucleus experiences only an averaged value. An effective relaxation time, T_{2m}, for these processes can be estimated from the contribution to the line widths using the uncertainty principle.

The resonant line width, therefore, is influenced by several factors, including high-frequency motions, characterized by the spin-lattice relaxation time, T_1; low frequency motions, characterized by the relaxation time T_{2d}; and the various contributions to magnetic field inhomogeneities caused by the external magnetic field and by the nature of the sample itself, characterized by the relaxation time T_{2m}. The total line width is related to the transverse relaxation time, T_2, by Equation 6:

$$\frac{1}{T_2} = \pi \Delta v = \frac{1}{2T_1} + \frac{1}{T_{2m}} + \frac{1}{T_{2d}} \tag{6}$$

where Δv for a Lorentzian line shape is the peak width at one-half maximum amplitude and for a Gaussian line shape is the peak width at maximum slope.

Experimental Methods (21, 23)

The most common method of obtaining an NMR spectrum is to use a spectrometer that operates at a fixed frequency, e.g., 60, 100, or 220 MHz for protons, and has a variable external magnetic field strength that is swept through the resonance condition defined by Equation 1. The result is a display of absorption versus field strength. Since the spin system is constantly being irradiated at one of the above frequencies throughout the experiment, this method is often called continuous wave or cw (21, 23).

An alternative method of obtaining an NMR spectrum is called pulsed NMR (21, 25, 26). In this method the spin system is irradiated by a short, intense pulse, and the resulting output is the free induction decay (21). By performing a mathematical operation called a Fourier transformation, an output can be displayed that is equivalent to that obtained by cw methods (21). The great advantage of the pulse method of experimentation is that the time required to obtain an NMR spectrum is often hundreds of times less than that required by cw methods. This type of experiment is called Fourier-transform NMR.

The sensitivity of an NMR experiment depends on the external field strength used and the computer facilities available (25). At 220 MHz a resonance can generally be seen with a 0.4-ml sample at 1–5 mM. This sensitivity can be further increased by time averaging several scans on a computer of averaged transients (CAT). After N scans the signal has increased by N, whereas the noise has increased by only $N^{1/2}$, and the resulting increase in signal-to-noise is $N^{1/2}$. When the total range of chemical shifts is large compared with the individual line widths, Fourier-transform NMR can give enhancements of 10–100 in signal-to-noise over cw methods (25, 26).

The value of the transverse relaxation time, T_2, may be easily measured from the line width obtained in a cw experiment. Equivalently this value may often be obtained from a pulsed experiment (21). The value of T_2 obtained by these methods includes contributions from T_1, T_{2d}, and T_{2m} processes. A spin-echo experiment can eliminate the contribution of T_{2m} to T_2 and therefore is a method of choice for obtaining the dipole contribution to the relaxation time (28, 29). There are several methods of obtaining the T_1 relaxation time (21, 28, 30, 31). The contribution of anisotropic chemical shifts or diamagnetic susceptibility to T_2 can often be assessed by the field dependence of the line width as well as by spin-echo experiments (21, 22). It should be pointed out that the data obtained from these parameters for nuclei in biological molecules, e.g. phospholipids or membranes, can give a great deal of information on the nature of the structural dynamics influencing the nuclei under investigation.

STUDIES ON DISPERSIONS OF PHOSPHOLIPIDS IN D_2O

Effect of Sonication on PMR Line Widths

If a thin film of phospholipids on the side of a flask is dispersed into water by gentle agitation, the result is a set of concentric bilayers (32, 33). The diameter of these large, coarse lipid globules is usually ~ 0.5–5 μ.

NMR study of these dispersions shows a broad (400–500 Hz), poorly resolved spectrum (34, 35). Sonication or extensive homogenization produces a well-resolved spectrum with relatively narrow resonances (Fig. 1) (34–36). The reason for the appearance of resolved resonances with sonication is not yet clear; the following evidence relates to this problem.

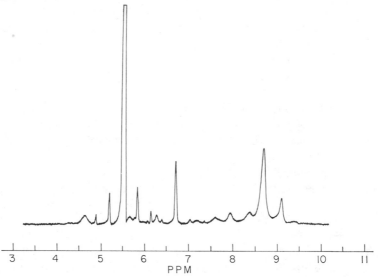

Figure 1

220-MHz proton magnetic resonance spectrum of sonicated egg yolk lecithin at 60°C in 0.15 M KCl in D_2O. The chemical shifts are with respect to TSS at $\tau = 10.00$. The resonances at $\tau = 9.1$, 8.7, 6.6, and 5.4 are from the fatty acid terminal methyl, fatty acid methylene, choline N-methyl, and HDO protons. The sharp resonances at $\tau = 5.8$, 6.2, 5.2, and 4.8 are spinning side bands from the water resonances. (See Table IV for the other assignments.)

Using results from model lipid systems and solid lipids, as well as studies on aqueous dispersions of phospholipids, several authors (19, 35, 37, 38) have suggested that the broad lines seen in coarse lipid dispersions of egg lecithin and other lipids are due in large measure to internal magnetic field gradients (37) and possible diffusion within these gradients (38). The evidence that the broad line widths in coarse lipid dispersions are indeed due primarily to non-dipolar contributions comes from the following findings. Penkett *et al.* (35) saw field-dependent line widths in solid egg yolk lecithin (80%), D_2O (20%) systems. Further, the line width went through sharp changes with temperature with no concomitant change in the thermal relaxation time. Steim (37) using pulsed PMR on coarse dispersions of erythrocyte lipids in D_2O, obtained, from the free induction decay, a T_2 of 0.56 ± 0.3 msec, and from the spin-echo sequence (28), a T_2 of 110 ± 10 msec. These corresponded to line widths of 1100 and 5.8 Hz, respectively. This work did not identify which resonances contributed to this relaxation time, nor did it indicate the fraction of the total protons that contributed to this relaxation time except to say that there seemed to be only one major relaxation time. Pulsed NMR studies on surfactants (38) indicated that in this system diffusion within local field gradients plays an important role in determining the relaxation time obtained from the free induction decay. Recently, evidence has been presented suggesting that the line width from coarse lecithin dispersions is largely dipolar in origin. Chan *et al.* (39) have described a preparation of lecithin dispersions in which the transverse relaxation rates are not field-dependent and are largely dipolar in origin. Further, Tiddy (40) has obtained results from a liquid crystal system that suggests that there is no rapid diffusion through gradients and that the observation made by other workers (38) of such a rapid diffusion may be an artifact.

If the line widths were due largely to local field gradients, then the effect of sonication could be to produce particles small enough that their tumbling time, $1/\tau_c$, would be greater than the maximum value of the magnetic susceptibility gradients, of the chemical shift anisotropy, or of any other factor that could contribute to the proposed local magnetic field inhomogeneities (34, 35). The effect of sonication could also be to produce a bilayer that is internally mobile so that the motion of the lipid protons themselves have correlation times short enough to produce the same effect (36). When the tumbling rate is much greater than the maximum value (in hertz) of the local magnetic field inhomogeneities, then that contribution to the line width is greatly reduced. Similarly, if the line width were mainly due to dipolar interactions, then tumbling rates much greater than the maximum dipole interaction would serve to reduce greatly the dipole contribution to the line width.

Estimates of the maximum line widths of the phospholipid proton resonances are about 7×10^4 Hz. Therefore, motions at rates greater than this number narrow greatly the PMR resonance lines. Table II shows estimates of the correlation time for spherical particles using Equation 4. Assuming that the maximum line width is largely dipolar and using Equation 5, it is readily seen that the tumbling of particles as small as 250 Å in diameter will not be very effective in narrowing the line width.

The early work of Attwood and Saunders (41) using light scattering showed that one effect of sonication was to break the large multilamellar structures found in the coarse lipid dispersions into smaller particles, and that after extensive sonication the resulting structures were small (~250 Å), single-lamellar structures (42). Chapman (34, 35) found an exponential increase in PMR signal intensity (with no change in line width) upon sonication

Table II

Inverse Correlation Time for Ideal Spheres of Different Radii Using the Stokes Equation

$a\ (cm \times 10^8)$	$1/\tau_c$
10,000	1.0
1,000	1.0×10^3
100	1.0×10^6

$n = 0.01$ poise; $T = 293°$ K.

that leveled off after 20 min for the resolved protons $(CH_2)_n$ and N^+—$(CH_3)_3$. Electron microscopy and x-ray studies on these sonicates showed that the particles resulting from 20 min of sonication were less than 1000 Å in diameter and still retained their closed lamellar structure (34). Sheard (36) also found that sonication produced an increase in intensity with a constant line width (suggesting more contributing protons), but claimed that it reached a maximum after 3 min which did not change with further sonication. He found further that he could account for only $88 \pm 5\%$ of the total choline and fatty acyl methylene protons in the narrow, resolved resonances. He concluded that the particle size after only 3 min of sonication was too large to permit averaging out the local field gradients presumably responsible for the broad, unresolved resonances in the unsonicated phospholipid dispersion. He suggested, however, that an internally mobile structure was acquired below a certain critical size and that these motions were responsible for the appearance of narrow, resolved resonances. Unfortunately, the work of Sheard does not give data on the sizes of, or the distribution of sizes of, particles he studied.

The effect of sonication has not yet been investigated adequately. Neither has it been adequately established what is the major contribution to the line width of unsonicated coarse lipid dispersions; and no satisfactory theoretical explanation of the phenomenon has been offered. An examination of a sample with homogeneous particle size like that described by Huang (42) may be useful in studying this problem. It appears that the phospholipid dispersions described by Chan *et al.* (39) may be of great value once the structure and properties of the system have been investigated.

The high-resolution spectra of sonicated dispersions of phospholipids in D_2O are characterized by relatively sharp resonances (Fig. 1). PMR studies at 60, 100, and 220 MHz indicated that these line widths were field-dependent (36). These field dependencies were interpreted as due to variations in diamagnetic susceptibility within the sample. To eliminate the effects of the diamagnetic susceptibility, the line-width data were extrapolated to zero field, giving the values shown in Table III. Thus, it appears that even in sonicated dispersions of egg yolk lecithin, the line widths are not entirely dipolar. Hopefully, more experimental work and a theoretical discussion of the line-width contributions will be forthcoming.

Table III

Line Widths of Lecithin Protons Extrapolated to Zero Field

PROTON	LINE WIDTH (Hz)
$-N^+(CH_3)_3$	2.2
$-(CH_2)_n-$	8.5
$-CH_3$	11.7

Data from Sheard, B. *Nature* **223**, 1057 (1969).

At this point, therefore, one must be cautious of discussions of line-width alterations due to various perturbations in sonicated lipid dispersions. It will be necessary to distinguish those contributions due to dipolar and those due to other factors, if any, to understand fully the meaning of the line-width changes.

Chemical Shift Assignments (43)

The fatty acid and glycerol methylene resonances were assigned by referring to studies on the triglycerides, tristearin and triolein. Ethanolamine resonances were assigned by referring to studies on *o*-phosphoethanolamine. Choline resonances were assigned by referring to studies with synthetic and natural phosphatidylcholines as well as studies with simple choline derivatives. Phosphatidylserine resonances were assigned by comparison with phosphatidylcholine and phosphatidylethanolamine. The results are shown in Table IV, and are expressed in parts per million with respect to TMS at $\tau = 10$ for studies in $CDCl_3$ or to TSS at $\tau = 10$ for studies in D_2O.

Continuous Wave Studies on Phospholipid Dispersions

Probably the most significant single result from cw PMR studies on phospholipid dispersions is that the fatty acid protons and N-methyl protons of the choline are in a highly mobile state (36, 44). This is based on the relatively narrow line widths seen for these resonances. Solid phospholipids at low temperatures have line widths of the order of 7×10^4 Hz, phospholipids sonicated in D_2O have line widths of about 15–30 Hz (36), unsonicated phospholipids in D_2O have line widths of about 1000 Hz, while fatty acids in organic solvents have line widths usually less than 1 Hz (43). The work of

Table IV

Chemical Shift Assignments (in Parts per Million) for Phospholipid Protons[a]

PROTON	TRIOLEIN	DIOLEOYLPHOS-PHATIDYL ETHANOLAMINE	EGG YOLK LECITHIN[b]	MIXED ACYL PHOSPHOTIDYL SERINE	O-PHOSPHO-ETHANOLAMINE[b]	O-PHOSPHO-CHLORINE CHLORIDE[b] Ca++ SALT
			Fatty Acid Protons			
—CH$_3$	9.11	9.12	9.11	9.12		
—CH$_2$—	8.7	8.7	8.74	8.7		
+CH$_2$CH=	~8.0	7.95	8.0	7.95		
—CH$_2$CO—	7.71	7.69	7.71	7.71		
—CH=CH—	4.63	4.60	4.63	4.65		
		Glycerol Methylene and Polar Headgroup				
—CH$_2$NH$_2$		6.70			6.71	
—CH$_2$N$^+$Me$_3$			~6.1			6.3
—N$^+$Me$_3$			6.65[c]			6.75
—CH$_2$OCO	5.74	5.89	~6.1	5.9		
—CH$_2$OPO		5.67	5.7	5.9		
—CHOCO	~4.8	~4.8	~4.8	~4.8		
N$^+$H$_3$		1.67				

[a] All assignments are in CDCl$_3$, TMS at $\tau = 10.0$, unless otherwise specified.
[b] In D$_2$O (TSS at $\tau = 10.0$).
[c] $\tau = 6.77$ in D$_2$O.
+CH=CH—CH$_2$—CH$_2$—CH=CH, 7.19 ppm.
From Chapman, D., and Morrison, A. *J. Biol. Chem.* **241**, 5044 (1966).

Chan *et al.* (39) is interesting in that they saw a heterogeneity in the line width in their preparation of unsonicated lecithin, suggesting that not all the protons in the hydrocarbon region of the bilayer undergo the same motion (Fig. 2). This is similar to the conclusions reached by McConnell and his coworkers (45) using spin labels. It remains clear that even though a theoretical description of the fatty acid chain motion that accounts for the observed NMR line shapes and widths has not been presented, at least some portion of the fatty acid chains are executing a relatively rapid motion (39). The line width provides (with the reservations mentioned above) a simple estimate of the mobility of and the interactions between the fatty acid chains. The following discussion illustrates some of the results obtained by studying changes in the PMR line width.

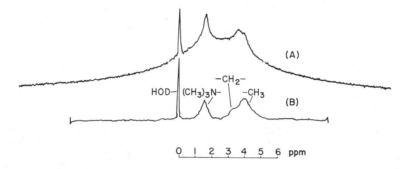

Figure 2

High-resolution proton magnetic resonance spectrum of a lecithin bilayer sample: *A*, 220-MHz spectrum at 18°C, and *B*, 100-MHz Fourier-transformed spectrum at 30°C. In *B*, any broad component, characterized by a T_2 shorter than 250 μsec (or a line width greater than ~1000 Hz) will not contribute to this spectrum. (From Chan, S.I., *et al. Nature 231*, 110, 1971.)

An important property of the lecithins studied so far is the endothermic transition that has been discussed and studied extensively (46, 47). These transitions have also been seen in plots of PMR line width versus temperature (49). For dipalmitoyl L-α-lecithin, the fatty acid methyl and methylene resonances exhibited a large broadening as the temperature was reduced below 41°C, the transition temperature seen by calorimetric techniques. In contrast, the methyl resonances of the choline headgroup exhibited a much smaller broadening at temperatures below 41°C (Fig. 3). These NMR studies at 220 MHz are consistent with the idea that the transition involves an alteration of the headgroup region as well as a melting of the fatty acid chains (45–48).

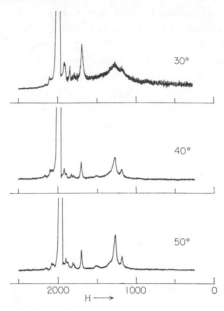

Figure 3

220-MHz proton magnetic resonance spectrum of sonicated dipalmitoyl L-α-lecithin at three temperatures. The abscissa is expressed in frequency units, Hz, with respect to an arbitrary zero.

A 1:1 mole mixture of sonicated egg yolk lecithin and cholesterol gave an NMR spectrum that showed a differential broadening of the resonances when compared with a sample without cholesterol (44). The N-methyl of the choline was broadened to a very small extent compared with that of the fatty acid methylene region (Fig. 4). This effect was interpreted in terms of one of several possibilities: an increase in solution viscosity, an inhibition of the hydrocarbon chain interactions, or an inhibition of chain motion. The latter is the most likely explanation.

Alamecithin, valinomycin, and gramicidin S are all cyclic antibiotics. The first two are believed to be cation carriers exhibiting their effect at nanomolar concentrations in liposomes as well as in mitochondria (50–52). In the absence or presence of salt, alamecithin and valinomycin greatly broadened the resonances of the hydrocarbon region and broadened the N-methyl resonances of the headgroup to a much lesser extent (53). This effect was also seen with phosphatidylserine at mole ratios of lipid to alamecithin as high as 600:1. Although gel filtration and ultracentrifugation studies showed that at lecithin-alamecithin ratios of 10:1 the particle size had increased by a

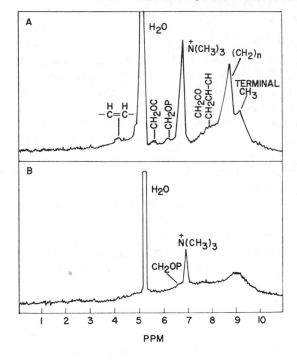

Figure 4

**High-resolution nuclear magnetic resonance spectrum at 60 Hz.
A. Egg yolk lecithin dispersed in D_2O. *B*. 1:1 molar egg yolk lecithin–cholesterol mixture dispersed in D_2O. (From Chapman, D.,
and Penkett, S.A., *Nature 211*, 1304, 1966.)**

factor of five, studies with phosphatidylserine and alamecithin at a ratio
(100:1) which produced a broadening similar to that seen with lecithin
showed no significant increase in particle size, suggesting that the broadening
was not caused by the increase in particle size. X-ray studies on the lecithin-
alamecithin system indicated that there was a gradual breakdown in the
lamellar structure with increasing concentrations of alamecithin. Gramicidin
S sonicated with egg yolk lecithin did not show any line broadening; the
x-ray evidence again indicated a breakdown in the bilayer structure. Further,
unlike alamecithin, the addition of gramicidin S to unsonicated lecithin dispersions produced resolved phospholipid resonances. These results were
interpreted as suggesting a different interaction between alamecithin and
gramicidin S with lecithin. The former produced a particle with hindered
mobility of the fatty acid region of the phospholipids, while the latter produced a smaller particle with a more mobile fatty acid region (53).

Pulsed PMR Results

Chan *et al.* (39) have investigated the T_1 and T_2 relaxation times on an egg yolk lecithin dispersion in D_2O. With this system, there was no field dependence of the line widths, and the free induction decay was used to measure the spin-spin relaxation time, T_2. They found that the free induction decay was not exponential, suggesting to the authors a distribution of T_2 values. In particular, for the methylene resonances, there was a distribution of relaxation times which was interpreted as a dynamic heterogeneity along the fatty acid chains. While nearly all the terminal methyl protons of the fatty acid chains had a T_2 longer than 250 μsec, only a small percentage of the methylene protons did.

The T_1 of this egg lecithin system was studied using a standard 180–90° pulse sequence (28). A T_1 of 220 msec at 30°C and 60 MHz was found for all the protons contributing to the bulk magnetization. To explain this result, a spin diffusion mechanism was proposed in which one part of the molecule acts as a heat sink to keep the spin system in thermal equilibrium by the efficient spin-spin coupling (22). Therefore, if this mechanism is correct, the T_1 relaxation time will provide information only on that part of the chain responsible for the relaxation. Studies by Chapman (35) on sonicated egg yolk lecithin, using recovery from saturation (30), also indicated one main T_1 relaxation time for the methylene region. The relaxation times at 60 MHz and 33.4°C are shown in Table V.

Table V

Spin-Lattice Relaxation Times of Egg Yolk Lecithin Protons at 60 MHz and 33.4°C

LIPID	SOLVENT	T_1 (msec)
Egg lecithin	Chloroform	1610
Egg lecithin	D_2O (sonicated)	300
Egg lecithin	Methanol	1390

From Penkett, S. A., *et al. Chem. Phys. Lipids* **2**, 273 (1968).

Using Fourier transform capabilities, the spin-lattice and transverse relaxation times of the choline N-methyl and of the fatty acid α-carbonyl, allyl, vinyl, methylene, and methyl protons of sonicated egg yolk lecithin have been determined (29). The values of T_1 were similar to but different from one another; this along with other evidence suggested that, in sonicated egg lecithin, the relaxation rates of the fatty acid protons are not determined solely by spin-diffusion to a heat sink. The methylene protons appeared, however, to relax according to a single exponential. Arrhenius plots of the

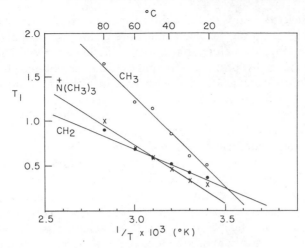

Figure 5

Variation of T_1 for the fatty acid CH$_3$, fatty acid CH$_2$, and choline N-methyl protons as a function of temperature in sonicated, EDTA-treated, deoxygenated samples of egg yolk lecithin dispersed in 0.15 M KCl in D$_2$O by 20 min of sonication.

T_1 values, which increased with temperature (see Fig. 5), gave an activation energy of 3.0 Kcal/mole for the relaxation of the methylene protons. This value is very similar to that seen for trans → gauche isomerizations in n-alkanes. These data implied that there is a rapid interconversion between the trans and the gauche isomers that is roughly uniform over much of the fatty acid chain. Unlike the spin-lattice relaxation rates, the transverse relaxation rates showed a relatively large variation among the proton resonances. Proceeding from the carbonyl end toward the methyl end of the fatty acid chain, the value of T_2 increased by a factor of about 2–3, and very near the methyl terminus there was an abrupt increase in the relaxation rates. For none of the resonances did $T_1 = T_2$, which implies (in the absence of exchange) that there are two classes of correlation times: one fast, resulting in T_1 relaxation; and one slow, resulting in T_2 relaxation. (This anisotropy in the motion also may be viewed as a rapid but restricted angular displacement; see below.) Similar results were obtained with dimyristoyl L-α-lecithin when studied above its transition temperature.

These data suggested a model in which extended displacements of a large segment of the chain caused by a gauche conformation are compensated for, in part, by a gauche rotation of the opposite sense about a bond β (or even farther down the chain, see below) toward the methyl end. Transitions of this sort would minimize collisional encounters with neighboring chains. Very near the methyl terminus the rotation of the methylene groups would be largely unrestricted, in this sense, and their motions would add onto those of the groups higher up on the chain, therefore giving rise to less efficient T_1 and T_2 relaxation and resulting in the high activation energy (4.2 Kcal/mole) observed for the methyl protons. Superimposed on the motions just described are the less probable extended displacements with the rotations (like those just discussed) of the groups that are on the segment. This will generate the gradual increase in T_2 seen as one proceeds down the chain. The relative probability of these latter motions remains to be established. Figure 6 illustrates a typical conformation at a given instant in time.

It is important to try to compare results obtained using different methods with the PMR data and interpretation presented above. At a superficial level of analysis the fatty acid conformations discussed above and illustrated in Figure 6 are consistent with the x-ray spacings ($\sim$ 4.6 Å) seen for lecithins above their transition temperature (54). If we assume that T_1 relaxation is primarily due to rotations about the long axis, and that T_2 relaxation is due to the more restricted angular displacements from the long axis (which are also caused by the isomerization), then the component of the anisotropic motion that corresponds to displacements from the long axis can be quite simply related to the order parameter discussed in this context by Seelig (55) and by Hubbell and McConnell (45). Following their arguments, it is readily shown that the value of the order parameter is linear in the line width [or $(1/\pi)T_2$]. The absolute value of the order parameter, in this approximation, is of the order of 6×10^{-4}. This value is about two orders of magnitude smaller than that seen using spin labels. Assuming that in this limit it is still valid to plot the log of the order parameter against the position of the substituent on the fatty acid chain, the slope of the approximately linear portion of the curve is in agreement with that of McFarland and McConnell (56) (i.e., within our error, which is much larger than that for spin labels). The data of McFarland and McConnell (56) show a deviation from linearity at carbon atom 16. This is also in agreement with the proton data which in our interpretation results from the abrupt increase in mobility at the terminal end of the molecule.

C^{13} magnetic resonance spectra and relaxation rates of sonicated dipalmitoyl L-α-lecithin have been reported by Metcalfe *et al.* (57). These studies have a distinct advantage over the proton studies in that more of the methylene carbon atoms than methylene protons can be resolved. In particular, with dipalmitoyl L-α-lecithin, they were able to obtain relaxation data for the fatty acid carbon atoms 2, 3, 4–13, 14, 15, and 16. These data have been interpreted as in accord with the proton T_1 and T_2 results discussed above.

NMR in the Presence of Paramagnetic Molecules

Using a paramagnetic analogue of lecithin in which one of the choline methyl groups is replaced by a spin label, Kornberg and McConnell (58)

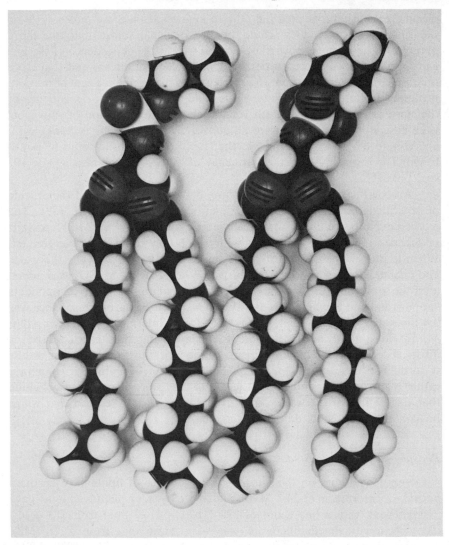

Figure 6

A typical structure at a given instant in time of the hydrocarbon region of sonicated dipalmitoyl L-α-lecithin. The conformation of the polar region is somewhat arbitrary and should not be taken seriously.

investigated the rate of lateral diffusion in sonicated lecithin. The effect of about 1 mole percent of spin-labeled dipalmitoyl lecithin in didihydrosterculoyl lecithin was to broaden the N-methyl proton resonances of the phospholipids. In the absence of any motion this resonance would be inhomogeneously broadened (i.e., the resonances from those molecules near the spin-labeled molecules would be broader than those that are more removed). The homogeneous line shape as well as the dependence of the line width on the added concentration of spin-labeled lecithin showed that there is considerable lateral motion. In an attempt to determine the rate of motion, proton and phosphorus relaxation rates and the temperature dependence of the slope of a plot of line width versus relative concentration of spin label were measured. These data were all consistent with the fast exchange case, and therefore only a lower limit to the lateral diffusion rate could be established. The lower limit for the frequency of the translation step is 3×10^3 sec^{-1} at $0°C$.

An NMR method for distinguishing between the inside and the outside of sonicated lecithin bilayers was described by Bystrov *et al.* (59). The method exploits the use of the paramagnetic ions Eu^{+++} or Mn^{++} to shift or broaden the choline resonances with which they come into close contact. When lecithin was sonicated in D_2O containing 10^{-3} M $MnSO_4$, the water and choline resonances were very broad. When 10^{-3} M $MnSO_4$ was added after the sonication, the choline resonance was composed of two components —one broad and one narrow—and the intensity of the narrow resonance was 0.4 times that of the original. These results were interpreted as showing that Mn^{++} does not readily penetrate the bilayer and thus can be used to distinguish the outside from the inside of the lecithin vesicles. Similar results were obtained using Eu^{+++}, which has the effect of shifting a resonance rather than broadening it. From their data they were also able to conclude that the lifetime of the lipids on a given surface is > 1 sec and that water has a lifetime on the inside of $< 10^{-2}$ sec.

Phosphorus Resonance Studies

Very few studies on the phosphorus resonance of lipids have been reported. In chloroform at 24.3 MHz, dioleoylsteroylphosphatidylcholine gave a single peak with a line width of 290 Hz at $+116.2$ ppm from P_4O_6; dioleoylphosphatidylmethanolamine was at $+115.6$ ppm from P_4O_6 (42). Aqueous dispersions of lecithin were shifted -1.0 ppm from 50% H_3PO_4 used an external reference (49).

An initial description of the phosphorus magnetic resonance (PhMR) spectrum and relaxation rates for sonicated and unsonicated lecithin have been presented (49). The absorption spectrum of unsonicated lecithin was interpreted as a superposition of two types of resonances, one broad (590 Hz) and one narrow (72 Hz). The effect of sonication was to increase the value of T_1 from 0.9 to 8.3 sec and to decrease the line width to a homogeneous line of 20 Hz width. This effect of sonication on the PhMR line width is qualitatively similar to that described above for the PMR line width. The PhMR spectrum from sonicated lecithin is similar to the PMR spectrum in that both have nondipolar contributions to the line width.

Studies of Membranes and their Lipids

Erythrocyte Ghosts

The PMR spectrum of intact erythrocyte membranes (ghosts) did not show well-resolved resonances below 30°C. Between 30 and 40°C weak resonances corresponding to methyl ($\tau = 9.1$), acetamido ($\tau = 7.88$), and choline ($\tau = 6.6$) were seen, and above 40°C there was an abrupt change in the methyl proton peak area and several new resonances appeared. As the temperature was increased further, the line widths did not change substantially, but the intensities continued to grow, indicating more contributing protons (61–64).

At very high temperatures there was a large absorption in the methyl region ($\tau = 9.1$), but without a concomitant absorption in the methylene region. This suggests that the protons contributing to the spectrum arise from the proteins rather than from the lipids. Using the chemical shifts for proteins in the random coil conformation and using the amino acid composition of erythrocyte ghosts, Glaser *et al.* (63) have computed the expected spectrum and found it to agree very closely with the observed spectrum (Fig. 7). The assumption that the proteins were in the random coil conformation was justified by circular dichroism results. It is quite interesting that even at a temperature of 82°C, the lipid fatty acid resonances apparently did not contribute significantly to the resolved proton resonance spectrum in this cholesterol-rich membrane.

Sonication of erythrocyte ghosts produced a reproducible spectrum that was qualitatively different from that of intact ghosts (Fig. 8) (61, 62). Between 30° and 40°C among the resonances seen were choline ($\tau = 6.71$), methylene ($\tau = 8.7$), methyl ($\tau = 9.1$), and acetamido ($\tau = 7.8$); no resonance due to vinyl protons ($\tau = 4.6$) was observed, however. As the temperature was increased, the choline resonance increased in intensity with little change in the peak width (suggesting more contributing protons), the resolved methylene and vinyl resonances began to grow, and the water shifted to higher τ. The appearance of the methylene and vinyl resonances above 40°C suggested that these may be due to the lipid protons (61).

One consequence of sonication was to reduce the ghosts to particles with a radius of 2000 Å or less (62). This size is not sufficiently small to permit tumbling at a rate adequate to average the magnetic field gradients or dipole interactions that are responsible for the lack of resolution and for the broad resonances from unsonicated erythrocyte ghosts (62). Therefore, the effect of sonication on erythrocyte membranes may be similar to that on

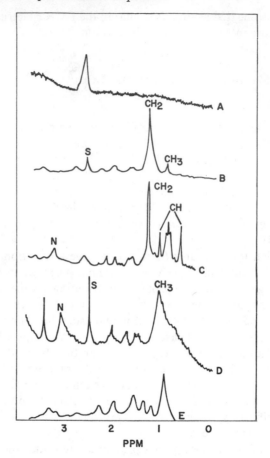

Figure 7

Proton magnetic resonance spectra in the region of the methylene and methyl proton resonances, in phosphate-buffered D$_2$O unless otherwise indicated. *A*. Untreated red blood cell membranes at 18°C. *B*. Phospholipid C–treated membranes at 18°C. *C*. CHCl$_3$-extracted lipids from untreated membranes, in CDCl$_3$ at 18°C. *D*. Untreated membranes at 82°C. *E*. Computed protein spectrum. The abscissa gives the chemical shifts in parts per million relative to the methyl protons of DSS. S, standard; N, choline methyl protons; CH cholesterol methyl protons. (From Glaser, M., *et al. Proc. Natl. Acad. Sci. U.S.A. 65,* 721, 1970.)

phospholipid dispersions. Some evidence for this has been presented by Steim (48), who compared the transverse relaxation time, T_2, obtained by a spin-echo sequence, with that obtained by the free induction decay. In these experiments one dominant relaxation time was seen and the objections

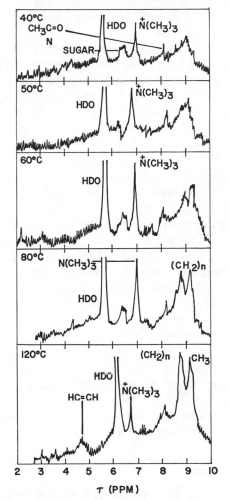

Figure 8

Proton magnetic resonance spectra at 60 MHz at different temperatures of a sonicated dispersion (5% w/v) of erythrocyte membrane fragments in D$_2$O (64 scans). (From Chapman, D., *et al. J. Mol. Biol. 31*, 101, 1968.)

discussed earlier in this chapter are applicable here. These results suggest that the apparent absence of resolved lipid fatty acid resonances in intact membranes may be due in part to magnetic field gradients perhaps similar to those proposed to explain the absence of relatively sharp resonances in unsonicated lecithin dispersions (see under "Chemical Shift Assignments").

This behavior was also seen with *Mycoplasma laidlawii* and *Escherichia coli* membranes (48).

PMR along with circular dichroism have been used to study erythrocyte ghosts. Glaser *et al.* (63) have suggested that there is a differentiation of the phospholipids in the membrane into at least two physical states. The evidence for this is as follows. With increasing temperatures, between 25° and 82°C, about 20% of the proteins changed from the α-helix to a random coil conformation. This was seen in the PMR spectrum; the estimates from the intensity of the methyl peak indicated that 20–40% of the membrane proteins contributed to this resolved line. Therefore, a significant fraction of proteins changed their conformation with temperature. The PMR spectrum did not show a change in the lipid methylene resonances, however. Conversely, when the lipids were treated with phospholipase C, there was an appearance of a lipid methylene resonance with no alteration in the protein circular dichroism spectrum. About 75% of the methylene lipid protons contributed to the resolved resonances after the phospholipase C treatment. These results were interpreted as evidence supporting a mosaic membrane structure.

In summary, PMR studies on erythrocyte ghosts have indicated that with intact membranes the lipid component apparently does not contribute significantly to the resolved resonances; and the lipid resonances appear to contribute to the resolved resonances only after the ghosts have been sonicated. Therefore, in sonicated erythrocyte membranes (as well as in sonicated lipid bilayers), the lipid component appears to be in a fluid state. The broad fatty acid resonances in unsonicated membranes are probably dipolar in origin, but may involve magnetic field gradients of some sort. The effects of sonication, high temperatures, and lyophilization on membrane functions and structure have not been clearly assessed.

Erythrocyte Ghost Lipids

Erythrocyte apoproteins obtained by a butanol procedure, lipoproteins obtained by a pentanol procedure, and lipids obtained by a chloroform methanol extraction were studied by PMR (64). In the spectrum of the sonicated lipids the N^+—$(CH_3)_3$ protons were broader than those in the sonicated membranes. The lipoprotein spectrum (not sonicated) was quite similar to that of the sonicated membranes, and the apoproteins themselves gave broad resonances similar to those seen with globular proteins.

The addition of cholesterol to sterol-free lipids produced a differential broadening primarily affecting the fatty acid protons. A great deal of motion remained, however, in that the resonances were still resolved. This is in contrast to lecithin (44), in which cholesterol appears to greatly broaden the fatty acid resonances. These results again suggest an effect of cholesterol on inhibiting the motion of the fatty acid chains.

Myelin and Myelin Lipids

At 65°C or higher a PMR spectrum of sonicated myelin with some resolved resonances was seen (65). As the sample was cooled below this temperature, there was a reversible loss of signal. Among the resolved resonances were vinyl, choline, methylene, and methyl, as well as galactolipid protons. Those assignments were not investigated in detail. The appearance of resolved resonances above a certain temperature parallels that seen with erythrocyte ghosts. With erythrocyte ghosts this temperature was at about 40°C and with myelin it was at about 65°C. The significance, if any, remains to be established.

The PMR spectra of sonicated myelin lipid extracts were qualitatively similar to those of sonicated myelin. The removal of cholesterol produced a sharpening of the hydrocarbon resonance and altered the relative peak intensities. As with erythrocyte lipids, the PMR spectrum of the total lipid extract (including cholesterol) still showed resolved fatty acid resonances. There was no major effect of the cholesterol on the choline resonances.

Microorganisms

Cerbon has studied the PMR of intact *Nocardia asteroides* and to a lesser extent *Mycobacterium smegmatis* (19, 66). He has observed a resonance at $\delta = 1.3$ (with respect to TMS at $\delta = 0$) that he believes is due to the aliphatic lipid component of the cell membrane. His assignment was based on a comparison with the resonances of the lipids extracted from the cells, on the observation that the signal was lost in defatted cells, and on the assignments from natural and synthetic lipids; it is suggestive at best.

This resonance was observed under hypotonic conditions, in some cases under hypertonic conditions, but was absent under isotonic conditions. However, if Mg^{++} or Ca^{++} were present, the signal was observed under a wide variety of ionic strengths and tonicities. With increasing pH or temperature, the line width decreased, which was interpreted as an increase in the mobility of the aliphatic region of the lipids.

Steim has studied the free induction decay of protons from isolated *E. coli* membranes (48). In this work he resolved the free induction decay into three different components; the longest of these had a T_2 of 45 msec. When the Y intercept of this component (proportional to the number of protons contributing to the resonance) was plotted against temperature, there was an increase, that began at about 20°C and ended at about 40°C, in the number of contributing (unassigned) protons.

Excitable Membranes

PMR spectra from rabbit sarcoplasmic reticulum and from rabbit sciatic nerve have been reported (68, 69). The spectrum from the sarcoplasmic reticulum (68), at 20°C, was interpreted as a superposition of two types of spectra: one type, which corresponds to an estimated 80% of the total lipid, was a broad 500-Hz resonance like that from unsonicated lecithin; the other, which accounts for the remainder of the phospholipid present, was relatively narrow and similar in appearance to that from sonicated lecithin. At 20° a narrow resonance corresponding to the N-methyl protons of the choline was not present. From these data the authors concluded that the membrane lipids were in two states of differing mobility. (There was no estimate of the distribution of particle sizes.)

On raising the temperature from 20° to 40°C, approximately 60% of the choline resonance appeared. If the membranes were previously treated with trypsin, the choline resonances were seen even at 25°C. The appearance of the choline resonance correlated with the temperature-dependent Ca^{++} efflux seen in these preparations.

The spectrum from rabbit sciatic nerve (69) was similar to that from sarcoplasmic reticulum in two respects: the resonances were sharp and well resolved like that from sonicated lecithin, and there was no sharp choline resonance at 20°C. The spectrum differed from that of sarcoplasmic reticulum in that no evidence was presented suggesting the presence of a broad resonance. The authors were unable to distinguish between resonances arising from the nerve and those from the myelin. Since isolated myelin shows virtually no narrow lipid resonances at this temperature, it is suggestive that these spectra are from the nerve itself.

It is interesting that these two excitable membranes show spectra that are so different from those discussed above from other membranes. The protein resonances from these two membranes are broad and unresolved, but the lipid resonances (at least in part) are relatively narrow and resolved. The other membranes discussed above, as well as those from *E. coli* (49), show virtually no narrow lipid signals at this temperature. That these spectra are a property of excitable tissue is a highly speculative but interesting possibility.

Summary and Conclusion

NMR studies on sonicated lecithin dispersions have provided strong evidence for, and an initial description of, the fluid nature of these bilayers. Analyses of the narrow line widths and of the long relaxation times support a fluid structure for the lipid molecules, and an initial description of the structure of the hydrocarbon region has been presented. Through the use of selectively protonated per-deutero fatty acids, selectively fluorinated fatty acids, and selectively enriched C-13 fatty acids as well as the use of oriented samples, more sophisticated resonance methods, and data from other physical techniques, a more detailed and quantitative theory of the structure of the interior region should be forthcoming.

NMR studies on unsonicated lecithin dispersions indicate that these bilayers are considerably less mobile than those from sonicated lecithin. An initial discussion of the structure of the interior region from one type of unsonicated dispersion has been presented, and undoubtedly a more detailed description will follow. The structural difference between sonicated and unsonicated lecithin dispersions is an important unanswered question. It is important to ask which type of lecithin dispersion, if any, is most relevant to the structure in a real membrane, and the answer is of course unknown. But based on the membrane spectra published so far, it is likely that both types may be relevant. An interesting example is the spectrum from the rabbit sarcoplasmic reticulum (68). This spectrum has been interpreted as a super-position of broad and narrow lipid resonances; the broad resonances are like those from unsonicated lecithin and the narrow are like those from sonicated lecithin.

Although we are beginning to understand the structure of the interior region of lecithin, the role of different fatty acids in determining the structure of the interior region, the function of the different headgroups, the structure of the headgroup itself, the structure of phospholipid dispersions composed of more than one class of phospholipid (e.g., do different classes of phospholipids form patches, and are different classes distributed symmetrically between the two bilayer surfaces?) are problems that are amenable to a systematic study using NMR.

There have been very few studies on the nature of lipid-protein inter-actions. A useful model system may be "functionally" reconstituted lipoproteins (70).

The NMR studies on membranes are not as well developed as those on isolated phospholipid bilayers. Among the problems in this area are unclear assignments and the apparent absence of resolvable lipid resonances in intact membranes. Studies using fluorine NMR with biosynthetically incorporated, selectively fluorinated fatty acids, like the studies of Keith and coworkers (Chap. 8) with spin-labeled fatty acids, may prove useful. The use of pulsed Fourier-transform techniques, gentle methods of breaking membranes into smaller structures, and homogeneous (in size) samples may be essential. In all this work it is extremely important to assess the fraction of the membrane proteins or lipids contributing to the resolved resonances.

REFERENCES

1. Chapman, D., and Salsbury, N. J. *Trans. Faraday Soc.* **62,** 2607 (1966).
2. Chapman, D., Byrne, P., and Shipley, G. G. *Proc. R. Soc.* **290,** 115 (1966).
3. Chapman, D., Williams, R. M., and Ladbrooke, B. D. *Chem. Phys. Lipids* **1,** 445 (1967).
4. Salsbury, N. J., and Chapman, D. *Biochim. Biophys. Acta* **163,** 314 (1968).

5. Veksli, Z., Salsbury, N. J., and Chapman, D. *Biochim. Biophys. Acta* **183**, 434 (1969).
6. Lecar, H., Menstein, G. E., and Stillman, I. *Biophys. J.* **11**, 140 (1971).
7. Steim, J. M., Edner, O. J., and Bargoot, F. G. *Science* **162**, 909 (1968).
8. Small, D. M., Penkett, S. A., and Chapman, D. *Biochim. Biophys. Acta* **176**, 178 (1969).
9. Leslie, R. B., Chapman, D., and Scanu, A. M. *Chem. Phys. Lipids* **3**, 152 (1969).
10. Chapman, D., Leslie, R. B., Hirz, R., and Scanu, A. M. *Nature* **221**, 260 (1969).
11. Cerbon, J. *Biochim. Biophys. Acta* **88**, 444 (1964).
12. Fritz, O. G., Jr., and Swift, T. J. *Biophys. J.* **7**, 675 (1967).
13. Chapman, G., and McLauchlan, K. A. *Nature* **215**, 391 (1967).
14. Cerbon, J. *Biochim. Biophys. Acta* **144**, 1 (1967).
15. Fritz, O. G., Jr., Scott, A. C., and Swift, T. J. *Nature* **218**, 1051 (1968).
16. Swift, T. J., and Fritz, O. G., Jr. *Biophys. J.* **9**, 54 (1969).
17. Cope, F. W. *Biophys. J.* **9**, 303 (1969).
18. Klein, M. P., and Phelps, D. E. *Nature* **224**, 70 (1969).
19. Cerbon, J. *Biochim. Biophys. Acta* **211**, 389 (1970).
20. Carrington, A., and McLachlan, A. D. *Introduction to Magnetic Resonance.* Harper & Row, New York, 1967.
21. Pople, J. A., Schneider, W. G., and Bernstein, H. J. *High Resolution Nuclear Magnetic Resonance.* McGraw-Hill, New York, 1959.
22. Abragam, A. *The Principles of Nuclear Magnetism.* Oxford University Press, London, 1961.
23. Becker, E. *High Resolution NMR.* Academic Press, New York, 1969.
24. Van Vleck, J. H. *Phys. Rev.* **74**, 1168 (1948).
25. Ernst, R. R. *Adv. Magn. Resonance* **2**, 1 (1966).
26. Ernst, R. R., and Anderson, W. A. *Rev. Sci. Instr.* **37**, 93 (1968).
27. Lowe, T. J., and Norberg, R. E. *Phys. Rev.* **107**, 46 (1957).
28. Carr, H. Y., and Purcell, E. M. *Phys. Rev.* **94**, 630 (1954).
29. Horwitz, A. F., Horsley, W. J., and Klein, M. P. *Proc. Natl. Acad. Sci. U.S.A.* **69**, 590 (1972).
30. Van Geet, A. L., and Hume, D. N. *Anal. Chem.* **37**, 983 (1965).
31. Vold, R. L., Waugh, J. S., Klein, M. P., and Phelps, D. E. *J. Chem. Phys.* **48**, 3831 (1968).
32. Bangham, A. D. *Progress Biophys. Mol. Biol.* **18**, 31 (1968).
33. Levine, Y. K., Bailey, A. I., and Wilkins, M. H. F. *Nature* **220**, 577 (1968).
34. Chapman, D., Fluck, D. J., Penkett, S. A., and Shipley, G. G. *Biochim. Biophys. Acta* **163**, 225 (1968).
35. Penkett, S. A., Flook, A. G., and Chapman, D. *Chem. Phys. Lipids* **2**, 273 (1968).
36. Sheard, B. *Nature* **223**, 1057 (1969).
37. Kaufman, S., Steim, J., and Gibbs, J. *Nature* **225**, 743 (1970).
38. Hansen, J. R., and Lawson, K. D. *Nature* **225**, 542 (1970).
39. Chan, S. I., Feigenson, G. W., and Seifer, C. H. H. *Nature* (1971).
40. Tiddy, G. J. T. *Nature* **230**, 136 (1971).
41. Attwood, D., and Saunders, L. *Biochim. Biophys. Acta* **98**, 344 (1965).
42. Huang, C. *Biochemistry* **8**, 344 (1969).

43. Chapman, D., and Morrison, A. *J. Biol. Chem.* **241,** 5044 (1966).
44. Chapman, D., and Penkett, S. A. *Nature* **211,** 1304 (1966).
45. Hubbell, W. L., and McConnell, H. M. *J. Am. Chem. Soc.* **93,** 314 (1971).
46. Phillips, M. C., Williams, R. M., and Chapman, D. *Chem. Phys. Lipids* **3,** 234 (1969).
47. Ladbrooke, B. D., and Chapman, D. *Chem. Phys. Lipids* **3,** 304 (1969).
48. Steim, J. M. *Liquid Crystals and Ordered Fluids.* Plenum Press, 1970.
49. Horwitz, A. F., and Klein, M. P. *J. Supramol. Struct.* In press. Also unpublished data.
50. Pressman, B. C. *Fed. Proc.* **27,** 1283 (1968).
51. Henderson, P. J. F., McGivan, J. O., and Chappell, J. B. *Biochem. J.* **111,** 521 (1969).
52. Pressman, B. C. *Proc. Natl. Acad. Sci. U.S.A.* **53,** 1076 (1965).
53. Finer, E. G., Hansen, H., and Chapman, D. *Chem. Phys. Lipids* **3,** 386 (1969).
54. Engleman, D. E. *J. Mol. Biol.* **47,** 115 (1970).
55. Seelig, J. *J. Am. Chem. Soc.* **92,** 3881 (1970).
56. McFarland, B. G., and McConnell, H. M. *Proc. Natl. Acad. Sci. U.S.A.* **68,** 1274 (1971).
57. Metcalfe, J. C., Birdsall, N. J. M., Feeney, M., Lee, A. G., Levine, Y. K., and Partington, P. *Nature* **233,** 199 (1971).
58. Kornberg, R. D., and McConnell, H. M. *Proc. Natl. Acad. Sci. U.S.A.* **68,** 2564 (1971).
59. Bystrov, V. F., Dubrovina, N. I., Barsukov, L. I., and Bergelson, L. D. *Chem. Phys. Lipids* **6,** 343 (1971).
60. Chapman, D., Kamat, V. B., de Gier, J., and Penkett, S. A. *Nature* **213,** 74 (1967).
61. Chapman, D., Kamat, V. B., de Gier, J., and Penkett, S. A. *J. Mol. Biol.* **31,** 101 (1968).
62. Kamat, V. B., and Chapman, D. *Biochim. Biophys. Acta* **163,** 411 (1968).
63. Glaser, M., Simpkins, H., Singer, S. J., Sheetz, M., and Chan, S. I. *Proc. Natl. Acad. Sci. U.S.A.* **65,** 721 (1970).
64. Kamat, V. B., Chapman, D., Zwaal, R. F. A., and Van Deewan, L. L. M. *Chem. Phys. Lipids* **4,** 323 (1970).
65. Jenkinson, T. J., Kamat, V. B., and Chapman, D. *Biochim. Biophys. Acta* **183** 427 (1969).
66. Cerbon, J. *Biochim. Biophys. Acta* **102,** 449 (1965).
67. De Vries, J. J., and Berendsen, H. J. C. *Nature* **221,** 1139 (1969).
68. Davis, D. G., and Inesi, G. *Biochim. Biophys. Acta* **241,** 1 (1971).
69. Dea, P., Chan, S. I., and Dea, F. J. *Science* **175,** 209 (1972).
70. Rothfield, L., and Finkelstein, A. *Annu. Rev. Biochem.* **37,** 463 (1968).

8

R. J. Mehlhorn
A. D. Keith

Spin Labeling of Biological Membranes

Spin labeling is one of the most recently developed methods used to derive information about membranes. Potentially, information can be derived relating to viscosity of local domains, polarity of local domains, states of molecular ordering, rotational or translational motion, precise measurement of molecular distances, interaction of molecular species, and no doubt, still other parameters.

Initially, the application of spin labels to membranes yielded general observations about the average lipid fluidity and orientation characteristics of the entire membrane. Current research, however, is largely concerned with the fine structure of membranes and with the restraint on structural alterations consistent with functional viability. This chapter provides an introduction to spin-labeling concepts together with an outline of past and current applications of the electron spin resonance (ESR) technique to the study of biological membranes. It deals with some fundamentals relating to resonance spectroscopy, discusses considerations relevant to deriving information from spin-label spectra, and presents a brief treatment of the spin-label literature relating to biological membranes.

Part of this work was supported by grants from the U.S. Atomic Energy Commission Project Agreement 194 and the U.S. Public Health Service (AM-12939).

MOLECULAR STRUCTURE

Nitroxide molecules used for spin-labeling studies share the structural feature:

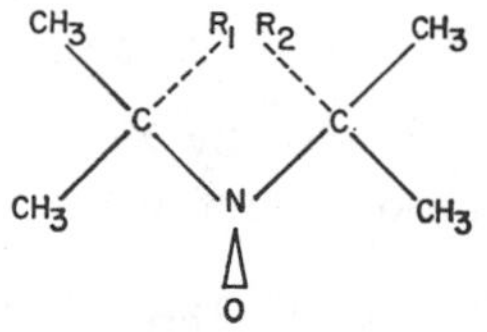

The magnetic properties of this molecule exploited in the spin-resonance experiment arise from the unpaired electron localized primarily on the nitrogen atom (1). The application of spin labels to biological problems requires an appreciation of the common chemical modifications of nitroxides which result in the cancellation of this "free" electron spin.

The usual mechanism for spin pairing is hydrogen donation by some other molecule, resulting in an OH bond. It is thought that deactivation of nitroxides by vitamin C proceeds in this manner (2). However, other hydrogen donors, like sodium borohydrate, lack the capacity to deactivate nitroxides. The critical parameter characterizing this type of hydrogen donation is not known.

Of particular importance for membrane research is the rapid destruction of spin labels by many organisms and organelles. Presently this type of spin destruction is only vaguely understood, and considerable research is being devoted to elucidating its mechanism.

The unpaired electron occupies a $2p\pi$ orbital which is confined largely to the nitrogen atom. Most of the electron density is concentrated about

an axis normal to the $\overset{\displaystyle C}{\underset{\displaystyle C}{\diagdown\!\!\diagup}}$ N—O plane and passing through the nitrogen.

Because interatomic Coulomb interactions are considerably larger than the spin-orbit interaction, the angular momentum of the unpaired electron is essentially quenched (3), and the dominant magnetic properties of a nitroxide molecule are those of a free electron spin.

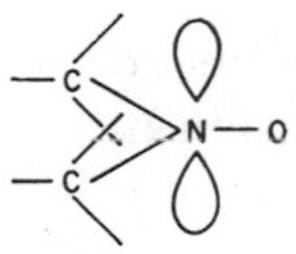

STATIC MAGNETIC PROPERTIES

A magnetic moment (μ) precesses in a constant magnetic field (H) with a characteristic frequency, known as its Larmor frequency, given by

$$\omega = \frac{\mu H}{\hbar}$$

where $\hbar$ is Planck's constant divided by 2π. This equation is rewritten for paramagnetic molecules and atoms in terms of the Bohr magneton (β) and the g value as

$$\omega = \frac{g\beta H}{\hbar} \tag{1}$$

Equation 1 describes the resonance condition for an ESR experiment. A properly oriented beam of microwaves of this frequency causes an ensemble of magnetic moments to precess coherently and induce a magnetization in the sample normal to the constant field, H. Loss of the coherence through thermal interactions is the dominant relaxation mechanism for nitroxide spin labels.

The usual resonance spectrometer operates at a fixed frequency, while the magnetic field is varied. Broad absorption of microwaves by electric dipoles and diverse paramagnetic molecules is effectively cancelled out by electronically taking the derivative of the power absorption with respect to the magnetic field. A simple type of ESR spectrum is illustrated by cigarette smoke, which contains a host of unstable free radicals having g values close to those of a free electron spin (Fig. 1).

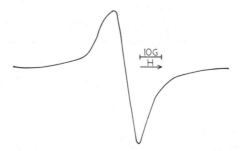

Figure 1
ESR spectrum of cigarette smoke radicals trapped on glasswool. We adhere to the convention that the magnetic field, *H*, increases from left to right.

Most of the interesting information to be obtained from spin labeling arises from the hyperfine interaction between the unpaired electron and the magnetic moment of the ^{14}N nucleus. This interaction has two contributions: an anisotropic dipole term associated with the p character of the unpaired electron and an isotropic Fermi contact interaction (4) arising from a slight *s* admixture into the *p*-wave function. The nuclear dipole field is given by the expressions

$$H_\parallel^D = \frac{\mu_N}{r^3}(3\cos^2\theta - 1)$$

$$H_\perp^D = \frac{\mu_N}{r^3}(3\sin\theta\cos\theta)$$

(2)

where $H_\parallel$ and $H_\perp$ are defined with respect to the orientation of the magnetic moment, μ_N. The nuclear spin of ^{14}N is 1, implying that the effective magnetic moment can have three values in a magnetic field given by μ_N, 0, and $-\mu_N$. If the isotropic hyperfine interaction is represented by the term $\mu \cdot H^c$, then the effective magnetic field seen by the electron spin is given by

$$'H' = H^L + [\sqrt{(H^c + H_\parallel^D)^2 + (H_\perp^D)^2}]M$$

(3)

where H^L refers to the laboratory magnetic field, and where $M = -1$, 0, $+1$. Consequently, there are three values of the laboratory magnetic field where the resonance condition of Equation 1 is met. For $M = \pm 1$, the field values of the resonances are orientation-dependent, as implied by Equations 2.

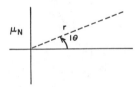

An early stable nitroxide to be synthesized for magnetic resonance experiments was di-*tert*-butyl nitroxide. Single crystals of tetramethyl-1,3-cyclobutanedione accept this molecule as a well-oriented impurity. Various orientations of the substituted crystal with respect to a laboratory magnetic field reflect the anisotropic character of the nitroxide hyperfine interaction as shown in Figure 2. The slight anistropy of the *g* value arises from the spin-orbit interaction but is not understood quantitatively. Numerical values of the *g* and hyperfine parameters found for this nitroxide are (5):

$$g_x = 2.00881 \qquad g_y = 2.00625 \qquad g_z = 2.00271$$

$$A_x = 7.59\ G \qquad A_y = 5.95\ G \qquad A_z = 31.78\ G$$

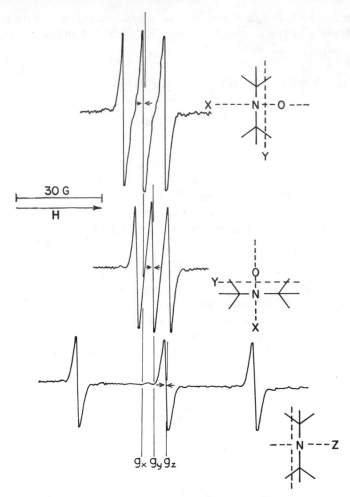

Figure 2

ESR spectra showing the crystal parameters derived from di-*tert*-butyl nitroxide. (Adapted from data kindly furnished by O. H. Griffith.)

Since the dipole interaction vanishes when averaged over all orientations, the contact interaction is given approximately by

$$H_c = \tfrac{1}{3}(A_x + A_y + A_z) = 15.1 \text{ G} \tag{4}$$

for the following nitroxide.

DI-TERT-BUTYL NITROXIDE

TETRAMETHYL-1,3-CYCLOBUTANEDIONE

INTERMOLECULAR INTERACTIONS

Polar solvents augment the polarity of the N—O bond, resulting in an increased density of the unpaired electron on the ^{14}N nucleus. The Fermi contact interaction is proportional to the s electron density at the nucleus so the isotropic hyperfine splitting increases with the polarity of the solvent. The g value changes as well, as shown in Table I. These parameters were observed in low-viscosity solvents where the anisotropic tensors average to zero, as explained later in this chapter.

Table I

Nitrogen Hyperfine Couplings (Gauss) and g Values for TEMPOL in Solvents of Different Dielectric Constant

SOLVENT	DIELECTRIC CONSTANT	HYPERFINE COUPLING	g VALUE
n-Hexane	1.9	15.2	2.0061
Ethyl ether	4.3	15.3	2.0061
t-Butanol	10.9	16.0	2.0059
n-Butanol	17.8	16.1	2.0059
Ethanol	24.3	16.1	2.0059
Methanol	32.6	16.1	2.0059
Ethylene glycol	37.7	16.4	2.0058
Water	78.5	17.1	2.0056

Polarity information gleaned from hyperfine couplings is of considerable importance in the spin labeling of membranes in that it allows the environment of the probe to be inferred. Taken together with the line width in a homogeneous environment, it can be used to estimate the range of local polarities where labels are dispersed.

The dipole field of a nitroxide spin is given by Equations 2, with $\mu = 10^4$ G and r expressed in Angstroms. Thus, appreciable interactions between nitroxides occur over distances of 10 Å or less. At smaller separations the wave functions of two nitroxide spins may overlap, giving rise to strong Coulomb interactions which lead to electron exchange.

Other paramagnetic species in a biological system, such as transition metal complexes, ions in solution, or dissolved O_2, produce magnetic dipole fields resulting in observable broadening of nitroxide spectra under appropriate conditions. Such intermolecular interactions provide a measure of the mean separation of paramagnetic species. In membrane studies, dipole and exchange broadening can be used as a geometric gauge; for example, the compartmentalization of various lipid classes can be inferred from the composite spectra of suitably chosen spin labels. Interactions between paramagnetic molecules will be treated in greater detail after a consideration of the effects of motion.

EFFECTS OF ROTATION

When 2,2,6,6-tetramethylpiperdinol-N-oxyl (TEMPOL) is dissolved in glycerol and observed over a broad temperature range, the spectra shown in Figure 3 result. These spectra can be interpreted in terms of a remarkably simple theoretical model proposed by Itzkowitz (6). The model assumes that the nitroxide molecule rotates while an energy transition occurs and that the effective energy, its expectation value, is given by the time average

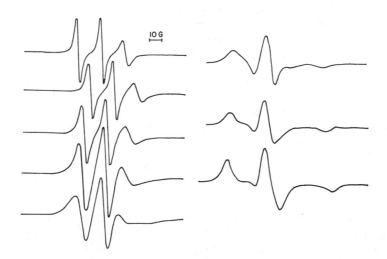

Figure 3
ESR spectra of the spin label, TEMPOL, in glycerol at a variety of temperatures. Spectra given illustrate the range of motion of interest to experimenters dealing with lipid or membrane systems.

of all the energies resulting from the various orientations sampled by the nitroxide during its tumbling motion. The time interval for this averaging is assumed to be the effective lifetime of the radiation process and is assumed to be the same for all three hyperfine lines.

TEMPOL

To demonstrate how the method works we can imagine a table of line positions for a large set of orientations described by the angles θ and ϕ. Such a table can be generated empirically with a single crystal. We give this table to a computer and have the computer select an arbitrary initial orientation for a nitroxide molecule. By means of another random number selection, a direction over the surface of the sphere is chosen, and new orientation angles corresponding to a fixed step size are generated. At the end of N steps the energies corresponding to all the orientations that were selected are averaged to give the effective energy of the nitroxide for this transition process.

It must be emphasized that the Itzkowitz time-averaging procedure makes no attempt to describe the mechanism whereby the nitroxide exchanges energy with the molecules of the solvent. Hence, the assumption of a single lifetime for all tumbling rates and for all the energy levels is not theoretically justified, and the Itzkowitz procedure is essentially a phenomenological theory whose validity rests on the excellent comparison between simulated and observed spectra.

As described by Itzkowitz, the time-averaging procedure is demanding of computer time. Also, to consider anisotropic motion, the technique has to be modified. In a new formalism three operators are employed to produce the tumbling motion of a nitroxide. These are rotation matrices which operate on the nitroxide orientation vector and rotate the nitroxide about any one of its three principal axes. Thus, the theory is readily adaptable to enhanced mobility about one of these axes through the appropriate step size adjustment. The actual procedure consists of selecting a random number between -0.5 and $+0.5$ and multiplying this by the step size about a principal axis. Three successive rotations about the x, y, and z axes of the nitroxide represent one step of the tumbling motion. Such a procedure with equal step sizes about these three principal axes produced the spectra shown in Figure 4.

The agreement with the observed spectra of TEMPOL in glycerol is good, even in the fast tumbling region. In the slow tumbling region the theoretical spectra show more structure than is observed, although subtle features of the experimental spectra may be obscured by noise.

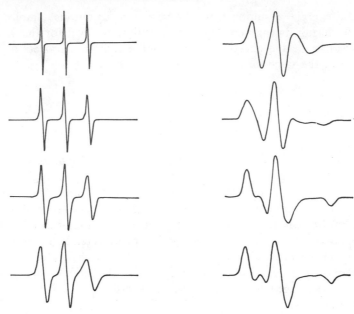

Figure 4

Simulated isotropic spectra ranging from the freely tumbling state to the fully immobilized limit. The maximum step sizes were chosen as $\Delta\theta_x = \Delta\theta_y = \Delta\theta_z = 0.4$ radian. Beginning with the upper left hand spectrum and reading down, the number of steps per lifetime and Gaussian line widths were: 16184, 1 G; 4096, 2 G; 1024, 3 G; 256, 4 G; 64, 5 G; 16, 5 G; 4, 5 G; 1, 5 G.

In the fast tumbling limit a more rigorous theory, which ascribes the relaxation process directly to the tumbling of the nitroxide, has been developed by Kivelson (7). This theory leads to an equation relating line-width differences to correlation times. If the line shapes are assumed to be single Lorentzian curves, the correlation time is given by

$$\tau_c = 6.5 \times 10^{-10} \; W_o(\sqrt{h_o/h_{-1}} - 1) \tag{5}$$

for the g- and A-tensor parameters of the nitroxide di-*tert*-butyl nitroxide (DTN). This equation has been used for some time to extract correlation times from ESR spectra. For some time we have used this expression in the tumbling range of 10^{-10} sec^{-1} to about 5×10^{-9} sec^{-1}. Since the Kivelson treatment is valid for tumbling times equal to or faster than about 10^{-9} sec^{-1}, we adopt the notation τ_0 as an empirical approximation of tumbling time for comparative purposes. The τ_0 notation will be used in this chapter.

DTN

An important attribute of the statistical averaging method is its adaptability to the treatment of nonisotropic tumbling. By defining different step sizes for rotation about the three principal axes, we can simulate motion expected for nonspherical molecules, or by imposing limits on the net excursion about some mean angle, we can deal with molecules that are anchored to large molecules or surfaces.

An anisotropic signal may be of two general types: *1*) when the population of nitroxides is immobilized in space and oriented, this sample may be mechanically rotated in space, thus rotating the different principal axes of anisotropy with respect to the applied magnetic field and achieving spectra such as those in Figure 2, and *2*) when the principal axes are moved in space but are time-averaged unequally. With anisotropic motion in non-oriented matrices the net position of all principal axes with respect to the applied magnetic field is random. The case where $x + y$ average faster than other principal axial combinations is referred to as "z-axis anisotropy." When we increase the step size for motion about the x axis in the nitroxide coordinate system near the fast tumbling limit, we obtain spectra A of Figure 5 (x-axis anisotropy). Experimental spectra from the elongated spin label 10:0-TEMPOL ester (B of Fig. 5) are in good agreement with this

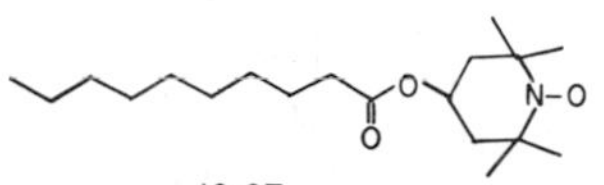

10:OT

simulation. Inspection of the spectra show that this spin label can be expected to exhibit enhanced mobility about its x axis. To further illustrate x-axis anisotropic motion, we refer to a commonly used spin label fatty acid, 12-dimethyl oxazolidine stearate (12NS), which approximates isotropic motion in most solvents. 12NS in the extended state rotates preferentially on its z principal axis. As another example, we take a spin-labeled fatty

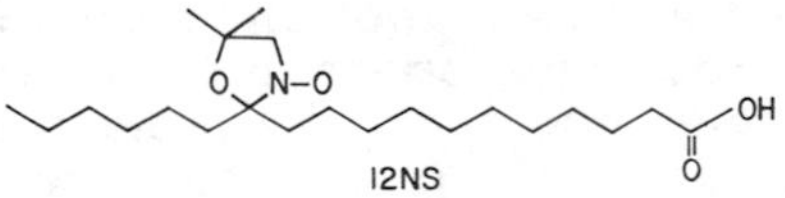

12NS

acid much like 12NS only with the oxazolidine ring turned 90° to the plane of the fatty acid where carbons 4 and 5 of the oxazolidine ring are part of the fatty acid chain.

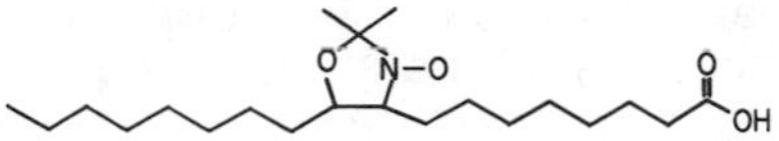

9,10NS

This spin label, 9,10-dimethyl oxazolidine stearate (9,10NS), in the extended state unlike 12NS would be expected to rotate preferentially on its x principal axis and would further be expected to give a signal like 10:OT (Fig. 5, spectrum B). The two spectra are shown in Figure 5, and can be seen to be very similar, with both having line heights of $h_1 > h_0 > h_{-1}$.

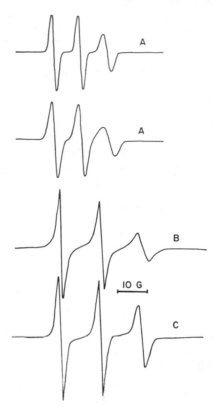

Figure 5

A. Simulated spectra where motion is enhanced about the *x* principal axis. For the upper spectrum the step size limits were $\Delta\theta_z = \Delta\theta_y = 0.8$ radian and $\Delta\theta_x = 3.2$ radians, with a line width of 3 G. For the lower spectrum the step size limits were $\Delta\theta_z = \Delta\theta_y = 0.4$ radian and $\Delta\theta_y = 1.6$ radians and the line width was 4 G. Both simulations involved 64 steps. *B.* Observed spectrum of spin label 10:OT dispersed in Asolectin at 30°. This spin label demonstrates enhanced motion about the *x* axis, as would be expected from its molecular geometry, shown in the text. *C.* 9,10NS, dispersed in Asolectin at 30°, shows an observed spectrum similar to that of 10:OT. Although this spin label is of a different type, it also would be expected to give enhanced motion about the *x* principal axis, as can be seen by its molecular geometry shown in the text.

The analogous simulated spectrum for facilitated tumbling about the y axis is given in Figure 6A. The spin label, 3-spiro-$[2^1$-(N-oxyl-4^1,4^1-dimethyl oxazolidine)] cholestane (3NC), while not expected to give precisely this kind of anisotropy, reflects the augmented midline height of the simulation (Fig. 6B).

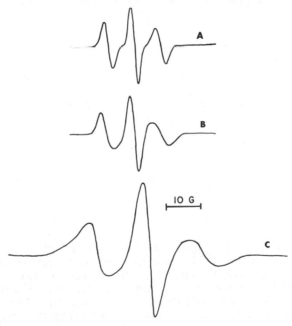

Since the spin Hamiltonian is nearly symmetrical about the z axis, it is predicted that z axis anisotropy will not differ appreciably from isotropic motion, and the simulations bear this out. Nevertheless, the spin

Figure 6
A. Simulated spectrum with motion enhanced about the *y* principal axis. The spectrum was generated with 64 steps using the maximum step limits $\Delta\theta_z = \Delta\theta_x = 0.4$ radian, $\Delta\theta_y = 1.6$ radians, and line width 4 G. *B.* Simulation analogous to *A*, with $\Delta\theta_y = 0.8$ radian, $\Delta\theta_z = \Delta\theta_y = 0.2$ radian, and line width 5 G. *C.* Observed spectrum of the spin label 3NC, dispersed in Asolectin at 30°. This spin label would be expected to give enhanced motion about the *y* principal axis, as can be seen by inspection of its molecular geometry, shown in the text.

label 4-nitroxide stearate (4NS) gives rise to a series of unusual spectra which appear so far to be unique to this system (4NS and related nitroxides).

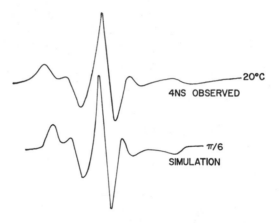

4NS

Seelig (8) has shown that spectra of a series of spin-labeled fatty acids like 4NS in smectic liquid crystals can be interpreted on the basis of a simple model in which the mean angular excursion of a CH_2 group in the fatty acid hydrocarbon tail is limited to some maximum angle related to the proximity of the carboxyl group. This notion is readily incorporated into the random step model by imposing a fixed excursion of the nitroxide z axis from its mean value (conical motion pattern). Assuming an excursion value of about 30° at 20°C as suggested by Seelig, the observed spectrum can be reproduced (Fig. 7). More generally, spectra of this type can be

20°C
4NS OBSERVED

$\pi/6$
SIMULATION

Figure 7

Observed and simulated spectra of the spin label fatty acid, 4NS. In the simulation, the molecular motion was approximated by a sequence of steps consisting of a random reorientation followed by a return to the initial orientation such that the effective excursion from the initial orientation was about 30°. The simulated spectrum closely approximates the observed spectrum of 4NS in Asolectin.

computed by assuming that each molecule of the ensemble has an identical tumbling trajectory. This would be the case, for example, if all the molecules obeyed the simple diffusion equation

$$\frac{\partial \Psi}{\partial t} = \text{const. } x\nabla^2\Psi \qquad (6)$$

often assumed for analytical convenience (9).

TRANSLATION

The translation of spin labels across biological and model membranes can be studied by means of various interaction effects. In this section several methods are outlined for deducing the permeability of intact cells and model membranes.

For liposomes the molecular translation rate across the liposome surface can be inferred for both directions by exploiting the spin destruction of nitroxides by vitamin C. In principle, the experiment consists of forming liposomes having nitroxides on one side of the surface and vitamin C on the other. The disappearance of the ESR signal depends on the translation rate of both types of molecules, assuming that the destruction of the nitroxide by vitamin C in water occurs much more rapidly than the membrane diffusion processes.

Nitroxides can be localized inside and vitamin C outside of liposomes as follows: Liposomes are formed in a nitroxide-containing medium and excess vitamin C is added to this product. The destruction of the nitroxides outside the liposomes proceeds very rapidly resulting in the desired partition (Fig. 8).

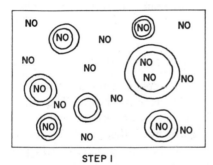 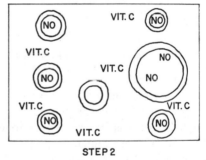

STEP I   STEP 2

Figure 8
Steps in localization of nitroxides (NO) inside, and of vitamin C outside, liposomes (see text).

The rate of spin destruction is given by

$$\frac{\Delta^c \text{ destruction}}{\Delta t} = \frac{\Delta^c \text{ vitamin C}}{\Delta t} + \frac{\Delta^c \text{ NO}}{\Delta t} \qquad (7)$$

The rate of appearance of vitamin C inside the vesicles (Δ^c vitamin C)/Δt is the product of the concentration of vitamin C molecules in the solution, C, and the rate of membrane diffusion, $R_{\text{vit.C}}$. Similarly, (Δ^c NO)/$\Delta t = C_{\text{NO}}R_{\text{NO}}$. Before appreciable spin destruction has occurred, the initial concentrations can be used for $C_{\text{vit.C}}$ and C_{NO}. By deducing the rate of spin destruction for two judiciously chosen concentrations, both rates $R_{\text{vit.C}}$

and R_{NO} can be calculated. At all times care must be exercised to avoid membrane alteration by excessive concentrations of spin labels and vitamin C.

The destruction of an ESR signal by vitamin C was used by Kornberg and McConnell to infer the rate of phospholipid "flip-flops" in dipalmitoyl phosphatidylcholine vesicles (12). In these vesicles there appears to be virtually no diffusion of vitamin C across the membrane.

Many biological cells have the capacity to destroy spin, as stated earlier. One method of estimating the transport of spin labels across the plasma membrane exploits this "spinase" activity. It is assumed that the rate of spin label translation into the cell is given by R_{NO}, and that the spin label concentration outside the cell is given by C_{NO}. The rate of appearance of spin labels within the cell is given by

$$\frac{dC_i}{dt} = R_{NO}C_{NO} - D_sC_i \tag{8}$$

where D_s is the rate of spin destruction and C_i the spin label concentration within the cell.

In general, the rate of spin destruction within yeast cells is quite large, so C_i never builds up appreciably. Hence, the spin concentration decreases linearly with time initially.

INTERACTIONS BETWEEN PARAMAGNETIC MOLECULES

The exchange interaction between nitroxides is complex, involving overlap integrals whose magnitude in liquids is not known. The dipole interaction depends on the molecular trajectories, often of more than two particles, and is not amenable to a straightforward theoretical treatment. In view of these difficulties, concentration effects will be presented in phenomenological terms—an approach that is adequate for most membrane applications.

Varying concentrations of TEMPOL in glycerol can be estimated visually with reasonable accuracy regardless of the glycerol viscosity, as shown in Figure 9. At very high concentrations the phenomenon of exchange narrowing is observed, while concentration broadening occurs in more dilute solutions of the spin label. Considerable broadening of ESR lines is observed for most spin labels in membranes, and rather high local concentrations are required before concentration effects become apparent. One possible application of these interaction effects utilizes two different kinds of nitroxides, preferentially ^{14}N and ^{15}N molecules, so that the hyperfine patterns do not overlap (Fig. 10). It is possible to introduce both spin labels into a membrane system and find that one exhibits concentration broadening by itself while the other does not. If both signals are broadened

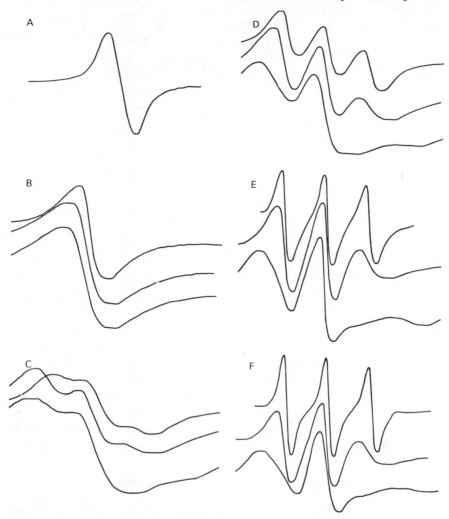

Figure 9

Concentration broadening of the spin label TEMPOL in glycerol at 70°C (top), 20°C (middle), and 5°C (bottom). The spectra A through F represent twofold dilutions beginning with an equimolar concentration of TEMPOL in glycerol and indicate that high spin label concentrations can be estimated.

when the combined spin labels are used, it follows that they are not compartmentalized in separate regions of the membrane. Such compartmentalization effects hold special promise in the study of lipoprotein systems and membrane mutants whose polarity patterns can be manipulated.

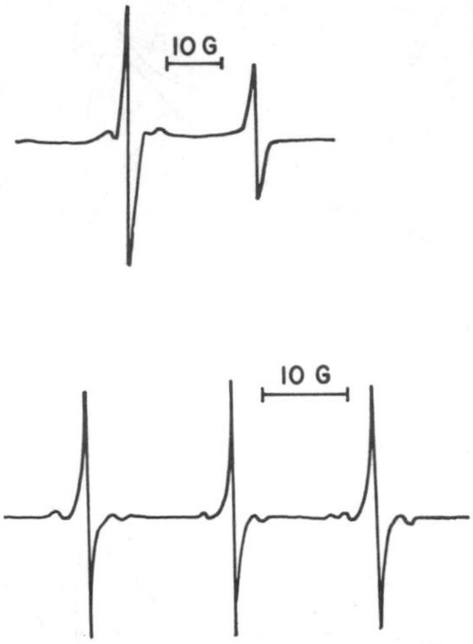

Figure 10

Observed spectra for ¹⁵N and ¹⁴N-containing nitroxides illustrating the two- and three-line character which is necessarily introduced by the different nuclear properties of the two isotopes.

Interactions of membrane lipids with divalent ions can be investigated by utilizing the paramagnetic effects of the transition metals. In aqueous solution these ions broaden nitroxide spectra by the mechanism of the dipole interaction which affects all three hyperfine lines equally.

An analogous effect is produced by molecular oxygen. The ground state of O_2 is a triplet, resulting in a magnetic moment which broadens the nitroxide spectrum at appropriate concentrations. As a rule, organic matrices contain sufficient dissolved O_2 to broaden ESR signals of spin labels appreciably (up to about 5 G).

ARRHENIUS PLOTS

Simulated spectra can be characterized in terms of line widths and heights to provide an approximate prescription for matching observed spectra with calculated ones. Such an analysis yields a motion parameter (the number of steps executed by the molecule during its lifetime) and the effective line width due to other broadening mechanisms.

When such data are used to compute motion parameters of spectra in glycerol taken over a wide temperature range, the results can be summarized in a concise manner, as shown in Figure 11. In this Arrhenius plot all points are seen to fall on one straight line, indicating that the rotational mobility is associated with an activation energy, E_a. The linearity of the points inspires confidence in the relative accuracies of the prescriptions for calculating rotational mobility.

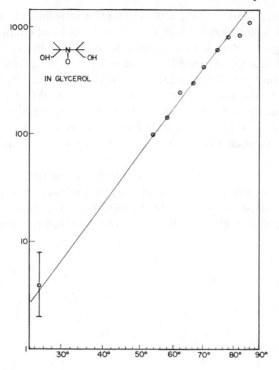

Figure 11

Arrhenius plot of Monte Carlo step parameters required to fit (on the basis of line heights and widths) observed glycerol spectra versus temperature. The low temperature error bar was estimated from simulations which clearly straddled the observed spectrum.

Experiments with a variety of model systems and biological preparations have shown that lipid melts can be inferred from such Arrhenius plots (13). As a rule the Kivelson formula is used to calculate correlation times, and these are used for phase transition studies.

For a sphere that is sufficiently large relative to the solvent molecules, the correlation time (τ_0) can be related to the bulk viscosity. Almost any nonpolar spin label will disperse in purified or heterogeneous fatty acid systems into nonaggregated states when the fatty acid matrix is a liquid. The solubility of the spin label may be greatly reduced, depending on the purity and other factors, when the fatty acid matrix is a solid. Consequently, line broadening due to concentration effects may occur in the range of temperatures where the matrix is a solid. When the spin label is maintained at concentrations where concentration effects do not occur, there is usually no striking difference between spectra just above or just below the melting point. One interpretation of the observed motion in solids is that the nitroxide is an impurity which prevents crystallization in its local environment. Thus, one can imagine the nitroxide molecules embedded in small liquid pools while the bulk matrix has solidified. The volume of these liquid pools varies with temperature, so the rotational mobility is affected by an additional parameter at temperatures below the melting point. While this picture explains the discontinuity of E_a, it is not capable of explaining the relative magnitudes of activation energies. More work with model systems will be necessary before such insight into the meaning of these numbers can be achieved.

Another method of observing phase transition utilizes spin labels which are preferentially solubilized in hydrocarbon matrices, but which also exhibit an aqueous signal. As shown in Figure 12, the partitioning between the two environments changes drastically over the region of the hydrocarbon melt.

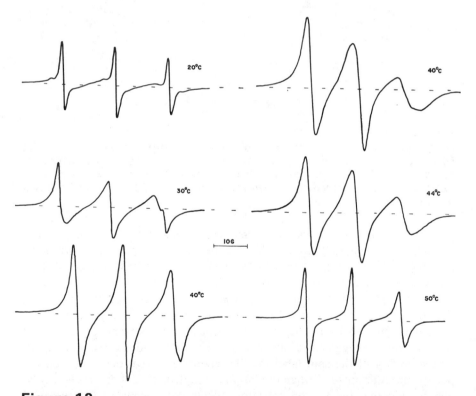

Figure 12
ESR spectra of the spin label 5-nitroxide decane in aqueous dipalmitoyl phosphatidylcholine. Left-hand side aqueous suspension, right-hand side 20% water.

Work with a variety of dilute solutions of nitroxide probes in homogeneous model systems has shown that the observed melting point is a property of the matrix itself and does not reflect a property of the combined system of spin label and solvent. This point is of considerable importance in the use of spin labels to explore biological environments, since the label is generally an impurity in the system whose perturbing influence is difficult to estimate.

The parameters that can be inferred about a solvent containing spin labels are summarized in Table II.

Table II

Parameters That Can Be Inferred About a Solvent Containing Spin Labels

ESR parameters

$\tau_{\|}$	Correlation time about minor axis of nitroxide "ellipsoid"
$\tau_{\perp}$	Correlation time about major axis
$g, \bar{A}_N$	Isotropic g value, hyperfine coupling; contains local polarity information
$\theta\bar{\psi}$	Spatial orientation, if any, of immobilized nitroxide molecules
J	Integrated intensity of entire ESR spectrum; uses include study of spin destruction by organism and detection of "hidden" immobilized spectrum
W	Line width; used to detect oxygen broadening, composite environments, etc.

Derived parameters

E_a	Activation energy; can be used to deduce melting point of solvent
D^T	Translational diffusion constant, inferred from interaction of multicomponent systems

Since spin label work dealing with lipid and membrane systems was initiated, a wide variety of useful compounds have been synthesized. The Keana synthesis has been largely responsible for much of this progress (14). Since then syntheses of sterols (15), fatty acids (16, 17), phospholipids (16), and a variety of other useful labels have been prepared. Aneja and Davis published the synthesis of a spin-labeled glycerol phospholipid containing the spin-label moiety on the phosphate polar head group (18). The Keana method for synthesizing nitroxides on ketone sites has been a major step in allowing experimenters to produce spin labels with desired structural shapes and other features essential for unique problems. Synthesis of long-chain fatty acids containing nitroxide at carbon 12 was carried out, and it was demonstrated that growing *Neurospora* hyphae incorporate the spin-labeled fatty acid into mitochondrial lipids (19). Resulting spectra from the isolated mitochondria demonstrated that the nitroxide-containing fatty acid is in a semifluid lipid environment. The same spin label was also useful in the study of phospholipid vesicles and other lipid systems (17).

12NS

Further spin-labeling experiments on *Neurospora* mitochondria showed that the signals originating from the isolated phospholipids and intact mitochondria were very similar, with the membrane preparation always showing slightly more restriction of the spin-label motion (20). This work also demonstrated that a spin-labeled fatty acid bound to a protein moiety such as bovine serum albumin results in an immobilized signal. No such immobilized signal was found at comparable temperatures originating from biological membranes or phospholipid vesicles using the same spin label.

Griffith *et al.* carried out spin-labeling studies at both 9.5 GHz and 35 GHz (21). They were able to show that the increased magnetic field and frequency required to perform spectra at 35 GHz result in the resolution of mixed spectra by several fold. This procedure is important where resolution of complex spectra due to mixed g values is of interest.

Huang *et al.* used a novel spin-labeled cholesterol linked to a thiol bridge to study phospholipid dispersions and showed that the saturation level of this spin label was about 1 spin label molecule per 10 phospholipid molecules (22).

One of the earliest studies employing spin labels to study biological membranes was carried out by Chapman *et al.* (23). These authors employed the much-used maleimide spin label which alkylates amines and sulfhydryl groups. These authors showed that the spin label alkylated on red blood cells resulted in a two-component signal: One was free in solution and the other demonstrated strong immobilization. Pretreatment of the red blood cells with maleimide, followed by treatment with spin-labeled maleimide, prevented the formation of any of the strongly immobilized component.

MALEIMIDE

The partitioning effect of a small nitroxide soluble in organic solvents and water was used by Hubbell and McConnell (24). This small nitroxide has different hyperfine coupling in organic and aqueous media. In a heterogeneous medium composed of an aqueous phase and a hydrocarbon phase, such as some biological membranes, it was possible to see two high-field lines: One was characteristic of the organic environment of the biological membrane, and the other was characteristic of the more polar aqueous environment. For example, if there is a comparable g-value increase (the g matrix increases with a decrease in polarity) to the A_N decrease caused by the nonpolar regions, then four lines are seen. The fourth line comes about largely because of the g shift so that the low-field and midfield lines still line up fairly well but the high-field line does not. Sometimes a relatively large A_N change may occur resulting in two low-field lines and two high-field lines. There is also a g-matrix shift but the magnitude of the hyperfine coupling shift still allows two low-field lines. A spin label can change its hyperfine coupling constant by 1.9 G as the polarity of the solvent changes from hexane to water while the g value changes by about 0.005 units at 3400 G which corresponds to about 0.85 G. Consequently, while both g-matrix values and hyperfine coupling values change in response to polarity of solvent at 3400 G, the magnitude of the hyperfine coupling changes more than twice that of the g matrix.

Other paramagnetic species such as many heavy and transition metal ions result in spin-spin interactions with nitroxides producing broadening of the nitroxide signal in a distance-dependent manner. If the other paramagnetic species is spaced appropriately from the nitroxide, the signal is broadened proportionate to the distance. Since these species interact with nitroxides in a distance-dependent manner, they only broaden a signal in the same environment. For example, some nitroxides partition in an oil–water emulsion, and if a transition metal paramagnetic ion such as copper is added in appropriate concentration, only the signal originating from nitroxides in the aqueous environment is broadened.

Gotto *et al.* (25) used the spin label maleimide to study the high- and low-density serum lipoproteins. This spin label serves as an alkylating agent where the alkyl group is the nitroxide. When the nitroxide was alkylated onto the free amines and sulfhydryls of the delipidated apoprotein of high-density serum lipoprotein, the ESR signal appeared relatively free. When the lipid moiety was added back to the apoprotein, the signal was largely immobilized. This gave some indication that the lipid elements of the high-density serum lipoprotein surround and place the surface of the apoprotein in hydrophobic zones. This would largely inhibit the free-flopping motion that might be expected to occur on a surface of proteins in aqueous media by moieties containing hydration spheres.

Chapman *et al.* (23) had previously shown that the maleimide spin label gave a two-component signal when red blood cell ghosts were treated, illustrating the heterogeneity of the binding sites. Holmes and Piette (26) used the maleimide spin label and iodoacetate containing a spin label to show at least two and probably three different types of binding sites on the sulfhydryls and amines of red blood cell proteins.

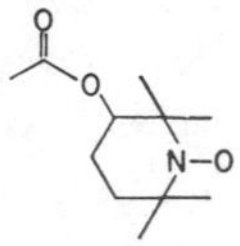

IODOACETATE TEMPOL

These authors used the effects of chlorpromazine and sodium dodecyl sulfate to refine their observations. Chlorpromazine added an immobilized component, and treatment with sodium dodecyl sulfate, in contrast, caused a two-component signal to become a free one-component signal.

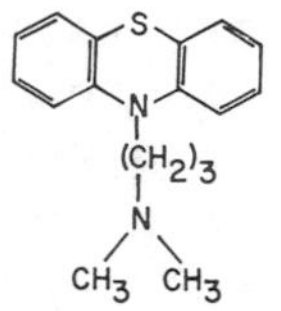

CHLORPROMAZINE

If the heterogeneity of a system, with respect to polarity or viscosity, is nonlinear, it may be possible to resolve three or more components. If the system is linearly heterogeneous, the lines broaden and resolution is not possible. A partially resolvable heterogeneous system is shown in Figure 13, where 2-nitroxide tetradecane (2NT) is dissolved in human high-density serum lipoprotein (27). Vitamin C has been shown to destroy nitroxide signals

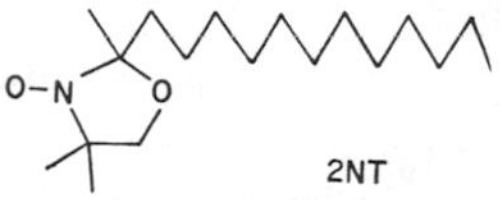

2NT

where molecular collisions are possible (the immediate proximity), probably by chemical reduction (2). In Figure 13, addition of ascorbate removed part of what appeared to be a two-component signal at $0°C$ and still left a two-component signal, showing that 2NT was really clustered in three domains having three different viscosities (27). Since ascorbate is water-soluble, its effect is reduced in hydrophobic zones and makes possible the selective destruction of signals originating from regions which contain water or have aqueous channels. The untreated sample at $30°C$ shows somewhat broadened lines and two third lines. After treatment with 0.3 equivalent of ascorbate, the lines become more narrow and homogeneous in appearance.

Hsia *et al.*, using red blood cell ghosts, tested the effect of eight different methods of membrane solubilization to determine the alterations reflected by spin-labeled protein in the membrane (28). Their results indicated that there was quite a degree of variation among the different methods and

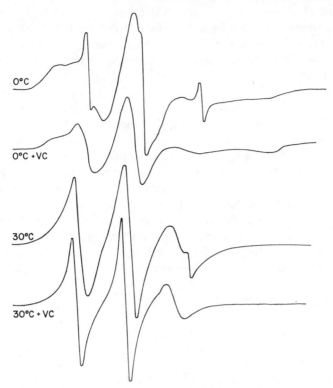

Figure 13

Spectra derived when a spin label, 2NT, is solubilized in high-density serum lipoproteins and dispersed in more than one local environment. Vitamin C can be used to selectively destroy parts of the signals which are in aqueous domains or which interface with aqueous domains.

that some methods resulted in a relatively native apoprotein compared with the protein in the intact red blood cell ghosts. Spin labels may be generally valuable for simple tests to establish the degree of molecular perturbation introduced by some treatment where other assays to detect perturbations may be more troublesome.

Recently, Hubbell *et al.* used red blood cells and a variety of spin labels to show perturbations in the spin label signals induced by foreign molecules (29). This line of research has high potential and will no doubt be of considerable interest to pharmacology and medicine since spin labels are sensitive to small changes in their local environment and reflect these changes in the spin label signal. Some of the chemicals which induced detectable alterations were benzol alcohol, tetracaine, butane thiol, and xylocaine.

Berger *et al.* demonstrated that 12NS bound to red blood cell ghosts was much freer than 12NS bound to the isolated red blood cell apoprotein fraction (30). Using 12NS and three other spin labels, they were also able to infer that some protein apopolar interactions exist in the intact red blood cell membrane.

Calvin *et al.* (31) exploited the use of biradicals in nerve membrane in an attempt to detect transient altered structural changes during the passage of the action potential. While the biradicals were not successful in detecting any of these changes, they have proved to be useful as structural probes. Since the character of the signal is dependent upon the spacing between the two radicals, signals arising from ordered and unordered matrices look quite different. When the matrix is completely ordered and the biradical is extended, a three-line signal is apparent. When the matrix is disordered and the two nitroxides move independently and flop around, there is a certain amount of electronic exchange between the two radicals resulting in a five-line signal. Phase transitions may also be detected in some systems, for at the temperature where a matrix goes from an ordered solid to a disordered liquid the five-line signal appears.

Jost *et al.* synthesized several spin-labeled long-chain fatty acids having an oxazolidine moiety at carbon 5, 7, 12, or 16 of stearic acid (32). These authors demonstrated that the motion of these spin labels in multilayers of phospholipids increased with degree of hydration and increased with an increasing number of methylene groups separating the nitroxide group from the carboxyl group. Long *et al.* showed that cholesterol increased the orientation of 12NS in phospholipid multilayers (33).

Libertini *et al.* showed that 3-nitroxide cholestanol (3NC) and 12-nitroxide stearic acid (12NS) would orient in an oriented egg lecithin multilayer (10). Under the conditions employed, the two spin labels were sufficiently immobilized that the three principal axes did not time average to

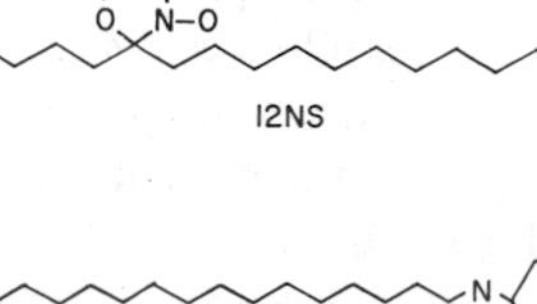

3NC

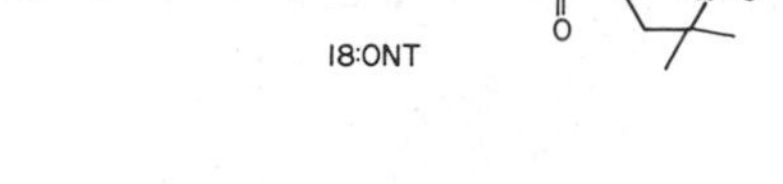

12NS

18:ONT

give an isotropic signal. Because of the way 3NC and 12NS intercalate into a phospholipid multilayer, the two z axes of symmetry are perpendicular to each other. Therefore, when 3NC yields its maximum hyperfine coupling, 12NS yields its minimum coupling. This convincing experiment showed how the spin-labeled sterol and spin-labeled fatty acid intercalated into a natural product, phospholipid. This study was made more convincing and more sophisticated by including computer simulations of the anisotropic spectra. Hsia *et al.* demonstrated that cholesterol alters the spectra of 3NC and the stearate nitroxide ester (18:ONT) in phospholipid dispersions (34). Later, Butler *et al.* demonstrated that sterols increased the ordering of polar lipids in phospholipids as detected by static anisotropy of 3NC (35). They further demonstrated that the hydroxyl group in the 3 position on cholesterol was required for the ordering effect in certain phosphatidylcholine multilayers. These authors also showed that an egg lecithin and cholesterol multilayer oriented 3NC whether hydrated or dehydrated. Hubbell and McConnell (11) showed orientation of spin-labeled fatty acids having the oxazolidine ring close to the carboxyl group in red blood cell ghosts and in the walking-leg nerve fibers of *Homarus americanus*. Nitroxides of this type all have the z principal axis of symmetry parallel to the long axis of the fatty acid. Hubbell and McConnell also demonstrated orientation of a spin-labeled sterol (the y principal axis of symmetry is parallel to the long axis of the sterol molecule) in red blood cells using hydrodynamic sheer as the mechanical method of orientation.

The last few paragraphs have shown that partially hydrated multilayers of egg lecithin (10), dehydrated egg lecithin and cholesterol (35), and red blood cell ghosts and nerve fibers (11) will orient certain spin labels with respect to the applied magnetic field when the structure containing the nitroxide is also oriented with respect to the magnetic field. Such orientation is static in nature and requires that the molecular motion of the spin label be slow enough not to time average away the anisotropic character of the ESR signal. Consequently, static anisotropy is not equivalent to anisotropic motion. As a further example of positional anisotropy, ESR spectra of a mixture of cytochrome c, egg lecithin, and water (1:1:1 by weight) doped with 4NS are shown in Figure 14. These clearly demonstrate that a protein (cytochrome c) and a phospholipid (egg lecithin) partially hydrated will orient under appropriate conditions. This is the same system that has been studied by x-ray diffraction (36).

4NS

12NS was used to probe the membranes of *Mycoplasma laidlawii* (37). The motion of 12NS in whole *Mycoplasma* cells and in extracted polar lipids

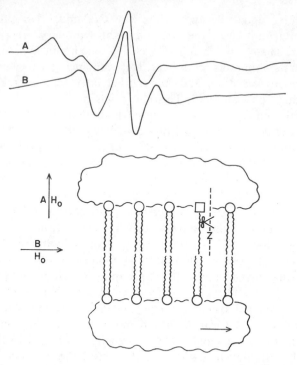

Figure 14

Spectra, at different mechanical orientations, of thick suspensions of egg lecithin, cytochrome c, and water (1:1:1), mixed together and spread in a thin film by pressure. The spin label 4NS was suspended in the emulsions in an approximate ratio of 1 spin label molecule to 500 phospholipid molecules.

showed a temperature dependency. This demonstrated that the fatty acid composition, as analyzed by gas-liquid chromatography, had a distinctively measurable effect on the rotational correlation time (τ_0). The spin label in *Mycoplasma* and in the extracted phospholipids gave very similar τ_0 values and varied in the same way with temperature changes for a given fatty acid composition. When the fatty acid composition was largely unsaturated, the motion was more free than when it was largely saturated. The activation energies were all quite similar. In every case the motion was more free in the extracted phospholipids than in the organism preparation, indicating that the protein elements of the membranes served in some way to restrict the motion of the substituent lipid elements. It was concluded that 12NS goes largely into the hydrophobic lipid zones of *Mycoplasma*, most of which are probably in the membranes. Since freeze-etch electron microscopy was also carried out on the same preparations, it was possible to state that the inclusion of the spin-labeled fatty acid into hydrophobic zones of *Mycoplasma* membranes did not cause gross abnormalities. Rottem *et al.* (38) labeled *Mycoplasma* with spin-labeled long-chain fatty acids and made approximately the same general conclusions as Tourtellotte *et al.*

It was also shown that gramicidin D dispersions immobilized 12NS and that when *Mycoplasma* was treated first with 12NS and then with gramicidin D, immobilization of the 12NS signal also occurred (37). This observation was not dealt with in detail but probably illustrates an important action by certain antibiotics. This mechanism of growth or function inhibition would be by binding membrane hydrophobic zones and rendering them nonfunctional by placing them in a state of rigid viscosity (see Chapter 14 for details).

Waggoner *et al.* showed that two organic, soluble spin labels were solubilized by sodium dodecyl sulfate concomitant with the micelle formation of sodium dodecyl sulfate (39). Later, Waggoner *et al.* showed that the freedom of motion and hyperfine coupling of spin-labeled fatty acids were proportionate to the degree of micelle formation in sodium dodecyl sulfate solutions (40). These studies showed that the nitroxide is solubilized in the micelles of sodium dodecyl sulfate concomitantly with the formation of micelles and that the motion of the spin label is much faster than that of the entire micelle. Therefore, there is much molecular motion independent of the micelle's total motion. These observations also constitute the first indication that spin labels can be used to monitor phase transitions in aqueous amphiphilic dispersions. A critical examination of the line shapes which occur during micelle formation was carried out by Oakes (41) using one of the nitroxides employed in the study by Waggoner *et al.* Oakes' analysis illustrated that τ_0 measurements are inaccurate when spin labels yield heterogeneous signals.

Barratt *et al.* used an intermediate size spin label with a favorable hydrophobic partitioning coefficient to probe aqueous dispersions of the synthetic phospholipid, dipalmityl phosphatidylcholine (42). When they plotted the log of τ_0 against temperature, a phase transition occurred at the expected temperature of 41°C which was detected by perturbation in the otherwise linear plot. These early observations, which indicate a non-linear transition as a function of temperature, are valid phenomenological observations, but do not justify the use of τ_0. Some relative parameters should be used instead.

Lyons and Raison (43, 44) showed that some homeothermic animals and chilling-sensitive plants have mitochondria which give nonlinear relations on an Arrhenius plot of oxygen uptake. These curves are all of the same type, having a straight-line relation above and below some "critical temperature." The temperature represented by the perturbation was thought to be the temperature where a membrane phase transition altered membrane-bound enzyme activity. These findings were made more interesting by the observations that a poikilothermic animal (trout) and a chilling-resistant plant (potato) had mitochondria which gave a straight-line relation for oxygen uptake on an Arrhenius plot over the same temperature range. Subsequent work was carried out on these same systems using the spin label 12NS added in vitro to mitochondrial suspensions (13). The homeothermic animal (rat) and chilling-sensitive plant (sweet potato) showed physical phase transitions at the same temperatures as previously recorded for oxygen uptake (Fig. 15). The poikilothermic animal and the chilling-resistant plant, as before with oxygen uptake, showed no nonlinearity on an Arrhenius plot for the rotational correlation time of 12NS. These data indicate that the respiratory enzyme activity of mitochondria is altered by the physical state of membrane lipids and that the ability of organisms to tolerate temperature extremes may be largely controlled by the physical state of membrane lipids.

The extracted phospholipids from sweet potato and rat liver mitochondria gave similar Arrhenius data, with the change in activation energy taking

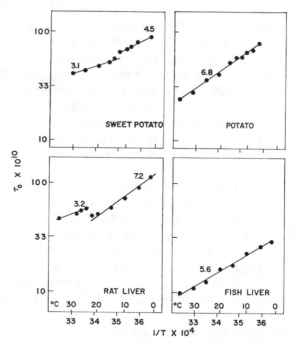

Figure 15

Arrhenius plots of relation between time of phase transition and temperature in different mitochondrial preparations labeled with 12NS.

place at the same temperature. A plot of τ_0 for 12NS in stearic acid is also shown to illustrate the same phenomenon; therefore, the observation is valid whether the nitroxide is localized in small domains of molecular dimensions or dispersed in a bulk phase.

One can imagine that phase transitions observable in isolated phospholipids, in yeast preparations, or other lipid preparations result from more than one cause. For example if the spin label, acting as an impurity, is randomly dispersed in the more liquid phase but is clustered into impurity pools in the more solid phase, we would expect to observe a change in line width at the temperature of phase transition. In yeast systems, using 12NS, no such broadening occurs, as in shown by the plot of log W_o against $°K^{-1}$ (Fig. 16). Instead, there is a change in the way h_0/h_{-1} responds to temperature changes.

A mutant of *Neurospora crassa* deficient in the synthesis of long-chain fatty acids has been probed with spin labels (45). This mutant (*cel*, for fatty acid chain elongation) grows well in the presence of such saturates as palmitate (16:0) or odd-chain components such as pentadecanoate (15:0) and desaturates these to the corresponding Δ^9 *cis*-unsaturate. This mutant is slightly "leaky" and will grow slowly without supplementation. Under nonsupplemented conditions, the addition of 12NS during growth followed by ESR analysis of the hyphae results in an immobilized signal compared with wild type under the same growth conditions. Supplementation of *cel* and wild type in this same way followed by treatment with 12NS and ESR analysis results in almost identical spectra over a temperature range as evidenced by h_0/h_{-1} (Fig. 17). The ratio, h_0/h_1, is a motion parameter which is proportionate to τ_0. Under unsupplemented and very slow growth conditions, the immobilized signals coming from 12NS in *cel* reflect a high viscosity of lipid zones. We suppose that this rigid viscosity of lipid zones is characteristic of poor physiological conditions for organismic function. The more fluid lipid zones of wild type and supplemented *cel* probably are indicative of a more functional system.

From the standpoint of this organism's membrane physiology the most important observation is that a saturated fatty acid is a requirement for growth. The organism is not deficient in desaturase activity and therefore can produce unsaturates when an appropriate saturate is available. The observation that growth cannot occur on pure Δ^9 *cis*-unsaturated fatty acids demonstrates, we believe, the requirement for a heterogeneous fatty acid composition where part of this composition must be supplied by saturated fatty acids.

Several palmitate requirers in yeast have recently been isolated in this laboratory (46). These mutants fall into several complementation groups and probably represent *cel* mutants for different sites in the fatty acid chain elongation enzyme complex. For present purposes, two aspects

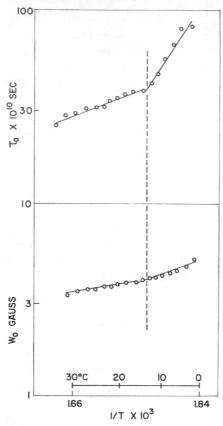

Figure 16
Ratio of the mid- to the high-field line as a function of temperature, and of the line width of the midline as a function of temperature. Local concentration effects of the spin label 12NS do not undergo drastic changes.

are of interest. First, these mutants will not grow on a highly purified Δ^9 *cis*-unsaturated fatty acid; consequently, a saturated fatty acid is a growth requirement. These yeast mutants and the *Neurospora cel* mutant represent the only reported cases of such observations. The biological aspect of these mutants will not be dealt with in detail here. The second aspect is the ESR analyses of these mutants. Since these yeast mutants all require saturated fatty acid it is possible to control the fatty acid composition to some extent. Supplementation with 15:0, 16:0, and 17:0 yields cells which show phase transitions at different temperatures. One of the yeast saturated fatty acid requirers (SH-1) has been examined by ESR using 12NS and several other spin labels. Figure 18 shows the changes in the temperature at which phase transition occurs as a function of fatty acid supplementation,

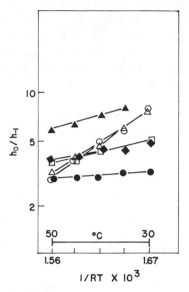

Figure 17

Analysis of ESR spectra of whole *Neurospora* hyphae of wild type and the mutant *cel* labeled with 12NS. ▲, unsupplemented *cel*; ●, unsupplemented wild type; ○ and △, *cel* and wild type supplemented with 17:0; □ and ◆, *cel* and wild type supplemented with a mixture of oleic and elaidic acids. The motion parameters shown on the ordinate are graphed in the standard manner of an Arrhenius plot.

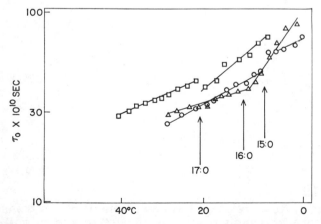

Figure 18

Effect of fatty acid supplementation on temperature at which phase transition occurs. A yeast fatty acid *cel* mutant (SH-1) was grown on three different fatty acid supplementations and then in vitro spin labeled with 12NS. □, growth on 17:0; △, growth on 16:0; ○, growth on 15:0.

and Figure 19 shows an Arrhenius plot of τ_0 where the yeast and lipid extracts have undergone several treatments. Heat treatment at 70°C for 15 min, heat treatment and freeze-thaw, and the extracted phospholipids all give the same temperature of phase transition; phospholipase C, pronase, sodium azide, and microbial lipase all show a phase transition at $13 \pm 1.5°C$. The slopes and absolute τ_0 change somewhat among the different preparations but probably not much more than among several different preparations with the same treatment. Since the temperature at which phase transition occurs is unchanged with different treatments, and τ_0 at various

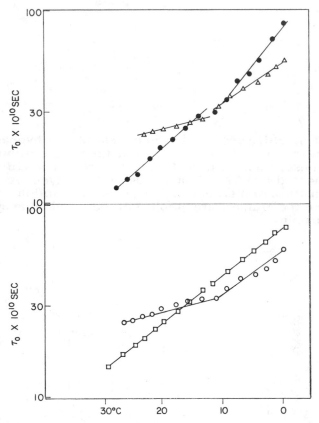

Figure 19

Arrhenius plots of relation between time of phase transition and temperature at which transition occurs in yeasts and lipid extracts subjected to various treatments. •, 12NS-labeled extracted phospholipids after yeast mutant SH-1 had grown on a supplementation of 16:0; △, 12NS-labeled heat-treated yeast; ○, 12NS-labeled heat-treated and freeze-thawed yeast; ▢, 12NS-C-labeled untreated mutant SH-1 grown on 16:0.

temperatures is not drastically changed among the different preparations, it appears that these properties are contained in the lipids and not in the proteins.

Figure 19 also shows the Arrhenius plot of 12-nitroxide stearate cholesterol (12NS-C) added in vitro to the SH-1 mutant grown on palmitate. This line is straight and shows no indication of a phase change in the temperature range. Possibly this indicates that the nitroxide is in an expanded impurity pool and is unable to sense changes in the membrane matrix or that the impurity effect is great enough to prevent a melt. There are numerous indications in the literature that cholesterol abolishes the phase change, and the observation reported here is consistent with others.

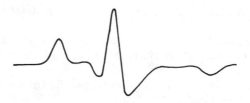

12NS–C

Intact yeast or yeast phospholipids bind several nitroxides in the same general way as reflected by the ESR spectra. A spin-labeled sterol (6NC) is highly immobilized in yeast, indicating that most of it goes into phospholipid domains (Fig. 20). This nitroxide is considerably more free in octadecane or triglycerides than in phospholipids.

Figure 20

ERS spectrum of the spin-labeled sterol 6NC in intact yeast at room temperature.

While the temperature of phase transition is the same for all preparations of yeast having the same fatty acid composition, the spectra are not identical. The spin labeled fatty acid 4NS illustrates this point.

In general terms, we postulate that membrane lipids make up the matrix for the deposition of membrane proteins. The physical dependency of function on physical parameters must partially reside in the lipids. The lipids impose physical limitations on the membrane proteins which may be expressed as restrictions in enzyme activity. For example, in the sweet potato and rat liver mitochondrial preparations discussed earlier, the enzyme-dependent oxygen uptakes have activity changes at the same temperatures where phase transitions are observed in their phospholipids and other mitochondrial lipids.

REFERENCES

1. Hamilton, C. L., and McConnell, H. M. *Structural Chemistry and Molecular Biology*. Ed. A. Rich and N. Davidson. Freeman, San Francisco, 1968.
2. Horwitz, A. Thesis. Stanford Univ., Stanford, Calif, 1969.
3. Pake, G. E. *Paramagnetic Resonance*. Benjamin, New York, 1962.
4. Fermi, E. *Z. Physik* **60**, 320 (1930).
5. Griffith, O. H., Cornell, D. W., and McConnell, H. M. *J. Chem. Phys.* **43**, 2909 (1965).
6. Itzkowitz, M. S. *J. Chem. Phys.* **46**, 3048 (1967).
7. Kivelson, D. *J. Chem. Phys.* **33**, 1094 (1960).
8. Seelig, J. *J. Am. Chem. Soc.* **72**, 3881 (1970).
9. Abraham, A. *The Principles of Nuclear Magnetism*. Clarendon Press, Oxford, 1961, p. 298.
10. Libertini, L. J., Waggoner, A. S., Jost, P. C., and Griffith, O. H. *Proc. Natl. Acad. Sci. U.S.A.* **64**, 13 (1969).
11. Hubbell, W. L., and McConnell, H. M. *Proc. Natl. Acad. Sci. U.S.A.* **64**, 20 (1969).
12. Kornberg, R. D., and McConnell, H. M. *Biochemistry* **10**, 1111 (1971).
13. Raison, J. K., Lyons, J. M., Mehlhorn, R. J., and Keith, A. D. *J. Biol. Chem.* **246**, 4036 (1971).
14. Keana, J. F. W., Keana, S. B., and Beetham, D. *J. Am. Chem. Soc.* **89**, 3055 (1967).
15. Griffith, O. H., and Waggoner, A. S. *Accounts Chem. Res.* **2**, 17 (1968).
16. Hubbell, W. L., and McConnell, H. M. *Proc. Natl. Acad. Sci. U.S.A.* **63**, 16 (1969).
17. Waggoner, A. S., Kingzett, T. J., Rottschaeffer, S., Griffith, O. H., and Keith, A. D. *Chem. Phys. Lipids* **3**, 245 (1969).
18. Aneja, R., and Davis, A. P. *Chem. Phys. Lipids* **4**, 60 (1970).
19. Keith, A. D., Waggoner, A. S., and Griffith, D. H. *Proc. Natl. Acad. Sci. U.S.A.* **61**, 819 (1968).
20. Keith, A. D., Bulfield, G., and Snipes, W. *Biophys. J.* **10**, 618 (1970).
21. Griffith, O. H., Libertini, L. J., and Birrell, G. B. *J. Phys. Chem.* In press.
22. Huang, C., Charlton, J. P., Shyr, C. I., and Thompson, T. E. *Biochemistry* **9**, 3422 (1970).
23. Chapman, D., Barratt, M. D., and Kamat, V. B. *Biochim. Biophys. Acta* **173**, 154 (1969).
24. Hubbell, W. L., and McConnell, H. M. *Proc. Natl. Acad. Sci. U.S.A.* **61**, 12 (1968).

25. Gotto, A. M., Kon, H., and Birnbaumer, M. E. *Proc. Natl. Acad. Sci. U.S.A.* **65,** 145 (1970).
26. Holmes, D. E., and Piette, L. H. *J. Pharmacol. Exp. Ther.* **173,** 78 (1970).
27. Keith, A. D., Mehlhorn, R. J., Freeman, K., and Nichols, A. V. In manuscript.
28. Hsia, J.-C., Schneider, H., and Smith, I. C. P. *Biochim. Biophys. Acta* **202,** 399 (1970).
29. Hubbell, W. L., Metcalfe, J. C., Metcalfe, S. M., and McConnell, H. M. *Biochim. Biophys. Acta* **219,** 415 (1970).
30. Berger, K. U., Barratt, M. D., and Kamat, V. B. *Biochem. Biophys. Res. Commun.* **40,** 1273 (1970).
31. Calvin, M., Wang, H. H., Entine, G., Gill, D., Ferruti, P., Haropold, M. A., and Klein, M. P. *Proc. Natl. Acad. Sci. U.S.A.* **63,** 1 (1969).
32. Jost, P., Libertini, L. J., Hebert, V. C., and Griffith, O. H. In manuscript.
33. Long, R. A., Hruska, F., Gesser, H. D., Hsia, J. C., and Williams, R. *Biochem. Biophys. Res. Commun.* **43,** 321 (1970).
34. Hsia, J.-C., Schneider, H., and Smith, I. C. P. *Chem. Phys. Lipids* **4,** 238 (1970).
35. Butler, K. W., Smith, I. C. P., and Schneider, H. *Biochim. Biophys. Acta* **219,** 514 (1970).
36. Gulik-Krzywicki, T., Schchter, E., and Luzzati, V. *Nature* **223,** 1116 (1969).
37. Tourtellotte, M., Branton, D., and Keith, A. *Proc. Natl. Acad. Sci. U.S.A.* **66,** 909 (1970).
38. Rottem, S., Hubbell, W. L., Hayflick, L., and McConnell, H. M. *Biochim. Biophys. Acta* **219,** 104 (1970).
39. Waggoner, A. S., Griffith, O. H., and Christensen, C. R. *Proc. Natl. Acad. Sci. U.S.A.* **57,** 1198 (1967).
40. Waggoner, A. S., Keith, A. D., and Griffith, O. H. *J. Phys. Chem.* **72,** 4129 1968).
41. Oakes, J. *Nature* **231,** 38 (1971).
42. Barratt, M. D., Green, D. K., and Chapman, D. *Biochim. Biophys. Acta* **152,** 20 (1968).
43. Lyons, J. M., and Raison, J. K. *Plant Physiol.* **45,** 386 (1969).
44. Lyons, J. M., and Raison, J. K. *Comp. Biochem. Physiol.* **24,** 1538 (1970).
45. Henry, S. A., and Keith, A. D. *J. Bacteriol.* **106,** 174 (1971).
46. Henry, S. A., and Keith, A. D. *Chem. Phys. Lipids* **7,** 245 (1971).

9

George Holzwarth

Ultraviolet Spectroscopy of Biological Membranes

The elucidation of protein structure through measurement of the optical activity of protein solutions probably began with Pasteur's studies on albumin, fibrin, and gelatin in the nineteenth century. Early in the present century it was observed that the optical rotation of proteins at the sodium D-line generally becomes more negative upon denaturation, but the molecular basis of these observations remained a mystery. With the synthesis of homopolypeptides and the elucidation of their structure by x-ray diffraction, light scattering, infrared spectrophotometry, and hydrodynamic measurements, it became possible in the 1950s and 1960s to associate in an empirical way observations of optical activity with known secondary structures. Moreover, a quantitative, molecular, quantum mechanical theory for the optical activity of specific structures, such as the α-helix, was developed by Moffitt, Kirkwood, Tinoco, and their coworkers. These empirical and theoretical foundations, coupled with the development of recording instru-

The analysis of membrane artifacts presented in this chapter, especially the results for absorption statistics and Mie scattering, is the product of collaboration with D. J. Gordon. I am also grateful to P. Urnes for his careful reading of this chapter and suggestions for improvement.

This work was supported by Grant NS-07286 from the National Institute of Neurological Diseases and Stroke, U.S. Public Health Service.

ments for the measurement of optical rotation and circular dichroism over a broad wavelength range, stimulated the utilization of optical activity as a structural probe for membranes.

What are the major advantages of this method?

1. The measurements are made in solution, so that pH, temperature, and solvent may be precisely controlled. The effects on membrane structure of compounds such as phospholipase or detergents are therefore readily studied.

2. The measurements are made in a short time upon samples as small as 10^{-4} g.

The information gained from measurements of optical activity is the average protein conformation for all the proteins in the membrane. Little is learned directly about lipids and carbohydrates.

What are the major shortcomings of this experimental technique?

1. The method yields conformational information averaged both for individual molecular species and, more seriously, for mixtures. It is not feasible to separate the contributions of different proteins to the spectra, unless the proteins can be physically isolated.

2. The measured spectra are distorted by light scattering and other artifacts which arise because membrane preparations are usually suspensions rather than true solutions. These artifacts, as will be shown below, are now understood quantitatively, but they misled some earlier workers.

The goal of this chapter is twofold: to describe what has been learned about membranes through optical activity measurements, and to provide a resumé of the basic physical chemistry of protein and lipid optical activity so the reader may grasp the strengths and weaknesses of this technique in sufficient depth to formulate his own experiments and understand further developments. General theory and terms of absorption spectroscopy, optical rotatory dispersion (ORD), and circular dichroism (CD) measurements are discussed first. The ultraviolet optical properties of proteins, polypeptides, and lipids in solution are then reviewed. Readers familiar with these areas may wish to proceed directly to the final two sections, which consider artifacts and special problems of measurements on membrane suspensions, in contrast to solutions, present CD and ORD data for a number of membrane preparations, and attempt interpretation on the basis of earlier discussions. There are three appendices, as follows: A, Some Units and Terms in Electronic Spectroscopy; B, Polarized Light; and C, Relation Between Circular Birefringence and Optical Rotation. The reader is cautioned that references are divided into two classes, General and Literature. The two lists adjoin one another.

ABSORBANCE, OPTICAL ROTATORY DISPERSION, AND CIRCULAR DICHROISM: GENERAL THEORY

Measurement Units

The measurement of absorption spectra is commonplace and need not be discussed in detail here. The measured quantity in ultraviolet spectra is usually the absorbance A; this may be converted to molar absorptivity ε, in liter $\cdot$ cm^{-1} $\cdot$ mole^{-1} (see Appendix A) to remove dependence upon path length and solute concentration.

The measurement of the optical activity of a molecule or a membrane involves the determination either of its optical rotatory dispersion (ORD) curve or of its circular dichroism (CD) spectrum. ORD is measured by passing a beam of monochromatic, linearly polarized light through a quartz cell containing the sample. If the sample is optically active, the direction of the plane of vibration of the electric vector of the linearly polarized light will be rotated. The extent of rotation can be measured by means of an analyzing prism; a plot of rotation versus wavelength is termed a rotatory dispersion curve. The molar unit of rotation $[\phi]$ is given by

$$[\phi] = \frac{\alpha \cdot M}{100 \cdot c \cdot d} \ \text{deg} \cdot \text{cm}^2 \cdot \text{decimole}^{-1} \tag{1}$$

where α is the rotation in degrees, M is the molecular weight of the solute, c is the solute concentration in g/ml, d is the path in decimeters. The Lorentz term $[3/(n^2 + 2)]$ is sometimes used to correct the rotation for local field effects arising from polarizability of the solvent. Here n is the refractive index of the solvent. The symbol $[\phi']$ is then used rather than $[\phi]$, and

$$[\phi'] = \frac{3}{n^2 + 2} [\phi] \tag{2}$$

It is shown in Appendix C that optical rotation α is a direct measure of the difference in refractive indices n_L and n_R for left and right circularly polarized light:

$$\alpha = \frac{\pi(n_L - n_R)l}{\lambda} \tag{3}$$

where α is the measured rotation in radians, l is the path length of the sample, λ is the wavelength of the radiant energy. This relation is derived in Appendix C.

CD, θ, although less familiar than optical rotation, is conceptually simpler; CD is just the difference in sample absorbance for left and for right circularly polarized light. It may be worthwhile here to describe right and left circularly polarized beams briefly; a more detailed description is given in Appendix B. For both right and left circularly polarized plane waves propagating in the direction $\hat{z}$ of a right-handed Cartesian coordinate system, the electric vector $\vec{E}$ at a point in space is of constant amplitude but time-dependent direction. For right circularly polarized light, $\vec{E}$ rotates clockwise in the plane normal to the $\hat{z}$ direction, when viewed toward the source. When examined over a region of space during an instant in time, the tip of the electric vector describes a right-handed helix whose axis corresponds to the $\hat{z}$ axis. This is illustrated diagrammatically in Figure 1. For left circularly polarized light, by contrast, the electric field moves counterclockwise with time about a point in space, when viewed toward the source, and the electric vector at an instant in time forms a left-handed screw.

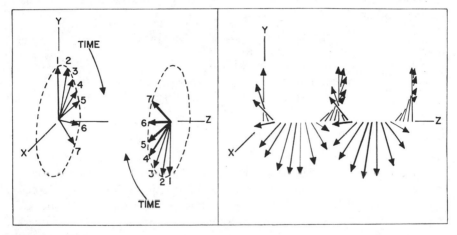

Figure 1
Right circularly polarized light in diagrammatic form. The light is propagating from left to right in the $+\hat{z}$ direction. *Left frame:* **Superposed sequential movie frames showing the field $\vec{E}$ at a two points in space along the $\hat{z}$ axis at times t_1, t_2, t_3, etc.** *Right frame:* **Snapshot showing the variation of $\vec{E}$ in space at the instant t_i.**

CD is today easily measured directly by the Grosjean–Legrand electro-optic method (57), in which a linear polarizer and Pockels cell are used to modulate sinusoidally the polarization of the beam at some audio frequency, f_o, from left to right circularly polarized light and back again. The beam *intensity* varies sinusoidally at frequency f_o only if the sample exhibits CD. The ratio of the intensity modulation at frequency f_o to the unmodulated light intensity, which is easily obtained electronically, gives a direct measure of this difference, $A_L - A_R$.

There are two molar units, $\Delta\varepsilon$ and $[\theta]$, in which CD is commonly reported. The first, $\Delta\varepsilon$, in liters $\cdot$ moles^{-1} $\cdot$ cm^{-1} is

$$\Delta\varepsilon = \varepsilon_L - \varepsilon_R = \frac{A_L - A_R}{c_M l} \tag{4}$$

Here ε_L is the molar extinction coefficient for left circularly polarized light, i.e., $\varepsilon_L = A_L/c_M l$; A_L is the optical density of the solute for left circularly polarized light; c_M is the molar solute concentration, and l is the optical path in centimeters. Corresponding units define ε_R. The second unit for CD, $[\theta]$, is termed the ellipticity and has the units deg $\cdot$ cm^2 $\cdot$ decimole^{-1}, just like optical rotation $[\phi]$. Ellipticity has its origins in the original method for detecting CD in the 1930s (33). The units $[\theta]$ and $\Delta\varepsilon$ are proportional to one another:

$$[\theta] = 2.303 \times 1800\,\Delta\varepsilon/4\pi \simeq 3298\,\Delta\varepsilon \tag{5}$$

One virtue of $[\theta]$, in comparison with $\Delta\varepsilon$, is that ORD and CD curves in these units are of comparable magnitude and identical dimensions. If CD is corrected for refractive index, $[\theta']$ is obtained; for the Lorentz local field one has

$$[\theta'] = \frac{3}{n^2 + 2}\,[\theta]$$

We have already seen that ORD is a manifestation of the difference in refractive indices $n_L - n_R$, whereas CD reflects the absorption difference $A_L - A_R$. It can be shown on very general grounds that ORD and CD are in fact intimately related to one another via the Kronig–Kramers integral transforms (5):

$$[\theta'(\lambda)] = -\frac{2}{\pi\lambda} \int_0^\infty [\phi'(\lambda')]\,\frac{\lambda'^2}{\lambda^2 - \lambda'^2}\,d\lambda' \tag{6}$$

$$[\phi'(\lambda)] = \frac{2}{\pi} \int_0^\infty [\theta'(\lambda')]\,\frac{\lambda'}{\lambda^2 - \lambda'^2}\,d\lambda' \tag{7}$$

Thus if the ORD curve is known for $\lambda = 0$ to $\lambda = \infty$, the CD curve can be calculated, and vice versa. The connection parallels that between the real refractive index and absorbance for an optically inactive solution. It reflects the fact that ORD and CD are both manifestations of the same term, β, in the complex polarizability of the medium. This term describes a linear response of bound electrons to a time-dependent electromagnetic field $\vec{E}_0 e^{iwt}$. The real part of β gives the ORD; the imaginary part of β gives the CD. Although the Kronig–Kramers integrals hold rigorously only with the limits of integration equal to zero and infinity, integration over a finite spectral range can give useful results.

Many absorption and CD bands arising from a single electronic transition are roughly Gaussian in shape. This makes it useful to examine the Kronig–Kramers transform for a Gaussian CD band (5, 33). Thus, suppose we have

$$[\theta(\lambda)] = [\theta_i^0]e^{-(\lambda-\lambda_i)^2/\Delta_i^2} \tag{8}$$

Here $[\theta_i^0]$ is the maximum value of the ellipticity, λ_i is the wavelength at which the maximum occurs, and Δ_i is a measure of the breadth of the band. Such a Gaussian band is plotted in normalized form in the upper part of Figure 2, using the universal coordinates $[\theta_i]/[\theta_i^0]$ and $(\lambda - \lambda_i)/\Delta_i$. The Kronig–Kramers transform of the Gaussian CD band yields the universal normalized ORD curve given in the lower half of Figure 2. The maximum

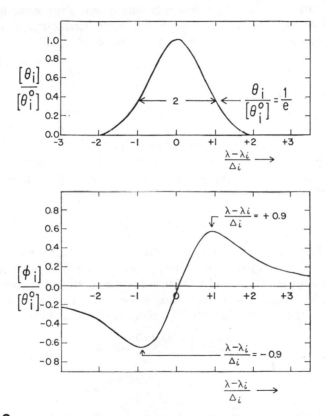

Figure 2

Circular dichroism (*upper graph*) and optical rotatory dispersion (*lower graph*) for a positive Gaussian CD band of the form $[\theta_i] = [\theta_i^0] \exp\{-(\lambda - \lambda_i)^2/\Delta_i^2\}$. The ordinate of both curves is normalized by dividing by $[\theta_i^0]$. The abscissa is normalized by the factor Δ_i.

and minimum values of $[\phi]$ are $+0.58$ $[\theta_i^0]$ and -0.64 $[\theta_i^0]$; these occur for $(\lambda - \lambda_i)/\Delta_i$ equal to $+0.9$ and -0.9, respectively (5, 33).

Dipole and Rotational Strengths

Absorption or CD curves generally exhibit a series of more or less Gaussian bands, each associated with a particular electronic transition in the molecule, e.g., the CD spectrum of a ketone between 200 and 400 nm shows two bands, one at 300 nm and another near 210 nm. The shape of each band reflects subtle vibrational effects which are very difficult to evaluate theoretically; the area (integral) of a band, by contrast, may be more easily evaluated because it requires knowledge of molecular electronic wave functions only. The theoretical quantity with which absorbance by a given band may be most easily compared is the dipole strength D_i of the band. This may be obtained from experiment as follows (5):

$$\left.\begin{aligned} D_i &= 2303 \cdot \frac{3hc}{8\pi^3 N_A} \int \frac{\varepsilon_i(\lambda)\, d\lambda}{\lambda}\ \text{erg} \cdot \text{cm}^3 \\ &= 0.92 \times 10^{-38} \int \frac{\varepsilon_i(\lambda)\, d\lambda}{\lambda}\ \text{erg} \cdot \text{cm}^3 \end{aligned}\right\} \tag{9}$$

where N_A is Avogadro's number. Sometimes the oscillator strength f_i, rather than the dipole strength D_i, is used to describe absorption intensity (see Appendix A). The oscillator strength is also defined by an integral over the absorption band:

$$f_i = 4.30 \times 10^{-9} \int \varepsilon_i \, d\bar{\nu} \tag{10}$$

where $\bar{\nu}$ is in wave numbers and f is a unitless number between zero and one.

The corresponding quantity allowing ready comparison between theory and experiment for ORD and CD curves is the rotational strength R_i of a transition. Rotational strengths are readily evaluated from CD curves by an integral over the partial ellipticity $[\theta_i]$ of the ith CD band (5):

$$\begin{aligned} R_i &= \frac{3hc}{144\pi^2 N_A} \int \frac{[\theta_i(\lambda)]\, d\lambda}{\lambda}\ \text{erg} \cdot \text{cm}^3 \\ &= 0.696 \times 10^{-42} \int \frac{[\theta_i(\lambda)]\, d\lambda}{\lambda}\ \text{erg} \cdot \text{cm}^3 \end{aligned} \tag{11}$$

For a Gaussian CD band, evaluation of the integral (5) in Equation 11 yields

$$R_i = 0.696 \times 10^{-42} [\theta_i^0] \Delta_i \sqrt{\pi}/\lambda_i \tag{12}$$

The Kronig–Kramers transform of the Gaussian band yields the ORD curve given in Figure 2.

For wavelengths far removed from the absorption band, the Kronig–Kramers transform yields (5)

$$[\phi_i(\lambda)] \cong \frac{2R_i}{0.696 \times 10^{-42}\pi} \cdot \frac{\lambda_i^2}{(\lambda^2 - \lambda_i^2)} \tag{13}$$

This equation is the Drude equation; it illustrates how the rotational strength enters into ORD curves in nonabsorbing spectral regions. For a molecule with several electronic transitions, there is the multiterm Drude equation:

$$[\phi(\lambda)] = \frac{2}{0.696 \times 10^{-42}\pi} \sum_i \frac{R_i\lambda_i^2}{\lambda^2 - \lambda_i^2} \tag{14}$$

Thus optical activity both within absorption bands and in nonabsorbing spectral regions is largely characterized by rotational strengths R_i and the wavelengths λ_i of the absorption maxima.

Quantum mechanical calculations show that D_i is in all practical cases given by $\vec{\mu}_i \cdot \vec{\mu}_i$, where $\vec{\mu}_i$ is the electric transition dipole moment of the ith electronic transition. For exact wave functions, $\vec{\mu}_i$ has the form

$$\vec{\mu}_i = \frac{e}{2\pi i v m_e c} \langle \psi_i | \vec{p} | \psi_0 \rangle \tag{15}$$

where $\vec{p}$ is the momentum operator $- i\hbar\nabla$. Here ψ_0 and ψ_i are the electronic wave functions of the molecule in the ground state and in the excited state characterizing the ith transition.

By contrast, the rotational strength R_i is given by

$$R_i = Im\{\vec{\mu}_i \cdot \vec{m}_i\} \tag{16}$$

where $\vec{m}_i$ is the magnetic transition dipole moment for the transition $i \rightarrow 0$ and $Im\{\ \}$ denotes the imaginary part of the bracketed quantity. The general form for $\vec{m}$ is

$$m_i = \frac{e}{2m_e c} \langle \psi_0 | \vec{r} \times \vec{p} | \psi_i \rangle \tag{17}$$

The form of the expression for rotational strength (Equation 16) immediately gives useful insight into the magnitude of R_i. Thus, R_i will be zero if $\vec{\mu}_i = 0$, $\vec{m}_i = 0$, or if $\vec{\mu}$ is perpendicular to $\vec{m}_i$.

It should be noted that, in practice, one knows very little about the wave functions of large molecules such as proteins or lipids. Nevertheless, it is possible, as explained below, to evaluate rotational strengths (Equation 16) approximately through judicious use of empirical information. For example, the amplitude and direction of μ_i for the 190-nm peptide band can be measured directly from polarized absorption spectra of crystals of simple amides. The magnetic transition moment $\vec{m}_i$ of a protein can then in favorable cases be approximated from information on $\vec{\mu}_i$ and the relative orientations of peptides in the protein.

ULTRAVIOLET OPTICAL PROPERTIES OF PROTEINS, POLYPEPTIDES, AND LIPIDS IN SOLUTION

Electronic Transitions of Proteins

The optical activity of a molecule is closely related to its electronic absorption spectrum, since both are manifestations of the same set of transitions between the molecular ground state and electronically excited states. The protein chromophores exhibiting electronic transitions between 180 and 600 nm, the spectral region readily accessible to instruments at present, can be divided into three groups: *1)* the peptide bond itself, *2)* amino acid side chains, and *3)* prosthetic groups such as porphyrins. In the paragraphs below the electronic transitions of these three groups are discussed in turn.

The molecular orbitals of the isolated planar peptide group fall into two classes: those orbitals exhibiting antisymmetry with respect to the amide plane and those which are symmetric with respect to this plane. In the former class are the π orbitals π^+, π^0, π^- of the NCO atomic framework; these exhibit respectively zero, one, and two nodes orthogonal to the amide plane, and are respectively bonding, nonbonding, and antibonding in first approximation. In the ground state there are four electrons occupying the π^+ and π^0 orbitals. An electronic transition corresponding to promotion of one electron from the π^0 orbital to the π^- orbital leads to strong absorption for $\lambda \cong 190$ nm. The molar extinction coefficient at band maximum is 5,000–10,000; the oscillator strength is 0.2–0.3. This transition is electric-dipole-allowed ($\vec{\mu} \neq 0$) and is polarized in the amide plane. For an isolated peptide group, $\vec{m}$ has no component in the amide plane for this transition, so the isolated peptide is optically inactive. For illustrative purposes, the peptide absorption spectrum of a simple polypeptide, poly-L-lysine, is shown in Figure 3 for three different conformations. The 190-nm $\pi^0 \to \pi^-$ transition dominates the spectra in all cases.

In addition to the $\pi^0 \to \pi^-$ transition, peptides show an $n \to \pi^-$ transition near 220 nm. This may be seen as a small but unmistakable long-wave tail in absorption spectrum of the α-helix (Fig. 3). The n orbital involved is a nonbonding orbital on the O atom; the orbital is approximately degenerate with π^0 and is symmetric with respect to the NCO plane. Absorption due to the $n \to \pi^-$ transition is weak; the ε_{max} is less than 300 and the oscillator strength is less than 0.01. Symmetry considerations suggest that the direction of $\vec{\mu}$ is perpendicular to the amide plane. The $n \to \pi^-$ transition has a large magnetic dipole transition moment $\vec{m}$ which is polarized in the amide plane. However, the transition of the isolated planar peptide group has zero rotational strength because $\vec{\mu}$ and $\vec{m}$ are orthogonal. The properties of the $n \to \pi^-$ transition in the NCO system are analogous to those of the $n \to \pi^*$ transition in the C=O system of small ketones, which is also weak and formally electric-dipole-forbidden. The $n \to \pi^*$ transition of ketones is magnetic-dipole-allowed by symmetry, a factor of major importance to its optical activity. The corresponding amide transition is largely localized on the C and O atoms. As a result, it is also profitably viewed as locally magnetic-dipole-allowed, electric-dipole-forbidden. The optical properties of simple amides have been intensively studied in the last few years (19).

The sensitivity of optical activity to conformation in proteins rests on the planar nature of the isolated peptide group. Since R_i depends upon

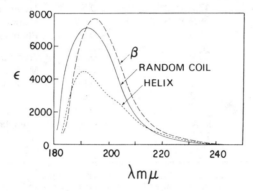

Figure 3
Ultraviolet absorption spectra of poly-L-lysine hydrochloride in aqueous solution: random coil, pH 6.0, 25°; helix, pH 10.8, 25°; β-form, pH 10.8, 52°. (From Rosenheck, K., and Doty, P. *Proc. Natl. Acad. Sci. U.S.A.* 47, 1775, 1961.)

$\vec{\mu}_i \cdot \vec{m}_i$, the rotational strengths of the isolated planar molecule are exactly zero. However, when two or more peptide groups have a fixed and chiral geometry with respect to one another, as in an α-helix or a globular protein the electric-dipole-allowed transitions acquire non-zero parallel $\vec{m}$, and the magnetic-dipole-allowed transitions acquire a non-zero component of $\vec{\mu}$ parallel to $\vec{m}$. The polymer therefore has non-zero optical activity.

The second major class of protein chromophores is the amino acid side chain. Several amino acid side chains possess electronic transitions between 180 and 300 nm which can add to the optical activity of the peptide chromophore. These side-chain transitions are listed in Table I in order of increasing transition energy. Approximate oscillator strengths are also given.

It is apparent from the wavelengths and oscillator strengths listed in Table I that amino acid side chains, and especially the aromatic side chains, can contribute to protein absorption spectra between 180 and 240 nm,

Table I

Ultraviolet Absorption Spectral Data for Amino Acid Side Chain Chromophores

WAVELENGTH	ε (liter · cm^{-1} · mole^{-1} × 10^{+3})	OSCILLATOR STRENGTH	CHROMOPHORE AND TRANSITION
280	5.6	0.1	Tryptophan $\pi \to \pi^*$
274	1.4	0.02	Tyrosine $\pi \to \pi^*$
257	0.2	0.003	Phenylalanine $\pi \to \pi^*$
250	0.3		$\frac{1}{2}$ (Cys)$_2$ $n \to \sigma^*$
223	8.0	~0.2	Tyrosine $\pi \to \pi^*$
219	47.0	0.7	Tryptophan $\pi \to \pi^*$
211	5.9	0.18	Histidine $\pi \to \pi^*$ (?)
206	9.3	~0.3	Phenylalanine $\pi \to \pi^*$
197	20.0	0.25	Tryptophan $\pi \to \pi^*$
195	1.8	0.05	Methionine $n \to \sigma^*$
193	48.0	0.55	Tyrosine $\pi \to \pi^*$
193	1.8	0.03	Cys-SH $n \to \sigma^*$
191	1.8	0.03	$\frac{1}{2}$ (Cys)$_2$ $n \to \sigma^*$
188	60.0	~0.3	Phenylalanine $\pi \to \pi^*$
185	14.0	0.1	Arginine $n \to \sigma^*$
<185			COO$^-$

Adapted from Sober. H. A. (ed.), *Handbook of Biochemistry*, Chemical Rubber, Cleveland, 1968, pp. B16–B18; Donovan, J. W., *Physical Principles and Techniques of Protein Chemistry, Part A*, Ed. S. J. Leach, Academic Press, New York, 1969; and Gratzer, W. B., *Poly-α-Amino Acids*, Ed. G. Fasman, Dekker, New York, 1967.

in spite of their low molar concentration compared with the omnipresent peptide group, whose absorption largely dominates the spectrum (1, 26, 42). More important for the understanding of protein optical activity than the oscillator strengths, however, are the rotational strengths of the side-chain chromophoric transitions. Several factors tend to minimize the contribution of side chains to protein optical activity curves in the far ultraviolet. First, all the chromophores except the disulfide bridge possess at least one mirror plane; several, such as tyrosine and phenylalanine, even possess two orthogonal reflection planes. Second, the strong transitions are all $\pi \to \pi^*$, which means that they are magnetic-dipole-forbidden. Third, the chromophoric groups are invariably isolated from the optically active α-carbon by a methylene group, in contrast to the peptide group itself, so that the chiral perturbation at the chromophore is reduced by distance. Fourth, the local concentration of a given side chain is generally too low to encourage the formation of chiral dimers or a helical array of chromophores which so effectively enhances the rotatory power of the peptide group through resonance excitation interaction (see below).

Some experimental support for the validity of these arguments for aromatic side chains may be extracted from the amplitudes of their rotational strengths near 280 nm. If for a particular side chain the 280 nm $\pi \to \pi^*$ rotational strength is weak or zero, the chromophore is most probably in an almost achiral environment. It is then probable that other $\pi \to \pi^*$ transitions of this chromophore are also optically inactive. One should note, however, that all these arguments are less convincing when applied to the $n \to \sigma^*$ transitions of cystine, cysteine, methionine, and arginine, all of which, though weak, are magnetic-dipole-allowed. Extensive data on side chain optical activity in proteins may be found in the work of Beychok (1).

The probable contributions to membrane optical activity of prosthetic groups such as porphyrins may be assessed in a manner parallel to that used for planar side chains. In most membrane preparations, there is no spectral evidence for these compounds in the form of visible absorption or CD bands. It is unlikely therefore that prosthetic groups are of major importance except in special cases.

Electronic Transitions of Lipids and Carbohydrates

Relatively little is known of the ultraviolet absorbance and optical activity of the glycerolipids, sphingolipids, sterols, and carbohydrates of membranes, because these compounds exhibit strong absorption bands only for wavelengths shorter than 220 nm. Experimental studies are therefore much more difficult than for substances with strong absorption in the visible or near ultraviolet. However, much can be predicted from the optical properties of simple molecules containing the same chromophoric groups. The major potential chromophoric groups for $\lambda > 180$ nm are ketones, dienes, esters, lactones, carboxylate ion, amines, alcohols, and conjugated enones. These are listed in Table II, along with approximate values of the wavelength and molar extinction coefficient at the absorption maximum.

The optical rotatory properties of saturated and unsaturated ketones and of steroids have been exhaustively studied, as have those of dienes; the other groups are largely neglected. This is less serious for understanding membrane optical activity curves than one might suppose, since, as we shall see below, the spectra for $\lambda > 190$ nm are predominantly those of protein constituents. Nevertheless, it is important to be aware of potential contributions by lipids.

Carbohydrates often constitute 5–10% of membrane weight and are invariably inherently chiral, as in D-glucose. Hence, one might expect substantial contributions by carbohydrates to membrane CD curves. For example, the lactone $n \to \pi^*$ transition near 210 nm may be important. However, data for sucrose suggest an exceedingly weak rotational strength.

Table II
Ultraviolet Optical Properties of Some Chromophores in Membrane Lipids

WAVELENGTH (nm)	ε_{max} (liter·cm^{-1}·mole^{-1} × 10^{+3})	CHROMOPHORE	TRANSITION TYPE
280–300	0.01	Ketone	$n \rightarrow \pi^*$
260–300	Variable	Conjugated enone	$n \rightarrow \pi^*$
230–260	3–10	Conjugated diene	$\pi \rightarrow \pi^*$
220–240	9–15	Conjugated enone	Charge transfer
190–230	0.1–1.0	Amine, unionized	$n \rightarrow \sigma^*$
210–220	0.02–0.1	Esters, lactone	$n \rightarrow \pi^*$
190–220	2–10	Ketone	$n \rightarrow \sigma^*$ (?)
170–200	< 4	Amine	$\sigma \rightarrow \sigma^*$
< 190		Alcohol	

Data from Crabbe, P., *Optical Rotatory Dispersion and Circular Dichroism in Organic Chemistry*, Holden-Day, San Francisco, 1965; Jaffe, H. H., and Orchin, M., *Theory and Applications of Ultraviolet Spectroscopy*, Wiley, New York, 1962; Snatzke, G. (ed.), *Optical Rotatory Dispersion and Circular Dichroism in Organic Chemistry*, Sadtler Research Laboratories, Philadelphia, 1967; and Suzuki, H., *Electronic Absorption Spectra and Geometry of Organic Molecules*, Academic Press, New York, 1967.

The other possibly important chromophore, the alcohol group, absorbs only for $\lambda < 190$ nm.

Optical Activity of Synthetic Polypeptides in Solution

STANDARD ORD AND CD CURVES FOR POLYPEPTIDES OF KNOWN CONFORMATION

Abundant evidence from x-ray diffraction, infrared spectrophotometry, light scattering, and hydrodynamic studies (for a review see ref. 21) demonstrates that polypeptides of high molecular weight such as poly-L-lysine and poly-L-glutamic acid can adopt the α-helical, disordered, and β-conformations under suitable solvent conditions. Poly-L-lysine, for example, is disordered at neutral pH because the side chains carry a charged —NH$_3^+$ group. At pH 10.8, the side chains are uncharged and the molecule is α-helical. Upon heating a solution of the α-helical polymer to 55°C, the β-form is obtained. The ORD and CD of these three conformations are profoundly different, as Figures 4 through 6 show.

It is immediately apparent that the optical activity of the two polymers depends profoundly upon conformation of the peptide groups. The side chains have little apparent effect on the optical activity except insofar as they determine the solvent conditions stabilizing the helix or disordered chain. The α-helix is characterized by two negative dichroic bands at 222 and 209 nm having peak heights for $[\theta]$ of approximately $-40,000$ and $-38,000$ for the glutamic polymer, and a positive dichroic band at 191 nm with $[\theta]$

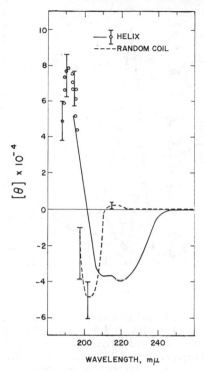

Figure 4

Circular dichroism of poly-L-glutamic acid in 0.1 M NaF. Helix, pH 4.3; random coil, pH 7.6. (From Holzwarth, G., and Doty, P. *J. Am. Chem. Soc. 87*, 218, 1965. Copyright 1965 by the American Chemical Society. Reprinted by permission of the copyright owner.)

equal to 80,000 at the peak. There is a distinct and characteristic "notch" in the negative CD at 215 nm. The α-helix spectrum can be resolved into three roughly Gaussian bands with rotational strengths -22, -29, and $+81$ erg $\cdot$ cm^3 (28). The ORD curve of the α-helix is characterized by a trough at 233 nm ($[\phi] \cong -19,000$), a crossover at 222–223 nm, an inflection point at 210 nm, a peak at 198 nm ($[\phi] \approx 100,000$), and a further crossover at 188–190 nm (13, 60). The ORD features can be generated by Kronig–Kramers transform of the CD curve (15, 28). The electronic origins of the observed rotatory power will be discussed below.

The disordered chain CD curve exhibits a small negative band at 238 nm, a somewhat more prominent but weak positive CD maximum at 217 nm ($[\theta] \cong 3000$), and a strong negative band ($[\theta] \cong -35,000$) at 198 nm. The corresponding ORD curve shows weak features in the 215- to

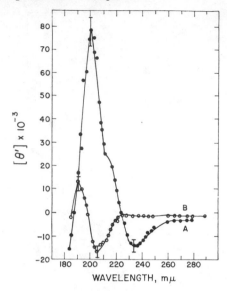

Figure 5
Rotatory dispersion of α-helical and randomly coiled poly-L-glutamic acid. *Curve A*, helix, pH 4.3; *curve B*, coil, pH 7.1. (From Blout, E. R., Schmier, I., and Simmons, N. S. *J. Am. Chem. Soc. 84*, 3193, 1962. Copyright 1962 by the American Chemical Society. Reprinted by permission of the copyright owner.)

240-nm region, a prominent trough at 205 nm ($[\phi] = -20,000$), and positive peak at 189 nm.

The β-form of poly-L-lysine shows features distinct from both the α-helix and random coil (Fig. 6). The CD curve shows a negative extremum at 218 nm with $[\theta] = -12,000$, and positive peaks at 197 and 189 nm with $[\theta] = 41,000$ and $37,000$ (7). The corresponding ORD curve shows a trough at 231 nm, a peak at 203 nm ($[\phi] = 36,000$), and crossovers at 243, 225, and 195 nm. The β-form in poly-L-serine shows similar features with a few differences in amplitudes and positions of wavelength maxima (40).

Poly-L-proline has been shown to adopt two different helical conformations, depending upon the solvent (34). In poor solvents such as propanol the molecule folds in a compact right-handed helix called polyproline I, in which peptide units are all *cis*. In good solvents, such as acetic acid or water, the molecule folds as polyproline II, which is a left-handed helix of *trans* peptide units. The CD curves of polyprolines I and II differ profoundly from one another (Fig. 7). The corresponding ORD curves are thus also radically different from one another. These curves may be used as empirical "fingerprints" of these structures.

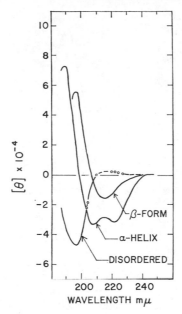

Figure 6

Circular dichroism of poly-L-lysine in the α-helical, disordered, and β-conformations. (Adapted from Holzwarth, G., and Doty, P., *J. Am. Chem. Soc.* **87**, 218, 1965, and Timasheff, S. N., *et al.*, *Conformation of Biopolymers,* Ed. G. N. Ramachandran, Academic Press, New York, 1967.)

ELECTRONIC ORIGINS OF POLYPEPTIDE CD AND ORD SPECTRA

The extraordinary sensitivity of the optical activity of polypeptides to conformation in the 180- to 240-nm spectral region has received intensive theoretical attention. It is known from measurements of absorption spectra and from theoretical calculations that a $\pi^0 \rightarrow \pi^-$ transition and an $n_1 \rightarrow \pi^-$ transition of the peptide group occur in this spectral region; these transitions are thus the most probable source of the optical activity. The theoretical basis of the optical activity of $\pi^0 \rightarrow \pi^-$ transition in the α-helix has been studied (35, 36, 51). The monomer units are sufficiently far apart that electron overlap between peptide units is negligible. However, resonance excitation interaction occurs among the energetically degenerate $\pi^0 \rightarrow \pi^-$ transitions in different peptide groups. This leads to splitting of the single excited state of the monomer into a band of N states, if there are N peptide units in the polymer. Transitions from the ground state described by the wave function ψ_0 to the N excited state wave functions ψ_K occur at slightly different photon energies and possess different rotational strengths.

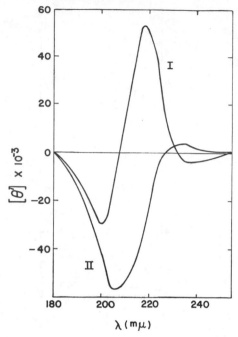

Figure 7

Circular dichroism [*θ'*] of polyproline I in *n*-propanol and poly-proline II in water. (Data from Timasheff, S. N., *et al. Conformation of Biopolymers*. Ed. G. N. Ramachandran. Academic Press, New York, 1967.)

The underlying idea is readily described by a first-order theory for a dimer, for which $N = 2$ (see ref. 50). The monomers, designated by L and R, are at fixed distance and in fixed orientation with respect to one another; the monomers are themselves optically inactive, like planar peptide groups. As shown in Figure 8, $\vec{r}_{12}$ is a vector from group L to group R. Now let the

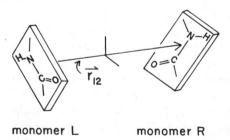

monomer L   monomer R

Figure 8

Diagram of an optically active dimer consisting of two peptide groups.

ground state electronic wave functions of the monomers be ψ_L^0 and ψ_R^0; and let the corresponding monomer $\pi^0 \pi^-$ excited states be ψ_L^σ and ψ_R^σ. Then ground state wave function Ω_0 of the dimer is given by $\Omega_0 = \psi_L^0 \psi_R^0$. The dimer excited state wave functions are not given by the simple product functions $\psi_L^\sigma \psi_R^0$ and $\psi_L^0 \psi_R^\sigma$, because these functions are degenerate. Resonance excitation interaction lifts this degeneracy. The 0th order wave functions of the dimer are therefore

$$\Omega_+^\sigma = \frac{1}{\sqrt{2}}(\psi_L^\sigma \psi_R^0 + \psi_L^0 \psi_R^\sigma) \tag{18}$$

and

$$\Omega_-^\sigma = \frac{1}{\sqrt{2}}(\psi_L^\sigma \psi_R^0 - \psi_L^0 \psi_R^\sigma) \tag{19}$$

The two transitions between the dimer ground state and the dimer excited states, $\Omega_0 \rightarrow \Omega_+^\sigma$ and $\Omega_0 \rightarrow \Omega_-^\sigma$ can then, following Tinoco (50), be characterized as to energy, dipole strength D, and rotational strength R. If $h\nu_0$ is the energy of the monomer transition $\psi^0 \rightarrow \psi^\sigma$ in ergs, and if $\vec{\mu}$ is the monomer electric transition dipole moment in esu $\cdot$ cm, then the energies $h\nu_+$ and $h\nu_-$ of transitions $\Omega_0 \rightarrow \Omega_+^\sigma$ and $\Omega_0 \rightarrow \Omega_-^\sigma$ are given, to first order, by

$$h\nu_+ = h\nu_0 + V_{LR} \tag{20}$$
$$h\nu_- = h\nu_0 - V_{LR} \tag{21}$$

where

$$V_{LR} \cong \frac{1}{r_{12}^3}\left\{\vec{\mu}_L \cdot \vec{\mu}_R - \frac{3(\vec{r}_{12} \cdot \vec{\mu}_L)(\vec{r}_{12} \cdot \vec{\mu}_R)}{r_{12}^2}\right\} \tag{22}$$

i.e., V_{LR} is the first-order correction to the energy due to the interaction of the transition moments. But for the $\pi^0 \rightarrow \pi^-$ transition, the magnitude and direction of $\vec{\mu}$ are known from experiments on crystals of simple amides. Thus the energy difference between ν_+ and ν_-, which equals $2V_{LR}$, is easily calculated from the geometry of the dimer.

The dipole and rotational strengths of the two transitions in the dimer may be evaluated from the electric and magnetic transition dipole moments of the dimer, $\vec{\mu}^+$, $\vec{\mu}^-$, $\vec{m}^+$, and $\vec{m}^-$. Because of the orthogonality of ψ^0 and ψ^σ for each monomer, the electric transition dipoles have a simple form:

$$\vec{\mu}^+ = (\vec{\mu}_L + \vec{\mu}_R)/\sqrt{2}$$
$$\vec{\mu}^- = (\vec{\mu}_L - \vec{\mu}_R)/\sqrt{2} \tag{23}$$

Moreover, if the magnetic transition moment of the monomer is zero, then the magnetic transition moments of the dimer also reduce to a simple form involving the electric transition moment of the monomer:

$$\vec{m}^+ = -i\pi v_0 [\vec{r}_{12} \times \vec{\mu}_L^* - \vec{r}_{12} \times \vec{\mu}_R^*]/\sqrt{8}c$$

$$\vec{m}^- = -i\pi v_0 [\vec{r}_{12} \times \vec{\mu}_L^* + \vec{r}_{12} \times \vec{\mu}_R^*]/\sqrt{8}c$$

(24)

It is then easily shown that

$$D^+ = \vec{\mu}^+ \cdot \vec{\mu}^+ = \tfrac{1}{2}\{\vec{\mu}_L \cdot \vec{\mu}_L + \vec{\mu}_R \cdot \vec{\mu}_R\} + \vec{\mu}_L \cdot \vec{\mu}_R$$

$$D^- = \vec{\mu}^- \cdot \vec{\mu}^- = \tfrac{1}{2}\{\vec{\mu}_L \cdot \vec{\mu}_L + \vec{\mu}_R \cdot \vec{\mu}_R\} - \vec{\mu}_L \cdot \vec{\mu}_R$$

(25)

and

$$R^+ = Im\{\vec{\mu}^+ \cdot \vec{m}^+\} = +\pi v_0 \vec{r}_{12} \cdot \{\vec{\mu}_L \times \vec{\mu}_R\}/2c$$ (26)

$$R^- = Im\{\vec{\mu}^- \cdot \vec{m}^-\} = -\pi v_0 \vec{r}_{12} \cdot \{\vec{\mu}_L \times \vec{\mu}_R\}/2c$$ (27)

where D and R are both in erg $\cdot$ cm^3. Thus, the absorption band at energy hv_0 is split into two bands at hv_+ and hv_-. The dipole strengths of these two transitions are in general unequal, but the total dipole strength of the dimer is just twice that of the monomer, independent of geometry. For the rotational strengths, by contrast, $R^+ = -R^-$, i.e., the rotational strengths are of equal intensity but opposite sign. This is true regardless of the geometry. Such a set of rotational strengths, which arises from a single monomeric transition and has a sum of zero, is termed conservative; this pattern is an automatic consequence of resonance excitation interaction. The energy levels and spectra of the dimer are presented in diagrammatic form in Figures 9 and 10.

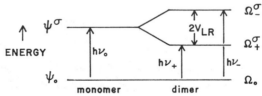

Figure 9
Energy level scheme for a monomer and a dimer.

As can be seen from Equations 26 and 27, the optical activity of a dimer formed from optically inactive monomers will be zero if the relative orientation of the two monomer units is random, since the average value of $\vec{\mu}_L \times \vec{\mu}_R$ will be zero. This is, of course, a necessary consequence of the symmetry argument first recognized by Pasteur in 1848: "Imagine a [right-handed] spiral stair whose steps are cubes, or any other objects with superposable images. Destroy the stair and the dissymmetry will vanish. The dissymmetry of the stair was simply the result of the mode of arrangement of the component steps."

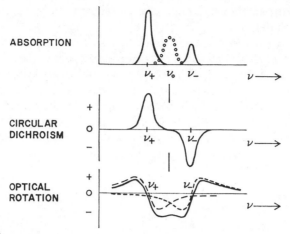

Figure 10

Schematic diagram of the absorption, circular dichroism, and optical rotatory dispersion spectra of an optically active dimer formed from optically inactive monomers. *Upper frame:* **The monomer absorption is shown dotted; the two bands of the dimer, centered at frequencies v^+ and v^-, are shown with solid lines.** *Center frame:* **Circular dichroism of the dimer showing bands of equal intensity but opposite sign. The monomer circular dichroism is zero.** *Lower frame:* **Optical rotation of the dimer. The rotation of the monomer is zero; the two circular dichroism bands of the dimer give separately the ORD curves shown by the dashed lines. The sum of the two dashed ORD curves gives the resultant ORD of the dimer, shown by the solid line.**

For a polymer with N residues in an α-helical array, there will be N excited state wave functions ψ_K^σ consisting of a linear combination of the N degenerate polymer excited states in which a particular residue, such as the jth, is excited.

$$\Omega_K^\sigma = \sum_{j=1}^{N} C_{jK} \frac{\psi_j^\sigma}{\psi_j^0} \prod_{i=1}^{N} \psi_i^0 \tag{28}$$

The energy of each member of this band of excited states, and the coefficients C_{jK}, can be calculated by diagonalization of an $N \times N$ matrix. The electric and magnetic transition dipole moments can then also be evaluated, and D_K and R_K found for each transition $\Omega_0 \rightarrow \Omega_K^\sigma$. For the $\pi^0 \rightarrow \pi^-$ transition of ordered polypeptides such as the α-helix, theory predicts splitting of 10 nm and very large rotational strengths. The sum of the rotational strengths of the $\pi^0 \rightarrow \pi^-$ transition is in theory equal to zero.

Calculations on the optical activity of the amide $n_1 \rightarrow \pi^-$ transition, which is magnetic-dipole-allowed, show that this transition acquires strong rotatory power through static electric field perturbation of the wave functions by highly polar neighboring amide groups rather than by resonance excitation interaction (43, 59).

Detailed calculations on the optical activity of the α-helix, β-forms, polyproline I and II, and recently, the disordered chain in the 180- to 240-nm region have been reported by Tinoco, Woody and coworkers, and Pysh and coworkers (for a review, see ref. 19). These calculations, which are all based upon the theory developed by Tinoco, are in remarkably good accord with experiment. For example, Figure 11 shows the experimental CD curve of largely α-helical poly-L-alanine together with the results of Woody's calculations for an infinitely long, rigid α-helix.

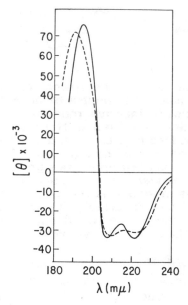

Figure 11

Theoretical and experimental CD curves for α-helical poly-L-alanine. The dashed line is experimental; the solid line, theoretical. (From Woody, R. W. *J. Chem. Phys. 49,* 4797, 1968.)

Analysis of Protein Conformation

The striking differences observed in the optical properties of synthetic polypeptides known to be in the α, β, disordered, polyproline I, and polyproline II geometries prompted the development of methods utilizing optical data to assess the conformation of polypeptides and proteins of unknown secondary structure.

Most extensively used between 1956 and 1965, before far ultraviolet spectra became measurable, was the Moffitt–Yang b_0 method for estimating the relative amounts of α-helix and disordered chains in conformational mixtures. Moffitt predicted that the right-handed α-helical array of planar peptide groups should show two strong rotatory bands which we designate $R_{\parallel}$ and $R_{\perp}$, one near 195 nm with strength -120×10^{-40}, the other near 185 nm with strength $+120 \times 10^{-40}$. The rotatory dispersion of two such bands in the visible spectra should be given, according to the Kronig–Kramers transforms, by a two-term Drude equation:

$$[\phi'] = \frac{2}{0.696 \times 10^{-42}\pi} \left\{ \frac{R_{\parallel}\lambda_{\parallel}^2}{\lambda^2 - \lambda_{\parallel}^2} + \frac{R_{\perp}\lambda_{\perp}^2}{\lambda^2 - \lambda_{\perp}^2} \right\} \qquad (29)$$

This four-parameter equation can be simplified, because $R_{\parallel} = -R_{\perp}$ and $\lambda_{\parallel} - \lambda_{\perp} \ll \lambda_{\parallel} + \lambda_{\perp}$. A series expansion of the four-parameter equation, in inverse powers of $(\lambda - \lambda_0)$, yields the three-parameter Moffitt–Yang equation (see refs. 8 and 60 for thorough reviews of this method).

$$[\phi'] = \frac{a_0\lambda_0^2}{\lambda^2 - \lambda_0^2} + \frac{b_0\lambda_0^4}{(\lambda^2 - \lambda_0^2)^2} \qquad (30)$$

Although Moffitt's original theory for the rotatory bands of the α-helix has been modified by inclusion of the $n_1 \rightarrow \pi^-$ transition and the "end effects" of Moffitt, Fitts, and Kirkwood, the Moffitt–Yang equation retains its value as an empirical means for describing visible ORD curves of helical polypeptides. Indeed the Moffitt-Fitts-Kirkwood terms would also be expected to fit a Moffitt–Yang equation, since they are also of the conservative type with two bands at almost equal energy but opposite sign. The ORD data for α-helical polypeptides between 300 and 600 nm, which cannot be fitted by the one-term Drude equation, can easily be fitted to the Moffitt equation. By plotting $[\phi'](\lambda^2 - \lambda_0^2)/\lambda_0^2$ against $\lambda_0^2/(\lambda^2 - \lambda_0^2)$ for various values of λ_0, it is found that for $\lambda_0 = 212$ nm, the plots are linear. The quantities a_0 and b_0 can then be evaluated from the intercept and slope respectively. The customary units of a_0 and b_0, degrees $\cdot$ cm^2 $\cdot$ decimole^{-1}, are often abbreviated to "degrees." It is found empirically that, although a_0 is quite variable for different side chains, b_0 has the value $-630° \pm 50$ for a variety of synthetic polypeptides which are α-helical in aqueous solution. This is not surprising, since a_0 reflects the averaged contributions of the $n_1 \rightarrow \pi^-$ transition and other nonconservative bands, whereas the b_0 value, by the nature of the Kronig–Kramers transforms, selectively emphasizes the contributions of contiguous bands with equal intensities but opposite sign.

The Moffitt–Yang analysis acquires real utility when the data for various disordered polypeptides are plotted according to the Moffitt–Yang equation with $\lambda_0 = 212$ nm. The values of a_0 are again quite variable, but b_0 is found to fall between $0°$ and $70°$. Since b_0 is very negative for α-helical polymers and essentially zero for disordered chains, the conformation of mixtures of α-helical and disordered regions can be estimated from their b_0 values by a linear interpolation:

$$f_H \approx -\frac{b_0^{obs}}{630} \qquad (31)$$

where f_H is the fraction of residues in the α-helical form and b_0^{obs} is the observed b_0 value of the unknown mixture.

This method and its refinements for analyzing the conformation of polypeptides from visible ORD data, as well as extensive supporting data, are discussed in detail in references 8 and 60. Attempts have been made to incorporate mixtures of α-helix, β-form, and random coil into the theory; these attempts have not been particularly successful. Finally there is some evidence that in organic solvents b_0 depends upon refractive index through factors other than the Lorentz local field correction (16). Such solvent effects could be important for proteins in a lipid environment (see below).

Once ORD and CD curves in the 185- to 240-nm spectral region could be readily measured, these data largely supplanted the Moffitt–Yang method for the analysis of protein secondary structure. The procedures adopted by various investigators are not standardized, but most fall into two categories depending on whether CD or ORD has been measured.

CD. A comparison of the CD curves of polypeptides with a unique conformation (Figs. 4 through 7) reveals that at 225 nm only the α-helix has large negative values of $[\theta]$, as shown in Table III. If the protein can be regarded as a mixture of α-helix and random coil only, then $[\theta]_{225}/40,000$ serves as an approximate measure of α-helix content, f_H. Of course, if $[\theta]_{225}$ lies between $+2,000$ and $-15,000$, contributions from the β-form, and perhaps other ordered structures, can contribute significantly to the measured $[\theta]$ value. The value of the ratio $[\theta]_{225}/40,000$ may then be a poor indicator of α-helical regions. Two features are pathognomic for the presence of β-form: *1*) a negative peak between 215 and 220 nm, where the α-helix has a distinct notch but the β-form has a trough, and *2*) a positive peak near 198 nm (the α-helix CD peak is nearer 190–192 nm). Additional conformational information can be obtained from the 208-nm region, where only the α-helix and polyproline II helices have negative peaks.

Table III

CONFORMATION	$[\theta]_{225}$
α-Helix	$-40,000$
Disordered chain	$+1,500$
β-Form	$-7,000$
Polyproline I	$+15,000$
Polyproline II	$\sim +1,500$

ORD. The characteristic features of the synthetic polypeptide ORD spectra can be used in the same empirical manner to judge conformation. In this case, the trough at 233 nm is a useful guide, as shown in Table IV. The polyproline I and II structures seem to occur only in polymers for which several adjacent amino acids are the special residues proline or hydroxyproline; they seem not to occur in proteins of less than 50% of these amino acids. Thus for most proteins, the two polyproline structures can be neglected. Therefore, if $[\phi]_{233}$ is more negative than -6000, significant α-helical content is indicated and can be estimated. For $[\phi]_{233}$ less negative than -5000, one can only conclude that most of the peptides are *not* in an α-helical array. Other features of the curves may then give useful clues. Attempts to automate such an analysis of ORD curves of proteins by computer fit to standard α-helix, β-form, and disordered chain spectra (27, 48) suggest that it is difficult to evaluate secondary structure in proteins with no one secondary structure predominant.

Table IV

CONFORMATION	$[\phi']_{233}$
α-Helix	$-15,000$
Disordered chain	$-2,000$
β-Form	$-3,000$ to $-4,000$
Polyproline I	$+4,000$
Polyproline II	$-15,000$

Four additional factors enter into the interpretation of protein optical activity curves: *1)* side-chain contributions, *2)* inequivalence between the random coil of synthetic polypeptides and the nonrepetitive secondary structure found in globular proteins, *3)* chain end effects, and *4)* distorted helices.

Inspection of Table I, listing the electronic transitions of side chains in the ultraviolet, makes it obvious that the side chains may make significant contributions to the CD and ORD between 180 and 240 nm. In the case of homopolymers such as polytyrosine, such effects are large. Studies on proteins in the 260- to 300-nm region clearly show that aromatic amino acid side chains can be in a chiral (dissymmetric) environment which becomes achiral upon denaturation. For example, carbonic anhydrase, myoglobin, hemoglobin, chymotrypsin, lactoglobulin, insulin, and trypsin all show weak optical activity in this region (for reviews, see refs. 1 and 7). Side chains may contribute significantly in the 180- to 240-nm spectral region in these cases. However, it seems more likely that the rotational strengths per peptide will be small. More serious contributions can arise from prosthetic groups. For example, the visible ORD spectrum of cytochrome c (53) reveals a wealth of strong bands clearly demonstrating the highly chiral nature of the heme group or its environment. Heme transitions are therefore likely to contribute significantly to ultraviolet CD and ORD data for this protein between 190 and 240 nm.

The second factor complicating the interpretation of protein ORD and CD curves is the lack of correspondence between the disordered chain of homopolymers in solution and the nonrepetitive secondary structure of globular proteins. Such nonrepetitive structure is clearly seen in solved crystal structures of ribonuclease and cytochrome c, for example. The dihedral angles in globular proteins are probably relatively fixed in time even in nonrepetitive dihedral angle sequences because of packing requirements. By contrast, the ORD and CD curves of the synthetic polymers are time averages of many local conformations weighted by Boltzmann factors (12); each individual dihedral angle is a function of time, and entropy considerations lead to dihedral angles quite different from those of the potential energy minimum. Thus, it may be misleading to use random coil ORD and CD data to predict the ORD and CD of structurally nonrepetitive regions of proteins.

The third complication in the utilization of homopolymer data as a standard for proteins lies in end effects. For example, do two short four-residue α-helices have the same optical activity per residue as does one long eight-residue α-helical sequence? Calculations by Tinoco and Woody on the absorption spectra and optical activity as a function of chain length suggest that for 10 residues the α-helix has attained most of its major features. A similar conclusion is suggested by comparison of the observed CD and ORD of myoglobin in solution with the helix content seen in the crystals. From $[\theta]_{225}$, $[\phi']_{233}$, and b_0 one estimates that f_H equals 0.65–0.75; the crystal structure shows f_H equal to 77% in helices containing 16, 16, 7, 7, 10, 10, 10, 19, and 26 residues in α-helical sequences. These data suggest that partial helix contents of proteins are underestimated by ORD and CD.

The fourth factor complicating the direct comparison of homopolymer data to protein ORD and CD curves is distorted helices in proteins, as observed, for example, in the lysozyme crystal. The magnitude of the ORD and CD changes wrought by small distortions has not been determined.

MEMBRANES: PROBLEMS WITH MEASUREMENTS UPON PARTICULATE SYSTEMS

Typical ORD and CD Data for Membrane Suspensions

The utility of optical activity curves of simple protein solutions prompted several bold efforts to measure and interpret the ORD and CD of membrane suspensions by the methods outlined above. Typical data for several membrane preparations are shown in Figures 12 through 15.

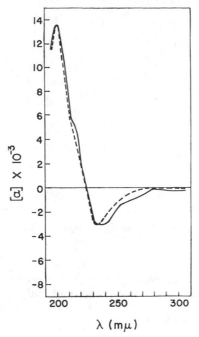

Figure 12

Optical rotatory dispersion of plasma membrane from Ehrlich ascites carcinoma cells in aqueous suspension. The dashed line is the rotatory dispersion of α-helical polyglutamic acid at pH 4.25, but with scale reduced by a factor of 4.5. (Data from Wallach, D. F. H., and Zahler, P. H. *Proc. Natl. Acad. Sci. U.S.A.* **56, 1552, 1966.)**

The observed curves, although reminiscent of a protein with perhaps 10–40% α-helix, show low amplitudes, red shifts of CD bands, and red-shifted ORD features corresponding to 233-nm trough and 199-nm peak. Such data were initially interpreted directly in terms of protein conformation by the methods developed for protein solutions. However, experimental evidence has now accumulated which shows that CD and ORD spectra of membrane suspensions, such as those shown in Figures 12 through 15, are distorted owing to their particulate nature.

The first evidence of systematic effects of particle size was provided by Ji and Urry (30), who showed that similar red shifts and distortions occur in the CD spectra of α-helical polyglutamic acid upon aggregation of the polymer by sonication for various periods of time. Their results are shown in Figure 16.

More recently, the effects of particle size on membrane CD curves were investigated by Schneider *et al.* (6). Sonication of red cell ghosts for

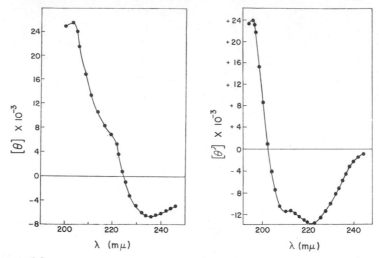

Figure 13

Optical rotatory dispersion (*left*) and circular dichroism (*right*) of aqueous suspensions of human red cell ghosts. (Data from Gordon, A. S., Wallach, D. F. H., and Straus, J. H. *Biochim. Biophys. Acta 183,* 405, 1969.)

various lengths of time breaks the ghosts down into small vesicles. As particle size decreases, the amplitude of the CD increases between 200 and 240 nm, and the red shift disappears, as shown in Figure 17. The distortions are believed to have two main origins, absorption statistics and the unequal scattering of left and right circularly polarized light. After discussing the nature and magnitude of these effects in the paragraphs below, we shall be in a better position to analyze the measured data and their significance.

Absorption Statistics

Absorption statistics (also termed bunching effect, Duysens' effect, concentration masking, and sieve effects) manifest themselves by a decreased amplitude of both absorption and optical activity curves in spectral regions of high absorption by the large particles when compared with the curves for the same number of chromophores in true solution. The origin of this flattening effect has been examined by Duysens (20) and by Rabinowitch (41), who were concerned about corrections to absorption spectra of suspensions of whole cells.

The importance of absorption statistics to CD and ORD measurements of suspensions was first pointed out by Urry and Ji (54). They applied to their CD and ORD curves the correction factors calculated by Duysens for absorption spectra. This procedure is not correct for optical activity

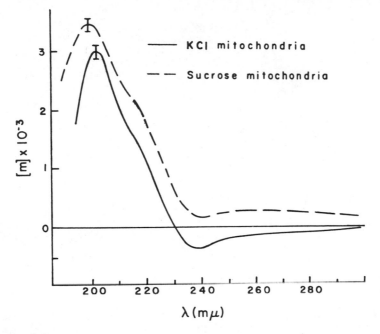

Figure 14
Differential optical rotary dispersion curves of KCl mitochondria and sucrose mitochondria. For calculation of the mean residue rotation, [*m*], it was assumed that all the optical activity was due to biuret protein. A mean residue weight of 115 was used. Note the low amplitude of the Cotton effect. (From Urry, D. W., Mednieks, M., and Bejnarowicz, E. *Proc. Natl. Acad. Sci. U.S.A. 57*, 1043, 1967.)

measurements. In the following discussion we adopt Duysens' model and extend it to the measurement of optical activity. We will find that for a given particle model, such as a sphere or a shell, a different correction factor is needed for CD and ORD curves than for absorption spectra (24).

Let the suspension be represented by a collection of homogeneous cubic particles of identical size distributed randomly in a transparent solvent. The particles, which constitute a fraction q of the total volume of the suspension, are composed of material having absorptivity α per centimeter, i.e., the transmission of z cm of this material is exp $\{-\alpha z\}$. Consider now the beam of light used in a spectrophotometer or spectropolarimeter to measure absorbance, CD, or ORD. Such a beam usually has an area of 10–100 mm². Let I_0 denote the light intensity incident upon the sample cell, whose path length is taken to be h. Now consider I_0 to be decomposed into many fine rays of intensity i_0 falling on different portions of the cell. In passing through

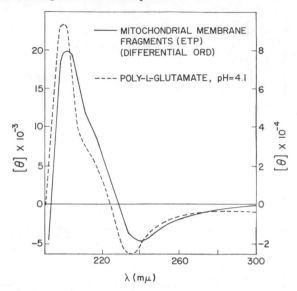

Figure 15

Differential optical rotary dispersion curve of electron transport particles from heavy beef heart mitochondria (left ordinate). For calculation of the mean residue rotation, [*m*], it was assumed that all the optical activity was due to biuret protein. A mean residue weight of 115 was used. Also included in the figure is poly-L-glutamic acid (right ordinate). (From Urry, D. W., Mednieks, M., and Bejnarowicz, E. *Proc. Natl. Acad. Sci. U.S.A.* 57, 1043, 1967.)

h cm of suspension, the ray will traverse z cm of absorber, so that its emergent intensity i will equal $i_0 \exp\{-\alpha z\}$. The total transmitted beam intensity I_{sus} for the suspension will then be the sum of many fine rays i:

$$I_{sus} = I_0 \int_0^h P(z)e^{-\alpha z}\,dz \qquad (32)$$

where $P(z)$ is the probability of finding a path with z cm of absorber. The absorption statistics artifact arises from the nonlinear (i.e., exponential) dependence of I upon absorbing path z. The absorbance A_{sus} of the suspension is given by $\ln[I_0/I_{sus}]$, i.e.,

$$A_{sus} = -\ln \int_0^h P(z)e^{-\alpha z}\,dz \qquad (33)$$

For a true solution, the average number of particles encountered by a ray is very large, e.g., 10^9. Random statistical variations in z deviate by only a very small fraction from the average z. Under these conditions, to a very good

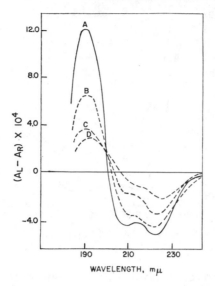

Figure 16

Circular dichroism of polyglutamic acid at various degrees of aggregation, prepared by sonication as follows:

Curve	pH	Sonication time (sec)	OD$_{700}$ (scattering)
A	3.9	0	0.000
B	2.4	0	0.0025
C	2.4	10	0.0040
D	2.4	60	0.0075

The optical density (turbidity) at 700 nm is a measure of particle size, greater turbidity corresponding to *larger* particles. (From Ji, T. H., and Urry, D. W. *Biochem. Biophys. Res. Commun. 34*, 404, 1969.)

approximation, $P(z) = 1$ when $z = qh$ and zero otherwise, i.e., each ray encounters the same path of absorbing material with $z = qh$. Thus the transmitted intensity I_{sol} and absorbance A_{sol} for the solution are given by

$$I_{sol} = I_0 e^{-\alpha q h} \qquad (34)$$

$$A_{sol} = \alpha q h \qquad (35)$$

For a suspension, the number of particles is small, e.g., $10^8/cm^3$, and $P(z)$ will have a broad relative distribution. One then finds that A_{sus} and A_{sol} are unequal; their ratio is termed the absorption flattening coefficient Q_A.

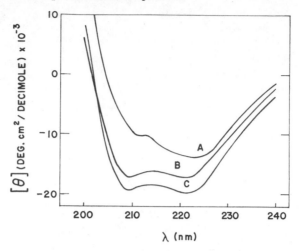

Figure 17

Circular dichroism of whole and of sonicated red blood cell membranes: *A*, whole membranes; *B*, sonicated membranes (total sonicate); *C*, sonicated membranes (supernatant after 1 hr centrifugation at 30,000 × g. (From Schneider, A. S., Schneider, M-J. T., and Rosenheck, K. *Proc. Natl. Acad. Sci. U.S.A. 66*, 793, 1970.)

In order to evaluate Q_A, P(z) must be known. But P(z) is easily found for the simple cubic model utilized by Duysens: a suspension of uniformly aligned cubes, each having edge d. Each fine ray of $d \times d$ square cross-section traverses a column of area d^2 and length h which can be divided along the path into $m = h/d$ cubic boxes of volume d^3. The probability that a given box contains a cube is just q, the volume concentration of absorbing material. The probability P_n that a path contains n particles is just the binomial probability,

$$P_n = \frac{m!}{(m-n)!\, n!}\, q^n (1-q)^{m-n} \tag{36}$$

This P_n corresponds to the P(z) in Equation 33 yielding, for $q \ll 1$ (a realistic assumption),

$$Q_A = \frac{(1 - e^{-\alpha d})}{\alpha d} \tag{37}$$

The quantity Q_A is always less than unity.

The effect of particle size on Q_A is shown diagrammatically in Figure 18, which illustrates the variations in optical path encountered by "fine" rays when particles have d equal to $h/2000$, $h/125$, and $h/20$. In each case the total amount of absorbing material is identical. If in Figure 18 the absorbance A_{sol} of the solution is 2, then Q_A is essentially 1.0, 0.96, and 0.80 for $d = h/2000$, $h/125$, and $h/20$, respectively. At another wavelength, A_{sol} might have the value 0.1 for the same suspensions shown in Figure 18. The quantity Q_A would then be essentially 1.0 for all three cases. Thus absorption statistics lead to a distortion of the absorption spectrum, since, for a given preparation, spectral regions of high absorptivity are more drastically affected than spectral regions of low absorptivity. The absorption bands are therefore flattened. The absorbance will be substantially lower in case C than in case A (referring to Fig. 18).

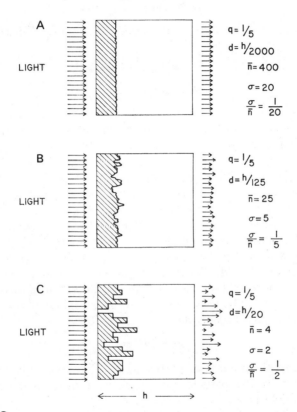

Figure 18

Diagrammatic representation of the variation in character of light transmitted by a solution (upper curve), a fine suspension (middle curve), and a coarse suspension (lower curve) of cubic particles. q, volume fraction of absorbing material; d, length of cube edge; h, optical path length; $\bar{n}$, average number of particles encountered by a ray; σ, root mean square variance in the number of particles encountered.

Duysens' model can easily be extended to CD θ and ORD ϕ (24). Let us consider CD first, and let θ_{sol} and θ_{sus} be the CD of the solution and suspension, respectively. The ratio $\theta_{sus}/\theta_{sol}$, which measures the absorption flattening of CD curves, will be termed Q_B. Imagine the circularly polarized light beam as broken up into a set of fine rays, as above. The particles now generate CD in direct proportion to their number. They also absorb the light, so that the transmission varies exponentially with the number of particles. The measured θ_{sus} is then an intensity weighted average of the CD of the fine rays:

$$\theta_{sus} = \frac{\int_0^h P(z)[z\theta_{sol}/qh] \exp\{-A_{sol}z/qh\}\,dz}{\int_0^h P(z) \exp\{-A_{sol}z/qh\}\,dz} \tag{38}$$

For the cubic model, with $P(z)$ in the form of the binomial distribution, this simplifies to

$$\theta_{sus} = \theta_{sol} \exp\{-\alpha d\} \tag{39}$$

$$Q_B = \exp\{-\alpha d\} \tag{40}$$

This ratio is always less than unity and is, moreover, smaller than Q_A. If the particle is large and highly absorbing, Q_B will be small. The flattening coefficient for optical rotation ϕ turns out to be identical to that for CD (24).

It is useful to evaluate and compare Q_A and Q_B for a range of cube sizes and absorptivities α. This is easily done since Q_A and Q_B depend only upon the parameter A_{sol}/qm. For cubic particles A_{sol}/qm is just the absorbance (base e) per particle. In Figure 19 both Q_A and Q_B are plotted against the parameter A_{sol}/qm. Examination of the figure shows that Q_A and Q_B change rapidly near A_{sol}/qm equal to 1.0, which corresponds to a particle transmission of $1/e$ or 37%. The flattening coefficients Q_A and Q_B have also been evaluated for more realistic models for membrane suspensions, such as spherical shells (24). The results are qualitatively similar to those for cubes, provided m is given a broader definition, as follows:

$$m = N\sigma_p h/q \tag{41}$$

where N equals the number of particles per cubic centimeter and σ_p is the area projected by the particle onto a plane normal to the propagation vector of the light beam.

Typical values of Q_A and Q_B for cubes, spheres, and shells are given in Table V for a range of particle sizes. In this table the parameter α, the absorption index per centimeter within the particle, is taken to be $2.3 \times 10^{+5}$ cm^{-1}. This is a reasonable number for material with a residue molecular weight of 100, molar extinction coefficient 10^4 liter cm^{-1} mole^{-1}, and density 1 g/ml, and corresponds roughly to proteins at 190 nm. For a lower molar extinction coefficient, the values of Q_A and Q_B would be more nearly unity.

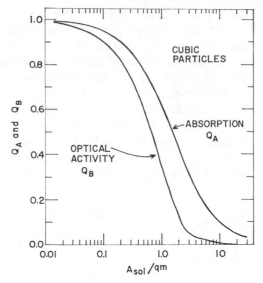

Figure 19

Calculated flattening coefficients due to absorption statistics in the cubic particle model. Note that the coefficient Q_B for CD and ORD is significantly less than the coefficient Q_A for absorbance. (From Gordon, D. J., and Holzwarth, G. *Arch. Biochem. Biophys. 142*, 481, 1971.)

Table V

	EDGE OR RADIUS (cm)	N (ml^{-1})	MW (daltons)	qm	Q_A	Q_B
Cube	10^{-7}	10^{16}	6×10^2	100.0	0.99	0.98
	10^{-6}	10^{13}	6×10^5	10.0	0.89	0.80
	10^{-5}	10^{10}	6×10^8	1.0	0.40	0.11
	10^{-4}	10^7	6×10^{11}	0.1	0.04	0.00
Sphere	10^{-7}	2.4×10^{15}	2.5×10^3	75.0	0.98	0.96
	10^{-6}	2.4×10^{12}	2.5×10^6	7.5	0.84	0.71
	10^{-5}	2.4×10^9	2.5×10^9	0.75	0.29	0.05
	10^{-4}	2.4×10^6	2.4×10^{12}	0.075	0.03	0.00
Shell	1.0×10^{-6}	2.4×10^{12}	2.5×10^6	7.5	0.84	0.71
	3.5×10^{-6}	8.8×10^{10}	6.9×10^7	3.4	0.64–0.70	0.43–0.47
	3.5×10^{-5}	6.7×10^8	9.0×10^9	2.6	0.57	0.34
	3.5×10^{-4} (3.5μ)	6.5×10^6	9.2×10^{11}	2.5	0.57	0.34

For $q = 10^{-5}$, $h = 1$ cm, $A_{sol} = 2.3$, $\rho = 1$ g/ml, and $\alpha = 2.3 \times 10^{+5}$ cm^{-1}. The shell thickness is taken to be 100 Å. From Gordon, D. J., and Holzwarth. G. *Arch. Biochem. Biophys 142*, 481 (1971).

Two points are especially noteworthy from the table. The first is that significant correction factors occur for cubic and spherical particles of molecular weight 10^6 or greater. The second point is the relative insensitivity of Q_A and Q_B to molecular weight for spherical shells. This implies that the flattening correction for membranes depends primarily on membrane thickness and only secondarily upon molecular weight of the particle.

It is instructive to calculate the influence of absorption statistics on CD and ORD spectra of two model systems. This is easily done for spheres of α-helical polyglutamic acid and for spherical shells in which the shell is composed of α-helical polyglutamic acid, but the region inside and outside the shell is transparent. In Figure 20 is shown the ellipticity calculated for

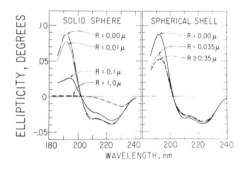

Figure 20

Calculated curves showing the influence of absorption statistics on CD spectra in the Duysens' model. The particles are assumed to be made of material having the absorbance and CD of α-helical poly-L-glutamic acid in solution. Solute concentration is in all cases 0.15 mg/ml; path length is 0.1 cm. *Left frame:* Solid spheres with density 1.5 g/ml, radii as shown. *Right frame:* Spherical shell with shell density of 1.5 g/ml, thickness 100 Å, and radii as shown. (Courtesy of D. J. Gordon.)

a solution and suspensions with concentration 0.15 mg/ml in all cases, but with various particle radii. The solid sphere (left side) shows substantial distortion near 190 nm even for $R = 0.01$ μ, and as R becomes larger, the CD rapidly approaches zero in regions of strong absorption. For spherical shells of fixed shell thickness (right side) large changes occur for the 190- to 200-nm region, but the ellipticity rapidly approaches a non-zero limit for all R greater than 0.35 μ. This is not surprising, since for large particles the membrane thickness, not radius, determines its optical behavior. Similar results for the ORD spectrum of spheres and shells are shown in Figure 21.

It is clear that absorption statistics probably explain, at least in part, the low amplitudes of the CD spectra of membranes at 208 nm relative to the values at 222 nm. However, the calculations provide no evidence for substantial red shifts in band positions, and the shifts in crossover points are, of course, not explicable. These shifts, as we shall see below, are more probably a result of light scattering by the particles.

Light Scattering

For purposes of optical study, red cell ghosts and other membranes are usually suspended in a transparent solvent. These suspensions appear turbid in visible light. This turbidity can be expected to differ for left and for right circularly polarized light, since the particles have different refractive

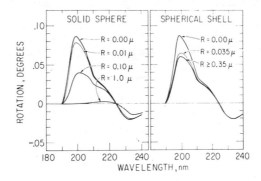

Figure 21
Calculated curves showing the influence of absorption statistics on ORD spectra in the Duysens' model. The particles are assumed to be made of material having the absorbance and ORD of α-helical poly-L-glutamic acid in solution. Solute concentration is in all cases 0.15 mg/ml; path length is 0.1 cm. *Left frame:* **Solid spheres with density 1.5 g/ml, radii as shown.** *Right frame:* **Spherical shells with shell density 1.5 g/ml, thickness 100 Å, and radii as shown. (Courtesy of D. J. Gordon.)**

indices, n_L and n_R, for the two polarizations. Several attempts have recently been made to calculate the probable magnitudes of the scattering effects and their wavelength dependence. Urry and his associates (9, 54) developed a semiquantitative theory for scattering effects in polyglutamic acid aggregates, but their analysis makes some incorrect assumptions about the manner in which light scattering contributes to measured ORD and CD curves (39). A calculation of the effects of scattering in the Rayleigh approximation was made by Ottaway and Wetlaufer (39). The Rayleigh approximation can be expected to give quantitative predictions of ultraviolet scattering effect in CD and ORD for particles smaller than 60 Å; red cell ghosts are probably about $7 \times 10^{+4}$ Å in diameter, and are thus well outside the region where Rayleigh theory is applicable. The problem can be solved rigorously by Mie theory, without any assumptions of particle size, for particles with spherical symmetry, e.g., spheres or spherical shells (23, 25). The theory, as we shall see below, predicts quantitatively the red shifts and distortions so commonly seen in membrane CD and ORD curves. Unfortunately, it is not yet possible to reverse this process, i.e., to calculate the solution properties from measurements on suspensions.

Mie's theory, which is based upon the exact solution of Maxwell's electromagnetic equations, yields the scattering of an incident plane wave by an isotropic spherical particle of known refractive index and absorptivity suspended in a transparent medium. The theory has been strikingly successful in explaining a variety of scattering phenomena in astronomy, physics, and chemistry (31, 56). The extension of the theory to isotropic spherical shells has been given by Aden and Kerker (11). Briefly, a spherical coordinate system $\{r, \theta, \phi\}$ is set up at the center of the spherically symmetric particle. The incident vector wave is expanded in an infinite series of spherical vector wave functions which are general solutions of the vector wave equation in spherical coordinates. The scattered vector waves in all directions are likewise expanded as an infinite series in the same spherical vector wave basis. The coefficients of each term in the series expansions are found from the boundary conditions of the problem, i.e., the field of the scattered waves must be finite everywhere, go to zero as $r \to \infty$, and satisfy continuity conditions at the boundaries between the particle and the transparent medium. The boundary conditions depend upon the complex refractive indices m_1, m_2, and m_3 in the external medium, the shell, and the inner sphere, respectively.

The present problem, to evaluate the effect of Mie scattering on CD and ORD, involves only the evaluation of forward scattering, since no other radiation reaches the detector. This forward scattering can be rigorously calculated provided one knows the particle radius, its thickness if it is a shell, and complex refractive indices m_1, m_2, m_3, for both left and right circularly polarized light. Let us suppose the shell radius is 3.5 μ and the shell thickness is 70 Å, reasonable values for a red cell ghost. For transparent, optically inactive solvents, the refractive indices m_1 and m_3 are real and are the same for left and right circular polarization. We therefore assume $m_1 = m_3 = 1.4$, an average value for water between 190 and 260 nm; dispersion is neglected for simplicity. The refractive indices m_{2L} and m_{2R} of the membrane material are unknown; if these were known, one would have no need to measure the absorbance, CD, and ORD of suspensions, since the optical properties of the solution are directly calculable from m_{2L} and m_{2R}. We have chosen to approximate m_{2L} and m_{2R} from the measured absorbance, CD, and ORD of red cell ghosts solubilized by 0.1% sodium dodecyl sulfate (SDS). One additional piece of information, the mean real part of m_2, is unknown; ultraviolet refractive indices are not readily measured. Let us assume a value 1.7, and neglect dispersion, as for the solvent. We may then write exact expressions for m_{2L} and m_{2R}:

$$m_{2L} = 1.7 - (iA_{\text{sol}}\lambda/4\pi Nh) + \lambda(\phi_{\text{sol}} - i\theta_{\text{sol}})/360hNV \qquad (42)$$

$$m_{2R} = 1.7 - (iA_{\text{sol}}\lambda/4\pi Nh) - \lambda(\phi_{\text{sol}} - i\theta_{\text{sol}})/360hNV \qquad (43)$$

where A_{sol} is the absorbance of the solution in base e, ϕ_{sol} and θ_{sol} are the solution ORD and CD in degrees, h is the path length, N is the number of particles per milliliter, and V is the volume of membrane material in the shell. With these refractive indices fully specified, the Mie scattering of the particles for left and for right circularly polarized light may be rigorously calculated from the equations of Aden and Kerker. From the absorptive loss and phase lag in the forward direction, and the total scattering in non-forward directions, one may evaluate the OD, CD, and ORD of the suspension (44, 56). The results are shown in Figures 22 through 24.

In Figure 22 are shown the CD curves of a red cell ghost suspension, for the same preparation solubilized by SDS, and calculated for a suspension in the manner described above. The dissolved ghosts exhibit a CD curve much like that of a protein with 40% α-helix and no contributions from other

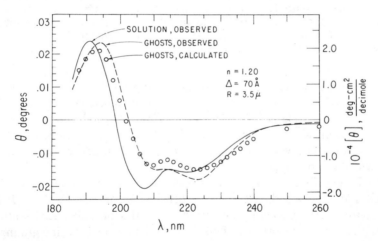

Figure 22

Circular dichroism of human red blood cell ghosts. The solid line is observed data for ghosts solubilized by 0.1% SDS. The dashed line is observed data for ghosts in suspension. The open circles are points calculated by Mie theory for ghosts in suspension, in the manner described in the text. For the calculations it is assumed that solvent refractive index equals 1.4; shell refractive index 1.7; shell thickness 70 Å; shell radius 3.5 μ; number of ghosts/ml 2.2 × 10⁸. (From Gordon, D. J., and Holzwarth, G. *Proc. Natl. Acad. Sci. U.S.A. 68*, 2365, 1971.)

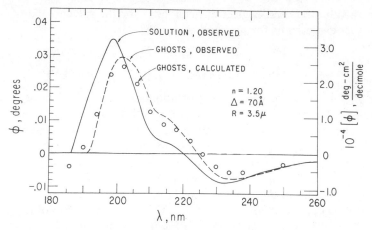

Figure 23

Optical rotation of human red blood cell ghosts. The solid line is observed data for ghosts solubilized by 0.1% SDS. The dashed line is observed data for ghosts in suspension. The open circles are points calculated by Mie theory for ghosts in suspension, in the manner described in the text. For the calculations it is assumed that solvent refractive index equals 1.4; shell refractive index 1.7; shell thickness 70 Å; shell radius 3.5 μ; number of ghosts/ml 2.2 × 10[8]. (From Gordon, D. J., and Holzwarth, G. *Proc. Natl. Acad. Sci. U.S.A. 68*, 2365, 1971.)

conformations; the suspended ghosts show typical 3- to 5-nm red shifts, a diminution of the 208-nm CD band, and an enhancement of the 225-nm CD band. The calculated CD of a suspension (open circles) differs from the solution CD in a manner similar to the observed suspension, i.e., the red shifts are quantitatively reproduced, and the amplitude changes are also similar to those observed.

In Figure 23 are shown the corresponding ORD curves. Again, calculated and observed data for the suspension are in good accord. It is of interest to ascertain whether absorption statistics or differential scattering causes the major distortions. This can be established by integrating the differential scattering for all non-zero angles, which one may call θ_S, subtracting it from the total ellipticity measured in the forward direction, termed θ_T, and call the difference, which must originate in absorption, θ_A. The curve for θ_A includes implicitly the effect of absorption flattening on the CD. In Figure 24 are shown curves of θ_T, θ_S, and θ_A for the red cell ghost model. One can see that θ_S resembles the ORD curve of the ghost, as Urry and coworkers and Ottaway and Wetlaufer have suggested, whereas θ_A resembles the CD of the solution upon correcting for absorption statistics. In the 190- to 200-nm spectral region, the effects of absorption statistics and

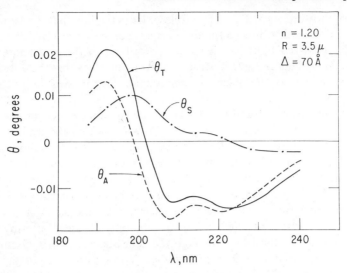

Figure 24

Scattering and absorptive contributions to the calculated CD of red cell ghost suspension. All conditions are identical to those in Figures 22 and 23. The total CD, θ_T, is a sum of an absorptive term θ_A, which appears to be a flattened version of the CD of dissolved ghosts, and a scattering term θ_S, which resembles the ORD of the dissolved ghosts. (From Gordon, D. J., and Holzwarth, G. *Proc. Natl. Acad. Sci. U.S.A. 68*, 2365, 1971.)

differential scattering nearly cancel. The shift in crossover is entirely a result of scattering. The 208-nm band is reduced in amplitude by both scattering and absorption statistics. The 225-nm band is essentially unaltered in amplitude by absorption statistics or by scattering, but it is red-shifted because θ_S, like the ORD curve, is positive for $\lambda < 222$ nm and negative for $\lambda > 222$ nm.

The limitations of the model are perhaps usefully mentioned at this juncture: *1*) multiple scattering is neglected; *2*) the detector solid angle is assumed to be zero; *3*) the membrane is assumed to be isotropic; *4*) dispersion in m_1, m_3, and the mean real part of m_2 is neglected; *5*) the particles are assumed to be spherical. None of these limitations is likely to alter seriously the conclusions derived from the model presented above (25).

Other Optical Artifacts

Several other optical artifacts have been suggested in the past, notably depolarization of the light beam and stray light. Let us consider depolarization first. Depolarization would cause the measured amplitude of the CD to decrease toward zero. This effect follows from the fact that the depolarized light would not contribute to the difference signal generated by left and right circularly polarized light, yet it would contribute to the mean intensity of transmitted radiant energy. Since the instrument measures the ratio of the difference signal to the mean signal, depolarization can only cause an amplitude decrease.

The simplest test for this artifact is the tandem cell technique. In this procedure one prepares two sample cells which we call *A* and *B*. Cell *A* contains the suspension; cell *B* contains a solution of a simple molecule with measurable CD in the same spectral region as the suspension. One then records the circular dichroism spectrum when *A* precedes *B* in the light beam, and also when *A* follows *B* in the beam. If the two spectra are identical, depolarization is negligible, since the spectrum of the *solution* would obviously be adversely affected by scattering only if the scattering occurs prior to passage of the beam through the cell. Such tests of CD curves for membrane suspensions (6, 9, 32) show that depolarization is not significant.

In the case of ORD, turbidity per se seems to pose no problem, since the modern spectropolarimeter measures rotation by a null technique. Slight depolarization of light then has no effect upon measurements except to increase noise. The absence of artifacts from this source has been verified by using the tandem cell method (32).

Stray light can also cause artifacts in optical measurements if sample optical density becomes too high. With the double monochromators now in general use, stray light figures of less than 0.05% are common, and no artifact is expected from this source for sample optical densities less than 2.0 as commonly employed.

Solvent Shifts and Local Field Effects

In addition to the factors cited above, solvent effects may cause membrane protein spectra to differ from the spectra of the identical protein conformation in aqueous solution. Two general mechanisms may be important: local field effects, which change the amplitudes of bands, and solvent shifts, which alter the spectral positions of bands.

The time-dependent electric field experienced by a molecule (the "local field") in a light beam will differ from that which would be expected from the intensity of light external to the sample cell because of time-dependent changes in the medium surrounding the molecule. The Lorentz–Lorenz model is often applied for this effect. The model is based upon a simple macroscopic picture of polarizable solvent surrounding the solute. According to this model, the dipoles induced in the solvent by the field generate at the solute molecule a field parallel to the applied field, so that the local field exceeds the vacuum field. The measured absorption should then exceed the vacuum value by the factor $(n^2 + 2)^2/9n$, whereas the measured CD and ORD should exceed the vacuum field by the factor $(n^2 + 2)/3$ (see ref. 37). The adequacy of the Lorentz–Lorenz model is seriously questioned by comparison with absorption spectral data (49). Its predictions are not obeyed for $\pi \rightarrow \pi^*$ transitions in conjugated hydrocarbons, although they appear to hold for certain other cases, e.g., the $n \rightarrow \pi^*$ transition of benzoquinone. It is found that, as a rule, the correction seems to be too large. Because of its erratic success, the Lorentz–Lorenz correction is rarely applied in absorption spectroscopy. Its use in optical activity is purely a matter of custom; the few studies which have been performed to test the Lorentz–Lorenz factor have shown it to be inadequate (see, for example, ref. 52). More sophisticated theories taking the polarizability of the solute into account have been developed (58), but adequate tests are lacking. How important are these local field effects to the ultraviolet spectroscopy of membranes? It is difficult to be certain at this time, but it seems unlikely that lipids will have a local field effect exceeding 5–10%.

Solvent shifts form the second broad class of solvent effects which may cause the spectra of proteins in contact with lipid to differ from the spectra in aqueous solution. Solvent shifts have been extensively investigated by absorption spectroscopy (49). For nonpolar molecules in nonpolar solvents, dispersion forces between solvent and solute generally lower the energy of the excited states with respect to the ground state because the excited state is more polar. This "dispersion red shift" can reach 300 cm^{-1} when a solute is transferred from heptane to benzene. Polar molecules in polar solvents, such as proteins in water, show strong solvent-solute interactions generating spectral shifts as large as 3000 cm^{-1}. In general, $\pi \rightarrow \pi^*$ transitions undergo red shifts, and $n \rightarrow \pi^*$ transitions undergo blue shifts as the solvent becomes more polar. The explanation is believed to lie in the fact that the dipole moment of the molecule increases during a $\pi \rightarrow \pi^*$ transition but decreases in an $n \rightarrow \pi^*$ transition. The energy of the solute-solvent dipolar complex is therefore lower in the solvated $\pi\pi^*$ excited state than in the unsolvated $\pi\pi^*$ state, and this lowering of the $\pi\pi^*$ state by the solvent exceeds the lowering of the ground state. The energy difference between ground and excited $\pi\pi^*$ states is therefore reduced by solvent, i.e., a red shift is observed. A comparable but reversed argument applies to the $n \rightarrow \pi^*$ blue shift. The energy of $n \rightarrow \pi^*$ transitions is also greatly influenced by hydrogen bonding. How important are such solvent shifts in causing differences between spectra of proteins in membranes and the same proteins in an aqueous environment? It seems likely that these effects are small, because the contribution of a given solvent dipole to the energy of a solute-solvent complex varies roughly as the inverse fourth power of the distance between solute and solvent (dipole-dipole interaction). Thus near-neighbor effects predominate; the near neighbors of a peptide group inside a globular protein are other peptide groups and side chains, not external solvent. Of course, groups on the surface interact with solvent, but their relative numbers are small. In addition to the general solvent effects discussed above, proteins may be subject to a unique type of "local field effect" because their optical properties are influenced by resonance excitation interaction. As shown earlier in this chapter, the spectra are greatly affected by the splitting of the $\pi \to \pi^*$ transition. The splitting energy, as given in Equation 22, will be reduced in magnitude by a factor proportional to $1/\chi$, where χ is an effective optical frequency dielectric constant. Presumably χ may be slightly smaller in membranes than in aqueous solutions; χ will also show dispersion. Thus small changes in χ could alter the CD pattern, because the $\pi^0 \to \pi^-$ bands of opposite sign are strongly overlapped. There exists suggestive experimental evidence for such an effect in measurements of the parameter b_0 of poly-γ-benzylglutamate in a series of α-helix-promoting solvents of different refractive index (16). It is found that b_0 varies between $-706°$ for $n = 1.364$ and $-505°$ for $n = 1.635$. The direction and magnitude of the change are consistent with a mechanism involving the splitting energy. Another interpretation may lie in solvent-induced spectral shifts of the more usual type, as discussed in the previous paragraphs.

Membranes: Results and Interpretation

It is now appropriate to consider ORD and CD data for several membrane preparations. We shall first consider the heavily studied red cell ghost and then discuss in turn plasma membranes of Ehrlich ascites carcinoma cells, mitochondrial membranes, and bacterial membranes.

Red Cell Ghosts

Because of its ready preparation, the human erythrocyte ghost is the most extensively studied membrane preparation (4, 6, 32). Typical CD and ORD data of whole, sonicated, and solubilized ghosts have already been shown in Figures 13, 17, 22, and 23.

Most striking is the close resemblance in shape to the CD and ORD of an α-helical polypeptide or protein; the molar amplitude of the ellipticity is about 40% of that of an α-helical polypeptide in aqueous solution. Some differences are noteworthy: *1)* the CD bands at 208 nm and 195 nm are of unusually small amplitude compared with the 225-nm band; *2)* the CD crossover point is red-shifted to 202 nm from its more usual 200 nm location; *3)* the ORD curve shows a peak at 202 nm of much lower amplitude than the magnitude of the 237-nm peak would lead one to expect; *4)* the ORD peaks, troughs, and crossovers are all red-shifted by about 3–5 nm from those of α-helical polypeptides.

As we have seen, all these spectral features are expected for large particles in suspension. Direct correction for the artifacts is not at present possible. However, the similarity between the CD of the ghost suspension and the CD calculated for spherical shells "prepared" from SDS-solubilized ghosts (Fig. 22) suggests that the CD and ORD spectra of the membrane material would, when properly corrected to solution conditions, be similar to the CD and ORD of the SDS-treated preparation. This implies that the membrane proteins, if solubilized but unaltered in conformation, would show $[\theta]_{192} \sim +22,000$, $[\theta]_{208} \sim -18,000$, and $[\theta]_{222} \sim -15,000$–17,000. The corresponding values for α-helical poly-L-glutamic acid are 89,000, $-39,000$, and $-40,000$ (17). Thus the ellipticities at 208 and 222 nm suggest that about 40% of the membrane protein peptides are in an α-helical conformation; the ORD data are also consistent with this interpretation. It is not possible to say whether the α-helical segments are localized in one of two protein species or are uniformly distributed; such questions can only be settled by resolving the protein mixture into its constituents. The 40% estimate of helix content should be tempered by all the caution one would normally apply to a comparable CD pattern for a single protein, based upon an appreciation of the ignorance of at least three points: *1)* end effects arising from the short helices commonly found in proteins, *2)* contributions from amino acid side chains, and *3)* contributions from the remaining 60% of the peptide residues. Additional uncertainties arise from the special features of membranes: *4)* local field effects by lipids, *5)* circular dichroism contributions from carbohydrates and lipids, *6)* inadequacies in the scattering correction. This is a formidable list; it is probably more formidable than it need be. For example, the absence of significant optical activity near 280 nm argues against the possibility that significant numbers of tyrosine, histidine, phenylalanine, and tryptophan side chains are rigidly held in a chiral environment of a unique handedness. Hence these side chains are unlikely to contribute to the optical activity between 190 and 250 nm (item 2).

Carbohydrates, which account for 7.5% of the weight of ghosts, probably contribute only to circular dichroism for $\lambda < 210$ nm, where alcoholic groups absorb (4).

Measurements by Lenard and Singer (32) on the optical activity of membrane lipids extracted from red cell ghosts showed no significant CD effects from free lipids for wavelengths longer than 215 nm. Of course bound lipid could act differently. Measurements of the ultraviolet circular dichroism of L-α-lecithin and phosphatidylethanolamine (55) showed a positive CD peak at 218 nm and a negative peak at 192 ± 2 nm, in tri-fluoroethanol solution. However, the absence of molar ellipticities for the data makes it difficult to judge the importance of the effects.

It should be noted that lipid-protein interactions probably proceed through the polar groups, such as esters, of lipids. Since the 210-nm ester absorption band is caused by an $n \to \pi^*$ transition, one cannot dismiss the possibility that lipid optical activity is different in solution and in a membrane. However, treatment of the membranes by various lipid perturbants, discussed below, gives a partial answer to the problem.

EFFECTS OF PHOSPHOLIPASE C

Treatment of ghosts by phospholipase C is known to release 60–70% of membrane phosphorus into solution while leaving fatty acid esters and cholesterol undisturbed. Does phospholipase treatment affect the conformation of membrane proteins? Measurements of the optical activity of red cell ghosts before and after treatment by phospholipase C show no change (3). Thus one may conclude that these phosphate groups have no effect on the conformation of protein in ghosts, nor do they contribute to the optical activity directly. Fatty acid esters may nevertheless still contribute in both ways. The absence of any effect of phospholipase C upon protein conformation contrasts sharply with the profound effects of this enzyme on the fatty acids remaining in the membrane; proton nuclear magnetic resonance studies show markedly increased methylene proton mobility upon enzyme treatment (3).

EFFECTS OF PHOSPHOLIPASE A, LYSOLECITHIN, DIGITONIN

Treatment of membranes with phospholipase A, lysolecithin, or digitonin is known to influence the binding of lipids in membranes. All these compounds lead to an increased intensity of the 208-nm CD band, an increase in intensity and a blue shift of the 190-nm $\pi^0 \to \pi^-$ CD band, and a blue shift of the crossover (4). These effects parallel those observed upon sonication of the membrane upon solubilization by SDS, suggesting that the ghosts are broken up into much smaller fragments. However, materials treated with phospholipase A remain large enough to be easily sedimented (4). Further experiments establishing particle size would be useful.

EFFECTS OF DETERGENTS, GLYCEROL, AND 2-CHLOROETHANOL

The influence of dilute SDS upon the CD and ORD of ghosts has been studied by Lenard and Singer (32) and Gordon and Holzwarth (25). Data are given in Figures 22 and 23. SDS effects on band intensities and positions are discussed above. The major changes arise from removal of scattering and absorption statistics artifacts, but it is certainly conceivable that some proteins are altered in conformation as well. For example, the reduced intensity of the 225-nm CD band in SDS-treated material (Fig. 22) compared with the ghost CD may reflect real conformation change.

Addition of glycerol to a ghost suspension increases $[\theta]_{225}$ from $-13,000$ to $-17,000$, probably by altering the scattering (32). A more striking change occurs when membranes are dissolved in 2-chloroethanol (32); $[\theta]_{225}$ changes to $-37,000$, which corresponds to a fully helical protein, as shown in Figure 25. The 2-chloroethanol probably acts in two distinct ways: *1*) it solubilizes the membrane proteins and thus removes artifacts, and *2*) it almost certainly causes a conformational change of the proteins to yield a large increase in the helix content. Such behavior has been observed for other proteins (8), but the extent of the change for membrane proteins is unusually large.

Plasma Membranes of Ehrlich Ascites Carcinoma Cells

The ORD and CD of plasma membrane vesicles 750 Å in radius, which can be prepared from Ehrlich ascites carcinoma cells, reveal band shapes remarkably similar to those of the red cell ghost (4, 10), as shown in Figures 12 and 26. However, the amplitudes of the bands are all approximately one-half those of red cell ghosts. No corrections for absorption statistics or scattering have been applied to these data. Since the vesicles are smaller than red cell ghosts, corrections are also likely to be smaller.

Gordon *et al.* (4) estimate the "helix content" of the membrane protein as 58%. This value is based upon "per cent helix" standards derived from the measured CD and ORD of hemoglobin, myoglobin, and lysozyme (48) rather than from synthetic polymers. The helix content for the standard proteins is taken from their x-ray structures in the crystalline state. An uncertainty of $\pm 20\%$ is assigned to the derived value. We have reservations about this method and suggest instead that the helix content of these vesicles is probably less than 25%, based upon the observed amplitudes. This value is so low that its uncertainty is probably $\pm 15\%$ of helix content.

EFFECTS OF PHOSPHOLIPASES A AND C, LYSOLECITHIN, AND 2-CHLOROETHANOL

CD and ORD are more usefully employed in the study of conformational change than for estimates of helicity. Wallach and his coworkers have studied the effects of several perturbants on their plasma vesicles.

In Figure 26 are shown their CD curves of lysolecithin-treated and phospholipase A-treated membranes. Phospholipase A causes an increase in the 208-nm and 193-nm band intensities and a blue shift in crossover from 206 to 204 nm; lysolecithin has the same effects but gives in addition a blue shift of the $n_1 \rightarrow \pi^-$ band. The observed changes are all consistent with a reduction in artifacts from absorption statistics and light scattering. It would be useful to know the absorption spectrum of the treated and untreated membranes; this could give independent evidence for changes in light scattering and absorption flattening.

Phospholipase C has also been added to Ehrlich ascites tumor cell

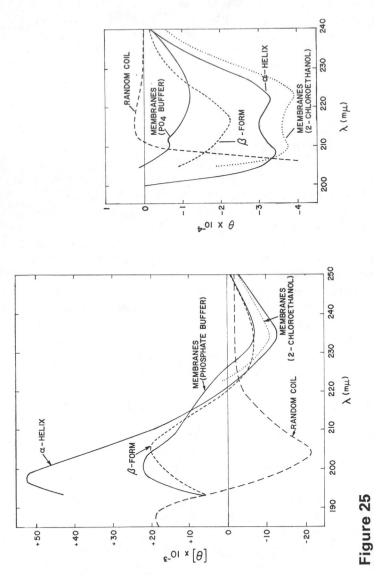

Figure 25

Left. ORD spectra of red blood cell membranes suspended in 0.008 M phosphate buffer, pH 7.7, and dissolved in 2-chloroethanol. The ORD spectra of poly-L-lysine in the α-helical, β, and random coil forms are plotted for comparison. *Right.* CD spectra of red blood cell membranes suspended in 0.008 M phosphate buffer, pH 7.7, and dissolved in 2-chloroethanol. The CD spectra of poly-L-lysine in the α-helical and β forms, and of random-coil poly-L-glutamic acid, are plotted for comparison. (From Lenard, J., and Singer, S. J. *Proc. Natl. Acad. Sci. U.S.A. 56,* 1828, 1966.)

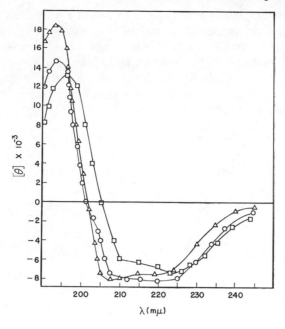

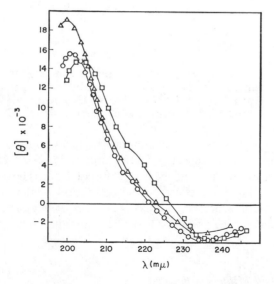

Figure 26

Observed CD and ORD spectra of plasma membrane fragments
(□—□), plasma membrane fragments *plus* lysolecithin (△—△),
and plasma membrane fragments *plus* phospholipase A (○—○).
(From Gordon, A. S., Wallach, D. F. H., and Straus, J. H. *Biochim.
Biophys. Acta 183,* 405, 1969.)

plasma membranes, with the CD results shown in Figure 27. The distortions characteristic of the particulate nature of suspensions are further enhanced by this treatment; the 208- and 196-nm bands are diminished and the CD crossover shifts even further to the red, to 206 nm. The absence of turbidity and absorption data again makes interpretation difficult. These effects of phospholipase C on the Ehrlich plasma membrane fragments differ from those found in red cell ghosts (see above).

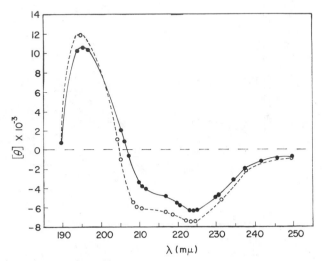

Figure 27

Observed CD spectra of plasma membrane fragments (O---O) and plasma membrane fragments *plus* phospholipase C (●—●). Solutions contain 1 mM Ca²⁺. (From Gordon, A. S., Wallach, D. F. H., and Straus, J. H. *Biochim. Biophys. Acta 183*, 405, 1969.)

Treatment of the plasma membrane fragments with 2-chloroethanol causes two- to three-fold increases in the amplitude of the ORD curve, as shown in Figure 28, with attendant blue shifts of 4 nm. The 2-chloroethanol again appears to have two effects: *1*) solubilization of the membrane (and thereby removal of artifacts), and *2*) induction of major conformational change in the membrane protein. These effects parallel those observed in the red cell ghost.

Mitochondria and Mitochondrial Proteins

The CD and ORD curves of rat liver mitochondria and of whole and fragmented beef heart mitochondria reveal features similar to those of the systems above (30, 55, 61). Whole mitochondria have the ORD curve shown previously in Figure 14. Most striking is the extremely low amplitude of the entire curve. This is due, as Urry and coworkers have suggested, to light scattering and the Duysens' effect. Mitochondria can be crudely represented by solid spheres of radius 0.5 μ; this gives values of Q_A and Q_B much less than unity (Table V) for reasonable values of absorption index α.

More useful information is obtained if the particles are sonicated and then separated into three sizes by centrifugation, as shown in Figure 29. In this figure, particle size as judged by turbidity increases from curve E

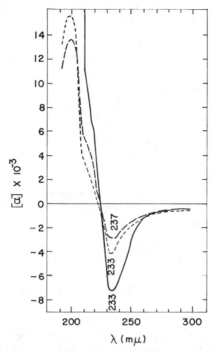

Figure 28

Optical rotatory dispersion of plasma membrane in aqueous suspension (short-dashed line), in an aqueous solution of lyso-lecithin (130 μg/ml) (long-dashed line), and in 9:1 2-chloro-ethanol:water (solid line). (From Wallach, D. F. H., and Zahler, P. H. *Proc. Natl. Acad. Sci. U.S.A. 56,* 1552, 1966.)

to F to G. The effect of particle size is unambiguous, as Ji and Urry have noted: The CD decreases in amplitude and the greatest changes occur in spectral regions of high absorbance. The changes in particle size were monitored by light scattering at 700 nm, as noted in the figure legend. Unfortunately, the $A_L - A_R$ values reported cannot be readily converted to $[\theta]$ values, so quantitative comparison with other membrane systems is not possible. However, the shape of the sonicated curves is once again remarkably reminiscent of the spectrum of an α-helical polymer.

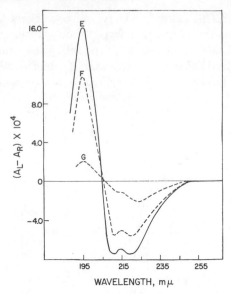

Figure 29

Circular dichroism of mitochondrial fragments in different stages of breakdown by sonication. Protein concentration, 2 mg/ml; path length, 0.1 mm.

Curve	Sample	OD_{700}
E	Supernatant, 20,000 × g	.000
F	2000–20,000 g pellet	.022
G	2000 × g pellet	.055

(From Ji, T. H., and Urry, D. W. *Biochem. Biophys. Res. Commun.* 34, 404, 1969.)

Mitochondrial "structural protein" has also been studied by ORD methods (47). It has been shown that the 233-nm ORD trough undergoes large red shifts, from 232 to 239 nm, upon aggregation of the protein. The crossover in ORD also shifts from 219 to 226 nm. These features, and the decreased amplitude of the over-all curve, are similar to those seen in so many other particulate systems. A mixture of absorption statistics and scattering is probably the cause of these spectral changes upon aggregation. The ORD of the unaggregated material shows that $[\phi']_{233}$ is only -4000. It is unlikely, therefore, that the α-helix plays a dominant conformational role in this protein fraction.

Other Membrane Systems

Sarcotubular vesicles have been studied by Mommaerts (38); they also show typical α-helical curves. Membranes of *Bacillus subtilis*, studied by Lenard and Singer (32), showed typically α-helix-shaped CD curves with $[\theta]$ equal to $-13,000$ at 225 nm in aqueous buffer, $-21,000$ in 75% glycerol. No corrections for absorption statistics are made, nor have absorption spectra been reported. The addition of sodium lauryl sulfate to the membrane preparation also causes $[\theta]$ to increase to $-20,000$. It seems likely, then, that the proteins of the membrane have a helix content near 50%.

SUMMARY

Let us attempt to summarize the definite information about membrane structure which ORD and CD measurements have already provided.

1. ORD and CD curves of membranes originate almost entirely in protein structure, not in lipid or carbohydrate structure.

2. Several membrane preparations appear to contain 30–50% of their peptide groups in an α-helical conformation.

3. Significant amounts of lipid phosphate can be removed without altering the protein conformation; e.g., phospholipase C treatment has negligible effect on the CD curves.

4. The averaged conformation of proteins in red blood cell membranes solubilized by 0.1% SDS is quite similar to that of these proteins in intact ghosts.

5. A significant fraction of the proteins of membranes is very sensitive to 2-chloroethanol, becoming substantially more α-helical in this solvent.

The major questions of membrane structure are still unanswered. It should be clear that ORD and CD experiments, by themselves, are often difficult to interpret in any absolute way. Parallel ultraviolet absorption measurements are most desirable; such measurements give an independent check of light scattering and absorption statistics. Moreover, measurement on the same sample of CD, ORD, infrared absorption, nuclear magnetic resonance, and electron spin resonance spectra, as well as x-ray scattering, is much more valuable than measurements by any one technique, since each method has different strengths and weaknesses. Finally, it is evident that CD and ORD measurements increase in reliability and usefulness when spectra are compared for one sample treated in several ways—e.g., with and without lipase digestion; at high and low temperature; with and without SDS, glycerol, or 2-chloroethanol. If these points of strategy are appreciated, CD and ORD measurements will provide further useful information on membrane structure.

APPENDIX A: SOME UNITS AND TERMS IN ELECTRONIC SPECTROSCOPY

Absorption

ABSORBANCE A, MOLAR ABSORPTIVITY ε, OPTICAL DENSITY OD

The radiant energy I_0 incident upon a sample may be attenuated both by absorption and by scattering. If I is the transmitted radiant energy in the absence of scattering, then absorbance A is defined by $A = -\log_{10}(I/I_0)$. It is implied here that I is corrected for reflectance losses, solvent absorption, and refractive effects by means of a suitable reference, e.g., a solvent-filled cell. The molar absorptivity ε in liter $\cdot$ mole^{-1} $\cdot$ cm^{-1} is then $\varepsilon = A/bc'$, where b is path in centimeters and c' is solute concentration in moles/liter. The term optical density (OD) is reserved for samples in which I is reduced both by absorption and by scattering. For these cases, as for absorbance, OD $= -\log_{10}(I/I_0)$.

DIPOLE STRENGTH D_i OF ABSORPTION BAND i
Units: esu$^2 \cdot$ cm^2 or debye^2 (1 debye $= 10^{-18}$ esu $\cdot$ cm)

$$D_i = \frac{3hc(2303)}{8\pi^3 N} \int_0^\infty \frac{\varepsilon_i \, dv}{v}$$

$$= 9.18 \times 10^{-39} \int_0^\infty \frac{\varepsilon_i \, dv}{v} \qquad \text{esu}^2 \cdot \text{cm}^2$$

where N is Avogadro's number, c is speed of light, h is Planck's constant, and v is frequency of light.

OSCILLATOR STRENGTH f_i OF BAND i
Units: unitless

$$f_i = \frac{2303 mc^2}{\pi e^2 N} \int_0^\infty \varepsilon_i \, dv$$

$$= 4.318 \times 10^{-9} \int_0^\infty \varepsilon_i \, dv \qquad \text{with } v \text{ in cm}^{-1}$$

where m is electronic mass, and e is electronic charge.

Optical Rotation

1. Specific rotation $[\alpha]_\lambda$

$$[\alpha] = \frac{\alpha_\lambda}{dc'}$$

where α_λ is the measured rotation in degrees at wavelength λ, d is the cell path length in decimeters, and c' is the weight concentration of solute in g/ml.

2. Molar rotation $[\phi]$ in degrees $\cdot$ cm$^2 \cdot$ decimole^{-1}

$$[\phi] = \frac{M[\alpha]}{100}$$

where M is the molecular weight of solute or, in case of polymers, the mean residue weight.

$$[\phi'] = \frac{3}{n^2 + 2} [\phi]$$

gives molar rotation corrected by the Lorentz factor for solvent refractive index, n.

Circular Dichroism

UNITS OF MEASUREMENT

1. Molar ellipticity $[\theta]$ in degrees $\cdot$ cm^2 $\cdot$ decimole^{-1}

$$[\theta] = \frac{\theta M}{100 \, dc'}$$

where θ is measured ellipticity in degrees, other symbols as above.

$$[\theta'] = \left(\frac{3}{n^2 + 2}\right) [\theta]$$

corrects the molar ellipticity for solvent polarizability, using the Lorentz factor.

2. Another unit sometimes used to report CD measurements is $\Delta\varepsilon$ in liter $\cdot$ mole^{-1} $\cdot$ cm^{-1}.

$$\Delta\varepsilon \equiv \varepsilon_L - \varepsilon_R$$

where ε_L is the molar absorptivity for left circularly polarized light and ε_R is the molar absorptivity for right circularly polarized light.

$$\Delta\varepsilon = \frac{\Delta A}{c'l}$$

where ΔA is the measured difference in absorbance of the sample for left vs. right circularly polarized light, c' is molar concentration of solute, and l is path in centimeters.

It turns out that

$$[\theta] = \{(2303)9/2\pi\}\{\varepsilon_L - \varepsilon_R\}$$

$$= 3298(\varepsilon_L - \varepsilon_R)$$

ROTATIONAL STRENGTH R_i OF CD BAND i

Units:

    1. erg $\cdot$ cm^3, or

    2. debye $\cdot$ magneton (with 1 Bohr magneton $= 0.927 \times 10^{-20}$ esu $\cdot$ cm and 1 debye $= 10^{-18}$ esu $\cdot$ cm).

$$R_i = \frac{3hc(2303)}{32\pi^3 N} \int_0^\infty \left[\frac{\varepsilon_L - \varepsilon_R}{\nu}\right]_i d\nu$$

where ε_L and ε_R are the molar absorptivities for left and for right circularly polarized light.

Evaluation of the constants yields

$$R = 2.3 \times 10^{-39} \int_0^\infty \frac{\Delta\varepsilon_i \, d\nu}{\nu}$$

$$\approx 0.696 \times 10^{-42} \int_0^\infty \frac{[\theta_i(\lambda)]}{\lambda} d\lambda \qquad \text{erg} \cdot \text{cm}^3$$

APPENDIX B: POLARIZED LIGHT

The material presented below is to be found in any classical optics textbook, e.g., Born and Wolf (14).

GENERAL DESCRIPTION OF A PLANE WAVE OF ARBITRARY POLARIZATION

Consider a monochromatic plane wave of electromagnetic radiation (e.g., visible light) propagating along the direction $\hat{z}$ in a Cartesian coordinate system. The electric vector $\vec{E}$ will then have components along the $\hat{x}$ and $\hat{y}$ directions. In the most general case,

$$E_x = a_1 \cos (\text{wt} - \vec{k} \cdot \vec{r} + \delta_1)$$
$$E_y = a_2 \cos (\text{wt} - \vec{k} \cdot \vec{r} + \delta_2)$$

Here $\vec{k}$ is the propagation vector of the wave, i.e.,

$$\vec{k} = \frac{2\pi \hat{z}}{\lambda} = \frac{2\pi \hat{z} n}{\lambda_0}$$

where n is the refractive index, λ is the wavelength of the wave, and λ_0 is the wavelength of the wave in vacuum. It is often convenient to adopt a complex notation rather than the real one above:

$$E_x = Re\{a_1 \exp [-i(\text{wt} - \vec{k} \cdot \vec{r} + \delta_1)]\}$$
$$E_y = Re\{a_2 \exp [-i(\text{wt} - \vec{k} \cdot \vec{r} + \delta_2)]\}$$

$$Re\{x\} \equiv \text{real part of } x$$

At a point in space, the electric vector $\vec{E}$ equals $E_x \hat{x} + E_y \hat{y}$; the tip of the $\vec{E}$ vector describes an ellipse in space, i.e.,

$$\left(\frac{E_x}{a_1}\right)^2 + \left(\frac{E_y}{a_2}\right)^2 - \frac{2E_x E_y}{a_1 a_2} \cos \delta = \sin^2 \delta$$

where $\delta = \delta_2 - \delta_1$.

LINEARLY POLARIZED LIGHT

Suppose $\delta = \delta_2 - \delta_1 = m\pi (m = 0, \pm 1, \pm 2, \cdots)$, then

$$\frac{E_x}{E_y} = \frac{a_1}{a_2}$$

The ellipse then reduces to a line, and we have *linearly polarized light*. Suppose we choose the direction of vibration by an angle α, as viewed toward the source

$$\cot \alpha = -\frac{a_1}{a_2} = \frac{-E_x}{E_y}$$

Two simple cases: $\alpha = 0$ ($\vec{E}$ vibrates in xz plane); $\alpha = \pm \pi/2$ ($\vec{E}$ vibrates in yz plane).

CIRCULARLY POLARIZED LIGHT

The ellipse described by $\vec{E}$ becomes a circle if $a_1 = a_2 = a$ and $\delta = m\pi/2$ ($m = \pm 1, \pm 3, \pm 5, \dots$).

1. Right circularly polarized light, E_R.

We consider $\delta = \pi/2$. Then

$$E_x = a \cos(\tau)$$

$$E_y = a \cos(\tau + \pi/2)$$

$$\vec{E}_R = E_x \hat{x} + E_y \hat{y}$$

where $\tau = \text{wt} - \vec{k} \cdot \vec{r} + \delta_1$.

The locus of $\vec{E}_R$ at a point in space is a circle; the circle is traced in a *clockwise* direction if one looks toward the source. The locus of the tip of $\vec{E}$ at an instant in time forms a *right-handed* helix. In complex notation $E_x/E_y = e^{-i\pi/2} = -i$. A drawing of right circularly polarized light is given in Figure 1.

2. Left circularly polarized light, E_L, is noted as follows:

$$a_1 = a_2 = a$$

$$\delta = -\pi/2$$

$$E_x = a \cos \tau$$

$$E_y = a \cos(\tau - \pi/2)$$

$$\frac{E_x}{E_y} = e^{+i\pi/2} = i$$

COMPLEMENTARITY OF (E_x, E_y) NOTATION AND (E_L, E_R) NOTATION

We have already written E_L and E_R in terms of E_x and E_y. We also note that, if the phase factors for E_L and E_R are identical, then

$$\tfrac{1}{2}(E_L + E_R) = E_x$$

$$\tfrac{1}{2}(E_L - E_R) = E_y$$

APPENDIX C: RELATION BETWEEN CIRCULAR BIREFRINGENCE
AND OPTICAL ROTATION

A medium is said to exhibit circular birefringence if $n_L \neq n_R$, where n_L and n_R are the refractive indices for left and right circularly polarized light. Consider a linearly polarized beam with $\vec{E} = \vec{E}_x$ incident upon a medium with $n_L \neq n_R$. At the interface (point A in Fig. 30) we may write, $\vec{E} = (\vec{E}_L + \vec{E}_R)/2$, with $\vec{E}_L$ and $\vec{E}_R$ in phase.

Suppose we define $z = 0$ at A, and set the phase factors for E_L and E_R equal to zero for convenience. Then at B we have for E_R:

$$E_x = Re\{a \exp[-i\text{wt} + i2\pi T(n_R/\lambda_0)]\}$$

$$E_y = Re\{a \exp[-i\text{wt} + i2\pi T(n_R/\lambda_0) + i\pi/2]\}$$

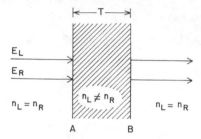

Figure 30
Relation between circular birefringence and optical rotation.

If $n_R = \bar{n} - \Delta$, $n_L = \bar{n} + \Delta$, this can be written, for $\tau = wt - (2\pi T\bar{n})/\lambda_0$, as

$$E_x = Re\{a \exp\left[-i\tau - 2\pi i\Delta T/\lambda_0\right]\}$$
$$E_y = Re\{a \exp\left[-i\tau - 2\pi i\Delta(T/\lambda_0) + 2\pi i/4\right]\}$$

Similarly, if $n_L = n + \Delta$, we have for E_L

$$E_x = Re\{a \exp\left[-i\tau + 2\pi i\Delta T/\lambda_0\right]\}$$
$$E_y = Re\{a \exp\left[-i\tau + 2\pi i\Delta(T/\lambda_0) - 2\pi i/4\right]\}$$

Now, the field of the emergent beam is $(E_L + E_R)/2$. The x component of this emergent field is given by:

$$E_x = Re(a \exp\left[-i\tau\right]\{\exp\left[+2\pi i\Delta T/\lambda_0\right] + \exp\left[-2\pi i\Delta T/\lambda_0\right]\}/2)$$
$$= \cos(2\pi\Delta T/\lambda_0)Re\{a \exp\left[-i\tau\right]\}$$

Similarly, the y component of the emergent field is given by:

$$E_y = Re(a \exp\left[-i\tau\right]\{\exp\left[2\pi i\Delta(T/\lambda_0) - 2\pi i/4\right]$$
$$+ \exp\left[-2\pi i\Delta(T/\lambda_0) + 2\pi i/4\right]\}/2)$$
$$= -\sin(2\pi\Delta T/\lambda_0)Re\{a \exp\left[-i\tau\right]\}$$

Note that E_x and E_y are in phase; the light is linearly polarized. To find the angle of rotation, we take the ratio of E_x to E_y:

$$\frac{E_x}{E_y} = -\cot\frac{2\pi\Delta T}{\lambda_0} = -\cot\left\{\frac{\pi(n_L - n_R)T}{\lambda_0}\right\}$$

This describes linearly polarized light with vibration direction defined by the angle α, where

$$\alpha = \frac{\pi(n_L - n_R)T}{\lambda_0}$$

i.e., circular birefringence implies optical rotation.

GENERAL REFERENCES

1. Beychok, S. *Poly-α-Amino Acids.* Ed. G. Fasman. Dekker, New York, 1967, chap. 7.
2. Donovan, J. W. *Physical Principles and Techniques of Protein Chemistry, Part A.* Ed. S. J. Leach. Academic Press, New York, 1969.
3. Glaser, M., Simpkins, H., Singer, S. J., Sheetz, M., and Chan, S. I. *Proc. Natl. Acad. Sci. U.S.A.* **65,** 721 (1970).
4. Gordon, A. S., Wallach, D. F. H., and Straus, J. H. *Biochim. Biophys. Acta* **183,** 405 (1969).
5. Moscowitz, A. *Optical Rotatory Dispersion.* Ed. C. Djerassi. McGraw-Hill, New York, 1960, chap. 12.
6. Schneider, A. S., Schneider, M-J. T., and Rosenheck, K. *Proc. Natl. Acad. Sci. U.S.A.* **66,** 793 (1970).
7. Timasheff, S. N., Susi, H., Townend, R., Stevens, L., Gorbunoff, M. J., and Kumosinski, T. F. *Conformation of Biopolymers.* Ed. G. N. Ramachandran. Academic Press, New York, 1967, p. 173.
8. Urnes, P., and Doty, P. *Adv. Protein Chem.* **16,** 401 (1961).
9. Urry, D. W., Hinners, T. A., and Masotti, L. *Arch. Biochem. Biophys.* **137,** 214 (1970).
10. Wallach, D. F. H., and Zahler, P. H. *Proc. Natl. Acad. Sci. U.S.A.* **56,** 1552 (1966).

LITERATURE REFERENCES

11. Aden, A. L., and Kerker, M. *J. Appl. Phys.* **22,** 1242 (1951).
12. Aebersold, D., and Pysh, E. S. *J. Chem. Phys.* **53,** 2156 (1970).
13. Blout, E. R., Schmier, I., and Simmons, N. S. *J. Am. Chem. Soc.* **85,** 644 (1963).
14. Born, M., and Wolf, E. *Principles of Optics.* Macmillan, New York, 1964, chap. I.
15. Carver, J. P., Shechter, E., and Blout, E. R. *J. Am. Chem. Soc.* **88,** 2250 (1966).
16. Cassim, J. Y., and Taylor, E. W. *Biophys. J.* **5,** 553 (1965).
17. Cassim, J. Y., and Yang, J. T. *Biopolymers* **9,** 1475 (1970).
18. Crabbe, P. *Optical Rotatory Dispersion and Circular Dichroism in Organic Chemistry.* Holden-Day, San Francisco, 1965.
19. Deutsche, C. W., Lightner, D. A., Woody, R. W., and Moscowitz, A. *Annu. Rev. Phys. Chem.* **20,** 407 (1969).
20. Duysens, L. N. M. *Biochim. Biophys. Acta* **19,** 1 (1956).
21. Fasman, G. (ed.). *Poly-α-Amino Acids.* Dekker, New York, 1967.
22. Glaser, M., and Singer, S. J. *Biochemistry* **10,** 1780 (1971).
23. Gordon, D. J. *Biochemistry* **11** (1971). In press.
24. Gordon, D. J., and Holzwarth, G. *Arch. Biochem. Biophys.* **142,** 481 (1971).
25. Gordon, D. J., and Holzwarth, G. *Proc. Natl. Acad. Sci. U.S.A.* **68,** 2365 (1971).
26. Gratzer, W. B. *Poly-α-Amino Acids.* Ed. G. Fasman. Dekker, New York, 1967, chap. 5.
27. Greenfield, N., Davidson, B., and Fasman, G. D. *Biochemistry* **6,** 1630 (1967).
28. Holzwarth, G., and Doty, P. *J. Am. Chem. Soc.* **87,** 218 (1965).
29. Jaffe, H. H., and Orchin, M. *Theory and Applications of Ultraviolet Spectroscopy.* Wiley, New York, 1962.

30. Ji, T. H., and Urry, D. W. *Biochem. Biophys. Res. Commun.* **34,** 404 (1969).
31. Kerker, M. *The Scattering of Light and Other Electromagnetic Radiation.* Academic Press, New York, 1969.
32. Lenard, J., and Singer, S. J. *Proc. Natl. Acad. Sci. U.S.A.* **56,** 1828 (1966).
33. Lowry, T. M. *Optical Rotatory Power.* Longmans, London, 1935; Dover, New York, 1964.
34. Mandelkern, L. *Poly-α-Amino Acids.* Ed. G. Fasman. Dekker, New York, 1967, chap. 13.
35. Moffitt, W. *J. Chem. Phys.* **25,** 467 (1956).
36. Moffitt, W., Fitts, D. D., and Kirkwood, J. G. *Proc. Natl. Acad. Sci. U.S.A.* **43,** 723 (1957).
37. Moffitt, W., and Moscowitz, A. *J. Chem. Phys.* **30,** 648 (1959).
38. Mommaerts, W. F. H. M. *Proc. Natl. Acad. Sci. U.S.A.* **58,** 2476 (1967).
39. Ottaway, C. A., and Wetlaufer, D. B. *Arch. Biochem. Biophys.* **139,** 257 (1970).
40. Quadrifoglio, F., and Urry, D. W. *J. Am. Chem. Soc.* **90,** 2760 (1968).
41. Rabinowitch, E. I. *Photosynthesis.* Interscience, New York, 1956, vol. 2, part 2, p. 1863.
42. Rosenheck, K., and Doty, P. *Proc. Natl. Acad. Sci. U.S.A.* **47,** 1775 (1961).
43. Schellman, J., and Oriel, P. *J. Chem. Phys.* **37,** 2114 (1962).
44. Schneider, A. S. *Chem. Phys. Letters* **8,** 604 (1971).
45. Snatzke, G. (ed.). *Optical Rotatory Dispersion and Circular Dichroism in Organic Chemistry.* Sadtler Research Laboratories, Philadelphia, 1967.
46. Sober, H. A. (ed.). *Handbook of Biochemistry.* Chemical Rubber, Cleveland, 1968.
47. Steim, J. M., and Fleischer, S. *Proc. Natl. Acad. Sci. U.S.A.* **58,** 1292 (1967).
48. Straus, J. H., Gordon, A. S., and Wallach, D. F. H. *Eur. J. Biochem.* **11,** 201 (1969).
49. Suzuki, H. *Electronic Absorption Spectra and Geometry of Organic Molecules.* Academic Press, New York, 1967.
50. Tinoco, I., Jr. *Radiat. Res.* **20,** 133 (1963).
51. Tinoco, I., Jr., Woody, R. W., and Bradley, D. F. *J. Chem. Phys.* **38,** 1317 (1963).
52. Urry, D. W. *Spectroscopic Approaches to Biomolecular Conformation.* American Medical Association, Chicago, 1970.
53. Urry, D. W., and Doty, P. *J. Am. Chem. Soc.* **87,** 2757 (1965).
54. Urry, D. W., and Ji, T. H. *Arch. Biochem. Biophys.* **128,** 802 (1968).
55. Urry, D. W., Mednieks, M., and Bejnarowicz, E. *Proc. Natl. Acad. Sci. U.S.A.* **57,** 1043 (1967).
56. van de Hulst, H. C. *Light Scattering by Small Particles.* Wiley, New York, 1957.
57. Velluz, L., Legrand, M., and Grosjean, M. *Optical Circular Dichroism.* Academic Press, New York, 1965.
58. Weigang, O. E., Jr. *J. Chem. Phys.* **41,** 1435 (1964).
59. Woody, R. W. *J. Chem. Phys.* **49,** 4797 (1968).
60. Yang, J. T. *Poly-α-Amino Acids.* Ed. G. Fasman. Dekker, New York, 1967, chap. 6.
61. Wrigglesworth, J. M., and Packer, L. *Arch. Biochem. Biophys.* **128,** 790 (1968).

III

MEMBRANE
FUNCTION AND
ASSEMBLY

10

Robert D. Simoni

Macromolecular Characterization of Bacterial Transport Systems

The emphasis of this chapter is on the function of biological membranes. The discussion is limited to recent work concerning attempts to define the macromolecular components involved in solute translocation in bacterial cells. Excitability, muscle contraction, intestinal absorption, and secretory activity, which are but a few of the many functions associated with the cell membranes of higher organisms, are not considered. It is, however, the hope of workers concerned with transport studies in microbes that the information obtained will aid examination of more complex systems and that the study of membrane function will promote a better understanding of membrane structure.

The choice of microorganisms for membrane transport studies is chiefly for reasons of technical simplicity. It is easy to maintain a pure culture of a particular species and to grow it on a large scale with strict control over the physiological state of the organism. Most important is the ability to alter gene expression through mutation. This approach, along with the use of radioisotopes, has been extremely useful for metabolic studies in microbes and is now being applied with equal success to the study of transport. The use of genetic techniques to study transport in bacteria has been reviewed (1). These methods have been invaluable for distinguishing the substrate specificity of various transport systems and for identifying the components involved, and they promise to be even more important in the study of the mechanisms of transport. In close conjunction with the genetic approach much information has been obtained using biochemical techniques. There has been a transition in transport studies from the kinetic toward an enzymological approach. The kinetic studies have done a great deal toward limiting speculation on the general mechanism of transport but of course cannot provide a direct demonstration of macromolecular participation in the transport processes.

The major emphasis of this chapter is on those systems for which a significant biochemical characterization has been carried out; historical developments are discussed only briefly.

The Concept of Carrier-Mediated Solute Movement

The cell membrane was once considered a static structure which simply isolated the cell from its environment. It was thought to allow communication with the environment by regulation of solute exchange according to the hydrophobicity, molecular size, and electrical charge of the solute. It has since been realized that the cell membrane is a far from static barrier and that it plays an active role in solute movement. In many cases solute movement either into or out of cells is due to "pumps" within the membrane. As a result, the mode of passage cannot be restricted to passive diffusion of solute through hydrophobic or hydrophilic pores.

The idea that an active membrane component might be involved in solute binding was suggested by Osterhout (2). Such a concept would allow higher efficiency and greater possibility for regulation than simple diffusion. The term *carrier* was introduced to describe a structure which binds solute, is not itself rigidly fixed within the membrane, and somehow facilitates entry or exit of a particular solute.

The following discussion of possible types of carrier-mediated solute movement is restricted to the three most general types: facilitated diffusion, active transport, and group translocation. Those interested in the kinetic description of the more complex variations of transport systems are directed to the reviews of Wilbrandt and Rosenberg (3) and Stein (4). The three general processes have several common features which serve to distinguish them from passive diffusion: *1*) the presence of an active membrane component provides for rates more rapid than can be predicted from the chemical nature of the solute; *2*) the rate of entry reaches a limiting value with increasing solute concentration (rate of entry as a function of solute concentration often resembles simple Michaelis-Menten enzyme kinetics); *3*) the carrier is usually stereospecific, which implies different carriers for different solutes, a fact which has been amply demonstrated; *4*) the processes are temperature-dependent. Thus the three general parameters which serve to distinguish carrier-mediated processes from passive diffusion are rapid rates, saturation kinetics, and stereospecificity.

Each type of system differs from the others in several important ways. *Facilitated diffusion* is the simplest and the most common type of carrier process. Such systems provide for solute movement *down* a concentration gradient until the solute concentration inside the cell equals that outside. The energy for translocation is provided by the solute concentration gradient, and no metabolic energy is required. The carrier only affects the *rate* of movement and not the final equilibrium. If metabolic energy is coupled to facilitated diffusion so that accumulation of solute results, the system is called *active transport*. This definition also requires that the solute remain unaltered during the over-all process. Active transport of amino acids and lactose in *Escherichia coli* exemplifies this process. In some cases the energy-coupling reactions can be dissociated from facilitated diffusion and each studied separately.

Group translocation, in contrast to both facilitated diffusion and active transport, involves solute modification during the transport process. This mechanism, first proposed by Mitchell and Moyle (5), suggests that the carrier molecules behave like enzymes in catalyzing group transfer reactions. Although in this case the solute is altered, it should be apparent that such a mechanism results in "accumulation" as does active transport. The phospho-enolpyruvate-glycose phosphotransferase system in many bacteria is the prime example of this transport mechanism. There has been no functional dissociation of translocation and energy coupling as yet, and they seem to be intimately connected. The remainder of this chapter is devoted to examples of each type of transport system. Emphasis is on those for which insight has been gained into the biochemical nature of the macromolecular participation.

TRANSPORT OF CARBOHYDRATES

Although carrier-mediated diffusion was an early concept, the demonstration of carrier specificity came much later. By the late 1950s it was clear that there existed specific permeation systems for different solutes. This was especially true for carbohydrates. Evidence was accumulated in two areas: selective permeability and accumulation of some solutes and selective crypticity of some mutants toward certain solutes.

The state of selective crypticity of certain cells toward selected substrates was particularly pertinent. Some mutants were unable to metabolize a given substrate even though they possessed the necessary metabolic enzymes. A particularly striking example of selective crypticity was revealed by the experiments of Doudoroff *et al.* (6). A mutant of *E. coli* was incapable of metabolizing glucose although it metabolized maltose (made of two glucose units) normally. It was shown that free glucose is liberated by the amylomaltase reaction and that the mutant cells metabolized both glucose moieties of the maltose molecule. Therefore it appeared that glucose could be used when liberated intracellularly, while free glucose in the external medium could not. The conclusion of these experiments seemed inescapable; the cells were impermeable to glucose. But a membrane impermeable to glucose could not be permeable to maltose except via a stereospecific permeation system.

The Lactose System in E. coli

A remarkable series of experiments from The Pasteur Institute provided the first demonstration of carrier specificity in the β-galactoside system in *E. coli* The review of Kepes and Cohen (7) provides a good background for bacterial transport systems, including the β-galactoside system in *E. coli.*

By studies on genetic control and specific inducibility, these workers showed the narrow range of structural specificity for the substrate, the indirect demonstration of the essential stereospecific macromolecular component of the system as a protein, and the proof of the uniqueness of the carrier. They also introduced the term *permease* that in spite of many objections has survived over the years and is still the most widely used, albeit nondescript, term in the transport field.

The classic genetic studies of the lactose operon of *E. coli* will not be discussed here in detail; those unfamiliar with this work are directed to Chapter 12 of this book and to the review by Jacob and Monod (8). *E. coli* can grow on lactose owing to the presence of a specific permease for β-galactosides and a hydrolytic enzyme, β-galactosidase, which cleaves the glycosidic bond to yield glucose and galactose.

In wild-type *E. coli*, β-galactosidase is found only when the cells have been grown on lactose. If the cells are transferred to media with another carbon source, i.e., succinate, the enzyme is diluted by growth until it is at a very low level. In such a culture addition of a nonmetabolizable analogue of lactose, methyl-β-D-thiogalactoside (TMG), again elicits the synthesis of β-galactosidase. Such inducers as TMG are said to be gratuitous.

Mutant strains were isolated which did not grow on lactose as sole source of carbon (but grew normally on glucose and galactose). Some of these mutants were found to be devoid of β-galactosidase when grown in the presence of a gratuitous inducer for the lactose operon, while others could synthesize a normal complement of this enzyme. The only explanation for nonutilization of lactose in the latter class is that in some way it has no access to the hydrolytic enzyme; i.e., the cells are cryptic toward lactose. With the use of radioactive TMG it was possible to show that induced wild-type cells could *accumulate* the substrate while the cryptic mutants or un-induced wild type could not. It should be emphasized that nonmetabolizable substrate analogues are extensively used for transport experiments, and they are invaluable for the determination of meaningful transport kinetics free of complicating metabolic reactions.

Thus it was possible through genetics and biochemical analyses to distinguish between two independent gene products, both of which are required for lactose utilization. Loss of the permease results in the absence of growth on lactose and loss of the ability to accumulate TMG. Loss of β-galactosidase affects only the ability to use lactose for growth; TMG accumulation is normal. It was noted, however, that in permease mutants a higher concentration of inducer was required for complete β-galactosidase induction, which is to be expected, since the inducer is concentrated intracellularly by the permease. Table I summarizes the phenotypes of various lactose-negative mutants of *E. coli*. This brief historical description of the development of an understanding of the lactose system in *E. coli* cannot adequately emphasize the great impact its characterization has had on the entire field of membrane transport.

It is useful at this point to consider the over-all active transport process as two distinct steps: *1*) the facilitated diffusion of solute across the membrane, and *2*) the coupling of metabolic energy to achieve and maintain a concentration gradient. In the *E. coli* lactose system these two processes are experimentally separable. The hydrolysis of *o*-nitrophenyl-β-D-galactoside (ONPG) by metabolically poisoned intact cells is routinely used to measure facilitated diffusion. ONPG is a substrate for the permease and is hydrolyzed as it enters the cell to yield *o*-nitrophenol (yellow) plus galactose.

Table I

Correlation of Phenotypes and Genotypes of Various Lactose-Negative Mutants of *E. coli* in Relation to Permease and β-Galactosidase Activities

STRAIN	PHYSIOLOGICAL TYPE	GENOTYPE	Galactoside Permease		β-Galactosidase	
			INDUCED	NON-INDUCED	INDUCED	NON-INDUCED
ML30, K12	Normal wild	$y^+z^+i^+$	+	0	+	0
ML3, K12, W2241	Negative cryptic	$y^-z^+i^+$	Trace	0	+	0
K12, W2244, K12, W2242	Absolute negative	$y^+z^-i^+$	+	0	0	0
ML308, K12, S$_4$	Normal constitutive	$y^+z^+i^-$	+	+	+	+
ML3088	Absolute negative	$y^+z^-i^-$	+	+	0	0
ML35	Constitutive cryptic	$y^-z^+i^-$	0	0	+	+

After Kepes, A , and Cohen, G. N. *The Bacteria*. Ed. I. C. Gunsalus and R. Y. Stanier. Academic Press, New York, 1962, vol. IV, p. 179.

β-Galactosidase is present in large excess in the cell, and thus the rate of ONPG hydrolysis is a measure of the rate of ONPG entry. Alternatively radioactive TMG can be used for both rate measurements and accumulation. However, it is less convenient and is usually used only when extent of accumulation is required. It has been demonstrated (9) that the rate of ONPG hydrolysis in whole cells is not significantly affected by treatment of the cells with metabolic poisons such as 2,4-dinitrophenol or sodium azide. The process measured by the accumulation of radioactive TMG is completely abolished by such treatment. Thus, it has been assumed that poisons inhibit the energy-coupling reactions while leaving the carrier fully functional. These two steps will be discussed separately.

DIRECT DEMONSTRATION OF LACTOSE CARRIER (M PROTEIN)

With the large amount of information available on the lactose system, Fox and Kennedy and their collaborators began what would ultimately be the first direct isolation and characterization of a transport protein, the product of the lactose *y* gene. This work has been excellently reviewed by Kennedy (10).

Kepes (11) observed that sulfhydryl reagents such as *p*-chloromercuribenzoate were effective inhibitors of the over-all transport reaction and that the compound thiodigalactoside (TDG) could protect the cell from the inhibition. This experiment suggested that the sulfhydryl inhibitor was reacting at some substrate-binding site for which TDG had a high affinity. Fox and Kennedy (12) proposed to use this substrate-protection phenomenon to label specifically the permease protein. The experiment in its simplest form involved the treatment of cells or usually isolated membranes induced for the lactose permease with nonradioactive N-ethylmaleimide (NEM) in the presence of TDG. NEM forms a covalent linkage with sulfhydryl groups and is highly specific. This treatment theoretically should block all sulfhydryl groups except those protected by TDG. The cells are then washed free of NEM and TDG and treated with radioactive NEM. With this technique Fox and Kennedy were able to demonstrate the following:

1. The M protein, as the *y* gene product is now called, was present in the membrane fraction of induced cells.

2. It was only present in cells induced or constitutive for the lactose operon and in cells with a normal *y* gene (Table II).

3. Temperature-sensitive (ts) revertants which were lactose-positive when grown and induced at 25°C but were lactose-negative at 42°C were isolated from y^- strains. This heat lability was exhibited when ONPG hydrolysis was measured in whole cells which had been grown and induced for proteins of the *lac* operon at 25°C. Most importantly, the M protein from a ts revertant was also shown to be more than 10-fold more sensitive to incubation at 42°C than was M protein from the y^+ parent strain.

Table II

Demonstration of the Product of the lac y Gene (M Protein) by the NEM Labeling Technique in Various Strains of *E. coli*

STRAIN	PROTEIN-BOUND NEM ($\mu\mu$MOLES/MG PROTEIN)		M PROTEIN ($\mu\mu$MOLES/MG PROTEIN)
	NO TDG	0.01 M TDG	
ML 308	200	86	111 ± 5
$(i^- z^+ y^+ a^+)$	188	87	
Uninduced	204	85	
ML 3	94	99	
$(i^+ z^+ y^- a^-)$	97	99	NS
Induced	105	102	
ML 308–225	250	126	
$(i^- z^- y^+ a^+)$	245	132	119 ± 5
Uninduced	260	129	
ML 35	89	97	
$(i^- z^+ y^- a^-)$	90	95	NS
Uninduced	99	95	
ML 30	235	110	
$(i^+ z^+ y^+ a^+)$	239	104	135 ± 3
Induced	236	103	
	246	109	
ML 30	124	123	NS
Uninduced	124	127	

NS, not significant.
After Fox, C. F., Carter, J. R., and Kennedy, E. P. *Proc. Natl. Acad. Sci. U.S.A.* **57**, 698 (1967).

These three lines of evidence demonstrate that the M protein is the product of the lactose *y* gene and is a component of the lactose permease system.

The radioactive NEM labeling technique has been extended by Jones and Kennedy (13) for use in the isolation and characterization of the M protein. It should be realized that this labeling technique offers both advantages and disadvantages for the final resolution of a transport mechanism. Since the labeling reagent is a covalently bound inhibitor, the product isolated is of little use for any study which requires the native protein, such as the determination of possible enzymatic function, mode of solute binding, properties of the protein in response to the binding of solute, and possible reconstitution experiments. This limitation is, however, being overcome by the development of a direct solute-binding assay which should be profitable in terms of isolating the "active" protein. The advantages of the NEM technique are that the protein can be easily followed through isolation procedures which would normally be too vigorous for maintenance of the native protein.

The M protein is an integral part of the membrane matrix, and the extraction procedure which renders it soluble involves use of the detergent sodium dodecyl sulfate and of mercaptoethanol. This procedure is extremely disruptive to multisubunit proteins and membranes from a variety of sources. In most cases the treated materials are dissociated to single polypeptide chains. It is obvious that only a labeling technique such as covalent attachment of NEM would allow the use of a rigorous isolation technique. The solubilized membrane extract containing the M protein has been treated by gel filtration and analyzed by sodium dodecyl sulfate gel electrophoresis. The polypeptide which reacts specifically with NEM has a molecular weight of about 30,000. This, of course, represents the monomer of the M protein which contains the active sulfhydryl group to which the NEM is attached. The current extension of this work with the direct binding assay, coupled with a less rigorous isolation procedure, hopefully will lead to the isolation of "native" M protein and permit an examination of its function. The significance of the approach used cannot be minimized, however, since it resulted in the first isolation of a permease protein.

ENERGY-COUPLING REACTION IN β-GALACTOSIDE TRANSPORT

The experimental dissociation of the intact active transport system into the energy-coupling and facilitated diffusion components has most clearly been demonstrated by Winkler and Wilson (9), who carefully measured *rates* of uptake of β-galactosides in cells with intact or inhibited energy coupling. Energy coupling can be effectively eliminated by treating the cells with the energy poisons sodium azide and iodoacetamide. Such an experiment is shown in Figure 1. The energy-uncoupled cells cannot accumulate solute, although they can equilibrate with the medium. Under such conditions the rate of ONPG hydrolysis is essentially the same as in uninhibited cells. Thus the transport and accumulation processes are at least functionally dissociable.

The nature of the metabolic energy which converts the facilitated diffusion process to an active transport mechanism is still obscure, although some information is now beginning to clarify this problem. From the inhibitor studies it has been assumed that adenosine triphosphate (ATP) was involved, since compounds known to uncouple oxidative phosphorylation from electron transport were potent inhibitors of β-galactoside accumulation. This has now been brought into serious question. Pavlasova and Harold (14) showed that TMG could be accumulated under anaerobic conditions, which in itself excluded the possibility that ATP generation from oxidative phosphorylation was involved. In addition, they showed that several inhibitors of oxidative phosphorylation and transport were inhibitory to the anaerobic accumulation process. They have suggested that the compounds tested affect a proton gradient which exists across the cell membrane. Mitchell (15) first proposed that respiratory carriers are arranged across the membrane

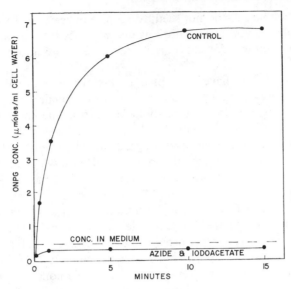

Figure 1

Effect of metabolic poisons on the accumulation of β-galacto-sides in *E. coli*. (After Winkler, H., and Wilson, T. *J. Biol. Chem.* *241,* 2200, 1966.)

in such a way that H⁺ and OH⁻ are deposited on opposite sides of the membrane. Thus, association of the carrier with the proton gradient could drive active transport. A more complete evaluation of this proposal awaits further work.

Additional evidence against the role of ATP in β-galactoside transport has come from experiments of Klein *et al.* (16), who showed that cells continue to accumulate TMG in the presence of 100 mM arsenate, a concentration which almost completely abolished ATP formation.

Barnes and Kaback (17) have shown that D-lactate stimulates the uptake of TMG in isolated membrane vesicles. This effect was also noted for amino acids and is discussed later in this chapter. The effect seemed restricted to a few compounds, D-lactate being the most effective, while DL-hydroxy-butyrate, succinate, and L-lactate were 80%, 50%, 25% as active, respectively. A wide range of other compounds showed no stimulation. The D-lactate-stimulated TMG uptake in these vesicle preparations was strongly inhibited by anaerobiosis and various uncouplers (proton conductors), but only slightly inhibited by arsenate and oligomycin. These authors suggest that energy coupling occurs via membrane-bound D-lactate, and succinate de-hydrogenases, and an electron transport chain, and that the coupling reaction itself may involve oxidation and reduction of the carrier. It is, however, curious that whole cells do not seem to require oxygen for TMG accumulation (14) while the vesicle preparations do. Whether this apparent inconsistency is meaningful will depend on a clearer understanding of the mechanism involved.

Wong *et al.* (18) have attempted to use the mutational approach to unravel the energy-coupling reaction. Using a rationale based on evidence that the coupling reaction can be dissociated from the entry process, they selected for mutants which were able to transport β-galactosides but unable to retain them against a concentration gradient. The *exit* of TMG from such mutant cells was much faster than from normal cells, which accounts for the inability of the cells to retain a high intracellular concentration of sugar. The levels of β-galactosidase and M protein in the mutant are actually higher than normal. The effect of the mutation on the solute accumulation seems to be specific for the β-galactoside system, since uptake of fucose, a substrate for the galactose permease, and of α-aminoisobutyrate, an amino acid permease substrate, appears normal in the mutant.

Wilson *et al.* (19) have recently been able to demonstrate this mutation is within the lactose *y* gene. Thus the M protein itself is involved in the terminal reaction of energy coupling.

Several models for galactoside transport have been proposed, all of which are quite similar conceptually. The model shown in Figure 2 is a composite taken from many sources.

In such a model it is proposed that the M protein binds β-galactosides with high affinity at the external surface of the membrane. This initial binding is followed by translocation to the inner face of the membrane. These two steps do not require energy and represent the facilitated diffusion process. At the inner surface energy coupling leads to a marked reduction in the affinity of the carrier for the substrate, resulting in solute displacement. The low-affinity carrier then returns to the outer surface where the high-affinity form is regenerated, and the system is able to recycle, resulting in the net accumulation of β-galactosides.

The Phosphoenolpyruvate-Glucose Phosphotransferase System

In contrast to the active transport system for lactose in *E. coli*, many sugars enter the bacterial cell by group translocation. In many bacteria this process has been shown to involve the phosphorylation of sugars as they pass through the membrane. An enzyme system, the phosphoenolpyruvate-glucose phosphotransferase system (PTS) mediates this uptake process. A complete description of the PTS can be found in a review by Roseman (20).

The PTS was first described by Kundig *et al.* (21) as an enzyme system from *E. coli* which can phosphorylate many sugars according to the general reaction shown in Figure 3. The reaction is specific for phosphoenolpyruvate,

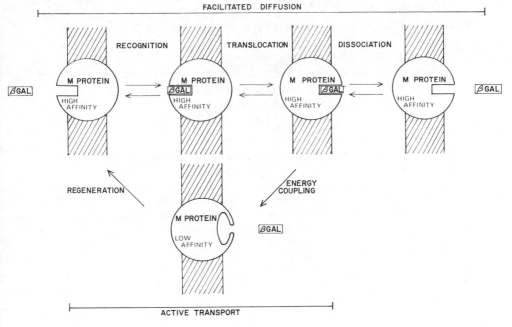

Figure 2
Model for β-galactoside transport in *E. coli*.

although a variety of sugars can serve as acceptors. Most of the mono-saccharide phosphates that have been characterized have been the 6-phosphate esters; the exception is fructose, which is the 1-phosphate ester.

The enzyme system has been demonstrated in many bacterial species including *E. coli*, *Salmonella typhimurium*, *Staphylococcus aureus*, *Streptococcus lactis*, *Aerobacter aerogenes*, *Bacillus subtilis*, and *Lactobacillus*

$$
\begin{array}{c}
\text{Sugar} \\
+ \\
CH_2 = C - CO_2^- \\
| \\
O - PO_3^=
\end{array}
\xrightarrow[\text{PTS}]{Mg^{++}}
\begin{array}{c}
\text{Sugar} - PO_3^= \\
+ \\
CH_3COCO_2^-
\end{array}
$$

Phosphoenolpyruvate
(PEP)

Pyruvate

Figure 3
Over-all reaction of the phosphoenolpyruvate-glycose phosphotransferase system (PTS). The PTS below the arrow represents a number of protein components as discussed in the text. (After Roseman, S. *J. Gen. Physiol. 54*, 138s, 1969.)

plantarum. The discussion following is largely limited to *E. coli, S. typhimurium,* and *S. aureus,* which have been most extensively studied.

PTS ENZYMOLOGY

The complexity of the system is shown in Table III. In the three organisms studied extensively to date four protein components, enzyme I, histidine protein (HPr), factor III, and enzyme II, have been identified.

Table III

Protein Components of the Bacterial PTS Required for the Over-All Reaction Shown in Figure 3

PROTEIN FRACTION	OCCURRENCE	SUGAR SPECIFIC	INDUCIBILITY
Enzyme I	Cytoplasm	−	−
HPr	Cytoplasm	−	−
Fraction or enzyme II	Membrane	+	+ and −
Factor III (*S. aureus*)	Cytoplasm	+	+

After Roseman, S. *J. Gen. Physiol.* **54.** 138s (1969).

The bacterial source of the various proteins is designated by a subscript and the sugar specificity, where applicable, by a superscript. Table III also gives the cellular location of these protein components as determined by cell lysis experiments. The bulk of enzyme I and HPr is found in the soluble fraction, while all of the detectable enzyme II is located in the membrane fraction and cannot be removed by washing or sonication. Factor III is soluble. Enzyme I and HPr are not sugar-specific components and are common for all sugars handled by the system. Factor III and enzymes II are sugar-specific and represent families of proteins each with activity toward a single sugar or group of closely related sugars. Enzyme I and HPr are generally thought to be produced constitutively by the cells, although the levels can sometimes be affected by altering growth conditions (22). Enzyme II and factor III are often inducible proteins (Table IV), although some are either constitutive or formed in response to an internally produced inducer or inducers. When *S. aureus* is grown under conditions inducing for the lactose operon and extracts are examined, a marked increase in the sugar-specific components enzyme II^{lac} and factor III^{lac} is noted while enzyme I and HPr remain essentially the same (22).

GENERAL REACTIONS OF THE PTS

It has been possible to show that the over-all reaction shown in Figure 3 is the sum of at least two separate reactions given in Figure 4. The first reaction is the enzyme I–catalyzed transfer of a phosphoryl group from

Table IV

Induction of the Sugar-Specific Components of the Lactose PTS in *S. aureus*

COMPONENT	Specific Activity in Extracts[a]	
	NONINDUCED	INDUCED
Enzyme I	52.0	58.0
HPr	6.8	6.7
Enzyme II^{lac}	0.1	38.0
Factor III^{lac}	0.1	2.5

[a] The cells were induced by addition of galactose to the medium. The activity for each component was determined by the rate of β-methylthiogalactoside phosphate formation.

After Roseman, S. *J. Gen. Physiol.* **54**, 138s (1969).

phosphoenolpyruvate to HPr. This reaction is general for *all* sugars and involves no sugar-specific components. The subsequent reactions of P-HPr introduce sugar specificity and are considered later in this chapter.

Properties of the first reaction. $HPr_{E. coli}$, $HPr_{S. typhimurium}$, and $HPr_{S. aureus}$ have been purified to homogeneity. They all have molecular weights about 9000. $HPr_{E. coli}$ and $HPr_{S. typhimurium}$ are similar as judged by their amino acid composition and electrophoretic behavior. $HPr_{S. aureus}$ is somewhat different in amino acid content. Also, while $HPr_{E. coli}$ and $HPr_{S. typhimurium}$ show good cross-reactivity in enzyme assays, $HPr_{S. aureus}$ reacts poorly in the *E. coli* system. Enzyme I has been purified to near homogeneity, but it has not been extensively characterized.

Enzyme I transfers 1 mole of phosphate to each HPr molecule via an enzyme I–phosphate intermediate. It has been suggested that the linkage

$$PEP + HPR \underset{Mg^{++}}{\overset{I}{\rightleftharpoons}} Phospho-HPR + Pyruvate$$

$$Phospho-HPR + Sugar \xrightarrow[III]{II} Sugar-P + HPR$$

$$PEP + Sugar \xrightarrow[II,III,HPR]{I,Mg^{++}} Sugar-P + Pyruvate$$

Factor III detected in induced *S. aureus*

Figure 4

Partial reactions of the bacterial PTS. The first reaction is general for all systems studied. The second reaction varies as is discussed in the text. I, enzyme I; II, enzyme II; III, factor III; HPr, histidine protein; PEP, phosphoenolpyruvate. (After Roseman, S. *J. Gen. Physiol.* 54, 138s, 1969.)

of the phosphoryl group to enzyme I may be to a histidine residue. The product, P-HPr, has been characterized, and the linkage of the phosphoryl group is to the N-1 position of a histidine residue (23, 24). The detailed sequence of the first reaction can be written as shown in Figure 5.

The enzyme II complex. While the mode of action of the phosphorylation of HPr by enzyme I seems to be the same in all organisms studied, the subsequent reactions of P-HPr → sugar-P seem to vary in different organisms and for different sugars. Three enzyme II systems have been studied in most detail: the constitutive enzyme II for glucose, fructose, and mannose in *E. coli* (25), the inducible enzyme II system for lactose in *S. aureus* (26), and the inducible β-glucoside system in *E. coli* (27).

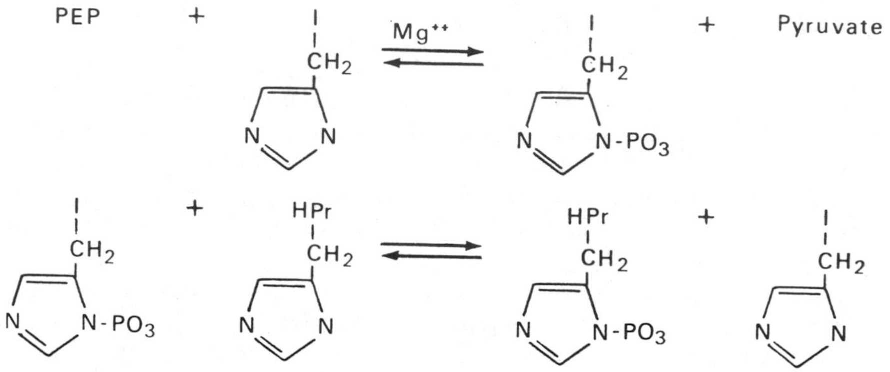

Figure 5
First reaction of the PTS system.

The most clearly defined (in terms of protein components) are the constitutive glucose, fructose, and mannose systems in *E. coli* characterized by Kundig and Roseman (25). The reaction

$$\text{P-HPr} + \begin{array}{c} \text{glucose} \\ \text{fructose} \\ \text{mannose} \end{array} \xrightarrow{\text{(washed membranes)}} \begin{array}{c} \text{glucose-6-}\textcircled{P} \\ \text{fructose-1-}\textcircled{P} \\ \text{mannose-6-}\textcircled{P} \end{array} + \text{HPr}$$

is catalyzed by washed membranes, with no requirement for an additional soluble component. When the membranes are solubilized and fractionated, three components are obtained which are essential for enzymatic activity, two proteins (II-A and II-B) and a lipid (phosphatidylglycerol). The II-A fraction can be further fractionated into separate sugar-specific components II-A^{glu}, II-A^{man}, and II-A^{fru}. The II-B fraction has also been fractionated to a point where the product gives a single band on sodium dodecyl sulfate gel electrophoresis with a molecular weight of about 36,000. If II-B is a single species it indicates that the II-B protein is a general component for at least glucose, fructose, and mannose. Furthermore it may represent about 10% of the total membrane protein. The question of the purity of II-B is a difficult one, however, since the protein aggregates, making the usual criteria for purity useless.

The requirement for phosphatidylglycerol, a relatively minor constituent of the *E. coli* membrane, is quite specific. The major lipid, phosphatidyl-ethanolamine, is not active. Several detergents tested have slight activity, although they are less than 10% as active as phosphatidylglycerol.

The reconstitution of enzyme activity requires a specific order of mixing of the three components and also divalent metal ion (Ca^{++} or Mg^{++}). Only when the order II-B + Ca^{++} + phosphatidylglycerol + II-A is followed, is an insoluble complex formed which is enzymatically active. A more detailed examination will hopefully lead to a clearer understanding of the functional relation of the complex to its membrane organization.

Although the following reaction can be written for the constitutive activities of the *E. coli* membrane, the precise mechanism is obscure:

$$\text{P-HPr + glucose} \xrightarrow[\quad Ca^{++} \qquad\qquad\qquad PG \quad]{\text{II-A} \qquad\qquad\qquad\qquad \text{II-B}} \text{glucose-6-} \textcircled{P} + \text{HPr}$$

or
(intact membranes)

The enzyme II for β-glucosides in *E. coli* is inducible and does not seem to require an additional soluble component. It has been partially resolved from the intact membrane by extraction with detergent and is essentially free of lipid. There is an absolute requirement for anionic lipid or detergent for the reaction. The isolated protein has a molecular weight of about 200,000, and no evidence has been obtained for an additional protein component. At this stage, however, the system has not been sufficiently resolved to speculate on the mechanism of the reaction.

The lactose enzyme II system in *S. aureus* has been studied in some detail. It differs from the *E. coli* constitutive system in that a soluble protein III^{lac} is required in addition to II^{lac} in the membrane. The reaction mechanism is as follows:

(1) $\qquad\qquad 3\ \text{HPr-P} + III^{lac} \rightleftharpoons P_3\text{-}III^{lac} + 3\ \text{HPr}$

(2) $\qquad\qquad P_3\text{-}III^{lac} + 3\ \text{lactose} \rightarrow 3\ \text{lactose-P} + III^{lac}$

The first reaction has been demonstrated with pure $\text{HPr}_{S.\ aureus}$ and pure III^{lac} and seems to be self-catalyzed (26). The phosphoryl linkages to III^{lac} appear to be the same as in P-HPr, and therefore are assumed to be to the N-1 position of two histidine residues. Reaction 2 is not well understood owing to great difficulty in resolving the II^{lac} from the membranes. It has been extensively purified and is essentially free of lipid, but at this point it is impossible to determine whether II^{lac} is composed of one or more components. Kinetic studies carried out on this reaction indicate that the reaction proceeds by an ordered mechanism with a ternary complex formed between the two substrates, P_3-III^{lac} and lactose and enzyme II^{lac}. This excludes a sequential (ping-pong) mechanism of transfer of the phosphoryl group from P_3-III^{lac} to enzyme II^{lac} (26). In addition, direct binding of lactose to II^{lac} has been demonstrated (28). It is therefore suggested that the functional permease for this system consists of II^{lac} and P_3-III^{lac} with II^{lac} functioning as the lactose carrier and P_3-III^{lac} providing the energy coupling.

It should be emphasized that the phosphorylation of III^{lac} and the direct binding of lactose to II^{lac} represent the first demonstrations of energy coupling and solute recognition in a transport system of known enzymatic function (26, 28).

A factor-III–like protein has been described by Hanson and Anderson (29) for the fructose system in *Aerobacter aerogenes*, although its function is not clear.

The apparent lack of analogy between the various enzyme II systems studied can be resolved by answering the following questions: Are the constitutive II-A proteins functionally analogous to factor III systems and do they differ only in cellular location or in affinity for the membrane? Is the II-B fraction sugar-specific as in the lactose system in *S. aureus*? Is the *S. aureus* II^{lac} component a single protein? Does the β-glucoside system in *E. coli* contain two protein fractions? If the answer to each of these questions is yes, then the following general reaction can be written for the PTS:

$$
\text{PEP} + \text{HPr} \xrightarrow[\text{I-P}]{\text{I}} \text{P-HPr} + \text{Pyr}
$$

$$
\begin{array}{cc}
\text{P-HPr} + \text{III}^x & \text{III}^x\text{-P} + \text{HPr} \\
(\text{II-A}^x) & (\text{II-A}^x)
\end{array}
$$

$$
\begin{array}{ccc}
\text{III}^x\text{-P} \quad + \text{X} & \text{II}^x & \text{X-P} + \text{III}^x \\
(\text{II-A}^x\text{-P}) & (\text{II-B}^x) & (\text{II-A}^x)
\end{array}
$$

THE PTS AND SUGAR TRANSPORT

A great deal of evidence suggests that the PTS plays an important role in sugar transport; only the major points will be discussed.

The first demonstration that sugars were accumulated as their phosphate esters was that of Rogers and Yu (30). They were able to show that in *E. coli* galactose, α-methylglucoside, and β-methylglucoside were accumulated as their phosphate esters. Subsequent work in other laboratories has shown that essentially all the α-methylglucoside (a glucose analogue) taken up in the first minute by *E. coli* was α-methylglucoside phosphate (31). In *S. aureus* it appears that all sugars are converted to their phosphate esters during transport (30).

The first direct evidence linking the PTS to sugar transport came from the demonstration by Kundig *et al.* (32) that cells subjected to the osmotic shock technique of Nossal and Heppel (33) lost the ability to transport sugar It was also demonstrated that the shocked cells had lost most of their HPr. When the shocked cells were incubated with purified HPr, transport ability could be restored. A variety of experiments showed that the restoration was specific for HPr. In addition to implicating the PTS in transport, these experiments have led to the wide use of the osmotic shock technique in the study of other transport systems, particularly those for amino acids.

The mutational approach has provided the most compelling evidence for the linkage of the PTS and sugar transport. From the enzymology of the system it is easy to predict the possible mutant phenotypes. Enzyme I and HPr are general components required for all sugars utilized by this mechanism, and a mutational loss of either should result in a cell unable to grow on a number of different sugars. Enzyme II and factor III systems, on the other hand, are sugar-specific, and loss of a particular enzyme II or factor III should affect only a single sugar or closely related sugars. The simple model in Figure 6 illustrates this point.

It is apparent from the model that the enzyme II systems form the carrier complex and the generation of P-HPr by enzyme I provides the energy-coupling reaction. Whether the carrier complex can function in facilitated diffusion depends on the precise mechanism involved and is discussed later.

Several enzyme I and HPr mutants have been isolated. The first such mutant to be recognized as a transport mutant was described by Egan and Morse in *S. aureus* (34). The mutant was called car⁻ and was unable to utilize 11 carbohydrates for growth because of a transport defect. It has since been shown to be an enzyme I mutant (35). Table V lists a few of the various PTS mutants obtained and their phenotypes. It is apparent that the

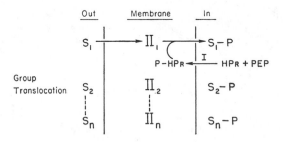

Figure 6
Simple model for sugar transport by the PTS. The enzyme II complex (see text) serves as the membrane carrier and the energy is provided from P-HPr. (After Roseman, S. *J. Gen. Physiol. 54,* 138s, 1969.)

number of sugars affected varies considerably. The possible reasons for this are discussed later. It is important to note, however, that either enzyme I or HPr deficiency causes a pleiotropic defect in the ability to use sugars for growth. It is, of course, critical to show that the PTS mutations actually affect transport, and the resultant phenotype is not simply due to the lack of a phosphorylating system which is essential for metabolism. Several lines of evidence have established that the PTS is required for transport and that concomitant phosphorylation is simply an energetic bonus which prepares the sugar for catabolism. An enzyme I mutant of *S. typhimurium* can be "tricked" into growing on glucose either by raising the glucose concentration in the medium or by inducing the cells for the galactose permease which allows glucose to enter (37). Under either of these conditions an ATP-dependent glucokinase can phosphorylate glucose at a rate sufficient to allow growth. Thus if sugar can enter the cell there are metabolic alternatives to the PTS.

The clearest demonstration is the direct measurement of rate of uptake of various sugars. In an enzyme I mutant of *S. typhimurium* uptake is severely decreased as shown in Figure 7. This figure compares the uptake ability of

Table V

Phenotypes of Various PTS Mutants

| | Defect in Carbohydrate Utilization | | |
	NO. OF CARBOHYDRATES NOT UTILIZED	TRANSPORT	REFERENCE
Enzyme I			
E. coli,			
A. aerogenes	5	?	36
S. typhimurium	9	Negative	37
E. coli	12	Negative	38
S. aureus	11	Negative	34
HPr			
A. aerogenes	5	?	36
S. typhimurium	9	Negative	37
Enzyme II			
A. aerogenes	Mtl^-	Mtl^-	39
	Fru^-	Fru^-	29
E. coli	β-Glucosides$^-$	β-Glucosides$^-$	38
S. aureus	Lac^-	Lac^-	40

Pleiotropic carbohydrates include glycerol, hexoses (Glc, Gal, Man), glucosides, GlcNAc. hexitols, keto-hexoses pentoses (?), disaccharides (Mal, Lac, Meli. Suc).

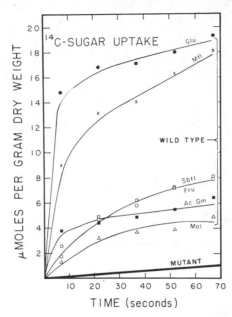

Figure 7

Inability of an enzyme I mutant of *S. typhimurium* to transport sugars. The wild type shows appreciable uptake for glucose (Glu), mannitol (Mtl), sorbitol (Sbtl), fructose (Fru), N-acetyl-glucosamine (AcGm), and maltose (mal). The mutant is represented by the dark line which represents the average uptake for the same six sugars. (After Roseman, S. *J. Gen. Physiol. 54,* 138s, 1969.)

parent and mutant for six sugars. The mutant uptake is represented by the heavy line which represents the average for all six sugars tested.

Although this experiment uses metabolizable sugars, it is clear that the defect is in the rate of uptake. The experiments have been carried out with α-methylglucoside and similar results have been obtained.

Mutants isolated in the enzyme II complex are also given in Table V. These mutants, in contrast to the pleiotropic enzyme I and HPr mutants, are only defective for a single sugar or closely related sugar. Figures 8 and 9 illustrate the transport ability of enzyme I, enzyme II^{lac}, and factor III^{lac} mutants of *S. aureus*. These data clearly show that the enzyme I defect affects both the glucose permease and the lactose permease, while the II^{lac} and III^{lac} mutations affect only the lactose system. It should also be noted that all three mutations render the cells essentially impermeable to sugar. The enzyme I mutant cannot carry out facilitated diffusion even though the carrier complex is intact.

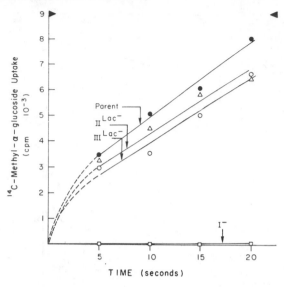

Figure 8

Uptake of ¹⁴C-labeled methyl-α-glucoside (a glucose analogue) by *S. aureus* PTS mutants. The parent and two sugar-specific mutations (II^{lac} and III^{lac}) show normal uptake, while the enzyme I mutant is defective. ▶ ◀, mark equilibration levels.

WHICH SUGARS ARE DEPENDENT ON THE PTS?

In *S. aureus* all sugars tested seem to be accumulated as their phosphate esters and are apparently dependent on the PTS for transport. In *E. coli* and *S. typhimurium* the situation is not so clear. A definite PTS dependence can be demonstrated by both phosphorylation in vitro and mutant phenotype for glucose, fructose, mannose, N-acetylglucosamine, mannitol, sorbitol, and glucosamine. However, several other compounds are curious.

The phosphorylation in vitro of lactose, glycerol, maltose, and melibiose has not been demonstrable in *E. coli* or *S. typhimurium*. Yet some enzyme I mutants are unable to utilize these sugars. There are several possible explanations for this apparent disagreement. It is possible that even though several sugars may not be phosphorylated by the PTS, the energy of P-HPr could still be required for transport. The difference between mutant phenotypes can be correlated with the levels of enzyme I (22). Mutants with no detectable enzyme I are negative on most sugars, including those mentioned above; those with detectable enzyme I are usually positive for these sugars. Thus the phenotypes could merely be a reflection on the different levels of enzyme I required for utilization of that particular sugar.

Another explanation has been presented for lactose and glycerol in *E. coli*. An examination of *E. coli* enzyme I mutants which are unable to use

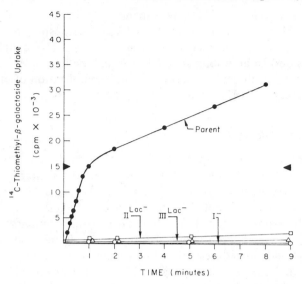

Figure 9

Uptake of ^{14}C-labeled-thiomethyl-β-galactoside (a lactose analogue) by *S. aureus* PTS mutants. Mutants of the sugar-specific components, II[lac] and III[lac], and a general component, enzyme I, are all defective. ▶◀, mark equilibration levels.

lactose or glycerol, in addition to the other sugars, has shown that the catabolic enzymes for these sugars are not induced normally (41–43). It has also been demonstrated that both enzyme I and HPr mutants are more sensitive to a phenomenon called catabolite repression (41, 43), which affects the induction of the catabolic enzymes. The lack of induction can be overcome by the addition of cyclic adenosine monophosphate (AMP) to the culture or conversion of the strains to constitutivity for the enzymes involved. For a thorough discussion of this phenomenon, the reader is directed to the review of Pastan and Perlman (41). Thus lactose- and glycerol-negative enzyme I mutants of *E. coli* can be converted to lactose- and glycerol-positive by treatment with cyclic AMP or by conversion of the cells to lactose or glycerol constitutivity. Neither of these treatments affects the enzyme I level of these mutants. It has been suggested that the lack of induction of catabolic enzymes can also explain why enzyme I mutants are negative on maltose and melibiose. In enzyme I and HPr mutants of *S. typhimurium* some disagreement is seen. Cyclic AMP does circumvent glycerol negativity in enzyme I mutants, although no effect is seen with other sugars including maltose and melibiose. Glycerol negativity is not reversed by cyclic AMP in an HPr mutant of *S. typhimurium*, and the cells are induced for normal levels of the glycerol catabolic enzymes (22). Thus the situation is far from clear. What should be remembered is that in spite of some minor uncertainties, there is a clear role for the PTS in the transport of many sugars such as glucose, mannose, fructose, sorbitol, mannitol, N-acetylglucosamine, and β-glucosides. In addition, the PTS in *E. coli* and *Salmonella* is in some way involved either directly or indirectly with the utilization of other solutes, such as lactose, glycerol, melibiose, and maltose.

Sugar-Binding Proteins

In the past few years much attention has been directed to a group of proteins which are not bound to isolated cell walls or membranes, but seem to be localized on the surface of the bacterial cell. Whereas the *precise* location of this class of proteins is not clear, the ease with which they can be removed from the cell strongly suggests localization in some compartment distinct from the cytoplasm. These proteins can be selectively released from the cell by relatively mild procedures. Either treatment of the cells with lysozyme and ethylenediamine tetraacetic acid (EDTA) or exposure of the bacteria to a form of osmotic shock causes selective release of this class of proteins. Some of the proteins released by these techniques are listed in Table VI. Included are nine hydrolytic enzymes, of which alkaline phosphatase was the first to be studied. Malamy and Horecker (44) showed that this enzyme is completely released into the sucrose medium when *E. coli* cells are converted to spheroplasts. The list in Table VI also includes binding proteins for leucine, sulfate, galactose, histidine, phosphate, arabinose, and arginine. These binding proteins are able to interact specifically with their respective "substrates" to form reversible complexes, but do not catalyze any change in the solute.

The osmotic shock technique was developed by Heppel (45) and co-workers. Washed *E. coli* cells are suspended in tris-EDTA-sucrose solution. The cells are then isolated and rapidly resuspended in dilute $MgCl_2$ solution. The cells are removed by centrifugation and the supernatant solution, "shock fluid," contains the various hydrolytic enzymes and binding proteins which in all represent about 5% of the total cell protein. The shocked cells are viable and grow normally when added to fresh growth media, although there is a short lag period before growth resumes. Heppel has pointed out that the shocked cells have lost their soluble nucleotide pool and have an increased permeability to compounds such as actinomycin D, so that the shock technique is not specific for just the so-called "shock" proteins.

The interest in shock proteins in relation to transport was prompted by the observation of Kundig *et al.* (32) that shocked cells had lost the ability to transport some sugars. This, plus the concept that some component of carrier-mediated transport must have the intrinsic ability to bind solute, prompted many workers to examine the shock fluid for binding

Table VI

Proteins Which Are or Are Not Released from *E. coli* Cells by the Osmotic Shock Procedure

PROTEINS RELEASED	PROTEINS NOT RELEASED
Alkaline phosphatase	β-Galactosidase
5'-Nucleotidase	Polynucleotide phosphorylase
Acid hexose phosphatase	Histidyl RNA (synthetase)
Nonspecific acid phosphatase	Adenosine deaminase
Cyclic phosphodiesterase	Thiogalactoside transacetylase
ADPG pyrophosphatase	Uridine phosphorylase
Ribonuclease I	Lactic dehydrogenase
Asparaginase II	Leucine aminopeptidase
Penicillinase	Certain other dipeptidases
Leucine-binding protein	DNA polymerase
Sulfate-binding protein	Ribonuclease II (phosphodiesterase)
Galactose-binding protein	Glucose-6-phosphate dehydrogenase
Arabinose-binding protein	Glutamic dehydrogenase
Phenylalanine-binding protein	Inorganic pyrophosphatase
Histidine-binding protein	Adenylic acid pyrophosphorylase
Phosphate-binding protein	Guanylic acid pyrophosphorylase
Arginine-binding protein	DNA exonuclease I
	5'-Nucleotidase inhibitor protein
	Glycerol kinase
	UDPG pyrophosphorylase

After Heppel, L. *J. Gen. Physiol.* **54,** 19s (1969).

components. As listed in Table VI many such binding proteins have been described.

GALACTOSE-BINDING PROTEIN

E. coli cells subjected to osmotic shock have a reduced ability to accumulate galactose. Anraku (46) purified a protein from shock fluid that bound radioactive galactose and glucose. The assay for binding activity usually involves equilibrium dialysis. The protein has been purified to homogeneity, has a molecular weight of about 35,000, and reversibly binds 1 mole of galactose or glucose per mole of protein with about the same affinity.

The evidence that the galactose-binding protein (GBP), and other binding proteins, is involved in transport is generally indirect. Shocked cells lose transport ability and GBP can be recovered in the shock fluid. Anraku (46) has reported experiments which indicate that transport activity of shocked cells can be partially restored by prior incubation of the cells with pure GBP. The restoration was about 25%. Essentially 100% restoration could be achieved by prior incubation of cells with a combination of

GBP and a crude ammonium sulfate fraction of shock fluid which contained no GBP. It should be pointed out, however, that these restoration experiments measured accumulation of galactose and not rate of transport. It is not difficult to imagine other alternations caused by the shock procedure which could effect the accumulation of a solute such as depletion of energy pools and alteration of general cell permeability.

Boos (47) studied the genetic relation between the GBP and the β-methylgalactoside permease. These studies have indicated that GBP activity correlates well with behavior of the β-methylgalactose permease (not to be confused with the lactose system). Three mutants have been isolated which have altered galactose transport activity. Of the three, only one has lost GBP activity. Thus it would appear that other components are involved in the transport process. In addition the mutant missing GBP does not revert, which indicates the lesion is something other than a single point mutation in the structural gene for the binding protein. It is hoped that a more detailed genetic characterization will clarify the role of this binding protein.

ARABINOSE-BINDING PROTEIN

An arabinose-binding protein (ABP) has been isolated and characterized by Hogg and Englesberg (48) and by Schleif (49). The substrate specificity of ABP closely resembles that of the arabinose permease, and ABP is under the same genetic control as the permease. Mutants have been isolated by Hogg and by Schleif which have about 10% of the normal amount of ABP and are not able to concentrate L-arabinose. However, other mutants with less than 10% ABP show near normal transport ability. ABP mutants do not map in the *araE* gene, which is the classic arabinose permease site. Again the transport function of ABP is far from clear, and a more extensive genetic analysis is required.

General Mechanism for Carbohydrate Transport in Bacteria

As the previous discussion indicates there are three major areas of investigation of carbohydrate transport in bacteria. The lactose system in *E. coli* is the best defined system of active transport. Several other sugars such as galactose, arabinose, maltose, and melibiose may be transported by an active transport mechanism as well.

The most general mechanism, however, seems to be group translocation mediated by the phosphoenolpyruvate-glycose PTS. A large number of sugars are transported by this system in a variety of bacteria. The examination of the PTS by Roseman and his collaborators has provided detailed information concerning the protein components involved in transport and the reactions which take place.

The third area of interest has been the isolation of binding proteins.

Although the evidence for their involvement in transport is not yet convincing, they must be considered potentially important.

Certainly in an organism such as *S. aureus* the evidence suggests that all sugars tested, including lactose, are transported into the cell as their phosphate esters. When one considers a sugar such as galactose in *E. coli*, however, a general concept of sugar transport in bacteria becomes difficult. This sugar is phosphorylated by the PTS in vitro and cells deficient in enzyme I have altered galactose transport activity (37). Whole cells, however, generally accumulate free galactose, suggesting an active transport system. In addition, a "shock" protein has been isolated which binds galactose and may participate in the transport process.

Further work is obviously required before such dilemmas can be resolved.

TRANSPORT OF AMINO ACIDS

The classic work of Gale and collaborators gave the first clear indication that microorganisms were able to accumulate amino acids without modification (50). These investigators found that when *S. aureus* was grown in the presence of a mixture of amino acids large amounts of glutamic acid and lysine could be extracted from broken cells. Later, free amino acids were also found to be accumulated by gram-negative bacteria.

The general mechanism for amino acid uptake is active transport. The model proposed for β-galactoside uptake in *E. coli* can serve equally well for amino acid uptake. There are many specific carriers implicated by both kinetic and genetic studies. The discussion in this section is limited to those amino acids for which some direct demonstration of the macromolecules involved has been attained.

Transport of Leucine, Isoleucine, and Valine

Much early work has indicated the presence of a transport system stereospecific for leucine, isoleucine, and valine (51, 52). This area has been greatly stimulated by Oxender and collaborators, who were able to demonstrate that, when *E. coli* cells were subjected to osmotic shock treatment, the accumulation of these three amino acids was reduced, while no change in the transport of alanine and proline could be detected. A binding protein was isolated from the shock fluid which bound the three-branched-chain amino acids (LIV protein) with binding constants which are similar to the K_m and K_i values for the uptake system.

The presence of leucine in the growth medium represses the LIV protein as well as the ability of the cells to transport leucine. This is analogous to the control of the leucine, isoleucine, and valine biosynthetic enzymes. There does, however, seem to be some functional separation in the control of the LIV protein and the biosynthetic enzymes. Oxender (53) has examined a leucine-requiring mutant of *S. typhimurium* which cannot make any of the biosynthetic enzymes, but has normal LIV protein activity and leucine transport. It is, however, curious that exogenous leucine represses the system required for its transport into the cell. Possibly LIV protein is part of a scavenging mechanism.

The LIV protein has been purified to homogeneity from *E. coli* K12 and crystallized. It has a molecular weight of about 36,000 and has a single binding site for leucine, isoleucine, or valine (54).

Furlong and Weiner (55) have found that there is a leucine-specific permease system in addition to the LIV system in *E. coli*. This system composes about 20% of the total capacity of the cell to take up leucine and is specifically inhibited by trifluoroleucine. They have also purified and crystallized a leucine-specific binding protein which also binds trifluoroleucine. This protein appears quite similar to the LIV protein, since they are immunologically indistinguishable.

Oxender has obtained indications that some strains of *E. coli* also contain a system specific for both isoleucine and valine and a corresponding binding protein (53). It therefore appears that *E. coli* K12 contains a common transport system for leucine, isoleucine, and valine and a minor leucine-specific system. *E. coli* B contains equal amounts of the shared system and the leucine-specific system. *E. coli* W contains the system specific for isoleucine and valine and the shared system. The corresponding binding proteins have been demonstrated in these organisms.

Anraku (46) has attempted to restore leucine transport in osmotically shocked cells with the LIV protein. LIV protein could restore about 25% of the transport activity, while a crude ammonium sulfate fraction containing no LIV protein restores nearly 50% of the activity. The combination of both fractions could restore leucine transport to near the level in non-shocked cells. These experiments are open to the same criticisms as discussed previously under "Galactose-Binding Protein."

Arginine Transport

Wilson and Holden (56) have demonstrated that *E. coli* W cells lose about 25% of their ability to transport arginine when they are osmotically shocked. They have fractionated the shock fluid and found four protein fractions which bind arginine. Two of the four proteins had some restorative power upon the ability of shocked cells to transport arginine.

Histidine Transport

Ames and collaborators (57–59) have studied histidine transport in *S. typhimurium* where at least four systems seem to be operative. One of these, the histidine-specific permease, has a high affinity and specificity (K_m for L-histidine: 3×10^{-8} M), while the other three have a much poorer affinity and specificity. For example, histidine is transported with poor affinity by the aromatic permease (*aroP*) which has a much higher affinity for the aromatic amino acids.

The high-affinity histidine permease is composed of at least three components distinguishable genetically, physiologically, and kinetically. One of these components has been demonstrated to be a histidine-binding protein (the J protein), thus confirming the involvement of this binding protein in transport (59). This group has now done considerable genetic work to identify the components required for histidine transport by the high-affinity system. They have described several classes of mutants with altered histidine transport activity.

1. *hisP* mutants. These are resistant to the histidine analogue 2-hydrazino-3-(4-imidazolyl)-propionic acid (HIPA) owing to an inability to transport this analogue into the cell. It has been shown that *hisP* codes for a protein by the isolation of amber nonsense mutants in this gene.

2. *dhuA* mutants. Histidine-requiring strains are not able to use D-histidine as a source of L-histidine, but mutations occur at high frequency at a site (*dhuA*) which allow D-histidine utilization. The *dhuA* gene is adjacent to the known *hisP* gene, and the *dhuA* mutation results in the increase in both D- and L-histidine uptake. The *dhuA* mutants also have an increased sensitivity to HIPA as a result of increased transport activity.

3. *dhuA-hisJ* double mutants. These have been selected from *dhuA* his⁻ strains for the loss of the ability to grow on D-histidine and the retention of HIPA sensitivity.

4. *dhuA-hisP* double mutants. The *hisP* mutation has been introduced into a *dhuA* strain by selecting for resistance to HIPA. The resulting double mutants have lost the ability to use D-histidine. Therefore the *hisP* product is required for transport of D- and L-histidine and HIPA. It has been demonstrated that *hisP*, *dhuA*, and *hisJ* are closely linked on the chromosome.

These mutant strains and the biochemical work on them has done much to clarify the macromolecular components involved in histidine transport. Ames and Lever (59) and Rosen and Vasington (60) have studied a histidine-binding protein. This binding protein (J protein) represents about 95% of the total binding capacity of the shock fluid while another protein (K protein) accounts for the remaining 5%. Ames and Lever have reported a binding constant of about 2×10^{-7} M for the J protein.

From analysis of the mutants it was demonstrated that *dhuA-hisJ* mutants are missing the major histidine-binding protein (J protein). The *dhuA* mutants have about five times the normal amount of J protein, and *hisP* mutants have normal J protein. None of these mutants has altered K protein.

The transport phenotypes of each are shown below:

	K_m TRANSPORT (L-HISTIDINE)
wild type	2.6×10^{-8}
dhuA	6.6×10^{-9}
hisP	10^{-6}
hisJ	2×10^{-7}

Although a complete rate study has not been done, it is apparent from the data presented that *hisP* mutants are quite defective in the high-affinity histidine transport, although they have a full complement of J protein. *hisJ* mutations also cause a defective uptake of histidine, though it is not nearly as low as in the *hisP* mutants. The residual activity in *hisJ* mutants has been attributed to the other permease component, K, although this component has not been genetically defined. The *dhuA* mutants have increased J protein and increased transport activity.

Ames and Lever have proposed the scheme shown in Figure 10 for histidine transport in *S. typhimurium*. Both the demonstrated J system and the proposed K system are coupled to the *hisP* product. The binding protein (J protein) appears to be an obligatory requirement for the high-affinity histidine transport system, but it is not sufficient for transport in the absence of *hisP* protein.

The complexity of this system is obvious. It is made up of at least three protein components which are genetically or kinetically distinguishable. One component is a binding protein which does play some role in the transport process.

As mentioned throughout this section, the evidence for the role of the binding proteins in transport has been generally indirect with the exception of the genetic data. The following correlations do exist:

1. Mutants which are missing binding protein and have defective transport systems have been isolated.

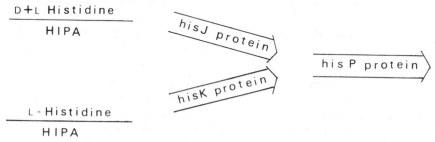

Figure 10
Proposed scheme for histidine transport in *S. typhimurium*. (After Ames, G. F., and Lever, J. *Proc. Natl. Acad. Sci. U.S.A. 66,* 1096, 1970.)

2. Cells subjected to the osmotic shock technique lose the ability to transport certain solutes. (Some solutes are, however, not affected by this procedure, i.e., alanine and serine.) Binding proteins for many of the affected solutes can be recovered from the shock fluid.

3. Binding constants for many of the binding proteins generally agree with the K_m values for transport.

4. Transport and binding activity seem to be under common genetic control.

5. Partial success has been obtained in restoration of transport in shocked cells by the isolated binding proteins. This effect has, however, been only partial in most cases, and some laboratories report a total lack of success.

If indeed binding proteins are a part of the permease system, it would seem most likely that they act prior to the membrane translocation step. They do not appear to be associated with the plasma membrane, as are the carrier components for the PTS and the M protein.

Energy-Coupling Reactions for Amino Acid Transport

The nature of the energy-coupling reaction in amino acid transport is unclear at present. Unlike the β-galactoside system, functional separation of facilitated diffusion and accumulation of amino acids has not been clearly established. Evidence does seem to be emerging to suggest that, as with the β-galactoside system of *E. coli*, energy coupling is dependent on electron transport rather than on oxidative phosphorylation. As discussed earlier, Pavlasova and Harold (14) proposed that a proton gradient across the membrane results from electron transport and that this is the driving force for solute accumulation. The effect of uncouplers is attributed to their ability to disrupt the proton gradient rather than to their inhibition of ATP synthesis.

Klein *et al.* (16), in addition to investigating β-galactoside transport, have studied the coupling of oxidative energy to proline transport. Using both isolated membrane vesicles and whole cells, they found that proline uptake was not markedly reduced by lack of exogenous substrate, but was greatly reduced by lack of oxygen. Arsenate at high concentration had little effect on transport, though it rendered the cells incapable of ATP synthesis. In these experiments, as with those of Pavlasova and Harold, uptake was sensitive to uncouplers, although the vesicles have no capacity for ATP synthesis. Klein *et al.* have suggested that active uptake of proline in *E. coli* may be directly coupled to a high-energy state or compound produced in the membrane by oxygen uptake.

Milner and Kaback (61) have studied the transport of many amino acids in *E. coli* membrane vesicles. These workers were the first to show that isolated membrane vesicles prepared from *E. coli* spheroplasts provide a suitable system for transport studies, and their work has been reviewed by Kaback (62). They have shown that proline, glutamic acid, glutamine, tryptophan, aspartic acid, asparagine, serine, glycine, alanine, and lysine can be accumulated by membrane vesicles. Phenylalanine, leucine, isoleucine, valine, and histidine can also be accumulated, but to a lesser degree. It is interesting that this latter group consists of amino acids for which binding proteins have been demonstrated. The low level of accumulation may result from the loss of binding proteins during vesicle formation.

The most intriguing information to come from the study of transport with isolated membrane vesicles is the effect of D-lactate on accumulation. Proline uptake, for example, is stimulated 20-fold by D-lactate, and the uptake of all the other amino acids tested is stimulated to varying degrees. α-Hydroxybutyrate, succinate, L-lactate, and DPNH all showed stimulation, but less than D-lactate, while a large number of other compounds showed essentially no effect. D-Lactate was converted almost solely to pyruvate in these experiments. The hypothesis offered for the D-lactate effect is that electron transport via a membrane-bound D-lactate dehydrogenase (or succinate dehydrogenase in the case of succinate) is coupled directly to the concentrative uptake process, possibly through a direct oxidation and reduction of the carrier.

If indeed an intact electron transport system is required for energy coupling, and oxygen is the terminal acceptor, can cells carry out active transport when growing under anaerobic conditions? In many bacteria some components of the electron transport system are repressed by anaerobic growth. It may prove interesting to study anaerobically grown cells in attempting to relate transport phenomena in vesicles to transport phenomena in intact cells. Since membrane preparations have lost the bulk of the soluble enzymes required for metabolism, they are restricted to the metabolism of compounds for which the enzymes remain attached to the membrane. It is possible, though probably not too likely, that the metabolism of D-lactate and succinate by their membrane-bound dehydrogenases simply results in maintaining a native membrane structure and that the resultant increase in transport activity is a reflection of the state of the membrane and not a direct effect on the transport process itself.

Ion Transport

The Sulfate Transport System

Ion transport is not discussed in any detail here, since it is covered in Chapter 11. It was, however, the isolation of the sulfate-binding protein from *S. typhimurium* by Pardee and his collaborators, the first binding protein described, which provided a stimulus to those involved in transport studies (63, 64).

Like the sulfate transport system, the synthesis of the sulfate-binding protein is repressed by the presence of cysteine in the growth medium. The protein has been purified and crystallized. It has a molecular weight of about 32,000. The binding protein differs from the permease system in one important respect: Thiosulfate is less effective in inhibiting sulfate binding to the binding protein than in inhibiting sulfate transport.

Like the other binding proteins studied, the sulfate-binding protein can be released by osmotic shock. There seem to be about 10^4 molecules of sulfate bound per bacterium. Pardee and Watanabe (65) developed methods to test whether a protein is located externally with respect to the plasma membrane and internally with respect to the cell wall. They found that diazo-7-amino-1,3-napthylene disulfonate can inactivate β-galactoside transport but no β-galactosidase of intact *E. coli* cells. This compound also inactivates sulfate binding and transport but not uridine phosphorylase of *S. typhimurium*. On the other hand, antibody to the sulfate-binding protein inactivates the protein when it is extracted and purified, but not when it is associated with whole cells. Thus, it would seem that the binding protein is indeed located somewhere between the cell wall and the plasma membrane, as expected from its ready release by osmotic shock.

It has been established that sulfate is bound to the binding protein by ionic forces. Binding can be abolished by treating the protein with amino group reagents, suggesting that positively charged groups on the protein might be so placed as to hold the negatively charged anion.

The analysis of mutants lacking the sulfate transport system has yielded some strains which lack the binding protein. It is, however, not clear if the transport defective mutants missing binding protein are stuctural gene mutants, or if they map in the *cysA* region which is known to be the permease site.

Phosphate Transport

The transport of phosphate in *E. coli* has been studied by Medveczky and Rosenberg (66). A phosphate-binding protein has also been isolated from shock fluid. It has been purified to homogeneity and has a molecular weight of about 42,000. There seem to be about 2×10^4 molecules of this binding protein per cell.

The pure phosphate-binding protein can restore some transport ability to both shocked cells and spheroplasts. In addition, two phosphate permease mutants have been examined. One mutant is missing the binding protein and can be stimulated to transport phosphate by incubation with the purified protein. The other mutant has normal binding protein and cannot be stimulated for transport by exogenous binding protein.

SUMMARY

The general field of solute transport in bacterial systems has progressed from an indirect kinetic examination to a genetic and biochemical dissection. This approach has in the past few years resulted in considerable progress toward an understanding of the macromolecular components involved in transport. The characterization of the phosphoenolpyruvate-glycose phosphotransferase system, the demonstration of the M protein, and the isolation of many binding proteins have been the major contributions to this effort.

In its simplest form solute translocation can be pictured as occurring in three steps: solute recognition, translocation across the membrane, and energy coupling to achieve accumulation. It is suggested that the various binding proteins are involved in solute recognition external to the plasma membrane. The M protein for lactose transport in *E. coli* seems to be involved in solute recognition, translocation, and energy coupling. The phosphoenolpyruvate-glycose PTS also functions in all three steps. The enzyme II complexes are necessary for solute recognition and translocation, and energy coupling occurs via factor III or P-HPr.

It is clear that the work to date has been largely descriptive in nature; a mechanistic understanding of membrane transport will be available within the foreseeable future.

REFERENCES

1. Lin, E. C. C. *Annu. Rev. Genet.* **4,** 225 (1970).
2. Osterhout, W. J. V. *Proc. Natl. Acad. Sci. U.S.A.* **21,** 125 (1935).
3. Wilbrandt, W., and Rosenberg, T. *Pharmacol. Rev.* **13,** 109 (1961).
4. Stein, W. D. *The Movement of Molecules Across Cell Membranes*. Academic Press, New York, 1967.
5. Mitchell, P., and Moyle, J. *Proc. Roy. Soc. Edinburgh* **27,** 61 (1958).
6. Doudoroff, M., Hassid, W. Z., Putnam, E. W., Potter, A. L., and Lederberg, J. *J. Biol. Chem.* **179,** 921 (1949).
7. Kepes, A., and Cohen, G. N. *The Bacteria*. Ed. I. C. Gunsalus and R. Y. Stanier. Academic Press, New York, 1962, vol. IV, p. 179.
8. Jacob, F., and Monod, J. *J. Mol. Biol.* **3,** 318 (1961).
9. Winkler, H., and Wilson, T. *J. Biol. Chem.* **241,** 2200 (1966).
10. Kennedy, E. P. *The Lactose Operon*. Ed. J. R. Beckwith and D. Zipser. Cold Spring Harbor Laboratory, Cold Spring Harbor, N.Y., 1970.
11. Kepes, A. *Biochim. Biophys. Acta* **40,** 70 (1960).
12. Fox, C. F., and Kennedy, E. P. *Proc. Natl. Acad. Sci. U.S.A.* **54,** 891 (1965).
13. Jones, T. D. H., and Kennedy, E. P. *J. Biol. Chem.* **244,** 5981 (1969).
14. Pavlasova, E., and Harold, F. *J. Bacteriol.* **98,** 193 (1969).

15. Mitchell, P. *Biol. Rev.* **41,** 445 (1966).
16. Klein, W. L., Dahms, A. S., and Boyer, P. D. *Fed. Proc.* **29,** 540 (1970).
17. Barnes, E., and Kaback, H. R. *Proc. Natl. Acad. Sci. U.S.A.* **66,** 1190 (1970).
18. Wong, P. T. S., Kashket, E. R., and Wilson, T. H. *Proc. Natl. Acad. Sci. U.S.A.* **65,** 63 (1970).
19. Wilson, T. H., Kusch, M., and Kashket, E. R. *Biochem. Biophys. Res. Commun.* **40,** 1409 (1970).
20. Roseman, S. *J. Gen. Physiol.* **54,** 138s (1969).
21. Kundig, W., Ghosh, S., and Roseman, S. *Proc. Natl. Acad. Sci. U.S.A.* **52,** 1067 (1964).
22. Saier, M., Simoni, R. D., and Roseman, S. *J. Biol. Chem.* **245,** 5870 (1970).
23. Anderson, B. *Studies on the Phosphoenolpyruvate-Dependent Phosphotransferase System of E. coli.* Doctoral thesis, University of Michigan, Ann Arbor, 1968.
24. Anderson, B., Kundig, W., Simoni, R. D., and Roseman, S. *Fed. Proc.* **27,** 643 (1968).
25. Kundig, W., and Roseman, S. *Fed. Proc.* **28,** 463 (1969).
26. Nakazawa, T., Simoni, R. D., Hays, J. B., and Roseman, S. *Biochem. Biophys. Res. Commun.* In press.
27. Rose, S. P., and Fox, C. F. *Fed. Proc.* **28,** 463 (1969).
28. Hays, J. B., and Simoni, R. D. *Fed. Proc.* (1971).
29. Hanson, T. E., and Anderson, R. L. *Proc. Natl. Acad. Sci. U.S.A.* **61,** 269 (1968).
30. Rogers, D., and Yu, S. H. *J. Bacteriol.* **84,** 877 (1961).
31. Winkler, H. *Biochim. Biophys. Acta* **117,** 231 (1966).
32. Kundig, W., Kundig, F. D., Anderson, B., and Roseman, S *J. Biol. Chem.* **241,** 3243 (1966).
33. Nossal, N. G., and Heppel, L. *J. Biol. Chem.* **241,** 3055 (1966).
34. Egan, J. B., and Morse, M. L. *Biochim. Biophys. Acta* **97,** 310; **109,** 172; **112,** 63 (1965).
35. Simoni, R. D., Smith, M., and Roseman, S. *Biochem. Biophys. Res. Commun.* **31,** 804 (1968).
36. Tanaka, S., and Lin, E. C. C. *Proc. Natl. Acad. Sci. U.S.A.* **52,** 913 (1967).
37. Simoni, R. D., Levinthal, M., Kundig, F. D., Kundig, W., Anderson, B., Hartman, P. E., and Roseman, S. *Proc. Natl. Acad. Sci. U.S.A.* **58,** 1963 (1967).
38. Fox, C. F., and Wilson, G. *Proc. Natl. Acad. Sci. U.S.A.* **59,** 988 (1968).
39. Tanaka, S., Lerner, S. A., and Lin, E. C. C. *J. Bacteriol.* **93,** 642 (1967).
40. Hengstenberg, W., Penberthy, W. K., Hill, K., and Morse, M. L. *J. Bacteriol.* **99,** 383 (1969).
41. Pastan, I., and Perlman, R. *Science* **169,** 339 (1970).
42. Berman, M., Zwaig, N., and Lin, E. C. C. *Biochem. Biophys. Res. Commun.* **38,** 272 (1970).
43. Epstein, W., Jewett, S., and Winter, R. H. *Fed. Proc.* **29,** 1986 (1970).
44. Malamy, M. H., and Horecker, B. L. *Biochemistry* **3,** 1893 (1964).
45. Heppel, L. *J. Gen. Physiol.* **54,** 95s (1969).
46. Anraku, Y. *J. Biol. Chem.* **243,** 3116, 3123, 3128 (1968).
47. Boos, W. *Eur. J. Biochem.* **10,** 66 (1969).

48. Hogg, R. W., and Englesberg, E. *J. Bacteriol.* **100,** 423 (1969).
49. Schleif, R. *J. Mol. Biol.* **46,** 185 (1969).
50. Gale, E. F. *Adv. Protein Chem.* **6,** 287 (1953).
51. Britten, R. J., and McClure, F. T. *Bacteriol. Rev.* **26,** 292 (1962).
52. Cohen, G. N., and Rickenberg, H. V. *Ann. Inst. Pasteur.* **91,** 693 (1956).
53. Oxender, D. L. *Metabolic Transport.* Ed. L. Hokin. Academic Press, New York. In press.
54. Piperno, J. R., and Oxender, D. *J. Biol. Chem.* **241,** 5732 (1966).
55. Furlong, C. E., and Weiner, J. H. *Biochem. Biophys. Res. Commun.* **38,** 1076 (1970).
56. Wilson, O. H., and Holden, J. T. *J. Biol. Chem.* **244,** 2743 (1969).
57. Ames, G. F. *Arch. Biochem. Biophys.* **104,** 1 (1969).
58. Ames, G. F., and Roth, J. R. *J. Bacteriol.* **96,** 1742 (1968).
59. Ames, G. F., and Lever, J. *Proc. Natl. Acad. Sci. U.S.A.* **66,** 1096 (1970).
60. Rosen, B. P., and Vasington, F. D. *Fed. Proc.* **29,** 545 (1970).
61. Milner, L., and Kaback, H. R. *Proc. Natl. Acad. Sci. U.S.A.* **66,** 1008 (1970).
62. Kaback, H. R. *Annu. Rev. Biochem.* **39,** 561 (1970).
63. Pardee, A. B., and Prestige, L. S. *Proc. Natl. Acad. Sci. U.S.A.* **55,** 189 (1966).
64. Pardee, A. B. *Science* **156,** 1627 (1967).
65. Pardee, A. B., and Watanabe, K. *J. Bacteriol.* **96,** 1049 (1968).
66. Medveczky, N., and Rosenberg, H. *Biochim. Biophys. Acta* **211,** 158 (1970).

11

P. D. Boyer
W. L. Klein

Energy-Coupling Mechanisms in Transport

The principal objective of this chapter is to furnish an analytical introduction to the energetics of active transport, predicated on the belief that satisfying explanations require detailed knowledge of chemical and molecular events. Such aspects, rather than kinetic or thermodynamic considerations, will be stressed. The latter eventually may provide a deeper level of understanding after the principal molecular components and reactions have been identified and described. As this chapter is not a comprehensive review article, the literature coverage is by no means inclusive, and some important, well-recognized contributions to our knowledge of active transport are not specifically cited. However, we have sought to call attention to certain aspects of this vigorous field which we feel may prove useful in developing energy-coupling hypotheses and hence have direct bearing on the design of pertinent experiments.

Preparation of, and experimental work from the authors' laboratory reported in, this chapter were supported in part by U.S. Public Health Service Grant GM 11094 of the Institute of General Medical Sciences. One of us (W. L. K.) gratefully acknowledges support by a Predoctoral National Science Foundation Fellowship.

ELUCIDATION OF ENERGY-COUPLING MECHANISMS

Present Level of Understanding

The coupling of metabolic energy to transport is one of the most basic and clearly one of the most complicated of primary biological processes. The long-range goal in the study of energy coupling is to establish the molecular nature of energy *source*, energy *conservation*, and energy *use* in the membrane system. More explicitly, the questions that must be answered in this field include from what compounds and reactions membranes derive their energy for transport, how and into what form the energy is converted and retained in the membrane, and in what manner the energy-rich compounds or states of the membrane are used to achieve a concentration gradient. Ultimately, the answers to such questions will require detailed comprehension of all molecules interacting with the transported substance, and further, the primary and secondary interactions of these molecules, in turn, with other membrane and cell components. The paucity of current knowledge becomes evident when the simple models representing transport are contrasted with the elegant description of molecular architecture and submolecular interactions of such enzymes as carboxypeptidase and lysozyme.

The investigator studying active transport confronts a spectrum of inherent difficulties, some obvious and some subtle. First of all, he is trying to understand a membrane process before the membrane itself is adequately described. Reactions and structural changes with ill-defined chemistry are involved. Also, he is seeking components whose existence appears required but whose precise function is a matter of conjecture. Further, a satisfying explanation of active transport in all likelihood will require an understanding of such related and complicated events as oxidative phosphorylation and contractility coupled to ATP cleavage.

Characterization of Energy Source, Conservation, and Use

Each of the myriad transport systems undoubtedly has uniquely individual aspects. However, we feel that organizing the present chapter into discussions of energy source, conservation, and use may be instructive. These divisions will be approached by examining features found among numerous systems. As the divisions merely represent breaking points in a continuous process, some overlap can be expected in the examination of experimental details.

Before these concepts are pursued further, it should be noted that energy-dependent accumulation of external solutes falls into two distinct categories: transport with chemical modification and transport without chemical modification. Accumulation with chemical modification is in a strict and important sense not active transport. Unless otherwise specified in this chapter, active transport is considered to represent accumulation of an unmodified substance against a concentration gradient.

Energy sources for active transport differ among cells and organisms. Excluding mitochondrial processes, most mammalian systems are likely to use ATP cleavage for driving active transport. On the other hand, with bacterial, mitochondrial, and other related systems, energy-yielding oxidative enzymes have been found closely associated with membrane components involved in transport; with these systems, evidence is accruing for direct oxidation-driven transport. In other words, energy from membrane-linked oxidations may be used directly for transport, without intervening ATP synthesis. A concept developed further in this chapter is that both ATP and oxidative reactions appear capable of generating an energized compound or state that is, in turn, coupled to ligand movement.

Present information suggests that energy conservation within membranes may involve one of three distinct primary events. The energy from ATP or oxidations may serve:

1. To give rise to conformational changes
2. To induce covalent bond formation in membrane components
3. To establish an electrochemical gradient

It must be emphasized that combinations of these possibilities are likely. For instance, covalent bond formation is likely to be accompanied by changes in conformation and ionization of protein groups and with consequent possible changes in membrane potentials. Thus, adequate definition of which events in energy conservation are primary and which are secondary is likely to remain a problem even if membrane conformation changes, covalent bond formation, and potential gradients are demonstrated.

With regard to utilization of conserved energy for specific ligand accumulation, one can readily envision two modes of action. Energy utilization may serve:

1. To modify a carrier involved in ligand translocation
2. To modify membrane sites interacting with carrier proteins

Evidence relevant to mechanisms is indirect and superficial, involving principally adaptations from better understood systems. For some time speculation will prevail as productive hypotheses are sought.

TRANSPORT ACCOMPANYING COVALENT MODIFICATION

The occurrence and properties of a phosphotransferase system in several bacterial species, as elucidated by Roseman and coworkers, has been described in other chapters of this volume. Uptake of sugars by this mechanism does not involve movement against a concentration gradient, but depends upon a vectorial phosphorylation, with phosphoenolpyruvate as the phosphoryl donor. The phosphorylated sugars may be metabolized as such, or cleaved by a phosphatase to give the free sugar, thus achieving an actual concentration gradient between extracellular and intracellular sugar. A similar concentrating system appears to be operative for the uptake of purines, with participation of purine nucleotide pyrophosphorylase (1).

Although phosphorylation of the sugar may provide the only means of energy transduction in these systems, the possibility that the membrane has additional mechanisms for concentrating the free sugar and that the accompanying phosphorylation is a useful initial metabolic process has not been eliminated. Observations on the uptake of α-methylglucoside, a nonmetabolizable analogue of glucose, are pertinent. Results showing that the ratio of intracellular α-methylglucoside phosphate to α-methylglucoside remains relatively constant with continued incubation (2, 3) are consistent with entry of a fraction of the sugar in a nonphosphorylated form. Action of an intracellular phosphatase should result in continuing decrease of phosphorylated sugar. The maintenance of a constant α-methylglucoside phosphate pool is also consistent with fortuitous balance of uptake, dephosphorylation, and exit. Haguenauer and Kepes (4) favor this view, although such a balance does not appear to be supported by their observations that energy-poisoned *Escherichia coli* maintain their α-methylglucoside phosphate pool even though further uptake is abolished and exit of the free sugar continues.

With mammalian systems, considerable experimental evidence has been accumulated that similar phosphorylation processes do not participate in sugar transport. Indeed, the transport of sugars and amino acids has not been demonstrated to involve any covalent modification.

COMPONENTS OF SYSTEMS FOR ACTIVE TRANSPORT

The characterization and isolation of the varied components of membranes will remain for some time as one of the principal research areas of active transport.* The emerging concept of membranes as dynamic structures with intermixed protein and lipid components, and with most of the structure arising from interaction of functional components, is of obvious importance to an understanding of the energization of transport. The evidence for a core "structural" protein for membranes has largely been discredited, and the usefulness of such a premature designation of function is doubtful.

Either of two types of membrane components serves to utilize the available energy sources. These components are the membrane-bound ATPases and the enzymes of the oxidative chain. Because anaerobic organisms, with limited or no oxidative enzymes present, can accumulate solutes effectively,

---

* Racker stated pertinently in a recent review (5) on the mitochondrial inner membrane: "However, I question whether it is possible to really 'see' the proverbial forest without an intimate knowledge of its trees."

it is clear that most or all of the oxidative capacities are not essential for active transport in some systems. Similarly, the occurrence of energy-linked transport in submitochondrial particles that contain no ATPase or contain inhibited ATPase demonstrates that active transport need not always be coupled directly to ATP cleavage. A concept that emerges is the formation, either through ATPases or through oxidative enzymes, of an energized state, designated as $\sim$,* in the simplified relations as follows:

$$\text{Oxidation} \leftrightarrow \sim \leftrightarrow \text{ATP}$$
$$\updownarrow$$
$$\text{Transport}$$

An extremely interesting group of membrane components that have been actively studied recently are the binding proteins that appear intimately linked to transport (Chap. 10 and ref. 7). Recent additions to an ever-expanding group include a glutamine-binding protein (8) and a basic amino acid–binding protein (9). The isolation of specific binding proteins is a logical development from earlier studies, since the existence of transport system components (very likely proteins) with specific binding sites has been strongly implied for many years by a number of different indirect approaches. The actual isolation of proteins appearing to fulfill the specific binding function offers enticing opportunities to elucidate the interplay with energy-yielding mechanisms.

Unfortunately, we do not have similar clear-cut theoretical or experimental indications about the nature of other protein or lipid components likely to be involved in the energy-conserving and energy-utilizing steps. Other chapters in this volume develop the probable function of lipid components. The participation of a variety of proteins with as yet unknown function is suggested by genetic studies (10, 11) as well as a number of other indications. The multicomponent interactions of the phosphotransferase system may serve as a pertinent example of expected complexities (12).

ENERGY SOURCES FOR ACTIVE TRANSPORT

ATP-Driven Transport

As mentioned above, metabolic considerations make it highly likely that in some cellular systems ATP cleavage at the membrane provides the energy for active transport. Such systems include anaerobically growing microorganisms, in which the only known mode of energy capture is formation of ATP in fermentations, and most mammalian cells, in which mitochondrial ATP synthesis or glycolytic reactions furnish ATP for use at cell

---

* The symbol $\sim$ was first used by Lipmann (6) to designate a high-energy phosphate bond, but its meaning since has been broadened to include high-energy membrane states as well as others.

surfaces. Any hydrolytic capacity for ATP exhibited by a membrane needs consideration for participation in active transport.

The wide occurrence of ATPase activity in fragmented membrane preparations and the lack of stimulation of such activity by the transported solute indicate that the solute is not essential as a stoichiometric reactant in the ATP cleavage reaction. Yet in the intact membrane, the presence of transported solute is likely to be a compulsory requirement for ATP cleavage. The "uncoupling" of ATP cleavage from the transport process may be akin to the "uncoupling" in oxidative phosphorylation.

The study of ATP cleavage stands as the best molecular analysis of an energy-transducing component linked to active transport. If other aspects of source, conservation, and use were so rigorously understood, we would be much closer to our goal of systematic knowledge at the molecular level. The more extensively studied ATPases are the Na^+,K^+-activated ATPases from nervous tissue, kidney tissue, and other sources; the Ca^{++}-activated ATPase of the sarcoplasmic reticulum; and the Mg^{++}-ATPase from *Streptococcus faecalis*.

Na^+,K^+-ATPase

Reviews by Whittam and Wheeler (13) and by Schultz and Curran (14), a volume edited by Armstrong and Nunn (15), and contributions of Christensen and colleagues (16) may be consulted for the considerable literature indicative of Na^+,K^+-ATPase involvement in transport of sugars, amino acids, and electrolytes in mammalian cells.

There is little evidence for any role of Na^+,K^+-ATPases in bacterial transport, as transport of many substances takes place adequately in medium lacking these ions. Of interest, however, is the demonstration by Mora and Snell (17) of K^+ stimulation and Na^+ inhibition of amino acid transport by protoplasts of *S. faecalis*, and more recently, observations reported in Roseman's laboratory of Na^+-dependent transport of thiomethylgalactoside by *Salmonella typhimurium* (18). The following discussion of Na^+,K^+-ATPase is based upon results from mammalian systems.

A diagram summarizing present information on the action of Na^+,K^+-ATPase is given in Figure 1.

A vital property of the Na^+,K^+-ATPases is the formation of a protein-bound acyl phosphate intermediate. A 1969 symposium volume gives good background information (20). The kinetic competency of the phosphorylated intermediate has been convincingly demonstrated by Kanazawa and colleagues (21) and further substantiated by observations of Neufeld and Levy (22). An important point is that the reversible phosphorylation of the enzyme takes place in the presence of Na^+ and Mg^{++}, thus producing an ADP $\rightleftarrows$ ATP exchange reaction.

A crucial part of the sequence is the conversion of "E_1-P" in an essen-

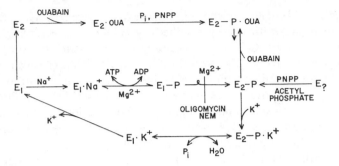

Figure 1

Na$^+$,K$^+$-ATPase cycle for ion transport and ouabain inhibition. (Adapted from Sen, A. K., *et al. J. Biol. Chem. 244*, 6596, 1969.)

tially irreversible step to "E_2–P" with Na$^+$ release, followed by hydrolytic formation of P_i in presence of K$^+$. Several experimental results suggest that conformational changes among protein components in the complex enzyme may occur during conversion of the E_1–P to the E_2–P form. The inhibitor ouabain can combine either with the free enzyme or the E_2–P form to give an inactive complex. Indications of more than one stable conformational form of the Na$^+$,K$^+$-ATPase are also shown by discontinuities of Arrhenius plots and lack of ouabain sensitivity below a critical temperature of 14°C (23).

Preparations retaining capacities indicated in Figure 1 show presence of several protein components (20). Purification of a protein catalyzing the ADP ⇄ ATP exchange reaction but not the over-all sequence has been claimed (24). More highly purified ATPase fractions such as those of Kyte (25) showing presence of only two polypeptide chains in contrast to the several components of less purified preparations may represent the hydrolytic capacity separated from associated proteins vital for conformational inter-actions with membrane and transport processes.

Recent observations show that the Na$^+$,K$^+$-ATPases have the capacity for catalyzing a K$^+$-dependent exchange of inorganic phosphate oxygens with water. This exchange occurs in the absence of ATP and is ouabain-sensitive (26). Na$^+$ inhibits the exchange, but ATP overcomes the Na$^+$ inhibition. Addition of detergent, which increases the potential ATPase activity, obliterates K$^+$-dependent exchange. This appears to be an additional type of "uncoupling," perhaps analogous to the uncoupling of oxidative phosphorylation.

Whether the ATPase puts its interesting chemistry directly to work for solute translocation or whether an intermediate state is formed via an energy conservation step is open to question. The first possibility, with the ATPase itself acting as a Na$^+$ carrier, is a hypothesis proposed by a number of investigators (14). The transport is regarded as driven by Na^+ flow from outside to inside, with energy supplied by ATP cleavage serving to pump Na^+ out of the cells. Caution in accepting this view is indicated by the data of Jacquez and Schafer (27), who showed that Ehrlich ascites cells appeared to accumulate α-aminoisobutyrate against a Na^+ concentration gradient. Observations of Kimmich (28) are even more convincing. He showed that intestinal cells can accumulate sugar for short periods against a distinct Na^+ gradient. Such results, as well as other facets of active transport systems discussed in this chapter, suggest that the Na^+,K^+-ATPases may share in common with a variety of membrane transport systems the capacity for generation of the energy-rich compound or state. Energy conserved in this manner is then used to drive transport without requiring stoichiometric movement of Na^+ or other ions.

Ca^{++}-ATPase

The uptake of Ca^{++} ions with ATP cleavage in vesicles of the sarcoplasmic reticulum suggests a fine system for present and future studies. Considerable insight into the ATPase has been obtained by various groups (29–31). Of interest is the recent demonstration that ATP can be synthesized by reverse of the sarcoplasmic Ca^{++} pump (32), indicative of the reversibility of the reactions making the energized state ($\sim$) available to transport.

Mg^{++}-ATPase

E. coli membranes have a prominent Mg^{++}-activated ATPase, but only a small amount of ouabain-sensitive Na^+,K^+-ATPase (33). The best studied bacterial ATPase, however, is the Mg^{++}-activated enzyme from membranes of *S. faecalis* (34, 35). The purified enzyme shows no ADP–ATP exchange activity and no apparent phosphoenzyme derivative. Of interest is the observation that N,N′-dicyclohexylcarbodiimide (DCCD) inhibits the membrane-bound ATPase and a number of energy-dependent transport processes (35, 36). Active ATPase can be isolated from the DCCD-inhibited membrane preparation, but ATPase added back to the DCCD-treated membrane loses activity. A combination of DCCD with components of the membrane responsible for energy coupling but distinct from the ATPase is thus indicated. The isolated ATPase is freed from its energy-coupling interactions. Such results are similar to earlier observations made with mitochondrial systems.

The establishment of a pH gradient across the cell membranes of *S. faecalis* is dissipated by action of DCCD as well as by uncouplers suggested to act as proton conductors. On the basis of this, Harold *et al.* suggest that transport by ATP cleavage is linked to establishment of a H^+ concentration gradient. They may take insufficient cognizance, however, of their own observations that cells grown in arginine do not give rise to a pH gradient

(37). Furthermore, *E. coli* cells appear capable of reversing the polarity of their membrane pH gradients, depending on the alkalinity of the growth medium (38).

Mention may be made that DCCD inhibition of other membrane systems has given interesting results. With red blood cell membranes, modification of a small portion of the total membrane carboxyl groups suffices for inhibition of the K^+-dependent p-nitrophenylphosphatase, an activity associated with the Na^+,K^+-ATPase (39). Perplexing complexities remain, however.

Some of the most pertinent observations regarding ATP-driven transport are those indicating mechanistic similarity to oxidative phosphorylation. For example, important implications for the mechanism of active transport come from the demonstration by Pavlasova and Harold (40) that uncouplers of oxidative phosphorylation inhibit anaerobic transport of thiomethyl-β-D-galactoside. This shows that ATP-driven transport may involve the same type of energized state as the elusive $\sim$ of oxidative phosphorylation.

Oxidative Mitochondrial Transport

The energy-linked transport of H^+ and of cations and anions by mitochondria has been intensively studied for the past several years. Increasing knowledge of a number of intricate, metabolically important interchanges of metabolites, together with descriptions of the inner and outer membranes, have given new vistas of mitochondrial function. Several recent reviews help give perspectives on the scope of the field (41–43). Most of the observed transport, however, has not been against concentration gradients, and active transport has been studied only with ionic substances, principally Ca^{++}, K^+, and H^+.

The most striking characteristic of the transport by mitochondrial systems is its ability to be energized by oxidations independent of participation of ADP or ATP (41–43). Both active transport of cations and energy-linked reductions can be observed with submitochondrial particles lacking the capacity to form ATP or blocked in phosphorylation of ADP by the action of inhibitors such as oligomycin (40–42). Oligomycin blocks the ability of mitochondria or submitochondrial particles to form ATP, but, unlike uncouplers such as 2,4-dinitrophenol, does not cause the continual dissipation of the mitochondrial high-energy state. Addition of 2,4-dinitrophenol uncouples ion transport and stimulates oxygen uptake by mitochondria in the presence of oligomycin. A further important point is that the transport is insensitive to the lack of phosphate or presence of arsenate. Such observations demonstrate that oxidative energy can be used directly in mitochondria for active cation transport without intervening phosphorylations.

As discussed later, the stoichiometry of K^+ transport is particularly relevant to how energy is conserved. In addition, the synthesis of ATP by efflux of K^+ down a concentration gradient (44) demonstrates the reversibility of the system, as implied in the simple scheme given on page 327.

An intriguing means of "uncoupling" mitochondrial oxidations from transport or ATP synthesis occurs with antibiotics such as valinomycin and related molecules that specifically bind K^+. These antibiotics, with cavities lined by carbonyl and hydroxyl groups, have lipophilic exteriors and facilitate penetration of K^+ through synthetic or natural membranes (45). In mitochondria, they make K^+ available to the matrix space, and oxidative energy is used for K^+ extrusion (46). Ion transport has been found to take precedence over ATP formation; thus valinomycin plus K^+ can uncouple oxidations from phosphorylations.

Oxidative Microbial Transport

The ability of yeast, bacteria, and other microorganisms to accumulate substances from low concentrations in the outer medium has long been recognized. With metabolites such as amino acids and sugars, it has been satisfactorily demonstrated that this is an accumulation against a concentration gradient. In addition to intact cells, vesicles formed from ruptured protoplasts provide valuable experimental systems (47).

The suggestion from data of Kepes (48) that thiomethylgalactoside transport involves hydrolysis of one ATP per molecule transported is one of the few attempts to assess the stoichiometry of transport. These and other studies led to establishment of the concept that oxidative energy was likely used first for synthesis of ATP which in turn was coupled through a hydrolytic reaction to active transport. As mentioned earlier, some sugars appear to be taken up through the vectorial phosphorylation system described by Roseman, but it is abundantly clear that galactoside uptake can occur without phosphorylation of the galactoside (49). Evidence for oxidatively driven uptake for galactosides and other substances is given in the following paragraphs.

Analogy with the mitochondrial systems suggests that bacterial transport driven by oxidation might occur independent of phosphorylations. Recent evidence gives strong support to this possibility. Bacterial vesicles with little or no capability for net oxidative phosphorylation accumulate proline in the presence of oxygen and substrate (50). D-Lactate appears to be a preferred substrate (51). Incubation with little or no phosphate but with high arsenate present, to further minimize any possible local phosphorylations, leaves transport intact (3). The arsenate-insensitive transport is inhibited by uncouplers of oxidative phosphorylation and is dependent on oxygen. Importantly, active uptake of several amino acids and sugars also can be demonstrated in intact cells having arsenate-depleted pools of ATP and phosphoenolpyruvate. However, in the same cells, anaerobically-driven uptake dependent upon metabolic phosphorylations is arsenate-sensitive.

In related observations, it has been demonstrated that vesicles from *E. coli* catalyze an ATP, energy-linked reduction of TPN by DPNH (52). Thus, both modes of energy input, ATP coupling and oxidative coupling, appear to be functional in the vesicles. Since only oxidative coupling has been linked to transport, several distinct high-energy states may possibly be generated within the membrane.

Earlier indications of transport dependent on oxygen but independent of phosphorylation were obtained with yeast by Halvorson and Cowie (53). They demonstrated arsenate-insensitive uptake of amino acids and recognized that this might reflect a fundamental difference in energy coupling between protein synthesis (requiring ATP) and transport.

ENERGY CONSERVATION MECHANISMS

The elucidation of the nature of the energized state or compound, $\sim$, long one of the central and unsolved problems of oxidative phosphorylation, now appears central to transport processes as well. Possible primary modes of energy conservation such as covalent bond formation, protein conformational changes, or establishment of electrochemical gradients have been mentioned earlier. Vigorous proponents of alternative modes of energy conservation have emerged. Regardless of how energy is conserved, the storage capacity of membrane systems for $\sim$ appears small. One of the interesting determinations is that of Chance and Azzi (54) based on uptake of Ca^{++}, measured using a bioluminescent jellyfish protein as an indicator of free Ca^{++}. Their data suggest that $\sim$ is equivalent to about 0.3 nmole of ATP per milligram in mitochondria and that it has a "lifetime" of 7 ± 2 sec at 25°C.

The participation of covalent bonds must remain hypothetical until and unless there is actual experimental demonstration of such bonds. The sensitivity and scope of searches still remain inadequate, but in absence of any demonstrable covalent chemistry, other alternatives must be considered.

Conformational coupling is one of the most attractive yet most vague of present suggestions (55–57). It has received support from such diverse observations as changes in electron micrographs of intact membranes (58) and changes in binding or environment of fluorescent probes. Evidence from the macromolecular structural observations of electron microscopy, although interesting, are definitively not convincing for establishing conformational changes as the basis of transport and phosphorylation at a molecular level.

Some of the changes reflected by fluorescent probes appear kinetically sufficiently rapid to be involved in primary coupling processes (59, 60). A recent observation of potential importance was reported by Jagendorf and Ryrie. They showed energy-linked conformational changes in an ATPase-type coupling factor of chloroplasts, as measured by exchange of hydrogens of the ATPase protein with solvent water (61). Whether the conformational changes observed with mitochondrial and chloroplast systems are primary or secondary, and to what extent the probes themselves may modify structures, are areas of uncertainty.

The suggestion that the carriers for transported substance may actually be allosteric electron transport components (51, 62) seems untenable at this stage. The abundant evidence that ATP cleavage can drive transport in membrane systems that lack oxidative components suggests that the oxidative enzymes provide an alternate means for the energization of, but not an integral component part of, transport systems.

The concept of primary coupling through electrochemical gradients, the most popular form of which is "the chemiosmotic" hypothesis of Mitchell, continues to gain considerable attention. The suggestion that transport may depend upon establishment of H^+ gradients across membranes seems refuted by various experimental observations (63). The postulated development of a potential gradient could be nearer the truth but much harder to check experimentally.

The ability of uncoupling agents to act as proton conductors has frequently been invoked as favoring a H^+ concentration gradient interconvertible with $\sim$. The interesting demonstrations of Wilson and colleagues (64) on the concentration dependency of different groups of uncoupling agents are pertinent in this regard. They show uncoupling versus concentration relations that are in accord with theoretical predictions of the simple Henderson-Hasselbalch relation. The carbonylcyanide phenylhydrazones and substituted benzimidazoles have maximum activity as acids, while the salicylanilide derivatives have maximum activity as bases. Such results appear incompatible with the H^+–potential gradient concepts and suggest participation of the uncouplers as general acid or general base catalysts, possibly for hydrolysis at critical sites.

As mentioned earlier, experimental findings on the stoichiometry of the active transport of K^+ ions by mitochondria appear incompatible with the development of an electrochemical potential as a primary basis for energy conservation by membranes. The maximum potential developed by passage of two electrons through a coupling site would be equivalent to a charge of 2^+ or 2^-. Similarly, energization of transport by cleavage of ATP dependent upon development of a potential could be expected to give rise to a maximum charge difference of 1 or 2. Hence on this basis, cleavage of one ATP molecule or passage of two electrons through a coupling site would be expected to, at the most, drive transport of two K^+ ions. The demonstration by Azzone (65) of at least four and by Pressman (66) of as many as seven K^+ ions transported per ATP molecule cleaved thus appears to necessitate some other or additional means of energy conservation.

ENERGY USE

Some General Characteristics

How energy conserved by the membrane is used to drive transport is the least understood aspect of active transport. The involvement of protein carriers specific for transported substances does appear well established and is discussed in Chapter 10. But beyond this, even the simplest questions remain unanswered. For instance, does energy output modify the carrier or does it alter a different membrane component? Does it act before the transported solute is released internally or after the solute is released? Or perhaps even independently of solute release? Furthermore, what is the nature of the membrane recognition sites demanded by the vectorial nature of the transport process?

Before discussing aspects of some models, brief mention of certain salient characteristics of solute movement appears appropriate. The kinetic and other characteristics of the galactoside transport system of *E. coli* established by the French investigators remain as landmarks in the field. These include such facets as the entry mechanism's being catalytic, showing saturation characteristics, being coupled with metabolic energy donors, and having passive exit not by free diffusion but rather by carrier mediation (67). More recent aspects of the lactose transport system have been usefully reviewed by Kennedy (68). Features to be considered in the present discussion include countertransport, known and purported kinetic characteristics, and effect of energy inhibitors on the phenomena associated with uptake, efflux, and exchange of transportable solutes.

When active transport systems are deprived of energy, they often still exhibit the phenomena of countertransport and exchange. Countertransport occurs when the transient uptake of one solute against a gradient is driven by the reverse concentration gradient of another solute. Participating solutes likely share the same binding site on a mobile carrier. An example is the uptake of glucose driven by a preestablished gradient of 3-0-methylglucose, or of lactose driven by a gradient of β-methylthiogalactoside. A recent contribution by Wong and Wilson gives a good experimental example and an adequate mathematical treatment (69). They demonstrated that bacterial cells exposed to azide or 2,4-dinitrophenol to uncouple energy linking exhibited countertransport followed by equilibration of internal and external galactosides.

The occurrence of countertransport and exchange when cells are deprived of energy sources demonstrates that energization is not essential for creation of specific binding sites and that energy use is likewise not required for exposure of binding sites to inner and outer sites of membranes. Indeed, lack of energy may promote access of mobile carrier proteins to the cell interior.

Exchange of internal and external solute may occur in one of two ways. One way is by nonspecific leaks, followed by reentry of the external solute by the specific energy-requiring system. Although such interchanges can occur, they are not likely to be metabolically important and do not give essential information about the transport process. Of more interest is the demonstration that exit of an accumulated solute from the cell is often dependent upon the same specific carriers as the entry. In addition, frequently the presence of external solute appears to promote the exit of internal solute. For example, bacterial cells with accumulated amino acids when washed on Millipore filters with buffer solutions retain most of the amino acids. However, washing with buffer containing the transported amino acid or a suitable analogue results in rapid loss of the corresponding internal amino acid.

Another intriguing observation of active solute interchange was first shown in studies of thiomethylgalactoside transport (70). Even after a near stationary internal concentration is reached, there is continued rapid uptake and efflux of the galactoside. Such recycling by some suggested mechanisms would appear to demand continued and perhaps wasteful use of energy. It is possible, however, that nature has evolved a means so that energy is not needed unless there is net gain in internal solute concentration.

The initial rate of uptake of a solute by an active transport system shows a concentration dependency akin to that of a simple enzymic reaction; i.e., the velocity soon becomes independent of further increase in the concentration of external solute. For this reason, the designation "apparent K_m" for uptake is often used, although a better term might be one that designates the solute concentration for the one-half maximal uptake rate, such as $S_{1/2}$, analogous to $t_{1/2}$ for half-time of a reaction. One reason for such a preference is that the concentration dependency of uptake may not follow a simple rectangular hyperbola as for an enzymic reaction. From theoretical considerations, deviations from simple saturation kinetics seem probable. Also, adequate experimental data are not yet readily available and may be difficult to obtain. Such factors as nonspecific leaks, differential permeability of solute to various cell compartments, and loss of accumulated solute during separation of cells for measurement make accurate uptake assays difficult. Nonetheless, the existence of an apparent saturation of uptake rate with increase in concentration has important implications. With high external solute, rate limitation results either from the rate of rotation or movement of the protein carrier to the inner surface or from the rate of energy-utilization reactions at the inner surface.

Another interesting transport characteristic is that the internal concentration achievable for galactosides has been reported to be largely independent of external concentration increase beyond a moderate level (67). However, data in this regard are not convincing. As noted later, some theoretical models predict that additional slow but continued increase in internal concentration may occur with time. Such has been noted experimentally in our laboratory for proline uptake. This gives rise to an increase in the $S_{1/2}$ with increase in the time period observations. This phenomenon is illustrated in Figure 2.

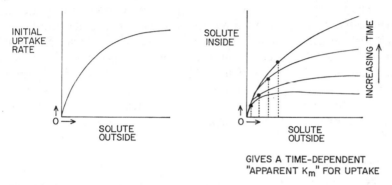

Figure 2
Some characteristics of conformational coupling models.

A difficulty in attempting to devise models from available data arises from the heterogeneity of transport systems that is sometimes overlooked. An example is the efflux of lactose, proline, and leucine from *E. coli*. Cells show a moderately rapid efflux of lactose, slow efflux of proline, and almost no efflux of leucine. However, when the substrates of each system are added to the external medium, an accelerated efflux of the intracellular substrates may be observed. Thus, three different transport systems exhibit three different patterns of efflux (3).

A summary of certain established features from various transport systems is an aid to further model building. Any given system is unlikely to include all the following features, but it may be expected to incorporate a large number of them. Thus, active transport models should be able to account for the following well-recognized characteristics:

1. A defined concentration gradient
2. Apparent saturation of uptake rate
3. Passive translocation

4. Exogenously stimulated efflux or exchange
5. Uncoupler stimulated efflux
6. Time dependency of $S_{1/2}$
7. Ability of efflux to regenerate $\sim$
8. Specific carrier participation

In addition to the characteristics listed above, we tentatively suggest another potentially important characteristic, namely, the likelihood that active transport systems always exhibit specific recognition sites at the inner membrane surface. Detailed appraisal of various suggested mechanisms for active transport shows that such recognition of the loaded or of the unloaded protein carrier is essential for the vectorial nature of the process. An example is discussed later. Such recognition, accompanied by a triggering of energy use or a removal of a barrier to access to the cell interior, appears essential to prevent unwanted exit, i.e., to provide the "one-way valve" for the active uptake. Various characteristics of transport systems previously allowed the confident prediction that protein molecules with specific binding sites participated in transport; subsequent isolation of transport proteins appears to fulfill this prediction. Similarly, it seems worthwhile to explore the possibility that a convincing argument might be constructed for the existence of membrane recognition sites.

A Carrier Conformational Change Model

Many suggestions for active transport systems are quite vague, perhaps wisely, about how energy use occurs. One of the more specific suggestions currently receiving considerable support is that energy is used to decrease the affinity of the transport proteins for the solute (72). Some support for such a possibility comes from the data of Boos *et al.* (73) indicating the existence of interconvertible forms of a galactose-binding protein, one form having considerably less affinity for galactose than the other. Also Winkler and Wilson (74) have interpreted their results as favoring a decreased apparent K_m for exit and have suggested that this may result from energy use to change the conformation of the binding protein.

A diagram of a transport system involving energy-linked change in the binding affinity of the carrier protein is shown in Figure 3. This is a simplified version and does not note membrane recognition sites. At the inner membrane, energy in some manner, possibly akin to the interactions of subunits of allosteric proteins, causes decrease in solute binding. Two types of membrane recognition would appear necessary. One type is that involved in the energy transfer. Protein components of the membrane may interact with those of the binding protein noncovalently, although transient covalent modification could be involved. The second type is recognition of the loaded in preference to the unloaded carrier. This seems essential to

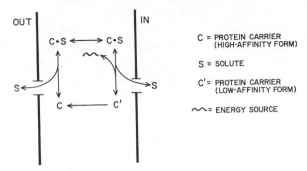

Figure 3
Oversimplified scheme for carrier-coupled conformational transport.

avoid wasteful energy-driven cyclization between the high- and low-affinity forms of the unloaded carrier.

Still a third type of membrane recognition also seems necessary. If, for example, the loaded carrier in the high-affinity form is allowed to have ready access to the inside as well as to the outside of the membrane, any gradient of accumulated solute would be dissipated. Membrane recognition of the loaded carrier to allow access to the interior solves this difficulty. Such recognition may require only a very minor shift in position of side groups of proteins. A model similar to that of Figure 3, but with this membrane recognition site depicted, is given in Figure 4.

Several facets of present information suggest caution in accepting the carrier conformation change model. The leucine-binding protein readily undergoes reversible denaturation, a fact regarded as favoring this model (75). A detailed appraisal of effects of various solvent characteristics and of temperature showed, however, that decreased binding affinity results only

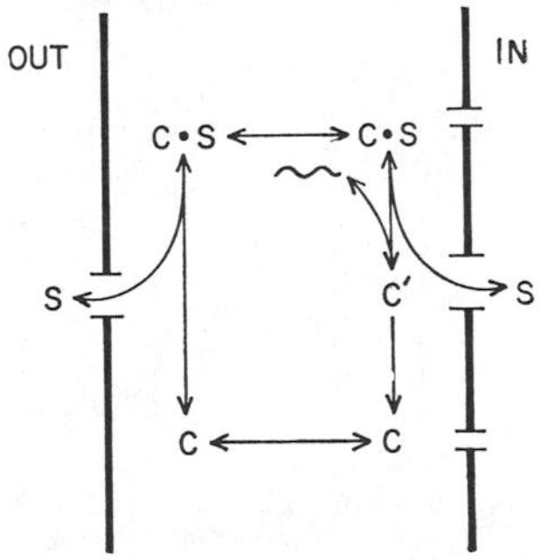

Figure 4
Scheme for carrier-coupled conformational transport.

when gross denaturation occurs (76). If conformational changes causing decreased ligand binding could be transmitted readily from the membrane to the carrier protein, such conformational changes would appear likely to be induced without gross denaturation. The change in apparent K_m for exit taken as supporting a carrier conformational change (74) can quite readily be explained by other models. Any system that allows efflux but maintains a higher internal concentration must increase the apparent K_m for exit. Another point is that it is difficult to explain the effect of uncouplers with the model. Entry of solute requires energy input in this model (Fig. 4), but exchange and countertransport occur readily in uncoupled systems. Perhaps more pertinent, evidence supporting this or any other mechanism is at present quite sketchy, and until better information is at hand, various alternative models need consideration.

A Specific Membrane Conformational Change Model

In consideration of how energy might be used at the inner membrane to capture a transported solute molecule, our attention has been directed recently to a more specific model of a membrane conformational change hypothesis. Essential features are that the "on" and "off" constants for solute binding and release, and thus the affinity, remain unchanged during the transport process, and that the inner membrane entry site, opened for the loaded carrier, is rapidly closed with energy use. Important stages for the cycle are indicated in Figure 5. As noted therein, the inner membrane site "opens" upon recognition of the loaded carrier, and after solute departs is "closed" by energy input. A simple competition between the rate of combination of solute with binding protein and the rate of the energy input reaction determines whether a given solute molecule is retained. A diagram of the transport cycle, like those previously presented, is given in Figure 6.

The model has a number of interesting features. Apparent saturation of entry rate would arise, as in other models, by sufficient external solute being

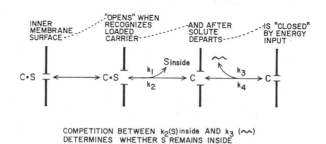

Figure 5
New mode of energy-coupling to transport.

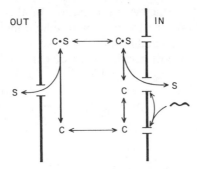

Figure 6
Boyer-Klein scheme for membrane-coupled conformational transport.

present to leave little or no unbound carrier at the outer membrane entry site. Apparent saturation of extent of uptake would arise when the internal concentration was sufficiently high so that rebinding of solute rather than closing of the entry site by energy use would result. Thus continued interchange after apparent saturation is reached would not require expenditure of energy. The model readily explains the need of carriers for exchange, but more importantly, readily explains exchange and countertransport in the presence of uncouplers. It allows a common energization mechanism, possibly even at the same site, for various protein carriers. It does not require transmission of energy from the membrane to the carrier.

A theoretical appraisal of this and related models shows some interesting characteristics (3). Both models depicted here (Figs. 3 and 5), although conceptually quite distinct, have some kinetic predictions in common. In both cases, the net rate of accumulation is given by an equation of the following form:

$$\frac{dS_i}{dt} = \frac{AS_e - BS_i}{CS_i + DS_e + ES_eS_i + F}$$

in which the terms A through F are functions of rate constants and energy input, S_e is the extracellular substrate concentration and S_i is the intracellular substrate concentration. They predict that the initial rate of uptake will saturate with increasing external solute. The final equilibrium concentration ratio between internal and external solute should be independent of the external solute concentration. The time required for this concentration-independent ratio to be established, however, could be considerable. Present information has been interpreted to indicate that internal concentration saturates at a certain value, as mentioned previously. However, results are not sufficiently precise and extensive to determine whether a continued slow uptake with increasing concentration may occur. Other factors may mask such uptake. Nonetheless, examination of the models suggests that more exploration of rates and extent of uptake could be useful. With the model of Figure 6 the rates and extent of uptake of a series of related solutes sharing a common transport carrier might be expected to be simply related to the rates of the association of the ligands.

The most searching assessment of the value of these or other models will come at a later time when their ability to generate useful experimentation has been tested.

REFERENCES

1. Berlin, R. D., and Stadtman, E. R. *J. Biol. Chem.* **241,** 2679 (1966).
2. Winkler, H. H. *Biochim. Biophys. Acta* **117,** 231 (1966).
3. Klein, W. L., and Boyer, P. D. Unpublished data.
4. Haguenauer, R., and Kepes, A. *Biochimie* **53,** 99 (1971).
5. Racker, E. *Essays Biochem.* **6,** 1 (1970).
6. Lipmann, F. *Adv. Enzymol.* **1,** 99 (1941).
7. Pardee, A. B. *Science* **162,** 632 (1968).
8. Berger, E. A., Weiner, J. H., and Heppel, L. A. *Fed. Proc.* **30,** Abs 1061 (1971).
9. Rosen, B. P. *J. Biol. Chem.* **246,** 3653 (1971).
10. Ohta, N., Galsworthy, P. R., and Pardee, A. B. *J. Bacteriol.* **105,** 1053 (1971).
11. Ames, G. F., and Lever, J. *Proc. Natl. Acad. Sci. U.S.A.* **66,** 1096 (1970).
12. Kundig, W., and Roseman, S. *J. Biol. Chem.* **246,** 1393 (1971).
13. Whittam, R., and Wheeler, K. P. *Annu. Rev. Physiol.* **32,** 21 (1970).
14. Schultz, S. G., and Curran, P. F. *Physiol. Rev.* **50,** 637 (1970).
15. Armstrong, W. W., and Nunn, A. S. (eds.). *Intestinal Transport of Electrolytes, Amino Acids, and Sugars.* Thomas, Springfield, Ill., 1971.
16. Christensen, H. N., Handlogten, M. E., and Thomas, E. L. *Proc. Natl. Acad. Sci. U.S.A.* **63,** 948 (1969).
17. Mora, J., and Snell, E. E. *Biochemistry* **2,** 136 (1963).
18. Stock, J., and Roseman, S. *Biochem. Biophys. Res. Commun.* **44,** 132 (1971).
19. Sen, A. K., Tobin, T., and Post, R. L. *J. Biol. Chem.* **244,** 6596 (1969).
20. Transport protein symposium. *J. Gen. Physiol.* **54** (Suppl.), 79s (1969).
21. Kanazawa, T., Saito, M., and Tonomura, Y. *J. Biochem.* **67,** 693 (1970).
22. Neufeld, A. H., and Levy, H. M. *J. Biol. Chem.* **245,** 4962 (1970).
23. Charnock, J. S., Doty, D. M., and Russell, J. C. *Arch. Biochem. Biophys.* **142,** 633 (1971).
24. Hossler, F. E., and Rendi, R. *Biochem. Biophys. Res. Commun.* **43,** 530 (1971).
25. Kyte, J. *J. Biol. Chem.* **246,** 4157 (1971).
26. Dahms, A. S. *Fed. Proc.* **30,** Abs 1169 (1971).
27. Jacquez, J. A., and Schafer, J. A. *Biochim. Biophys. Acta* **193,** 368 (1969).
28. Kimmich, G. A. *Biochemistry* **9,** 3669 (1970).
29. Kanazawa, T., Yamada, S., Yamamoto, T., and Tonomura, Y. *J. Biochem.* **70,** 95 (1971).
30. Kanazawa, T., Yamada, S., and Tonomura, T. *J. Biochem.* **68,** 593 (1970).
31. Pucell, A., and Martonosi, A. *J. Biol. Chem.* **246,** 3389 (1971).
32. Mackinose, M., and Hasselbach, W. *FEBS Letters* **12,** 271 (1971).

33. Hafkenscheid, J. C. M., and Bonting, S. L. *Biochim. Biophys. Acta* **178**, 128 (1969).
34. Abrams, A., and Baron, C. *Biochem. Biophys. Res. Commun.* **41**, 858 (1970).
35. Schnebli, H. P., Vatter, A. E., and Abrams, A. *J. Biol. Chem.* **245**, 1122 (1970).
36. Harold, F. M., Baarda, J. R., Baron, C., and Abrams, A. *J. Biol. Chem.* **244**, 2261 (1969).
37. Harold, F. M., Pavlasova, E., and Baards, J. R. *Biochim. Biophys. Acta* **196**, 235 (1970).
38. Kashket, E. R., and Wong, P. T. S. *Biochim. Biophys. Acta* **193**, 212 (1969).
39. Godin, D. B., and Schrier, S. L. *Biochemistry* **9**, 4068 (1970).
40. Pavlasova, E., and Harold, F. M. *J. Bacteriol.* **98**, 198 (1969).
41. Pressman, B. C. *Membranes of Mitochondria and Chloroplasts.* Ed. E. Racker. Van Nostrand Reinhold, New York, 1970.
42. Klingenberg, M. *Essays Biochem.* **6**, 119 (1970).
43. Lehninger, A. C., Carafoli, E., and Rossi, C. S. *Adv. Enzymol.* **29**, 259 (1967).
44. Cockrell, R. S., Harris, E. J., and Pressman, B. C. *Nature* **215**, 1487 (1967).
45. Eisenman, G., Szabo, G., McLaughlin, S. G. A., and Ciani, S. M. *Symposium on Molecular Mechanisms of Antibiotic Action on Protein Biosynthesis and Membranes.* Ed. D. Vasquez. Springer, New York, 1971.
46. Lardy, H. *Fed. Proc.* **27**, 1278 (1968).
47. Kaback, H. R., and Stadtman, E. R. *Proc. Natl. Acad. Sci. U.S.A.* **55**, 920 (1966).
48. Kepes, A. *Biochim. Biophys. Acta* **40**, 70 (1960).
49. Pastan, I., and Perlman, R. L. *J. Biol. Chem.* **244**, 5836 (1969).
50. Klein, W. L., Dahms, A. S., and Boyer, P. D. *Fed. Proc.* **29**, Abs 341 (1970).
51. Barnes, E. M., Jr., Lombardi, F. J., Kerwar, G. K., and Kaback, H. R. *Fed. Proc.* **30**, Abs 1061, 52 (1971).
52. Fisher, R. J., Lam, K. W., and Sanadi, D. R. *Biochem. Biophys. Res. Commun.* **39**, 1021 (1970).
53. Halvorson, H. O., and Cowie, D. B. *Membrane Transport and Metabolism.* Ed. A. Kleinzeller and A. Kotyk. Academic Press, New York, 1961.
54. Chance, B., and Azzi, A. *Biochim. Biophys. Acta* **189**, 141 (1969).
55. Boyer, P. D. *Oxidases and Related Redox Systems.* Ed. T. E. King, H. S. Mason, and M. Morrison. Wiley, New York, 1965, vol. 2, p. 994.
56. Boyer, P. D., Klein, W. L., and Dahms, A. S. *Proceedings, International Colloquium on Bioenergetics, Puguochiuso, Italy, September 1970.* In press.
57. Penniston, J. T., Harris, R. A., Asai, J., and Green, D. E. *Proc. Natl. Acad. Sci. U.S.A.* **59**, 624 (1968).
58. Green, D. E., and MacLennan, D. H. *Bio-Science* **19**, 213 (1969).
59. Chance, B. *Proc. Natl. Acad. Sci. U.S.A.* **67**, 560 (1970).
60. Nordenbrand, K., and Ernster, L. *Eur. J. Biochem.* **18**, 258 (1971).
61. Ryrie, I. J., and Jagendorf, A. T. *J. Biol. Chem.* **246**, 3771 (1971).
62. Gordon, A. S., and Kaback, H. R. *Fed. Proc.* **30**, 1061 Abs (1971).
63. Lardy, H. A., and Ferguson, S. M. *Annu. Rev. Biochem.* **38**, 991 (1969).
64. Wilson, D. F., Ting, H. P., and Koppelman, M. *Fed. Proc.* **30**, 1245 Abs (1971).
65. Azzone, G. F. *Proceedings, International Symposium on Biochemistry and Biophysics of Mitochondrial Membranes, Bressanone, Italy, June 1971.* In press.

66. Pressman, B. C. *Proceedings, International Symposium on Biochemistry and Biophysics of Mitochondrial Membranes, Bressanone, Italy, June 1971.* In press.
67. Cohen, G. N., and Monod, J. *Bacteriol. Rev.* **21,** 169 (1957).
68. Kennedy, E. P. *The Lactose Operon.* Ed. J. R. Beckwith and D. Zipser. Cold Spring Harbor Laboratory, New York, 1970, pp. 49–92.
69. Wong, P. T. S., and Wilson, T. H. *Biochim. Biophys. Acta* **196,** 336 (1970).
70. Kepes, A., and Cohen, G. N. *The Bacteria.* Ed. I. C. Gunsalus and R. Y. Stanier. Academic Press, New York, 1962, vol. 4, pp. 179–221.
71. Robbie, J. P., and Wilson, T. H. *Biochim. Biophys. Acta* **173,** 234 (1969).
72. Fox, C. F., and Kennedy, E. P. *Proc. Natl. Acad. Sci. U.S.A.* **54,** 891 (1965).
73. Boos, W., Gordon, A., and Hall, R. E. *Fed. Proc.* **30,** 1062 Abs (1971).
74. Winkler, H. H., and Wilson, T. H. *J. Biol. Chem.* **241,** 2200 (1966).
75. Penrose, W. R., Zand, R., and Oxender, D. L. *J. Biol. Chem.* **245,** 1432 (1970).
76. Berman, K., and Boyer, P. D. Unpublished observations.

12

C. Fred Fox

Membrane Assembly

Membranes are perhaps the most complex of the cellular supramolecular structures from the standpoint of their assembly. Cell walls, for example, are assembled largely by the formation of a prescribed set of covalent bonds. Ribosomes, though not held together by covalent linkages, are at least primarily made up of macromolecular components which are present in definite proportions. A given type of membrane, on the other hand, may be composed of hundreds of molecular species which are neither held together by covalent attachment nor present in defined molar ratios. Indeed, the gross composition of a given membrane may be markedly altered by a simple variation in the condition for cellular growth, with no apparent gross concomitant alteration in either membrane structure or membrane function. In addition to the obvious problems encountered in dealing with a structure which can undergo gross alterations without apparent functional change, progress in the area of membrane assembly has been impeded by a number of technological problems and by gaps in the understanding of pathways for the biosynthesis of membrane constituents. Recently, however, the bio-

Work in the author's laboratory was supported by U.S.P.H.S. research grant GM-18233. The author is the recipient of Research Career Development Award GM-42359 from the United States Public Health Service.

345

synthetic pathways have been largely resolved, as have many of the technical problems such as membrane isolation (Chap. 3) and the separation and identification of individual membrane lipids (Chap. 1) and proteins (Chap. 2).

ASSEMBLY OF MAMMALIAN CELL MEMBRANES

This chapter deals mainly with the biogenesis of membranes of microorganisms and viruses, since these systems offer definitive technical advantages for study. A great deal of information has been obtained in mammalian systems, however, and is briefly outlined here. For a more thorough treatment of the subject, the reader may consult Chapter 13, a recent review (1), or discussions of the topic in pertinent research treatises (2–5).

Studies on the assembly of mammalian cellular membranes consist largely of comparisons of the rates of synthesis and degradation of *1)* total membrane lipids and proteins, *2)* different portions of complex lipid molecules, or *3)* different individual proteins of a given species of membrane. The bulk of this information has been obtained using rat liver microsomes and plasma membranes, and these studies are directed to the question of whether a membrane is assembled and degraded as a unit, or whether selected membrane constituents may be synthesized and independently intercalated into the membrane matrix. All the investigations on the incorporation into and fate of radioactive labels in rat liver plasma membranes and endoplasmic reticulum indicate that membrane constituents turn over rapidly, the bulk of the protein and lipid components of the membranes turning over with a half-life of a few days. This indicates that these membranes are constantly being degraded and renewed. The proteins and lipids of rat liver microsomes turn over at different rates, the lipids turning over slightly faster than the proteins (4). Differences in the rate of turnover of the glycerol and fatty acid portions of the rat liver microsomal lipids have also been noted, the fatty acids having a longer half-life than glycerol. Differences in the rates of turnover of different individual membrane proteins range from as little as 60 hr to as much as 17 days (3; see also Chap. 13). Though the actual mechanisms of the various turnover processes are not known, the variations in rates of turnover of different membrane components suggest that the microsomal membranes are neither synthesized nor degraded as a unit. On the other hand, some differences in turnover rates for two given membrane components might be caused by an unequal distribution of the two components in dissimilar classes of hepatocytes. The microsomal preparations used in the turnover studies are mixtures of the total liver microsomes. By cytochemical techniques, the activities of many enzymes have been shown to vary from one area of the liver to another. Thus, differences in turnover rates for two given membrane components might reflect, at least to some extent, the turnover of membranes as a unit in classes of hepatocytes having dissimilar proportions of the membrane components, the membranes of each class of hepatocyte turning over as a unit but at different rates for each class. This criticism could best be resolved by the use of cell culture techniques, since these would provide a homogeneous population of cells.

Studies of the effects of phenobarbital administration to rats show that the rates of incorporation of some proteins into microsomes can be increased relative to the rates of incorporation of others (6). This indicates that even if the membrane is synthesized as a unit, the composition of that unit may vary considerably. A similar conclusion follows from experiments on the synthesis and incorporation of glucose-6-phosphatase into the endoplasmic reticulum of rat hepatocytes (7, 8). Near the time of birth, liver glucose-6-phosphatase activity increases 30-fold during a period of 5–6 days. By cytochemical techniques, glucose-6-phosphatase has been shown to be incorporated asynchronously into different hepatocytes, the cells having either a full complement of the enzyme or none. In addition, in cells which exhibited glucose-6-phosphatase activity, the enzyme activity developed simultaneously within *all* the rough endoplasmic reticulum membranes. Within the limits of the technique, there was no stage at which some cisternae of the rough endoplasmic reticulum of a given cell were positive for enzyme activity while other cisternae were not. These studies indicate that the endoplasmic reticulum consists of a mosaic of old and newly formed membrane components and that there exist no large regions of endoplasmic reticulum composed *entirely* of old or of new membrane. They do not, however, negate the possibility that this and other membranes might consist of regions composed *primarily* of old or new components.

CLASSES AND PROPERTIES OF MUTANTS IN LIPID BIOSYNTHESIS

The primary advantages of microorganisms as systems for the study of membrane assembly are *1*) they provide homogeneous populations of cells, and *2*) genetic techniques can be applied to construct mutants which allow the investigator to vary the quantity, composition, or order of synthesis of specified membrane components practically at will. Mutants in L-glycerol-3-phosphate or phosphatidic acid synthesis permit studies in which membrane proteins can be synthesized during periods of cessation of lipid synthesis, and mutants in fatty acid synthesis can be used to generate cells having grossly altered physical properties in the membrane lipids.

Glycerol Auxotrophs

In both gram-negative and gram-positive bacteria, the first step in complex lipid synthesis is the acylation of L-glycerol-3-phosphate. Since bacteria form L-glycerol-3-phosphate as the first intermediate in glycerol catabolism, a mutation which interrupts the normal anabolic pathway for the synthesis of L-glycerol-3-phosphate should make the cells dependent upon an exogenous supply of glycerol or L-glycerol-3-phosphate for lipid biosynthesis. This provides a rationale for the isolation of mutants dependent upon external glycerol or L-glycerol-3-phosphate for lipid biosynthesis and growth.

MUTANTS OF *ESCHERICHIA COLI*

An enzyme which catalyzes the $NADPH_2$-dependent reduction of dihydroxyacetone phosphate to yield L-glycerol-3-phosphate has been isolated and characterized (9). This enzyme is a "regulatory" enzyme and is feedback-inhibited by L-glycerol-3-phosphate. This information led to the hypothesis that the dihydroxyacetone phosphate reductase is the source of L-glycerol-3-phosphate for complex lipid biosynthesis. A number of *E. coli* mutants which require glycerol for growth have been isolated and all these are defective in the $NADPH_2$-linked reduction of dihydroxyacetone phosphate (10). This evidence supports the conclusion that anabolic L-glycerol-3-phosphate synthesis for complex lipid formation is mediated by this enzyme.

When the glycerol auxotroph is starved for glycerol in the presence of an otherwise suitable carbon source, an immediate cessation of complex lipid synthesis is observed. In contrast, protein synthesis and DNA synthesis continue at an unaltered rate for one-third of a generation time, but decrease to approximately 30% of this rate after one generation time. The fraction of newly synthesized protein which is incorporated into membrane is not altered significantly during glycerol starvation, indicating that de novo lipid synthesis is not required for the adsorption of membrane proteins by the membrane. Whether such proteins become functional, however, is another matter (see "Induction of Transport and Other Membrane Systems in Glycerol Auxotrophs"). Glycerol starvation in *E. coli* mutants does not cause an accumulation of unesterified fatty acids, and no increase in neutral lipid synthesis is observed (11). Extended periods of glycerol starvation produce no significant decrease in viability.

MUTANTS OF *BACILLUS SUBTILIS*

As is the case for *E. coli*, mutants of *B. subtilis* can be obtained which cease to grow or synthesize complex lipids when glycerol is removed from the growth medium (12, 13). The enzymatic defect is unknown. The *B. subtilis* mutants have properties similar to those of the *E. coli* mutants with regard to decreases in the rates of synthesis of proteins and nucleic acids and the retention of viability during glycerol starvation, and glycerol starvation does not affect the extent of incorporation of newly synthesized membrane proteins into membrane. In contrast to the *E. coli* mutants, the *B. subtilis* mutants and glycerol auxotrophs from *Staphylococcus aureus* continue to synthesize fatty acids and accumulate them in an unesterified form (14). The initiation of new rounds of DNA replication can proceed during glycerol starvation in the *B. subtilis* auxotrophs.

Acyltransferase Mutants of E. coli

In *E. coli*, the sequence of reactions leading to complex lipid biosynthesis proceeds from glycerol or L-glycerol-3-phosphate to phosphatidic acid is shown in Figure 1. In the case of mutations affecting L-glycerol-3-phosphate synthesis, lipid synthesis can be controlled by regulating the supply of glycerol or L-glycerol-3-phosphate in the growth medium. Mutants blocked in later steps in the sequence that leads to phosphatidic acid synthesis, i.e., the acylation of L-glycerol-3-phosphate, might be impossible to obtain as auxotrophs since the *E. coli* membrane is not known to be permeable to phosphatidic acid. For this reason, acyltransferase mutants of *E. coli* were isolated as temperature-sensitive mutants, i.e., mutants that can synthesize phosphatidic acid at low, but not at high, temperature. The first class of temperature-sensitive mutants in phosphatidic acid synthesis to be reported is defective at high temperature in the enzyme catalyzing the acylation of L-glycerol-3-phosphate, but not in the enzyme that catalyzes the acylation of monoacylglycerol-3-phosphate (15). At the nonpermissive temperature for L-glycerol-3-phosphate acylation, the properties of this temperature-sensitive mutant are identical to those of the glycerol auxotrophs of *E. coli* with regard to the effects of cessation of complex lipid synthesis on cell viability and the rates of macromolecule synthesis.

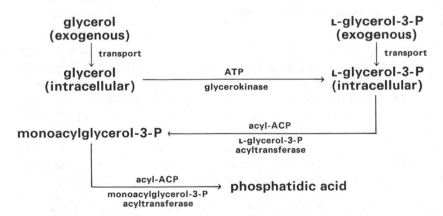

Figure 1
Scheme of events leading to phosphatidic acid synthesis in *E. coli*. ATP, adenosine triphosphate; ACP, acyl carrier protein.

A second class of mutant in phosphatidic acid biosynthesis is temperature sensitive in the enzyme catalyzing the second transacylation, the acylation of monoacylglycerol-3-phosphates (16). This class of mutant is similar to the glycerol auxotrophs and mutants in the first transacylation reaction in that an elevation in temperature sufficient to inactivate monoacylglycerol-3-phosphate acyltransferase brings about a reduction in the cellular capacity to synthesize DNA and protein. The reduction in rate of DNA synthesis, however, is far more pronounced than the reduction in protein-synthesizing capacity. In further contrast to the other classes of mutants defective in phosphatidic acid synthesis, a precipitous drop in cell viability occurs when cells temperature sensitive for monoacylglycerol-3-phosphate acyltransferase are held at the nonpermissive (elevated) temperatures for an hour, though 20% of the cells remain viable even upon heating for extended periods. The cause of this loss in viability has not been determined, nor is it known if it is directly related to the defect in monoacylglycerol-3-phosphate acyltransferase.

An additional, novel acyltransferase mutant has been described (17). This mutant was isolated as an auxotroph for L-glycerol-3-phosphate. The parent strain lacked both glycerokinase and the catabolic glycerophosphate dehydrogenase, but was constitutive for glycerophosphate transport. Thus the requirement for glycerophosphate could be supplied by exogenous L-glycerol-3-phosphate. This mutant is characterized by an L-glycerol-3-phosphate acyltransferase with a dramatically decreased (about 10-fold) affinity for L-glycerol-3-phosphate. Since the activity of the dihydroxyacetone phosphate reductase which catalyzes L-glycerol-3-phosphate biosynthesis is regulated by L-glycerol-3-phosphate concentration, the reductase is restrained from maintaining an increased steady-state level of L-glycerol-3-phosphate sufficient to compensate for the properties of the mutant acyltransferase.

Unsaturated Fatty Acid Auxotrophs of E. coli

The elucidation of the first distinct step in the pathway for unsaturated fatty acid synthesis in *E. coli* (Chap. 1) provided a rationale for the selection of mutants auxotrophic for unsaturated fatty acids (18). Two distinct genetic classes of unsaturated fatty acid auxotrophs have been identified (19). These have identical phenotypic properties with regard both to their response to unsaturated fatty acid supplements and to conditions of starvation for an essential fatty acid. These mutants enable the investigator to vary the structure of the essential fatty acid moiety of membrane lipids over a wide range.

STRUCTURE OF FATTY ACID SUPPLEMENTS WHICH PERMIT GROWTH

Unsaturated fatty acid auxotrophs of *E. coli* grow in media supplemented with *cis*-monoenoic fatty acids (Δ^9-$C_{14:1}$, Δ^9-$C_{16:1}$, Δ^6-$C_{18:1}$,

Δ^9-$C_{18:1}$, Δ^{11}-$C_{18:1}$, Δ^{11}-$C_{20:1}$); *cis,cis*-dienoic fatty acids ($\Delta^{9,12}$-$C_{18:2}$, $\Delta^{11,14}$-$C_{20:2}$); and *cis,cis,cis*-trienoic fatty acids ($\Delta^{9,12,15}$-$C_{18:3}$, $\Delta^{11,14,17}$-$C_{20:3}$) (20, 21). Fatty acids such as 9,10-methylene-octadecanoic acid (a cyclopropane fatty acid) can satisfy the growth requirement. This indicates that it is the structure imparted by the double bond, and not the double bond per se, which is the determinant of essentiality. The unsaturated fatty acid auxotrophs have a plating efficiency of only 10^{-6} when plated on solid medium containing a *trans*-unsaturated fatty acid (22). If cells from bacterial colonies which grow in a *trans*-unsaturated fatty acid medium are inoculated into a medium containing a *cis*-unsaturated fatty acid, grown for a number of generations, and then retested for growth on solid medium supplemented with a *trans*-unsaturated fatty acid, a plating efficiency of one colony formed per cell plated is obtained. This indicates that a mutation must occur if the growth requirement is to be satisfied by a *trans*-unsaturated fatty acid. In the first study on the nature of fatty acids which satisfied the auxotrophic requirement, the auxotroph was grown in a medium supplemented with *trans*-unsaturated fatty acid without the intermediary selection of a fatty acid mutant permissive for growth with a *trans*-unsaturated fatty acid (18). This can be explained by the fact that the fatty acid auxotrophs have a reversion frequency to prototrophy of approximately 10^{-7}, whereas the mutation which yields elaidate mutants* proceeds at a frequency of approximately 10^{-6}. The elaidate mutants have properties identical to those exhibited by the auxotrophs from which they were derived with respect to growth with *cis*-unsaturated and cyclopropane fatty acid supplements. They are distinguished from the parent auxotrophs by their growth in media supplemented with *trans*-unsaturated fatty acids (22–24) and in a medium supplemented with a mixture of 9- and 10-bromostearic acids (23). Neither the parent auxotrophs nor the elaidate mutants grow in media supplemented with 9,10-dibromostearic acid (23), nor with hydroxy substituted or keto substituted fatty acids (25). Branched-chain fatty acids do not satisfy the growth requirement of the parent auxotroph (20).

DIFFERENCES IN MEMBRANE FUNCTION AND PHYSICAL
PROPERTIES RESULTING FROM ALTERATION OF FATTY ACID
STRUCTURE OF MEMBRANE LIPIDS

The lipid-solvent extractable material of membranes of *E. coli* K12 consists almost entirely of the complex lipids phosphatidylethanolamine (85–90%), phosphatidylglycerol (7–12%), and cardiolipin (0.5–2%). The fatty acid composition of *E. coli* membranes is likewise simple, consisting almost entirely of palmitic acid ($C_{16:0}$), palmitoleic acid (*cis*-Δ^9-$C_{16:1}$), and

---

* Since the mutants which grow in medium supplemented with *trans*-unsaturated fatty acid are selected utilizing solid media containing elaidic (*trans*-Δ^9-octadecenoic) acid, they will be referred to here as elaidate mutants.

cis-vaccenic acid (*cis*-Δ^{11}-$C_{18:1}$) in cells grown in exponential phase. In cells grown to late exponential or stationary phase, a portion of the unsaturated fatty acids is converted to cyclopropane fatty acids at the level of the intact phospholipids. The functional importance of this conversion has not been established, and the conversion from monoenoic unsaturated to cyclopropane fatty acids produces no detectable change in the properties of monolayers of the phospholipids at the air-water interface (26). The relatively simple lipid composition of *E. coli* membranes makes the *E. coli* auxotrophs excellent systems for the study of the effects of fatty acid composition and structure on membrane function and physical properties.

Differences in saturated/unsaturated ratio as a function of growth temperature. The ratio of saturated to unsaturated fatty acids in the phospholipids of an *E. coli* unsaturated fatty acid auxotroph has been studied for three essential fatty acid supplements over a 15°C range of growth temperatures (21). Since the only major saturated fatty acid in *E. coli* complex lipids is palmitic acid, this ratio is essentially the ratio of palmitic acid to unsaturated fatty acids. The data in Figure 2 show that in all cases there was a 30–40% increase in the saturated/unsaturated fatty acid ratio as the temperature was increased from 27° to 42°C. Similar findings have been made using wild-type bacteria rather than the auxotrophic strains (26, 17). The increase

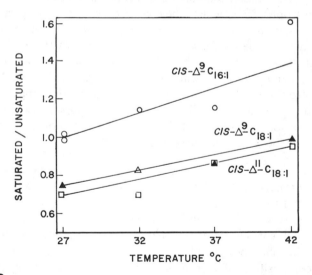

Figure 2

Effect of temperature on the ratio of saturated to unsaturated fatty acids in phospholipids of an essential fatty acid auxotroph grown with the indicated fatty acid supplements. (From Esfahani, M., et al. Proc. Natl. Acad. Sci. U.S.A. 64, 1057, 1969.)

in the proportion of unsaturated fatty acid as a function of decreased temperature for growth is thought to reflect a requirement for increased fluidity in the hydrocarbon side chain portion of the membrane lipid bilayer (Chap. 1).

Differences in saturated/unsaturated ratio as a function of structure of the essential fatty acid. Comprehensive studies of the effects of chain length, degree of unsaturation, and type of unsaturation is consistent with the idea that the cell has a regulatory mechanism for maintaining the proper fluidity of the fatty acid side chains in the membrane complex lipids (20, 21). Table I shows the effect of chain length of a homologous series of cis-Δ^9-unsaturated fatty acids in determining the ratio of saturated to unsaturated fatty acids in the auxotroph. As the chain length of the essential fatty acids in the series is increased, the proportion of unsaturated fatty acids in membrane lipids also increases by 29–53%. Table II shows the effect of increasing the degree of unsaturation of a homologous series of cis-unsaturated C_{18} essential fatty acid supplements. As the number of cis-olefinic bonds is increased from one to three, the proportion of saturated fatty acids in membrane increases. This

Table I

Effect of Chain Length of a Homologous Series of *Cis*-monoenoic Fatty Acids on Membrane Fatty Acid Composition

FATTY ACID COMPOSITION OF PHOSPHOLIPIDS	Fatty Acids Added to Growth Medium		
	cis-Δ^9-$C_{14:1}$ (%)	cis-Δ^9-$C_{16:1}$ (%)	cis-Δ^9-$C_{18:1}$ (%)
Saturated	70	52	45
Unsaturated	29	46	53
Unknown	1	2	2

From Esfahani, M., *et al. Proc. Natl. Acad. Sci. U.S.A.* **64,** 1057 (1969).

Table II

Effect of Increasing Degree of Unsaturation of a Homologous Series of *Cis*-unsaturated Fatty Acids on Membrane Fatty Acid Composition

FATTY ACID COMPOSITION OF PHOSPHOLIPIDS	Fatty Acids Added to Growth Medium		
	cis-Δ^9-$C_{18:1}$ (%)	All cis-$\Delta^{9,12}$-$C_{18:2}$ (%)	All cis-$\Delta^{9,12,15}$-$C_{18:3}$ (%)
Saturated	45	60	66
Unsaturated	53	37	30
Unknown	2	3	4

From Esfahani, M., *et al. Proc. Natl. Acad. Sci. U.S.A.* **64,** 1057 (1969).

indicates that the organism adjusts to the increase in fluidity imparted by an additional double bond in the unsaturated fatty acid supplement by elevating the saturated/unsaturated fatty acid ratio in the membrane lipids.

Trans-unsaturated fatty acids have physical properties intermediate between those of saturated fatty acids and *cis*-unsaturated fatty acids. The melting points for stearic acid ($C_{18:0}$), elaidic acid (*trans*-Δ^9-$C_{18:1}$). and oleic acid (*cis*-Δ^9-$C_{18:1}$) decrease in the order given. Similarly, the temperatures at which these fatty acids undergo a transition from a condensed to an expanded phase at the air-water interface also decrease in the order given. These properties are a reflection of the ability of *cis* double bonds to disrupt hydrocarbon side-chain stacking more effectively than the corresponding *trans* double bonds. This explains why the ratio of unsaturated to saturated fatty acids in membrane lipids is approximately 1.2 when the essential fatty acid supplement is *cis*-Δ^9-$C_{18:1}$ and increases to 2.4 when the essential fatty acid supplement is *trans*-Δ^9-$C_{18:1}$.

Membrane phase transitions as revealed by x-ray crystallography. The hydrocarbon side chains of fatty acids in lipids of membrane undergo a transition from a more ordered to a less ordered structure with increasing temperature. This transition can be detected by a change in the x-ray diffraction pattern of membrane suspensions (Chap. 4). Below the transition there is a sharp diffraction ring at 4.2 Å indicating hexagonal close packing of the hydrocarbon side chains. Above the transition, the diffraction ring contributed by the fatty acid hydrocarbon side chains is diffuse, with a maximum at 4.6 Å indicating the formation of a less ordered liquid crystalline structure. The transition from a sharp diffraction ring at 4.2 Å to a diffuse ring at 4.6 Å occurs over a broad temperature range of approximately 7–10°. Studies with membranes prepared from *Mycoplasma laidlawii* grown with different fatty acid supplements provide evidence that the midpoint of the thermal transition detected by x-ray diffraction is apparently determined by the degree of unsaturation of the fatty acid supplement employed (Chap. 14). A recent report on the x-ray thermal transitions of membranes from an *E. coli* unsaturated fatty acid auxotroph, however, indicates that the situation is more complex (Table III). Though the melting point of linoleic acid is approximately 40° lower than that of elaidic acid, the midpoint of the x-ray thermal transition contributed by the fatty acid side chains of the lipids in membranes of the auxotroph grown with linoleic acid is actually higher than that of lipids in membranes of the auxotroph grown with elaidic acid (28). This is the inverse of what would be expected if the transition detected by x-ray diffraction were a function of the degree of unsaturation of the essential fatty acid supplement, since the physical properties of *trans*-unsaturated fatty acids, such as elaidic acid, are more similar to those of saturated fatty acids than to those of cis-unsaturated fatty acids.

Table III

X-ray Transitions of Membranes from an *E. Coli* Unsaturated Fatty Acid Auxotroph Grown with Different Lipid Supplements

FATTY ACID SUPPLEMENT FOR GROWTH	TEMPERATURE RANGE OF X-RAY THERMAL TRANSITION (°C)
Linoleate (*cis,cis*-18:2)	37–45
Myristoleate (*cis*-14:1)	37–45
Oleate (*cis*-18:1)	20–27
12-Bromostearate	20–27
Elaidate (*trans*-18:1)	32–40

Data from Esfahani, M., *et al. Fed. Proc.* **30,** 1120 abs. (1971).

Transitions in rate of transport processes. An influence of membrane lipid fatty acid composition on the temperature dependence of transport rate was detected independently in two laboratories (22, 29). The influence of temperature on the rate of β-glucoside transport by an *E. coli* auxotroph grown at 37°C in media supplemented with either elaidic acid or linoleic acid is shown in the Arrhenius representation in Figure 3. The Arrhenius

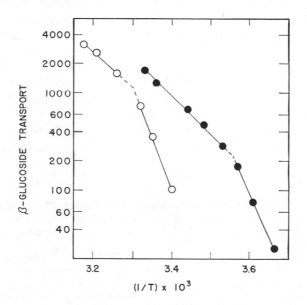

Figure 3

Arrhenius plots for the effect of temperature on the rate of β-glucoside transport by an essential fatty acid auxotroph grown at 37°C with linoleic acid (black circles) or elaidic acid (white circles) supplementation. (From Wilson, G., and Fox, C. F. *J. Mol. Biol. 55,* 49, 1971.)

plots for transport are biphasic, intersecting at transition temperatures which are unique for each fatty acid supplement. In the examples shown here, the transition temperature for cells grown in a medium supplemented with linoleic acid is 7°C, and the transition temperature for cells grown in a medium supplemented with elaidic acid is 30°C. The transition temperatures for transport are the result of a transition which occurs in the lipid phase, rather than the result of some lipoprotein interaction unique for a given transport system. Transition temperatures for transport mediated by two transport systems, different from one another in all aspects but their localization in membrane, have been determined after growth of the auxotroph with a wide number of essential fatty acid supplements (Table IV). Without exception, the transition temperatures in transport rate are identical for both transport systems for cells grown with a given fatty acid supplement (24). One further point which can be derived from these data is the apparent correspondence between the degree of unsaturation of the fatty acid used as a growth supplement and the relative position of the transition temperature for transport. Lipid containing elaidic acid, oleic acid, and linoleic acid would be expected to increase in fluidity in the order given. Elaidic acid would impart the least fluidity since the *trans* double bond causes less structural deformation and resultant interference with close packing in monolayers at the air-water interface than the *cis* double bond of oleic acid (see the following section of this chapter and Chap. 5). A similar increase in fluidity can be anticipated going from oleic acid (*cis*-Δ^9-$C_{18:1}$), which has one double bond, to linoleic acid (*cis,cis*-Δ^9-$C_{18:2}$), which has two. The transition temperatures for transport for cells grown in media supplemented with oleic and *cis*-vaccenic (Δ^{11}-$C_{18:1}$) acids are tightly grouped with the transition temperature for transport for cells grown in a medium supplemented with the cyclopropane fatty acid (dihydrosterculic acid). This last result might be expected, since lipids con-

Table IV

Transition Temperatures in Rates of Transport Mediated by the β-glucoside and β-galactoside Transport Systems in a Fatty Acid Auxotroph Grown with Different Essential Fatty Acid Supplements

	Transition Temperature (°C)	
FATTY ACID SUPPLEMENT IN GROWTH MEDIUM	β-GALACTOSIDE TRANSPORT	β-GLUCOSIDE TRANSPORT
Elaidate	30	30
Bromostearate	22	22
Oleate	13	13
Dihydrosterculate	11	11
cis-Vaccenate	10	10
Linoleate	7	7

taining cyclopropane fatty acids are similar in their properties in monolayers at the air-water interface to the precursor lipids containing the corresponding unsaturated fatty acids (26).

One of the transport systems used in the study described in Table IV, the β-glucoside transport system, functions by a mechanism termed "vectorial phosphorylation," in which phosphorylation of the glycoside is an integral part of the transport process (Chap. 10). The phosphorylation reaction can be assayed in broken-cell preparations under conditions where the activity of the membrane-bound component of the phosphotransferase system (the enzyme II) is rate limiting. When the effect of temperature on enzyme II activity is determined using membrane fragments prepared from cells grown with different fatty acid supplements as the source of enzyme II, the dependence of transport rate on temperature is independent of lipid composition (29). Unlike the case with transport, no transitions in the slopes of Arrhenius plots for enzyme II activity are observed. The preparative procedure for enzyme II yields membrane fragments which have both the inner and outer membrane surfaces exposed to the cofactors in the assay system. In the in vitro system which uses membrane fragments for which a compartmentalization between the inner and outer membrane surfaces does not exist, phosphorylation can evidently occur without the substrate having to penetrate the membrane barrier. This suggests that in intact cells, the rate-limiting step for vectorial phosphorylation below the transition temperature is the actual transport of the glycoside substrate across the membrane barrier.

Membrane phase transitions as revealed by monolayer techniques. A typical force-area diagram for the formation of a condensed lipid monolayer at the air-water interface is shown in Figure 4 of Chapter 5, and the various portions of the diagram are identified in Table II of that chapter. The transition in this force-area diagram which correlates with the transport transition temperatures is the transition from a liquid-expanded to a condensed lipid film. Figure 4 of the present chapter describes the force-area measurements obtained with phosphatidylethanolamine isolated from an essential fatty acid auxotroph grown in medium supplemented with elaidic acid (30). At 31°C and 25°C, there is a clear transition from the liquid-expanded to a condensed lipid film. No clear transition of this type is observed when the force-area measurements are made at 41°C, and the force-area measurements at 44°C show a transition from the liquid-expanded film directly to a collapsed monolayer. From these data, it appears that the transition from a liquid-expanded to a condensed film no longer occurs above 40°C. This temperature corresponds well with the transition temperature for β-galactoside transport obtained with this same mutant grown under identical conditions. The transition temperature for transport obtained in this study is higher than that given in Table IV for cells grown with elaidate. This difference is a result of the use of an essential fatty acid auxotroph with β-oxidation potential in the experiments described in Table IV, and the use of an essential fatty acid auxotroph incapable of β-oxidation in the study described in Figure 4. Excellent agreement between the temperatures above which a condensed lipid film can no longer be formed and the transition temperatures for transport was also observed for cells grown with oleic acid, linoleic acid, and a cyclopropane fatty acid. The good correlation between the transitions in monolayers and in transport suggests that the latter can be accounted for by the former. This conclusion is also consistent with the fact that condensed

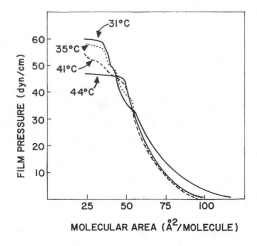

MOLECULAR AREA ($\mathring{A}^2$/MOLECULE)

Figure 4
Force-area diagram of phosphatidylethanolamine isolated from an essential fatty acid auxotroph grown with elaidic acid supplementation. (From Overath, P., *et al. Proc. Natl. Acad. Sci. U.S.A. 67*, 606, 1970.)

lipid films can be formed below a given temperature, but not above it. This phenomenological aspect is easily reconciled with the sharpness of the transition temperature for transport rate and should be contrasted with the relatively broad transitions revealed by x-ray diffraction. Furthermore, the x-ray thermal transitions for cells grown with linoleic or elaidic acid are the reverse of the order of the transition temperatures for transport (Tables III and IV). Apparent phase transitions in the hydrocarbon side chain portion of the membrane lipids have been observed by electron spin resonance (Chap. 8) and should soon be detectable by nuclear magnetic resonance techniques (Chap. 7). It would be of considerable interest to have the results of a comprehensive study where many biological and physical transitions are compared using the same cells and membrane preparations. The fact that certain of the x-ray transitions correspond to the transport transitions, where-as others do not, makes it imperative that as large a number of samples as possible be used in such a study.

EFFECTS OF STARVATION FOR AN ESSENTIAL FATTY ACID

When the accessibility of cells of an *E. coli* unsaturated fatty acid auxo-troph to an unsaturated fatty acid is blocked, lipid synthesis continues for a brief period with no reduction in rate. The lipids synthesized during periods of essential fatty acid deprivation are entirely saturated (31, 32). Longer periods of essential fatty acid deprivation lead to reductions in the rates of lipid, protein, and nucleic acid synthesis, and eventually to gross morpho-logical changes and cell death (33). There is no direct correlation between the loss of any given cellular function and the onset of loss of viability.

Fatty Acid Auxotrophs of Yeast

A number of fatty acid auxotrophs have been described in selected yeast strains. In yeast, it has been possible to obtain not only unsaturated fatty acid auxotrophs, but also auxotrophs which can synthesize neither saturated nor unsaturated fatty acids. Fatty acid substitution studies with a yeast unsaturated fatty acid auxotroph indicate that, as in the case of the *E. coli* auxotrophs, the requirement for an essential fatty acid can be satisfied by fatty acid supplements which vary widely in structure (34). The effects of variation of structure of the essential fatty acid supplements on the assembly of yeast mitochondria are described in Chapter 13. One of the more interest-ing observations with the fatty acid auxotroph which can synthesize neither saturated nor unsaturated fatty acids is its requirement for a saturated fatty acid. This class of mutant does not grow at any temperature in a medium supplemented solely with unsaturated fatty acids (35).

REQUIREMENTS FOR LIPID IN ASSEMBLY OF FUNCTIONAL MEMBRANE SYSTEMS

Two classes of lipid biosynthetic mutants with unique phenotypic properties have been described in the previous section. One class, glycerol or transacylase mutants, allows the investigator to turn lipid synthesis on or off at will with only minor initial changes in the rate of protein synthesis. With the other class of mutants, the essential fatty acid auxotrophs, the fatty acid composition of membrane lipids can be varied markedly. This variation can cause a concomitant alteration in the properties of membrane systems which have catalytic functions, and the altered catalytic properties allow tests to determine if certain proteins synthesized de novo associate preferentially with the lipids synthesized at the same time. A major incentive for studying the interrelations between protein and lipid biosynthesis in the assembly of catalytic membrane systems stemmed from observations on the induction of the components of the lactose operon of *E. coli* in strains which were haploid or diploid for the lactose operon (36). Induction of the lactose operon in diploid strains leads to approximately twice the level of synthesis of β-galactosidase, the lactose transport protein (M protein, see Chap. 10), and thiogalactoside transacetylase, as in the haploid strain. In some haploid-diploid pairs, a doubling of transport activity is also observed, but in others, the transport activity increases either much less than expected from the increased synthesis of the transport protein or not at all. Since plots of transport activity versus cell mass are linear in all cases, these data indicate that some substance or substances other than the lactose transport protein can be rate limiting for the expression of transport function, and that these substances and the protein must be synthesized simultaneously to permit transport function. Among the membrane components whose rates of synthesis could limit the expression of a fully functional lactose transport system are the cellular lipids.

Induction of Transport Systems in Unsaturated Fatty Acid Auxotrophs

INDUCTION OF LACTOSE TRANSPORT DURING STARVATION FOR AN ESSENTIAL FATTY ACID

The lactose transport system of *E. coli* provides a convenient experimental model for testing for interrelations between lipid and protein synthesis in membrane assembly. This transport system is inducible, and the transport protein is coordinately induced with the other proteins of the lactose operon. The three structural genes of the lactose operon are sequentially aligned on the chromosome in the order *lac* Z (which codes for the synthesis of the soluble enzyme β-galactosidase), *lac* Y (which codes for the synthesis of the transport protein), and *lac* A (which codes for the synthesis of the soluble enzyme thiogalactoside transacetylase). Assays of β-galactosidase and thiogalactoside transacetylase activities give reliable estimates of the actual amounts of the enzyme proteins. Since the structural genes are translated and transcribed in the order *lac* Z → *lac* Y → *lac* A, the synthesis of the lactose transport protein can be quantified simply by measuring the activities of β-galactosidase and thiogalactoside transacetylase. This facilitates comparisons of the extent of induction of the transport protein with the extent of induction of functional lactose transport without measuring directly the amount of synthesis of the transport protein, the assay for which is both cumbersome and insensitive.

The effect of essential fatty acid starvation on induction of functional lactose transport is described in Figure 5 (36). In this experiment, a culture of the auxotroph was first grown in medium supplemented with oleic acid, and the cells were washed free of oleic acid and divided into two portions.

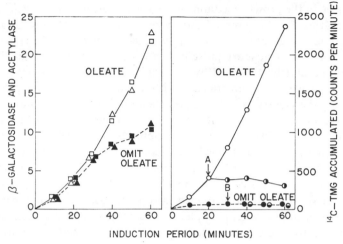

Figure 5

Essential fatty acid requirement for induction of β-galactoside transport in *E. coli*. Transport assays (right-hand figure) measured the ability of cells to accumulate radioactive methyl-β-thio-galactoside (TMG). Cells of an unsaturated fatty acid auxotroph were grown in an oleic acid supplemented medium, processed to remove oleic acid, and suspended in fresh medium containing oleic acid ($\square$, $\triangle$, $\bigcirc$) or in fresh medium containing no fatty acid supplement ($\blacksquare$, $\blacktriangle$, $\bullet$). The two cultures were then induced for the proteins of the lactose operon. At various times, portions of the cultures were removed and assayed for β-galactosidase ($\square$, $\blacksquare$), galactoside acetylase ($\triangle$, $\blacktriangle$), and transport ($\bigcirc$, $\bullet$). At point A in the right-hand figure, a portion of the culture was removed, and the cells were rendered free of oleic acid and inducer and incubated further to test for the stability of the induced transport system during growth in the absence of the essential fatty acid (◓). At point B, a portion of the culture, grown and induced in the absence of the fatty acid supplement, was removed; the cells were rendered free of inducer and then grown further in the presence of oleic acid (◑). See the text for further details. (From Fox, C. F. *Proc. Natl. Acad. Sci. U.S.A.* **63**, 850, 1969.)

One portion was induced for synthesis of the proteins of the lactose operon in the presence of oleic acid, and the second portion was induced in the absence of oleic acid. For both the essential fatty acid supplemented and unsupplemented cultures, the extent of synthesis of β-galactosidase and thiogalactoside transacetylase was nearly identical during the first 30 min of induction, indicating that the lactose transport protein was induced to a similar extent in both cases. The synthesis of lactose transport activity, however, displayed a strict requirement for the presence of an essential fatty acid supplement in the growth medium. A control experiment (arrow A) indicates that starvation for the essential fatty acid for periods of up to 30 min does not result in the inactivation of a preformed, functional transport system. Furthermore, the transport protein synthesized during essential fatty acid starvation cannot be activated by subsequent addition of the essential fatty acid to the growth medium after the withdrawal of the inducer for transport protein synthesis (arrow B). Since essential fatty acid containing lipids synthesized either before or after transport protein induction do not allow expression of the functional transport system, these data may indicate that the newly synthesized transport proteins associate primarily with lipids synthesized during transport protein induction. The transition temperatures for transport are unique for cells grown with different essential fatty acid supplements (Fig. 4). Thus induction of transport proteins after shifting from growth with one essential fatty acid supplement to growth with another provides a means to test for preferential association of newly formed transport proteins with newly formed lipids.

INDUCTION OF TRANSPORT AFTER A FATTY ACID SHIFT

The ratio of transport rate determined at a temperature below the transition temperature for transport, to transport rate determined at a temperature above the transition temperature is a constant unique for and determined by the essential fatty acid supplement. In the case of β-galactoside transport, for example, the $10°C/28°C$ ratio for transport rate is approximately 0.06 for cells grown with oleic acid supplementation, and 0.14 for cells grown with linoleic acid supplementation. These ratios remain constant throughout the course of growth in exponential phase at 37° and are referred to as transport temperature profiles (22).

When the β-galactoside transport system is induced during a 20% increase in cell mass after shifting from growth in oleate-supplemented medium to growth in linoleate-supplemented medium, the transport temperature profile is that of cells grown entirely in a medium supplemented with linoleic acid. On the other hand, when the β-galactoside transport system is first induced during growth in oleate-supplemented medium, and then grown further for a 20% increase in cell mass after shifting from oleate- to linoleate-supplemented medium, the transport temperature profile characteristic of cells grown exclusively in oleate medium is retained. This result is consistent with the hypothesis that the transport protein associates preferentially with lipids synthesized at the time of transport protein synthesis. A second type of essential fatty acid shift experiment employs induction for two transport systems (Table V). During the period of growth with the first fatty acid supplement, β-glucoside transport is induced. The β-galactoside transport system is then induced during a 20% increase in cell mass following the shift to growth with the second fatty acid supplement. When the shift is

Table V

Temperature Characteristics of Transport Obtained when β-glucoside Transport is Induced During Growth in Medium Supplemented with One Fatty Acid Followed by Induction of β-galactoside Transport in Medium Supplemented with Another Fatty Acid

FATTY ACID PRESENT DURING FIRST GROWTH PERIOD (β-GLUCOSIDE INDUCTION)	FATTY ACID PRESENT DURING SECOND GROWTH PERIOD (β-GALACTOSIDE INDUCTION)	β-GLUCOSIDE TRANSPORT 5°C/28°C	β-GALACTOSIDE TRANSPORT 10°C/28°C
Oleate	Oleate	0.030	0.062
Oleate	Linoleate	0.030	0.140
Linoleate	Oleate	0.066	0.138
Linoleate	Linoleate	0.064	0.144

During the first growth period, cultures of the auxotroph were induced for β-glucoside transport for four doubling times. The cells were then harvested, washed quickly, and suspended in medium for the second growth period. After induction of the *lac* operon during a 20% increase in cell mass, the cells were processed for transport assays. From Wilson, G., and Fox. C. F. *J. Mol. Biol.* **55**, 49 (1971).

from oleate to linoleate supplementation, the transport temperature profile of β-glucoside transport is unaltered after the shift, but the transport temperature profile of β-galactoside transport is that of cells grown only with linoleic acid. When the order is reversed so that linoleic acid is present during the first growth period when β-glucoside transport is induced, and oleic acid during the second when β-galactoside transport is induced, both transport systems exhibit the transport temperature profile characteristic of cells grown in linoleic acid. The results of the experiment described in Table V where the shift was from oleate- to linoleate-supplemented medium are consistent with the view that the newly synthesized lactose transport system protein associates primarily with newly synthesized (linoleate derived) lipids. The data obtained in the linoleate to oleate shift experiment, however, do not support this view.

In more recent studies, Arrhenius plots for transport rate were derived from measurements with cells induced for lactose transport after a shift from growth at 37° with one unsaturated fatty acid supplement to growth with another at the same temperature. The data from these studies indicate that the transition temperatures reflect the physical properties of the average membrane phospholipid fatty acid composition (37–38). For example, if cells grown with an unsaturated fatty acid supplement, which by itself produced cells giving a transport transition temperature of 30°, were grown and induced for transport during one generation of growth with a second unsaturated fatty acid supplement, which, were it the sole supplement for growth, would yield cells with a transport transition of 10°, the observed transport transition temperature would be approximately 20°. This is the type of observation expected if the membrane lipids are sufficiently mobile to randomize throughout the membrane by a lateral diffusion process. A markedly different result is obtained, however, if induction after the essential fatty acid shift proceeds below the transition temperature for one of the two essential fatty acid supplements. In the experiment described in Figure 6, cells grown initially with an oleic acid supplement were shifted to medium containing an elaidic acid supplement and then induced for β-galactoside transport during approximately one-fifth of a generation of growth. When both the initial growth and

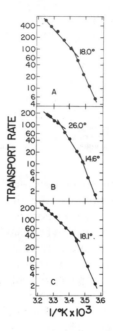

Figure 6

Properties of the *β*-galactoside transport system formed during induction at 37° and 25° after shifting an unsaturated fatty acid auxotroph of *E. coli* from growth in medium supplemented with oleic acid to medium supplemented with elaidic acid. Cells of the auxotroph were first grown in medium supplemented with oleic acid at 37°, processed to remove oleic acid from the medium, and suspended in medium supplemented with elaidic acid and containing an inducer for *β*-galactoside transport. One portion of these cells was then grown further at 37° (*A*) and another (*B*) at 25°. The times of growth at these two temperatures were regulated so that equivalent amounts of elaidic acid were incorporated into the membrane phospholipids in both cases. After growth and induction of the *β*-galactoside transport system in medium supplemented with elaidic acid at the indicated temperatures, the cultures were rapidly chilled to ice-bath temperature, and the cells were processed for assay of transport rate at the temperatures indicated in the figure. In *C*, cells from *B* were incubated for 10 min at 37° in the absence of inducer and then rapidly chilled to ice-bath temperature before assays of transport rate were begun (38).

subsequent growth during induction proceeded at 37°, the transition temperature for transport was 18°. This is essentially the result expected if the lipids synthesized before and after the shift can randomize by a diffusion process at 37°, since the transport transition temperature for cells grown entirely in oleate-supplemented medium is 13°, and that for cells grown entirely in elaidate-supplemented medium is 30°. When transport was induced at 25° (5° below the transport transition temperature for cells grown with an elaidate supplement) after the same fatty acid shift, two entirely different transition temperatures were observed (Fig. 6B). One of these was close to the transport transition temperature of cells grown with oleic acid and the other to that of cells grown with an elaidic acid supplement. When cells from the experiment described in Figure 6B were rendered free of inducer and merely incubated at 37° for 10 min, the two transitions disappeared, and a single new transport transition temperature was observed. This new transport transition temperature fell between those shown in Figure 6B and was identical within the limits of experimental error to that observed when induction proceeded at 37°. When the cells from Figure 6B were incubated for only 1 min at 37° and then rapidly cooled before transport rate was assayed, the Arrhenius plot of transport rate resembled that shown in Figure 6B indicating that the disappearance of the two transport transition temperatures and the appearance of the new one (Fig. 6C) proceed slowly. When the experiment described in Figure 6 was repeated with the essential fatty acid supplements supplied in the reverse order, i.e., where an elaidic acid supplement was used during the initial growth period and an oleic acid supplement used during subsequent growth and induction, qualitatively identical results were obtained. These experiments indicate that the transport protein can be incorporated into membrane with and influenced by the lipids formed during transport system induction below the transition temperature characteristic of cells grown entirely with the lipid supplement yielding the higher of the two transport transition temperatures. They further indicate that the rate of diffusion of phospholipids in biological membranes may, at least in certain instances, be well below the rate of lateral diffusion of phospholipids detected in artificially formed phospholipid bilayer vesicles (51).

Induction of Transport and Other Membrane Systems in Glycerol Auxotrophs

Experiments described in the preceding section indicate that some transport proteins may associate primarily with lipids synthesized at the time of transport protein synthesis. This nevertheless leaves open the question whether de novo lipid synthesis is or is not necessary for the formation of a functional transport system. This question is not answered by the failure to form a functional lactose transport system in an essential fatty acid auxotroph of *E. coli* when transport is induced during essential fatty acid starvation. In response to essential fatty acid starvation, the synthesis of complex lipids continues, though only saturated fatty acids are available for synthesis of the complex lipids formed during the starvation period. Since binding of the protein components of the transport system to abnormal (totally saturated) lipids might lead to abortive transport system assembly, additional experiments are needed to test for a requirement for de novo lipid synthesis during the assembly process. The glycerol auxotrophs described earlier in this chapter provide the means to test for the de novo incorporation of proteins into functional membrane systems during periods of complete cessation of complex lipid synthesis.

The removal of glycerol from the medium for growth of glycerol auxotrophs of *E. coli* leads to a cessation of synthesis of complex lipids. In the presence of an otherwise suitable carbon source, however, cells starved for glycerol can be induced for β-galactosidase and thiogalactoside transacetylase, and also for the synthesis and incorporation into membrane of the β-galactoside transport protein as determined by a technique which measures its binding ability for substrates for transport (10). During the initial periods of glycerol starvation, the newly formed transport proteins are functional in mediating transport. Glycerol starvation for periods of 30 min or longer, however, leads to the synthesis of transport proteins which are active in binding galactosides, but nonfunctional for transport. Furthermore, if glycerol is added after the induction period, the transport protein formed during glycerol starvation remains nonfunctional. Since glycerol starvation for periods of 1 hr or more does not lead to extensive inactivation of transport, simultaneous synthesis of transport proteins and lipids is required for assembly of a functional system. The lag in establishing the requirement for de novo lipid synthesis indicates that a pool of lipid which can satisfy the requirements for transport system assembly is initially present in the cell, but is exhausted during glycerol starvation. This suggests the presence of two classes of lipids in the cell: those available for structural assembly and those "fixed" and unavailable for the assembly of new structures.

The induction of a system which catalyzes the transport of lactose by vectorial phosphorylation (Chap. 10) has been studied in a glycerol auxotroph of *S. aureus* (39). This system catalyzes the transport and concomitant phosphorylation of β-galactosides and is coordinately induced with a soluble enzyme which catalyzes the hydrolysis of the phosphorylated galactosides (a phospho-β-galactosidase). In response to glycerol starvation, the membrane protein component of the transport system, as measured by phosphorylation activity in extracts, is both synthesized and incorporated into membrane to the same extent relative to synthesis of phospho-β-galactosidase as it is in the condition where induction occurs in the presence of glycerol. The ratio of transport activity to phosphotransferase activity measured in extracts, however, is drastically reduced during the later periods of glycerol starvation. This result is similar to that described for the induction of lactose transport in a glycerol auxotroph of *E. coli* during glycerol starvation, in that the transport proteins are incorporated into the membrane in a fashion which allows some expression of transport protein activity (substrate binding in the case of the *E. coli* system and phosphorylation in the case of the *S. aureus* system), but which does not permit optimal function of the transport system. In contrast to the properties of the *E. coli* lactose transport system formed during cessation of lipid synthesis, the nonfunctional *S. aureus* lactose transport system synthesized during glycerol starvation can be rendered functional when lipid synthesis resumes under conditions which permit no additional protein synthesis. Thus some membrane systems require concomitant synthesis of both lipid and protein for expression of function, whereas others do not.

Influence of Temperature and Structure of Essential Fatty Acid Supplement on Membrane Assembly

The structure of the essential fatty acid supplement is a determinant of the temperature below which the *E. coli* unsaturated fatty acid auxotrophs do not grow. With elaidic acid supplementation, for example, growth essentially ceases below the temperature at which the lipids extracted from cells grown with elaidic acid form a condensed phase in monolayers at the air-water interface (30). Cells supplemented with oleic or linoleic acids continue to grow at temperatures well below that which arrests growth in elaidic acid supplemented medium. One *E. coli* strain which is auxotrophic for unsaturated fatty acids not only ceases to grow, but lyses when the temperature is lowered from 38° to 27°C during growth in elaidic acid–supplemented medium (19). This, however, is not a general property of the *E. coli* essential fatty acid auxotrophs. The viability of another auxotrophic strain decreases only slightly after a shift from 37° to 25°C during elaidate supplementation, though growth essentially ceases after the temperature downshift (38).

The influence of temperature on the induction of the lactose transport system has been studied as a function of the fatty acid supplement for growth (38). These experiments use an *E. coli* auxotroph which does not lyse upon shifting from 37° to 25°C during incubation in elaidate medium. A downshift from 37° to 25°C has no influence on the efficiency of induction of lactose transport during growth in oleate-containing medium. This same downshift in temperature during incubation in elaidate-supplemented medium leads to a 30-fold reduction in the ratio of transport induction to β-galactosidase induction, but the ratio of the rate of elaidic acid incorporation into phospholipids to the rate of β-galactosidase induction is not grossly changed. Therefore, the failure to observe transport system induction, after the downshift from 37° to 25°C during incubation in elaidate-containing medium, indicates that the physical state of the membrane lipids is an important determinant of effective assembly of the lactose transport system. The induction of lactose transport as a function of temperature has also been studied in bromostearate- and oleate-supplemented media. Without exception, the temperatures below which transport system biogenesis becomes abortive are identical with the transition temperatures given in Table IV. This indicates that the assembly of this and other membrane systems may require a temperature above that which permits the formation of a condensed lipid phase at the air-water interface.

An additional experiment indicates that the physical state of the *newly* synthesized lipids is the determinant of effective assembly of the lactose transport system and of membrane assembly which permits cell growth. When the auxotroph is shifted from growth in elaidate-containing medium at 37°C to oleate medium at 25°C, growth and cell division continue, and induction for proteins of the lactose operon after the temperature and fatty acid shift leads to the formation of a normally functional lactose transport system.

Topography of Bacterial Membrane Assembly

Models for Membrane Assembly

LOCALIZED AND NONLOCALIZED ASSEMBLY

Experiments with an *E. coli* unsaturated fatty acid auxotroph indicate a preferential association of newly synthesized lipids and proteins. Two different types of models, localized and nonlocalized assembly, are advanced to account for this association. The term localized assembly is used to denote a process whereby the membrane grows by lateral extension, with the newly synthesized membrane components being inserted into the membrane matrix at a fixed focus or along a fixed perimeter. Nonlocalized assembly refers to a process whereby newly synthesized lipids and proteins are incorporated into the membrane at many points which have no fixed cellular location. To account for the preferential interaction of newly synthesized proteins with newly synthesized lipids in the model for nonlocalized assembly, the newly synthesized proteins would be incorporated into islets of newly synthesized lipids, or newly synthesized proteins and lipids would interact to form lipoprotein complexes, which would then be intercalated randomly into the membrane.

MEMBRANE GROWTH AS A DETERMINANT OF NUCLEAR SEGREGATION

In gram-negative and gram-positive bacteria, the cellular DNA is apparently attached to membrane both at the origin (terminus) of replication and at the replicative site. (The literature on membrane-DNA attachment has been recently reviewed; see ref. 40.) Hypotheses based on DNA-membrane attachment have been advanced to provide a rationale for investigating chromosome segregation in bacteria. The first hypothesis of this type stated that the chromosomes were attached to membrane at the equatorial perimeter of the cell (41). As the membrane grew by a localized assembly process, with newly synthesized proteins and lipids being inserted between the DNA-attachment sites and the equatorial perimeter, the DNA attachment sites for the two daughter chromosomes would be propelled in opposite directions. Cell division at the equatorial perimeter would then result in the orderly segregation of the daughter chromosomes. A similar mechanism can be envisaged where the DNA-attachment site resides at a pole of the cell, and where localized assembly results in growth of the membrane by linear extension from the pole. Here, one DNA-attachment site and the bound chromosome would remain stationary, and the second would be propelled away from it. The possible interrelation between localized membrane assembly and orderly segregation of daughter chromosomes in cell division has provided considerable impetus for the study of membrane assembly in bacteria.

Distinguishing New Membrane from Old

A test of the stated models for localized and nonlocalized assembly requires techniques which can clearly distinguish newly synthesized membrane from old. The first experiments of this type utilized a choline auxotroph of *Neurospora crassa* in an investigation on the origin of mitochondria in this organism (Chap. 13). A radioautographic study showed that labeled choline incorporated into cellular lipids was equally distributed in all mitochondria after three mass doublings during growth in a medium containing unlabeled choline. An additional study was based on the observation that mitochondria derived from cells grown with a low choline level in the medium have a different density than mitochondria from cells grown with a high choline level. When cells were shifted from low to high choline levels and the density of mitochondria determined, only mitochondria of an intermediate density were observed after growth for a period of time sufficient for a 60% increase in mitochondrial mass. These experiments can be interpreted to indicate a model for nonlocalized membrane assembly, but should be considered in terms of more recent information which indicates that in the cells of at least some higher organisms, a mechanism exists for the exchange of phospholipids between the different cellular organelles (Chap. 1).

MORPHOGENESIS OF *E. COLI* CYTOPLASMIC MEMBRANE

The cell envelope of *E. coli* contains two membranes: the cytoplasmic membrane, which is in continuity with the cytoplasm and constitutes the permeability barrier of the cell, and an "outer membrane," which is localized external to the cell wall. Separation of the cytoplasmic and outer membranes has been achieved by isopycnic banding in sucrose density gradients (23, 43–44). The "inner" (i.e., cytoplasmic) membrane contains the carrier proteins for transport (23), the cytochromes (23, 43), and most of the enzymes which catalyze the synthesis of complex lipids (45). The single exception is the enzyme which catalyzes phosphatidylserine synthesis (46). This enzyme is found primarily in the ribosomal fraction after cell disruption. No enzymes have so far been identified which are distinct for the outer membrane, and no proteins identifiable by acrylamide gel electrophoresis are common to both membranes (44). The lipid and lipopolysaccharide components present in the outer membrane are evidently synthesized on the inner membrane and are then transported to the outer membrane. Experiments designed to identify a protein which could serve to carry lipids from the inner to the outer membrane have been negative, and the mechanism for transport of lipids from their site of synthesis in the inner membrane to their site(s) of deposition in the outer membrane is unknown (47).

The assembly of the cytoplasmic membrane of *E. coli* has been studied with a density-labeling technique to test for the distinction between localized and nonlocalized processes. These studies make use of the lipid density label bromostearic acid (23). The cytoplasmic membrane fraction prepared from an unsaturated fatty acid auxotroph grown with bromostearic acid supplementation is approximately 0.06 g/cm^3 more dense than the cytoplasmic membrane fraction prepared from cells of the auxotroph grown with oleic or elaidic acids (Fig. 7). This density difference is that predicted by the fatty acid composition of the lipids from the density-labeled membranes, which contain approximately 60% of the total fatty acids as bromostearic and bromopalmitic acids.

Models of localized membrane assembly which could account for orderly chromosome segregation are depicted in Figure 8. In Model A, which depicts localized growth of the membrane from a single pole, one generation of growth of newly divided daughter cells would, after a density shift and subsequent cell division by fission at the equatorial perimeter, produce two populations of cells—one with the cytoplasmic membrane density labeled, and the other with the cytoplasmic membrane containing no density label. Models B and C depict localized assembly where membrane growth occurs simultaneously at both poles or at the equatorial perimeter. In contrast to Model A, two generations of growth and cell division after the density shift are required for the detection of two populations of cytoplasmic membrane on the basis of density. These models were tested (48) by first growing the auxotroph in medium supplemented with oleic acid (light) in the presence of a ^{3}H-amino acid. A population of the ^{3}H-labeled, newly divided daughter cells was isolated, inoculated into bromostearate-supplemented (heavy) medium containing a ^{14}C-labeled amino acid, and samples of the cells were removed after

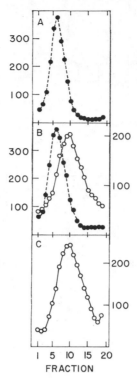

Figure 7

Isopycnic banding in a preformed sucrose density gradient of cytoplasmic membranes derived from cells of an *E. coli* essential fatty acid auxotroph grown in oleic acid and bromostearic acid. Fraction 1 is the fraction of highest density. *A*. Bromostearic acid membranes. *B*. Bromostearic acid and oleic acid membranes banded in the same gradient. *C*. Oleic acid membranes. Black circles, bromostearate membranes (^{3}H-labeled amino acid, cpm); white circles, oleate membranes (^{14}C-labeled amino acid, cpm). (From Tsukagoshi, N., and Fox, C. F. *Biochemistry 10*, 3309, 1971.)

either one or two rounds of synchronous cell division and processed for the isolation of the cytoplasmic membrane fraction. When the cytoplasmic membranes derived in this fashion were subjected to equilibrium centrifugation in sucrose density gradients, no separation of membranes with a bias toward light and ^{3}H-labeled or heavy and ^{14}C-labeled was observed. This clearly indicates that the cytoplasmic membrane of *E. coli* is not assembled by one of the localized processes depicted in Figure 8, with concomitant conservation of membrane synthesized before and after the density shift. In an extension of this experiment the density and radioactive label shifts were combined with controlled membrane fragmentation in a more general test

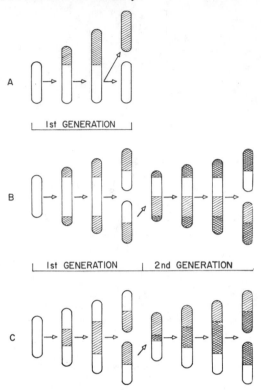

Figure 8

Models for localized growth of the membranes of rod-shaped bacteria at one (A) or both (B) poles, or at the equatorial perimeter (C). The newly formed membrane is indicated by the cross-hatched regions. (After Tsukagoshi, N., *et al. Biophys. Biochem. Res. Commun. 44*, 497, 1971.)

for localized membrane assembly. Uncontrolled fragmentation of a mixture of density-labeled and unlabeled membranes by sonic irradiation leads to the formation of a species of membrane of hybrid density. A procedure was developed which allowed fragmentation of a mixture of light and density-labeled membranes so that fragments of approximately 500 Å in linear dimension (about one-four hundredth the size of an intact *E. coli* membrane) could be produced with only limited hybrid formation (49). This permitted density shift experiments employing the lipid density label to test for localized membrane assembly at a greatly increased level of resolution (50). These experiments show that there are no areas of localized synthesis at the limit of resolution of the technique, i.e., 500 Å in lateral dimension. The tests for localized membrane assembly by lipid density labeling, and by other techniques such as radio-autography, assume that there is limited lateral diffusion of the lipid and protein components of the membrane. At the time of this writing, the lateral diffusion of lipids has been detected only in artificial bilayers (51). It is possible that the lipids of membranes from natural sources could be subject to restraints in movement to which the lipids in artificial bilayers are not. An experiment which obviates the possibility of lateral diffusion of lipids in a test for localized assembly of the membrane protein matrix is described below, under "Minicell-Producing Mutants." Though no definitive evidence for the lateral diffusion of proteins in membrane has been reported, the diffusion in membranes of chemically undefined antigens has been detected by immunofluorescence after the Sendai virus–induced fusion of two antigenically different lines of mammalian cells (52).

MORPHOGENESIS OF MEMBRANES OF *BACILLUS* SPECIES

Unlike gram-negative bacteria such as *E. coli*, gram-positive bacteria have a cytoplasmic membrane, but no outer membrane. The possibility that the cytoplasmic membrane of *B. subtilis* is assembled by a localized process has been studied by radioautography and by density labeling utilizing a shift from deuterated medium to a H_2O-containing medium (14). In the experiments which employed radioautography, a glycerol auxotroph was pulse labeled for 1, 2, or 4 min with radioactive glycerol, and the distribution of the isotope in the cell was determined. The label, which is incorporated only into the membrane lipids, was found to be uniformly distributed over the length of the cell with no differential labeling at the poles, the equatorial perimeter, or the sites of septum formation. Membranes from cells grown in deuterated medium are separable from membranes derived from cells grown in a H_2O-constituted medium. Cells were first grown in deuterated medium, shifted to a medium where H_2O was substituted for D_2O, and growth was continued for three cell divisions. Only hybrid membrane was observed after equilibrium density gradient centrifugation. Neither of these experiments supports the models for localized membrane assembly depicted in Figure 8. A radioautographic study of fatty acids taken up during a pulse labeling by *B. megaterium* KM, on the other hand, indicates that the label was distributed with a bias toward the poles of the cells (53). This raises the possibility of localized membrane assembly in this strain. Since the evidence with other organisms indicates a nonlocalized assembly process, it is important to determine if other techniques for distinguishing new membrane from old support the view that the membrane of *B. megaterium* KM grows by localized assembly.

Mutants with Defects in Cell Division and Nuclear Segregation

MINICELL-PRODUCING MUTANTS

A unique class of cell division mutants of *E. coli*, the minicell-producing mutants, produces small enucleate cells by fission occurring only at the poles (Fig. 9). Daughter nucleate cells are produced by normal fission at the

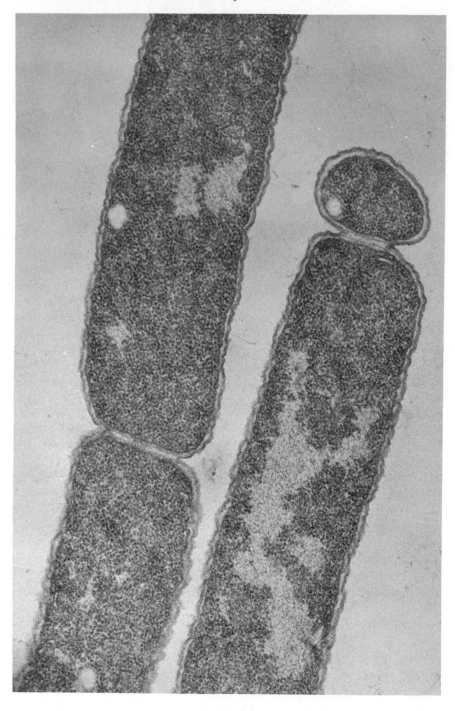

Figure 9
Electron micrograph of a section from the minicell-producing *E. coli* strain P678-54, showing a normal cell division and minicell formation at the pole in the same field ×53,000. (Courtesy of Drs. Howard Adler and David Allison, Biology Division, Oak Ridge National Laboratory.)

equatorial perimeter, but normal cell division and minicell production are not observed simultaneously in the same cell (54, 55). Though the molecular basis for the defect resulting in minicell production is unknown, the minicell-producing strains provide an additional system to test the models for localized membrane assembly described in Figure 8. These tests complement the studies which employ the density-labeled lipids, since they permit an examination of the possibility that newly synthesized proteins can be incorporated into the membrane matrix by a localized process without concomitant localized assembly of the membrane lipids (56). Figure 10 depicts the formation of a newly synthesized minicell membrane under conditions where assembly proceeds by a localized process. If growth of the membrane occurs at the poles of the cell, the membranes of the newly formed minicells should contain a higher relative proportion of newly synthesized membrane than the membranes of the cells from which the minicells are derived. If, on the other hand, the membrane grows by lateral extension from the equatorial perimeter of the cell, the membranes of newly formed minicells should contain virtually none of the newly synthesized membrane components. In one experimental test, the minicell-producing strain was first grown in a medium containing a ^{3}H-labeled amino acid for three generations of growth, and the cells were then separated from the minicells in the population and inoculated into a medium containing a ^{14}C-labeled amino acid. At various points, the newly produced minicells were removed from the population, and they and the nucleated cells from which they had been isolated were tested for the ratio

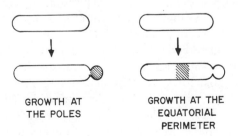

GROWTH AT
THE POLES

GROWTH AT THE
EQUATORIAL
PERIMETER

Figure 10
Schematic representation of localized processes of membrane growth for minicell formation.

of $^{14}C/^3H$ in the proteins of the cell envelope. This ratio was the same for both the cells and minicells isolated at periods after the isotope shift from 0.3 to 1.6 generations of growth. In a second experimental test, the minicell-producing strain was induced for one transport system, and the cells separated from the minicells and inoculated into a second growth medium for induction of a different transport system. At various periods of time after the induction shift, the minicells were separated from the cells, and the ratio of the two transport activities was compared in the minicells and the cells from which they had been isolated. Identical ratios of the two transport system activities were observed in cells and minicells sampled at periods of 0.3–2.5 generations of growth after the induction shift. Neither experiment supports the models given in Figure 10, indicating that the assembly of the protein portion of the *E. coli* membrane does not proceed by either of the stated localized assembly processes.

OTHER CELL DIVISION MUTANTS

In addition to the minicell-producing mutants, a variety of temperature-sensitive mutants in cell division has been obtained for *E. coli* (57). These include mutants which are defective in septation and which form multi-nucleate filaments, and mutants which divide to form full-sized, but enucleate, cells. Though the mutations which cause these defects have been mapped and are apparently single-site mutations; the molecular bases for the defects are unknown.

The Purple Membrane of Halobacterium halobium

Though there is no really convincing evidence for a localized process which could account for membrane assembly as a whole in bacteria or in higher cells, certain cells contain regions of membrane which are distinct from the membrane as a whole. One excellent example of this is the so called "purple membrane" of the halophilic bacterium, *H. halobium* (58, 59). Halophilic bacteria require high concentrations of inorganic salts for growth and for the maintenance of structural integrity. When membranes of *H. halobium* are subjected to conditions of low ionic strength, the membranes disintegrate into fragments, and one characteristic fragment, the purple membrane, can be isolated from other membrane. The purple membrane contains but a single protein species of molecular weight 26,000 to which the pigment retinal is bound in a 1:1 molar ratio. Freeze fracture studies indicate that the patches of purple membrane are continuous with the cell membrane. Thus in certain instances, at least, the protein matrix of defined membrane regions of considerable size can be assembled by a localized process and the structure conserved independent of that of other regions of the membrane.

ASSEMBLY OF VIRAL MEMBRANES

The viruses enveloped by lipoprotein membranes include: *1*) RNA viruses such as Sendai virus, Newcastle disease virus, and SV5 (the paramyxovirus group); *2*) influenza virus (a member of the myxovirus group); *3*) the causative agents in encephalitis (the arbovirus group); and *4*) the RNA tumor viruses, which have been shown to induce tumors and leukemia in rodents and fowls. Enveloped RNA viruses are thought to be cancer-inducing agents in humans as well, making the study of the assembly and function of viral membranes one of the most exciting and important areas of biology today. The herpes viruses, which contain DNA, share a common mechanism of envelopment with those named above in that the viral membrane is assembled on a host cell membrane template, and the viral capsid becomes enveloped by an exocytosis of the modified host nuclear membrane. Viruses of the herpes group are the causative agents in fever blisters, shingles, and chickenpox, and have been linked to the production of tumors in man.

Vaccinia virus, a DNA-containing poxvirus, and the enveloped bacterial virus PM2 which infects *Pseudomonas* BAL-31 may share a common means of development distinct from that of the previously mentioned examples. Vaccinia is produced in cytoplasmic "virus factories" which have no apparent continuity with any of the cytoplasmic membranes. Though the mechanism of PM2 assembly is not yet clear, it appears not to be enveloped by exocytosis from the cell membrane. This treatment of viral membrane assembly will deal primarily with the most thoroughly studied systems. SV5 is presented as an example of the viruses which are assembled on host membrane templates, and vaccinia and PM2 as examples of viral membrane assembly apparently at a site distinct from a host membrane template.

Assembly on Host Cell Membrane Templates (60)

The paramyxovirus SV5 consists of a ribonucleoprotein capsid surrounded by a lipoprotein envelope which has an electron microscopic appearance similar to that of the plasma membrane of the cell. The envelope is, in addition, covered with surface projections (spikes) which are approximately 100 Å in length. The SV5 virion contains six proteins, five of which are present in the envelope, the sixth being present solely in the nucleocapsid. No proteins of the host cell cytoplasmic membrane are present in the lipoprotein envelope of the virus in quantities detectable by acrylamide gel electrophoresis. The order of the steps in the assembly of the virus has been determined by electron microscopy. The first observable step is the assembly of the nucleoprotein capsids in the cytoplasm. Once the nucleoprotein capsids are formed, they can be detected adjacent to regions of the plasma membrane characterized by spikes present on the outer membrane surface. The spikes are not observed on regions of membrane where no capsids are aligned, nor are capsids seen adjacent to regions of membrane which contain no spikes, suggesting that the formation of the surface projections is a result of the attachment of the capsid to the plasma membrane. Other evidence indicates that regions of the host cell plasma membrane are altered prior to contact with the capsids, and that the alteration of the host cell membrane may be a prerequisite for attachment of the viral capsid. Ferritin-labeled antibodies specific for viral antigens attach to regions of the host cell plasma membrane which have no viral capsids aligned adjacent to the inner membrane surface. These studies indicate that the formation of the viral capsid and the assembly of the virus-specified regions of the host cell cytoplasmic membrane proceed independently before the attachment of the preformed capsids to specific viral determined regions of the plasma membrane. During the attachment process or shortly thereafter, the plasma membrane is modified further as shown by the appearance of spikes on the outer surface of plasma membrane regions which have capsids aligned beneath them. The final step in virus production (Fig. 11) is the "budding" of the virions by exocytosis from the plasma membrane (a budding-out process).

Biochemical data on the lipid composition of SV5 and the host cell plasma membrane afford further evidence for the assembly of the SV5 envelope on a plasma membrane template (Table VI). The relative proportions of phospholipids, triglycerides, cholesterol, and cholesterol esters in the virus reflect the lipid composition of the plasma membrane of the cell line in which the virus developed. One of the most striking similarities is the molar ratio of cholesterol to phospholipid in the host cell plasma membrane and viral membrane. This ratio is high in MK (monkey kidney) plasma membranes and in SV5 which developed in MK cells, and low in HaK (hamster kidney) plasma membranes and in SV5 derived from HaK cells. The relative proportions of the individual phospholipids in the viruses and the plasma membranes of the host cells from which the viruses are derived are also similar, but for a single exception (Table VII). The plasma membranes of MK cells and the envelopes of SV5 derived from MK cells are rich in phosphatidylethanolamine and phosphatidylserine, whereas the content of these lipids in BHK (baby hamster kidney) and HaK plasma membranes and in SV5 from BHK and HaK cells is relatively low. In these examples, the phospholipid composition of the virus reflects that of the plasma membrane of the host cell. MDBK plasma membranes, on the other hand, have a high content of phosphatidylcholine and a low content of phosphatidylethanolamine relative to the phospholipid composition of SV5 from MDBK (bovine kidney) cells. Though the reason for this exception is not known, it should be pointed out that the composition of the MDBK plasma membrane was not determined at the time of assembly of the viral envelope. Nor is it known if the lipids of the viral envelope are derived primarily from those synthesized

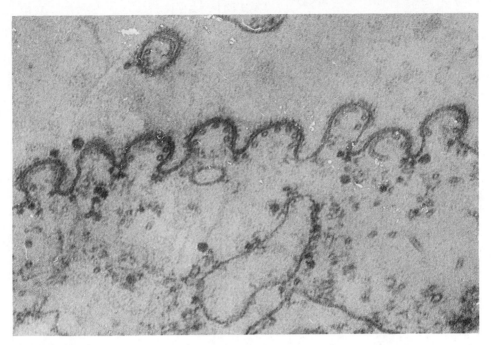

Figure 11

Electron micrograph of a section showing a row of SV5 virions budding by exocytosis from the plasma membrane ×100,000. (From Choppin, P. W., *et al. Perspectives in Virology*. Ed. M. Pollard. Academic Press, New York, 1970, vol. 7. The reproduction of the electron micrograph was provided by Dr. Purnell Choppin.)

prior to or during the time of viral envelope development. It is thus possible that in certain cell lines (such as MDBK) the pattern of host cell membrane lipid synthesis could be altered during the course of viral infection. The lipids of the virus might then reflect the composition of the nascent (newly synthesized) lipids, rather than the gross lipid composition of the host cell membrane which gives rise to the viral envelope.

The data on SV5 morphogenesis point to a localized mechanism of assembly of the proteins of the viral envelope at distinct sites on the host cell plasma membrane. Though the membrane lipids of the virus are apparently derived from the host cell plasma membrane, it is not clear whether the lipids which are incorporated into the viral envelope are scavenged primarily from those which are present in the host cell membrane prior to infection, or are primarily derived from those synthesized during the course of viral development.

Table VI

Lipid Content of Plasma Membranes of MK, BHK21-F, and HaK Cells, and of SV5 Virions Grown in These Cells

SOURCE	TOTAL LIPID (% DRY WEIGHT)	Percent of Total Lipid				MOLAR RATIO: CHOLESTEROL/ PHOSPHOLIPID
		PHOSPHOLIPID	TRIGLYCERIDES	CHOLESTEROL	CHOLESTEROL ESTERS	
MK membranes	28.5	55.0	5.1	23.0	0.3	0.81
SV5 from MK cells	20.0	50.9	5.0	29.0	0.9	0.89
BHK21-F membranes	30.7	60.0	2.8	20.5	1.3	0.68
SV5 from BHK21-F cells	21.0	57.0	3.0	18.6	2.7	0.64
HaK membranes	38.0	65.5	6.0	17.0	1.7	0.51
SV5 from HaK cells	19.0	65.0	4.6	20.0	2.0	0.60

Table VII

Phospholipid Content of Plasma Membranes of MK, BHK21-F, HaK, and MDBK Cells and of SV5 Virions Grown in These Cells

SOURCE	Percent of Total Phospholipid				
	SPHINGOMYELIN	PHOSPHATIDYL-CHOLINE	PHOSPHATIDYL-INOSITOL	PHOSPHATIDYL-SERINE	PHOSPHATIDYL-ETHANOLAMINE
MK membranes	11.8	32.1		17.2	38.8
SV5 from MK cells	12.2	25.2	2.9	17.9	40.3
BHK21-F membranes	24.2	49.5	10.0	5.1	11.2
SV5 from BHK21-F cells	30.0	38.5	10.5	5.2	15.6
HaK membranes	24.4	46.8	11.7	5.0	13.0
SV5 from HaK cells	25.8	43.8	8.5	5.0	17.1
MDBK membranes	22.8	44.5	2.9	2.2	27.2
SV5 from MDBK cells	27.3	23.8	5.2	2.0	40.9

MK, monkey kidney cells; BHK21-F, baby hamster kidney cells; HaK, hamster kidney cells; MDBK, bovine kidney cells. Both tables from Choppin, P., et al. Perspectives in Virology. Ed. M. Pollard. Academic Press, New York, 1970, vol. 7.

Assembly in Cytoplasm

Electron microscopic investigations of the course of development of vaccinia virus indicate that the viral envelope is formed by a self-assembly process in the cytoplasm at sites distinct from any membrane of host cell origin (61). The lipids which compose the viral envelope appear to be synthesized primarily during the course of virus assembly. Cells were pulse labeled with radioactive choline prior to or after infection, and the amount of radioactive label in viruses purified from the two batches of cells was determined. The viruses obtained from cells labeled during the course of viral development contained 3.5–7.6 times more radioactivity in lipid than did the viruses obtained from cells labeled prior to infection. This phenomenon could be explained either by de novo synthesis of lipids in situ in the developing virions or by transport of newly synthesized lipids from their site(s) of synthetic origin to the developing viral particles (see Chap. 1). If the viral lipids are synthesized at sites distinct from the developing virions and then transported to the developing virions for subsequent assembly, these data would indicate that there are two functionally distinct pools of lipid in the host cell membranes: lipids intimately associated in membrane structures and inaccessible for transport between cellular organelles, and lipids which, having been recently synthesized, are available for transport between cellular organelles.

Of the bacterial viruses, PM2, which infects the marine bacterium *Pseudomonas* BAL-31, is unique in that it is enveloped by a membrane which has a typical bilayer appearance by electron microscopy (62). Though it is premature to classify PM2 with viruses which develop at sites distinct from a host cell membrane, there is no morphological evidence to suggest that PM2 is enveloped by budding from the host cell membrane. Furthermore, the phospholipid composition of PM2 is quite distinct from that of either uninfected BAL-31 cells or that of the viral lysate (Table VIII) (63). It should

Table VIII

Phospholipid Content of BAL-31 Cells, PM2 Virus, and the Viral Lysate

SOURCE OF LIPID EXTRACT	Phospholipids as a Percent of Total		
	PA[a] AND LPE	PG	PE
BAL-31 cells	0.6	22.5	75.7
PM2 virus[b]	3.0	65.1	29.9
Viral lysate	22.5	33.8	40.0

[a] Only PA is present in the virus.
[b] Average of three independent viral preparations.
PA, phosphatidic acid; LPE, lysophosphatidylethanolamine; PG, phosphatidylglycerol; PE, phosphatidylethanolamine.
From Braunstein, S. N., and Franklin, R. M. *Virology* **43,** 685 (1971).

be recalled, however, that the phospholipid composition of SV5, which does receive its envelope by budding from a host cell membrane, sometimes displays a phospholipid composition quite different from that of the host cell membrane on which it develops (Table VII).

Viruses and Cell Fusion

The infection of animal cells by certain of the enveloped viruses gives rise to a phenomenon termed cell fusion. Cell fusion is also a normal developmental process and is responsible, for example, for the formation of multinucleate skeletal muscle cells by fusion of mononucleate myoblast precursors. Pathologically, multinucleate cells have been detected in association with the lesions of such viral diseases as chickenpox and smallpox. In tissue culture, the formation of multinucleate syncytia by cell fusion has been detected after infection with viruses of the paramyxovirus group such as Sendai virus, Newcastle disease virus, and SV5. Though the molecular events which give rise to virus-induced cell fusion are not yet known, studies in tissue culture have shown that it is in many cases a process associated with the membrane of the infecting virus. From a mechanistic standpoint, two types of fusion processes have been observed: *1*) fusion from without, which does not require an infectious virus particle or transcription of the viral genome, and *2*) fusion from within, which does require an infectious virus particle and transcription of the viral genome.

Fusion from without (64–66). Fusion from without can be caused not only by an infectious virus particle, but also by virus particles which have been treated with ultraviolet light or β-propiolactone to inactivate the viral nucleic acid. Such treatment has little influence on the fusion efficiency of a given number of virus particles. Though sonication reduces the efficiency of fusion by Sendai virus preparations, fragments of the viral envelope nevertheless retain the ability to cause fusion. The lipids of the viral envelope are an important determinant of the ability to induce fusion. The fusion ability of Sendai virus is destroyed by treatment with lipid solvents, and that of Newcastle disease virus by treatment with phospholipase. The first step in fusion is the adsorption of the virus particle to the cell. This proceeds rapidly at low temperature (0–4°C). Fusion can then be initiated by raising the temperature to 37°C. The next step in cell fusion could be the fusion of the viral envelope with the plasma membrane of the cell, with the establishment of continuity between the cellular and viral membranes. Cell-virus fusion has been observed with paramyxoviruses such as Sendai virus (67). As in the case of cell-cell fusion, the fusion of the viral and cellular membranes is critically temperature-dependent. Following virus adsorption or cell-virus fusion, the site of cell-virus fusion (or adsorption) might adsorb to a second cell, thus initiating cell-cell fusion. The final step would then be the establishment of continuity between the membranes of the two cells at the fusion site. This final step may be energy-dependent, since inhibitors of oxidative phosphorylation, such as dinitrophenol, are potent inhibitors of cell-cell fusion, but not of virus attachment. The ability of agents such as inactivated Sendai virus to induce cell fusion has proved a useful tool in animal cell genetics and tumor virology (Chap. 15).

Fusion from within. Fusion from within differs from fusion from without in that it requires viral development. This process has been studied in tissue culture largely with viruses of the herpes group (68) and with certain strains of Newcastle disease virus (66). The requirement for host cell macromolecule synthesis is believed to indicate that fusion from within results from a modification of the host cell plasma membrane. Whereas the herpes viruses can only induce fusion from within, the paramyxoviruses induce both fusion from within and fusion from without. With Newcastle disease virus, for example, the ability to induce fusion from without is strain-dependent in that certain strains can produce fusion only from within. The molecular determinants for both fusion processes are currently not known.

CONCLUSIONS AND PROJECTIONS

Though there is good evidence that newly synthesized proteins interact with newly synthesized lipids in the formation of a few defined membrane systems, membrane assembly processes in general do not proceed by growth with linear extension of new membrane from one or a few foci. Certain specialized regions of membrane, however, and the membranes of enveloped viruses are assembled by localized processes, at least with respect to the formation of their protein matrices. Though most membrane assembly occurs on a "template" of preexisting membrane, certain enveloped viruses are assembled at cellular sites distinct from any preexisting cellular membrane.

At the present time, little is known about the precise mechanism of membrane assembly or about the dynamics of the lipid and protein moieties of biological membranes. The experiments and experimental approaches described here, however, should provide a rationale for the design of in vitro approaches to the study of the assembly of membranes and specific membrane systems. The ability of animal cell viral envelopes to fuse with cells and to induce cell-cell fusion provides an excellent system for the determination of the lateral diffusion properties of membrane components. With these techniques, it is possible to establish a continuity between membranes which differ markedly in their chemical, physical, and antigenic properties, and the rates of mixing of chemically defined species can be studied after continuity has been established.

REFERENCES

1. Korn, E. D. *Annu. Rev. Biochem.* **38**, 263 (1969).
2. Arias, I. M., Doyle, D., and Schimke, R. T. *J. Biol. Chem.* **244**, 3303 (1969).
3. Dehlinger, P. J., and Schimke, R. T. *J. Biol. Chem.* **246**, 2574 (1971).
4. Omura, T., Siekevitz, P., and Palade, G. E. *J. Biol. Chem.* **242**, 2389 (1969).
5. Kuriyama, Y., Omura, T., Siekevitz, P., and Palade, G. E. *J. Biol. Chem.* **244**, 2017 (1969).
6. Orrenius, S. *J. Cell Biol.* **26**, 713 (1965).
7. Leskes, A., Siekevitz, P., and Palade, G. E. *J. Cell Biol.* **49**, 264 (1971).
8. Leskes, A., Siekevitz, P., and Palade, G. E. *J. Cell Biol.* **49**, 288 (1971).
9. Kito, M., and Pizer, L. I. *J. Biol. Chem.* **244**, 3316 (1969).
10. Hsu, C. C., and Fox, C. F. *J. Bacteriol.* **103**, 410 (1970).
11. Hsu, C. C., and Fox, C. F. Unpublished observations.
12. Mindich, L. *J. Mol. Biol.* **49**, 415 (1970).
13. Mindich, L. *J. Mol. Biol.* **49**, 433 (1970).
14. Mindich, L. *J. Bacteriol.* **110**, 96 (1972).
15. Cronan, J. E., Jr., Ray, T. K., and Vagelos, R. P. *Proc. Natl. Acad. Sci. U.S.A.* **65**, 737 (1970).
16. Hechemy, K., and Goldfine, H. *Biochem. Biophys. Res. Commun.* **42**, 245 (1971).
17. Kito, M., Lubin, M., and Pizer, L. I. *Biochem. Biophys. Res. Commun.* **34**, 454 (1970).
18. Silbert, D. F., and Vagelos, P. R. *Proc. Natl. Acad. Sci. U.S.A.* **58**, 1579 (1967).
19. Cronan, J. E., Jr., Birge, C. H., and Vagelos, P. R. *J. Bacteriol.* **100**, 601 (1969).
20. Silbert, D. F., Ruch, F., and Vagelos, P. R. *J. Bacteriol.* **95**, 1658 (1968).
21. Esfahani, M., Barnes, E. M., Jr., and Wakil, S. J. *Proc. Natl. Acad. Sci. U.S.A.* **64**, 1057 (1969).
22. Schairer, H. U., and Overath, P. *J. Mol. Biol.* **44**, 209 (1969).
23. Fox, C. F., Law, J. H., Tsukagoshi, N., and Wilson, G. *Proc. Natl. Acad. Sci. U.S.A.* **67**, 598 (1970).
24. Wilson G., and Fox, C. F. *J. Mol. Biol.* **55**, 49 (1971).
25. Wilson, G., and Fox, C. F. Unpublished observations.
26. Haest, C. W. M., De Gier, J., and van Deenen, L. L. M. *Chem. Phys. Lipids* **3**, 413 (1969).
27. Marr, A. G., and Ingraham, J. L. *J. Bacteriol.* **84**, 1260 (1962).
28. Esfahani, M., Limbrick, A. R., Knutton, S., Oka, T., and Wakil, S. J. *Proc. Natl. Acad. Sci. U.S.A.* **68**, 3180 (1971).
29. Wilson, G., Rose, S. P., and Fox, C. F. *Biochem. Biophys. Res. Commun.* **38**, 617 (1970).
30. Overath, P., Schairer, H. U., and Stoffel, W. *Proc. Natl. Acad. Sci. U.S.A.* **67**, 606 (1970).
31. Kass, L. R. *J. Biol. Chem.* **243**, 3223 (1968).
32. Silbert, D. F. *Biochemistry* **9**, 3631 (1970).
33. Henning, U., Dennert, G., Rehn, K., and Deppe, G. *J. Bacteriol.* **98**, 784 (1969).
34. Wisnieski, B. J., Keith, A. D., and Resnick, M. R. *J. Bacteriol.* **101**, 160 (1970).
35. Henry, S. A., and Keith, A. D. *Chem. Phys. Lipids.* **7**, 284 (1971).
36. Fox, C. F. *Proc. Natl. Acad. Sci. U.S.A.* **63**, 850 (1969).

37. Overath, P., Hill, F. F., and Lamnek-Hirsch, I. *Nature New Biol.* **234,** 264 (1971).
38. Tsukagoshi, N., and Fox, C. F. Manuscript submitted to *Biochemistry.*
39. Mindich, L. *Proc. Natl. Acad. Sci. U.S.A.* **68,** 420 (1971).
40. Goulian, M. *Annu. Rev. Biochem.* **40,** 855 (1971).
41. Jacob, F., Brenner, S., and Cuzin, F. *Cold Spring Harbor Symp. Quant. Biol.* **28,** 329 (1963).
42. Donachie, W. D., and Begg, K. J. *Nature* **227,** 1220 (1970).
43. Miura, T., and Mizushima, S. *Biochim. Biophys. Acta* **150,** 159 (1968).
44. Schnaitman, C. *J. Bacteriol.* **104,** 890 (1970).
45. Machtiger, N., and Fox, C. F. Manuscript in preparation.
46. Raetz, C. R. H., and Kennedy, E. P. *J. Biol. Chem.* **247,** 2008 (1972).
47. Tsukagoshi, N., and Fox, C. F. Unpublished observations.
48. Tsukagoshi, N., Fielding, P., and Fox, C. F. *Biochem. Biophys. Res. Commun.* **44,** 497 (1971).
49. Tsukagoshi, N., and Fox, C. F. *Biochemistry* **10,** 3309 (1971).
50. Tsukagoshi, N., and Fox, C. F. *Fed. Proc.* **30,** 1120 (1971).
51. Kornberg, R. D., and McConnell, H. M. *Proc. Natl. Acad. Sci. U.S.A.* **68,** 2564, (1971).
52. Frye, L. D., and Edidin, M. *J. Cell Sci.* **7,** 319 (1970).
53. Morrison, D. C., and Morowitz, H. J. *J. Mol. Biol.* **49,** 441 (1970).
54. Adler, H. I., Fisher, W. D., Cohen, A., and Hardigree, A. A. *Proc. Natl. Acad. Sci. U.S.A.* **57,** 321 (1967).
55. Adler, H. I., Fisher, W. D., and Hardigree, A. A. *Trans. N.Y. Acad. Sci.* (Series II) **31,** 1059 (1969).
56. Wilson, G., and Fox, C. F. *Biochem. Biophys. Res. Commun.* **44,** 503 (1971).
57. Hirota, Y., Ryter, A., and Jacob, F. *Cold Spring Harbor Symp. Quant. Biol.* **33,** 677 (1968).
58. Oesterhelt, D., and Stoeckenius, W. *Nature New Biol.* **233,** 149 (1971).
59. Blaurock, A. E., and Stoeckenius, W. *Nature New Biol.* **233,** 152, 1971.
60. Choppin, P. W., Klenk, H. -D., Compans, R. W., and Caliguiri, L. A. *Perspectives in Virology.* Ed. M. Pollard. Academic Press, New York, 1970, vol. 7, p. 127.
61. Dales, S., and Mosbach, E. H. *Virology* **35,** 564 (1968).
62. Silbert, J., Salditt, M., and Franklin, R. M. *Virology* **39,** 666 (1969).
63. Braunstein, S. N., and Franklin, R. M. *Virology* **43,** 685 (1971).
64. Okada, Y. *Curr. Top. Microbiol. Immunol.* **48,** 102 (1969).
65. Harris, H. *Cell Fusion.* Harvard University Press, Cambridge, Mass., 1970.
66. Bratt, M. A., and Gallaher, W. R. *In Vitro* **6,** 3 (1970).
67. Morgan, C., and Howe, C. *J. Virol.* **2,** 1122 (1968).
68. Roizman, B. *Viruses Affecting Man and Animals.* Ed. M. Sanders. Green, St. Louis, 1971.

13

Godfrey S. Getz

Organelle Biogenesis

This chapter is concerned with the biogenesis of membranous organelles in eukaryotic cells. The membranes of such cells have two major functions in the cell economy:

1. They have a boundary function enabling the cell or organelle to enclose and regulate an environment separate and distinct from that surrounding it. The permeability properties of such a membrane are attributable to its content of particular discriminating transport mechanisms which facilitate and control the transport of appropriate metabolites and cofactors between the two environments it separates. A special case of this boundary function is seen in the packaging within membrane-limited vesicles of exportable or digestive enzymes, as in secretory granules or lysosomes and their related particles.

2. They have an organizational function enabling specific molecular assemblages of enzymes and related molecules to accomplish, with startling efficiency, vital biochemical functions. This is perhaps best exemplified by the organization and consequent efficiency of terminal respiration and oxidative phosphorylation in the inner mitochondrial membrane.

Work of the author has been supported by grants GM 13048 and CA 10463 from the U.S. Public Health Service, and by grants 68-26 from the American Cancer Society (Illinois Division) and P-520 from the American Cancer Society.

Though the major general function of intracellular membranes can be readily categorized, the detailed nature of the transport and organizational mechanisms responsible for the attributes of any particular membrane is poorly understood. The unique properties of particular membranes are reflected in differences in chemical structure and composition, about which current knowledge has been recently reviewed (1–3). Indeed evidence is now accumulating for the heterogeneity of structure and function within conventional subcellular membrane fractions (see Chap. 3 and ref. 4).

The diversity of structure and function of various membranes may well be mirrored in diversity of biosynthetic mechanisms. Nevertheless, several related general questions have been posed with respect to membrane biogenesis:

1. Do membranes or membranous organelles arise de novo from a catalyzed or self-assembly of their component molecules? Up to the present there seems no well-documented evidence for an affirmative answer to this question.

2. Is a membrane composed of subunits and manufactured so that its lowest common subunits are unitarily regulated, or is each subunit or membrane formed from individual macromolecular species whose biosynthesis is independently regulated? At present no clear answers can be provided to this question, though it seems that the answer may vary for different membranes. The designation of the regulation of the synthesis and degradation of membranous elements as synchronous depends upon the time scale over which these processes are considered. If the time units are small enough, few if any processes would be thought synchronous. To answer this question requires work with a highly homogeneous membrane system, which has not often been done in the studies hitherto made on eukaryotic membranes.

The membranes of eukaryotic cells are composed largely of protein and lipid. There is little evidence for the unbalanced accumulation of either membrane lipids or membrane proteins under physiological circumstances. This suggests a close coordination between the synthesis of membrane lipid and protein, which has been shown to exist in at least one case. The formation of a functional galactoside transport system in *Escherichia coli* following induction of the *lac*-operon is dependent upon simultaneous lipid synthesis (Chap. 12).

Nothing is known about the mechanisms responsible for this coordination, though the coupling between the synthesis of these two major membrane constituents is neither quantitatively absolute nor always obligatory. For example, the proportion of lecithin found in the mitochondrial membranes in choline auxotrophs of *Neurospora crassa* can be readily modified by the manipulation of the choline content of the growth medium (5). However, these observations do not exclude the possibility that within the mitochondrial membrane may be subunits whose relative lipid and protein composition is more tightly regulated. It appears that the respiratory enzymes of *Bacillus subtilis* and the lactose permease of *Staphylococcus aureus* can be incorporated into their respective membranes in the absence of lipid synthesis (see also Chap. 12 and refs. 6 and 7). This chapter is concerned with the origin, biosynthesis, and control of the assembly of proteins and lipids to form the functioning membranes of the intracellular organelles of eukaryotic cells.

The discovery in 1963 of chloroplast DNA (8, 9) and in 1964 of mitochondrial DNA (10) allowed experimentalists to apply the powerful and successful tools of the biochemical geneticists to the problems of organelle biogenesis, and a great deal of effort has since been expended in the study of mitochondrial biogenesis. Of all the membr nous organelles in eukaryotic cells, most is known about the biogenesis of mitochondria, though even in this case knowledge is very incomplete. Since the genetics of mitochondria are complex and poorly understood, the rewards of the genetic approach have not been as rapidly forthcoming as might have been anticipated. However, progress now being made will undoubtedly greatly improve the understanding derived from this approach. Much of this chapter is devoted to a review of our knowledge of mitochondrial and chloroplast biogenesis.

MITOCHONDRIAL AND CHLOROPLAST BIOGENESIS

Autonomy of Mitochondria and Chloroplasts

The existence within each of these organelles of unique genetic systems suggested that they may be autonomous intracellular symbionts. It will become clear, however, in the succeeding discussion that the genetic information content of these organelles is inadequate to determine the structure of all the components uniquely localized within mitochondria or chloroplasts. They are semiautonomous at best. As discussed below, the synthesis of mitochondria and chloroplasts, and indeed of relatively restricted enzyme complexes within them, is dependent upon the cooperative interaction of nuclear and organelle gene products.

It is furthermore evident that not even the total biosynthetic capacity of these organelles derives from their own genetic systems. Thus the mitochondrial biosynthetic systems for nucleic acids, lipids, and heme pigments are almost certainly under the control of structural genes within the nuclear genome. It is conceivable that at least some of the proteins made within the mitochondria may be translated from RNA messages originating in the nucleus.

A clear distinction must therefore be made between the genetic and the biosynthetic semiautonomy of mitochondria and chloroplasts.

Genetics of Mitochondria

Knowledge of the phenomenon of cytoplasmic inheritance long predated (11) the discovery of mitochondrial and chloroplast DNA. In chromosomal or Mendelian inheritance, wild and mutant genes are equally distributed from a heterozygous parent to the haploid daughters of meiotic division. On the other hand, heterozygosity of cytoplasmic markers does not apparently exist as a stable condition, and such markers segregate among the four haploid products of a meiotic division on an all-or-none basis, i.e., 4:0 or 0:4.

For plants and *Neurospora* cytoplasmic inheritance is maternal. In other words, when male and female gametes unite to form a diploid zygote, the female cytoplasmic determinants are retained in all meiotic products, while the male markers are either physically or physiologically lost. The mechanism of this physiological exclusion of male cytoplasmic markers, which probably also operates in mammalian cells, is unclear. On subsequent meiosis of these diploid cells, the derived haploid gametes contain only the cytoplasmic markers derived from the maternal parents. Yeast is a particularly suitable organism for studies of mitochondrial genetics, since it is able to grow in the absence of terminal respiratory activity. *Saccharomyces cerevisiae* can survive and grow using only glycolytic pathways. In yeast, which may exist and divide mitotically either as a stable haploid or as a diploid, the transmission of cytoplasmic genes is somewhat more complex than for plants. In yeast, zygotes are formed by the copulation of haploid cells of each mating type, *a* and *α*. Following meiosis or sporulation, the non-Mendelian segregation pattern still obtains. Mitochondrial mutants are generally referred to as petites. They owe this name to the fact that in their complete lack of respiratory capacity, they yield only small colonies when grown on plates containing limiting amounts of a fermentable carbon source (0.1% glucose) and larger amounts of a nonfermentable carbon source (2% ethanol or glycerol). Wild-type cells yield large colonies on this medium. There are two classes of cytoplasmic petites: neutral and suppressive. If a haploid neutral petite cell is mated with a haploid respiratory-competent cell, the resultant diploid cells and all their mitotic and meiotic descendants will be normal or respiratory-competent regardless of the mating type of the petite. If a highly suppressive petite is crossed with a normal haploid cell, most of the resultant diploid cells will be petite. Petite strains may have different degrees of suppressiveness. The phenomenon of neutral and suppressive respiratory deficiency or petiteness is poorly understood. One attractive recent postulation suggests that highly suppressive petites (12) have a high frequency of recombination of their abnormal mitochondrial genomes with those of their mating partners. On this basis, the tendency of neutral petites to undergo cytoplasmic genetic recombination with their normal mating partners would be of a much lower order of magnitude. An alternative explanation for suppressiveness (13) postulates the elevation in petite strains of a cytoplasmically coded repressor of the initiation of mitochondrial DNA replication.

The respiratory deficiency of all cytoplasmic petites is associated with a complete multiple molecular defect of several mitochondrial elements. Among these is the absence of demonstrable cytochromes aa_3, b, and c_1, all of which are normally very tightly bound to the inner mitochondrial membrane. The synthesis and integration of these respiratory pigments in the inner mitochondrial membrane are dependent on the normal cooperative functioning of the nuclear and the cytoplasmic genomes. This is indicated by the isolation of a large series of chromosomally determined respiratory-deficient mutants (14), all of which have a phenotype similar to that of the cytoplasmic petites. The chromosomal mutants are differentiated from the cytoplasmic petites by the pattern of their inheritance, which shows classic Mendelian segregation, and hence they are designated segregational petites. The biochemical mechanisms responsible for these phenotypes are unknown.

It is considered almost certain that mitochondrial DNA is the cytoplasmic genetic determinant (see below). Each eukaryotic cell contains many mitochondria, each of which probably contains several molecules of mitochondrial DNA. The mitochondrial DNA is therefore multiply redundant, making it difficult to obtain the point mutations which would be most valuable for the application of genetic analysis to investigation of mitochondrial biogenesis. However, several authors have recently obtained yeast mutants in which resistance to antibiotics, especially erythromycin and chloramphenicol, are cytoplasmically determined apparently by point mutations (12, 15). Despite the prediction, based upon the patterns of cytoplasmic inheritance in yeast, that the cytoplasmic genome of one haploid parent is inactive, careful examination of the matings of these mutants has indicated that recombination between their cytoplasmic genomes does occur (12). An extensive quantitative analysis of recombination patterns between these various mutants has been made (12). Mutants can be isolated which exhibit varying degrees of recombination or transmission of antibiotic resistances. The frequency of recombination or of transmission of antibiotic resistance is a unique and invariable property of each mutant. This property is termed polarity. In other words, the mitochondria may be thought of as possessing a sexual character which is independent of the mating type of the whole cell. The conversion of these mutants to cytoplasmic petites by treatment with ethidium bromide can frequently be accomplished with retention of the cytoplasmic marker for antibiotic resistances, although extensive changes in the mitochondrial genomes result in a loss of these markers too. The careful quantitative analysis in these strains of the patterns of recombination as well as of the patterns of petite induction by ethidium bromide has led Slonimski and his colleagues to suggest that only one or at most a few of the multiple copies of the yeast mitochondrial genome in any zygote or its immediate progeny is involved in genetic recombination or in the determination of the cytoplasmic genotype of the cell. Should these suggestions be substantiated and elaborated, important fresh insight and understanding will have been brought to the complexities of mitochondrial genetics.

Genetics of Chloroplasts

The phenomenon of uniparental inheritance was first discovered in relation to the inheritance patterns of leaf variegation in studies on *Mirabilis jalapa* by Correns (16) and on *Pelargonium* by Baur (17). The inheritance of certain cytoplasmic traits exclusively from the maternal parent has been extensively studied in algae and higher plants. Genetic mapping of the non-Mendelian linkage group requires the elimination of uniparental inheritance in order that recombination studies may be carried out. Sager and Ramanis (18) have recently accomplished this by the ultraviolet irradiation of the maternal parent, following which a high proportion of zygotes accept cytoplasmically inherited factors from both parents. Using this device these workers have constructed a tentative map of the non-Mendelian linkage group in *Chlamydomonas reinhardi*.

Mitochondrial DNA

The structure, properties, and importance of mitochondrial DNA for mitochondrial biogenesis has been extensively reviewed recently (19–22). Only those features of mitochondrial DNA essential to the understanding of mitochondrial biogenesis are summarized here. References to the studies documenting these features may be found in the cited reviews.

Filaments within the matrix of the mitochondria of many species of cells have been demonstrated in a large number of electron microscopic studies (19). These filaments are removed by DNAse treatment and almost certainly represent the cytological demonstration of mitochondrial DNA. This is further supported by the autoradiographic demonstration of the incorporation of tritiated thymidine into DNAse-removable material within intact mitochondria. In one such study the analysis of the distribution of silver grains over *Tetrahymena* mitochondria (23) suggests that all mitochondria in the cell contain DNA, which can be labeled throughout the cell cycle. Many workers have isolated and characterized mitochondrial DNA from a wide range of species and cells (see Table I and refs. 19–22). The isolation of mitochondrial DNA is facilitated by two of its properties: *1*) in many species the bouyant density and thus the base composition of mitochondrial DNA is distinct from that of corresponding nuclear DNA; these differences have been summarized (19, 20, 22); *2*) in some species the mitochondrial DNA has

Table I

Features of Mitochondrial DNA

SPECIES	CONTOUR LENGTH[a] (μ)	ESTIMATED MW[b]	ESTIMATED GENOME SIZE[c]	REFERENCES
Higher eukaryotes (man,[d] mammals, birds, amphibians, worms, insects, echinoderm)	4.5–5.9[e] (circular)	9.0–11.8×10^6	10.0–11.0×10^6	19, 20, 22, 24
Higher plants (red bean)	19.5 (linear)	40×10^6	?	25
Euglena gracilis	40.0 (circular)	80×10^6	90×10^6	187, 188
Neurospora	Up to 25 (linear) 20 (open circles)	50×10^6 40×10^6	66×10^6	26, 189, 190[f]
Saccharomyces carlsbergensis; S. cerevisiae	Up to 26 (circular?)	50–52×10^6	$\pm 50 \times 10^6$	27[g]
Tetrahymena pyriformis	17.6	35×10^6	?	28

[a] Obtained by measurement of Kleinschmidt preparation of mitochondrial DNA by electron microscopy.
[b] Obtained by extrapolation from contour length measurements.
[c] Determined from the rate constant of renaturation.
[d] A recent estimate of the estimated genome size for human mitochondrial DNA yields a figure of 1.8×10^6, barely enough to code for ribosomal RNA (184).
[e] This range in sizes represents real differences as suggested by mixing of mitochondrial DNA of different species origins (185, 186).
[f] For the isolation of *Neurospora* mitochondrial DNA circles, a slime mutant free of cell wall was employed (190).
[g] The genome size of yeast mitochondrial DNA must remain suspect in view of the recent finding (191) that it contains significant segments with an A-T content higher than 90%.

higher, and in others lower, guanine plus cytosine content than nuclear DNA (19). However, in most mammals the bouyant densities of mitochondrial and nuclear DNA are barely distinguishable. In these species, as well as in all other vertebrates and some invertebrates (19), mitochondrial DNA is usually and perhaps invariably present as closed circular molecules having a contour length of about 5 μ. In all higher animals mitochondrial DNA has a double-stranded, closed, circular configuration. If the mitochondrial DNA exists as double-stranded, covalently linked, closed circular molecules with no breaks, the molecules are seen as twisted supercoils. Open circles are the result of one or a few breaks in one of the two DNA strands. These classes of circular molecules can be readily separated from one another and from linear molecules by density gradient centrifugation, especially on gradients containing ethidium bromide (19). Their circularity and their relatively small molecular size result in a very rapid renaturation of denatured mitochondrial DNA under conditions where nuclear DNA barely renatures at all. The separation of native mitochondrial DNA and denatured nuclear DNA can readily be accomplished after such a renaturation cycle. The circularity of mitochondrial DNA is a convenient property which facilitates the accurate measurement of its molecular size by electron microscopy. Each micron contour length corresponds to a polynucleotide of 3000 base pairs, so that 5-μ mitochondrial DNA molecules contain 15,000 base pairs or a coding ability for 5000 amino acids or 30 peptides of molecular weight 20,000. It is likely that all the circular mitochondrial DNA molecules isolated from a given cell species are identical or closely similar. This is based upon the narrow distribution of bouyant densities and of contour lengths of mitochondrial DNA of a given species. The most important evidence for identity of mitochondrial DNA molecules from a given species rests upon the rate of renaturation. The rate of renaturation of denatured DNA under standard conditions is proportional to the genetic content or number of base sequences it contains. The number of base pairs in chicken liver mitochondrial DNA (24) as determined from rate of renaturation corresponds closely to the number determined from electron microscopic size measurements and suggests that only a single predominant molecular species is present.

As summarized in Table I, mitochondrial DNAs from *Neurospora crassa* and from yeast are different from DNA of vertebrate cells. They are much larger—about five times the molecular weight of mitochondrial DNA in higher animals and birds. In the case of yeast, both linear and circular molecules of mitochondrial DNA have been observed. Recently (27) a few 26-μ circles have been observed, and this length corresponds with the molecular weight calculated from the kinetics of renaturation of yeast mitochondrial DNA. It is not clear whether the linear or circular forms in yeast are preparative artifacts and what their functional relation may be.

The most striking evidence implicating mitochondrial DNA as the cytoplasmic genetic determinant responsible for cytoplasmic genetic phenomena referred to in the preceding section, comes from the study of the mitochondrial DNA of the cytoplasmic petite. Mounolou *et al.* (29) first showed that several cytoplasmic petites have mitochondrial DNA of altered bouyant density and therefore profoundly altered base composition. This has been confirmed and extended (30). No alteration in mitochondrial DNA of segregational petites was observed. In those cytoplasmic petites in which this alteration has been observed, the density of the mutant mitochondrial DNA is usually less than that of the wild type. This indicates that mutant molecules have a higher content of adenine and thymine, in some cases as high as 96% (31). Even when almost no change in mitochondrial DNA bouyant density is demonstrable, DNA-DNA hybridization reveals a substantial change in DNA base sequence, probably as a result of scrambling of normal sequences (32). The molecular mechanism of petite induction is unclear. It seems likely that in all cytoplasmic petites the mitochondrial DNA is either partly or completely nonfunctional. Reference has already been made to the fact that the genetic determinants for antibiotic resistance are sometimes retained in cytoplasmic petites.

Mitochondrial DNA is synthesized many times more rapidly than nuclear DNA. This synthesis apparently occurs within the mitochondria and, especially in mature mammalian cells, is independent of nuclear DNA synthesis (19). In synchronized yeast, mitochondrial DNA is apparently synthesized separately and earlier than nuclear DNA in some strains (33), though not in others (13). The mode of replication appears to involve a semiconservative mechanism (34). In rat liver, Kirschner *et al.* (35) have observed forked double-stranded circular mitochondrial DNA molecules which are suggestive of replicating molecules.

Recent elegant electron microscopic studies have provided the basis of a model for the replication of mitochondrial DNA. A substantial proportion of supercoiled mitochondrial DNA molecules demonstrate a displacement loop (D-loop) representative of about 3% of the molecular length. It contains a short single-stranded DNA chain complementary to the light strand and therefore thought to be the initial replicative piece of motochondrial DNA heavy strand (192), whose replication appears to be initiated at a unique site. Further studies (193) suggest that after limited nicking of the molecule the displacement loop is expanded until the whole heavy strand is replicated. Only when it is approximately 60% replicated, does replication of the light strand commence. In other words, the two strands are asynchronously replicated. Biochemical and electron microscopic investigations of chicken liver mitochondria have suggested a mechanism of replication of their DNA (194, 195) consistent with the model proposed above.

Chloroplast DNA

It is thought that the non-Mendelian genes which control photosynthetic activity reside on chloroplast DNA. However, the categorical association of these genes with chloroplast DNA is as yet unproved (36). The mechanism of maternal inheritance in plants is unclear. Simple exclusion of the paternal cytoplasm is not the mechanism (18). Sager and Ramanis claim that the paternal chloroplast DNA is destroyed early in zygote history (18). However, Chiang (37) has presented compelling evidence that the paternal chloroplast DNA in *Chlamydomonas reinhardi* is replicated at the same rate as the maternal chloroplast DNA, at least in the initial phases of zygote multiplication. Chiang and Sueoka (38) had earlier shown in synchronously dividing *Chlamydomonas* that the chloroplast DNA was semiconservatively replicated and that its replication was at a different time in the cell cycle than that of the nuclear DNA. Chloroplast DNA is readily separable from nuclear DNA in both algae and higher plants by virtue of many of the same properties described above for mitochondrial DNA. Chloroplast DNA frequently differs in bouyant density from the nuclear DNA of the plant cell. The bouyant densities of a wide range of chloroplast DNAs relative to those of their respective nuclear DNAs are listed by Swift and Wolstenholme (39; their Table II). From this it can be seen that in the case of the protists and algae, chloroplast DNA is usually of lower density than nuclear DNA, while the reverse is the case for higher plants. Chloroplast DNA can be demonstrated cytologically. The exact size of chloroplast DNA is presently uncertain, but on the basis of genetic size as determined by the kinetics of renaturation in a number of species of chloroplast DNA, the size is approximately $1–2 \times 10^8$ daltons. The correspondence between the genetic size and the physical size of any single species of chloroplast DNA is presently uncertain. A figure of 1.14×10^8 daltons has been obtained for the genetic size (40) of tobacco chloroplast DNA, which is very close to the 1.2×10^8 daltons obtained for lettuce chloroplast DNA (41). The corresponding figure for *Euglena gracilis* chloroplast DNA is 1.8×10^8 daltons (42) and for *Chlamydomonas* chloroplast DNA 1.94×10^8 daltons (43). *Chlamydomonas reinhardi* has a single chloroplast which in the haploid cell contains 5.16×10^9 daltons, or 26 repetitions of the unique nucleotide sequence (43). Similar multiplicities have been suggested for other chloroplasts (40).

Transcription and Translation Within Mitochondria and Chloroplasts

For the expression of the genetic information contained within mitochondrial and chloroplast DNA, their presence must be coupled to transcriptive and translational systems. Indeed, in both mitochondria and chloroplasts all the components necessary for DNA transcription to RNA and its translation into proteins have been found to be uniquely associated with these organelles. This information and the documentation thereof are summarized in Tables II and III. Protein synthetic capacity of isolated mitochondria and chloroplasts appears to be completely independent of translational mechanisms present within the cell sap or cytoplasm. Protein synthesis of mitochondria (15, 21, 22, 64) and of chloroplasts (61) has been recently reviewed. That protein synthesis in these organelles is dependent upon the continual provision of RNA, presumably transcribed from organelle DNA, is indicated by the fact that, under appropriate conditions, synthesis is substantially inhibited by the presence of actinomycin D. Furthermore, the information presented in the tables suggests that protein synthesis in these organelles takes place on ribosomes smaller than those encountered in the extramitochondrial or extraplastid cytoplasm, and that these ribosomes contain smaller and unique molecules of ribosomal RNA. The protein synthetic mechanisms in these two organelles can be readily distinguished from those in the extramitochondrial or extraplastid cytoplasm by their sensitivity to a different range of antibiotics (15, 21, 64). Thus protein synthesis in mitochondria is inhibited by acriflavin, puromycin, chloramphenicol, oxytetracycline, and in the case of yeast mitochondria, also by erythromycin,

Table II

Unique Features of Transcriptive and Translational Mechanisms in Mitochondria

COMPONENT	SPECIES	MOLECULAR SIZE(S)[a] AND OTHER PROPERTIES	REFERENCES
Ribosomes	Animals (man, rat, hamster, rabbit, beef, pig, locust, *Xenopus*)	45–60 (78–83) *33–45 (60)* 25–35 (40)	44 196 197
	Neurospora, yeast	73–74 (77–80) 50–58 (60) *35–40 (37–38)* Inhibited by chloramphenicol, erythromycin, etc.	
Ribosomal RNA	Animals (man, rat, mouse, hamster, *Xenopus*, locust)	12–13 (18) 16–19 (28)	44
	Primitive eukaryocytes (*Neurospora*, yeast, *Aspergillus*, *Tetrahymena*)	14–16 (17) 21–23 (26)	
Ribosomal proteins	*Neurospora* Yeast Locust	Distinct from cytoplasmic ribosomal proteins by disc-gel electrophoresis	45 198 199
tRNA	*Neurospora* Yeast *Tetrahymena* Rat liver	Differ from cell sap tRNA chromatographically and in response to acylating enzymes	19
n-Formyl methionyl tRNA	Yeast Rat liver HeLa cells	Absent from cell sap; in HeLa cell mitochondria, initiator function demonstrated	46, 47 48
Aminoacyl synthetases	*Neurospora* Rat liver	Differ chromatographically from cell sap enzymes	19
mRNA	Yeast Rat liver HeLa cells		49, 50
RNA polymerase	*Neurospora* Rat liver	Sensitive to actinomycin D only when mitochondria disrupted	19, 21, 22
DNA polymerase	Yeast Rat liver Chick liver	Differ chromatographically in size and response to templates and inhibitors from nuclear enzyme	19
Elongation factors	Yeast	React with *E. coli*; mitochondrial ribosomes but not cytoplasmic ribosomes	200, 201

[a] Figures in parenthesis refer to parameters of cytoplasmic ribosomes; italicized figures refer to sizes of large and small subunits of mitochondrial and cytoplasmic ribosomes, respectively.

Table III

Unique Features of Transcriptive and Translational Mechanisms in Chloroplasts

COMPONENT	SPECIES	MOLECULAR SIZE(S)[a] AND OTHER PROPERTIES	REFERENCES
Ribosomes	*Chlamydomonas*	68 (80); *33 (43)*; *28 (30)*	51
	Euglena	70 (88); *50 (67)*; *30 (46)*	52–54
	Higher plants	67–68 (80)	55
Ribosomal RNA	*Chlamydomonas*	22.4 (25.1)	51
		16.7 (17.6)	
	Euglena	23.5 (24–26)	52–54
		16.5 (20–22)	
	Higher plants	22–23 (25)	55
		16–17 (16–18)	
Ribosomal proteins	Higher plants	Chloroplast, cytoplasmic and ribosomal proteins of mung bean are electrophoretically distinct from one another	202
tRNA	*Euglena*	Isoleucyl and phenylalanyl tRNA separable from extraplastid tRNA chromatographically and by response to chloroplast acylating enzymes	57
	Tobacco		40
n-Formyl methionyl tRNA	*Euglena*	Shown to be initiating tRNA	58
	Wheat	Methionyl tRNA of chloroplast is formylatable by plastid enzyme, while initiator methionyl tRNA of extraplastid cytoplasm is not	59
	Bean		60
Amino acyl synthetase	*Euglena*	Chromatographic and acylation properties for isoleucyl and phenylalanyl synthetases differ from extraplastid enzymes	57
	Pea		56
DNA polymerase	Tobacco		40
RNA polymerase	*Chlamydomonas, Euglena,* maize, tobacco, spinach, bean	RNA synthesis by isolated chloroplasts in *Chlamydomonas,* differentially inhibited by rifampicin	40, 61–63
mRNA	Tobacco	Hybridizes with chloroplast DNA not competed out by rRNA or tRNA of chloroplast	40

[a] Figures in parenthesis refer to parameters of cytoplasmic ribosomes; italicized figures refer to sizes of large and small subunits of chloroplast and cytoplasmic ribosomes, respectively.

lincomycin, oleandomycin, spiramycin, carbomycin, and mikamycin (15), although many of these latter also inhibit rat liver mitochondrial protein synthesis provided the mitochondria are swollen (65). Mitochondrial protein synthesis is insensitive to cycloheximide, which is a powerful inhibitor of cytoplasmic protein synthesis. Chloroplast protein synthesis is inhibited by puromycin, chloramphenicol, streptomycin, and spectinomycin (61, 63, 66).

Gene Products of Organelle DNA

RIBOSOMAL RNA

It has been clearly shown in a wide range of species (Table IV) that the ribosomal RNA of both mitochondria and chloroplasts hybridizes and is therefore genetically determined by the organellar DNA. In *Neurospora* and yeast, which apparently have larger mitochondrial genome molecular weights (Table I), saturation levels of hybridization suggest three or four ribosomal RNA cistrons per unit genetic molecular weight (but see 203, 204), while in the higher forms, *Xenopus laevis* and HeLa cells, there is apparently only one copy of each ribosomal RNA cistron. Chloroplast DNA, which has a larger genome molecular weight than yeast or *Neurospora*, apparently contains two or three copies of each ribosomal cistron (Table IV). This difference may well reflect a much greater complexity of the chloroplast relative to the mitochondrion. Fauman (32) has shown that even though the mitochondrial DNA of a particular petite of *Saccharomyces cerevisiae* was altered by 50% of its base sequences, its mitochondria nevertheless synthesized ribosomal RNA molecules which were complementary to this altered DNA and were themselves altered in 50% of their sequences. It is likely, however, that different petites vary greatly in the extent to which they retain mitochondrial ribosomal-type RNA. In one strain the absence of this type of RNA has been claimed (67). Indeed, petite strains have been described in which no mitochondrial DNA is demonstrable (80).

TRANSFER RNA

Transfer RNA (tRNA) sequences have also been shown to be complementary to mitochondrial or chloroplast DNA by the bulk hybridization of total organellar tRNA, as shown for HeLa cell and frog egg mitochondria (68, 70) or for tobacco chloroplast (40). For tobacco chloroplast DNA the saturation levels of hybridization are sufficient to account for the presence of at least one copy of each of 20–30 tRNA cistrons within each genome unit of the chloroplast DNA. However, two hybridization studies suggest, on the basis of saturation levels of hybridization, that only 12–15 tRNA cistrons are represented in mammalian mitochondrial DNA (68, 70). One cannot be certain of the absence of degradation products of heterogeneous RNA in the labeled samples of RNA used for these studies. Specific tRNA species can be followed into molecular hybrids with organellar DNA by loading the indi-

Table IV

Gene Products of Mitochondrial and Chloroplast DNA

COMPONENT	SPECIES	COMMENTS	REFERENCES
		MITOCHONDRIAL DNA	
Ribosomal RNA	Yeast	Both subunits 1 cistron[a]	67, 203
	Neurospora	Both subunits 1 cistron[a]	26, 204
	Xenopus laevis	Both subunits 1 cistron[a]	68
	HeLa cells	Both subunits 1 cistron[a]	69, 82
tRNA	*Xenopus laevis*	12 cistrons	68
	HeLa cells	12–15 cistrons (3 on L strand)	70, 82
	Rat liver	Leucine	71
		Phenylalanine	72
		Serine (on L strand)	
		Tyrosine (on L strand)	
	Yeast	f-Methionine	73
		Valine[b]	74, 75
		Leucine[b]	75
		Phenylalanine[b]	205
		Glycine[b]	205
		Alanine[b]	205
		Tyrosine[b]	205
		Isoleucine[b]	205
		CHLOROPLAST DNA	
Ribosomal RNA[c]	*Chlamydomonas reinhardi*	Both subunits 2–3 cistrons	77
	Euglena gracilis	Both subunits 1 cistron	78, 79, 187
	Tobacco	Both subunits 2–3 cistrons[d]	40
tRNA	Tobacco	20–30 cistrons	40

[a] Earlier estimates using stable mitochondrial RNA, separated as such into subunit RNA, yielded higher saturation levels of hybridization (equivalent to 2 or 3 cistrons; 26, 67), while more recent estimations using RNA extracted from purified ribosomal particles have yielded the lower saturation levels presented in this table (203, 204).

[b] All these aminoacyl tRNAas have been found represented on the mitochondrial DNA of selected cytoplasmic petites (76, 205).

[c] In all cases of chloroplast ribosomal RNA, numbers of cistrons are calculated per unit of unique nucleotide sequence. Since each chloroplast in each of these species contains 20–30 copies of the unique nucleotide sequence, the total number of ribosomal RNA cistrons per chloroplast is 20–30 times the number cited in the table. Similar arguments can be applied to the multiplicity of ribosomal RNA cistrons per mitochondrion and of transfer RNA cistrons per mitochondrion and chloroplast. In the case of mitochondria, the numbers of DNA copies per organelle have not been unequivocally defined.

[d] Corrected for heavy strand only as template.

vidual tRNA species with highly labeled amino acids and conducting the hybridization at low temperature and low pH in the presence of formamide, in order to minimize the deacylation of the amino acyl tRNA (81). As can be seen from Table IV, a number of different amino acyl tRNAs have been shown to hybridize with mitochondrial DNA. Most transcription of mitochondrial RNA employs the heavy strand of mitochondrial DNA as template (49), but it has been reported that in contradistinction to several other tRNAs, the seryl and tyrosyl tRNAs hybridize with the light strand of rat liver mitochondrial DNA (72). It will be of interest to learn which if any tRNA species is not represented in the mitochondrial DNA of mammals. As mentioned earlier, genetic information, in the form of genetic determinants of antibiotic resistance, is sometimes retained in the petite cytoplasmic genome. Some petite mitochondrial DNAs also retain the information for functional tRNA molecules (76). Indeed, in some cases, specific tRNA cistrons may be reduplicated in petite mitochondrial DNA (73). This offers a useful opportunity for combined genetic and molecular mapping of mitochondrial DNA.

ORIGIN OF MITOCHONDRIAL PROTEINS

It is quite clear from a consideration of the total genetic content of mitochondrial DNA in particular and perhaps also of chloroplast DNA that there is insufficient genetic information encoded in these molecules to account for all the proteins uniquely associated with these organelles. Indeed, it is unlikely that the primary structure of more than a relatively few proteins is determined in the genetic information content of animal mitochondrial DNA. The nature of these proteins is presently completely unknown. It is clear, therefore, that the majority of the proteins associated with animal mitochondria must have their primary structure determined by nuclear DNA. This has been proved for cytochrome c (83). From this and the existence of chromosomally determined mutants of organellar function, we can conclude that the synthesis and assembly of these organelles depend upon the cooperative functioning of both the nuclear and organellar DNA.

The most compelling evidence for the determination of the structure of a protein by the organellar genome would derive from genetic experiments. In no case is there yet evidence for point mutations in the organellar DNA which result in structural alterations of any single well-characterized protein contained uniquely within either mitochondria or chloroplasts. The best evidence that such proteins exist derives from the knowledge that the antibiotic sensitivity or resistance of organellar protein synthesis is frequently determined by non-Mendelian genes (12, 15). Reasoning by analogy with what is currently known about the determining factors of antibiotic sensitivity and resistance in bacteria, it is likely that this characteristic of organellar protein synthesis is determined at the level of one or another ribosomal protein. It is believed, therefore, that at least a few of the ribosomal proteins are structurally determined by the organellar DNA genetic information, but, no direct and concrete evidence of this is available. Indeed, the situation appears much more complex. Inhibition of mitochondrial protein synthesis by chloramphenicol during yeast growth does not interfere with the capacity of mitochondria isolated from these cells to manufacture protein in vitro (15). Thus mitochondrial protein synthesis is apparently not necessary for the manufacture of the mitochondrial protein-synthesizing mechanism. Several antibiotic-resistant yeast strains, in which resistance is determined cytoplasmically, exhibit anomalous behavior when resistance is assayed in vivo and in vitro (15). Also, labeling of mitochondrial ribosomal proteins in the presence of chloramphenicol suggests that, in *Neurospora* at least, proteins are not made by mitochondrial ribosomes (45). However, in this species the poky mutant, a cytoplasmic mutation, at certain stages of its growth is associated with a differentially low rate of synthesis of the small subunit of the mitochondrial ribosome (84). A reduced synthesis of ribosomes, in this case in chloroplasts, is associated with a nuclear mutation of *Chlamydomonas reinhardi*, ac-20 (36, 63). This dilemma is yet to be resolved. It is conceivable that the crucial one or two ribosomal proteins are made by the mitochondria and that their synthesis has escaped detection in the experiments so far reported, or that they are made in the extramitochondrial cytoplasm using as messenger RNA (mRNA) a transcript of mitochondrial DNA. Perhaps chloramphenicol inhibition of mitochondrial protein synthesis is selectively incomplete.

Three other approaches have been employed to identify the proteins whose primary structure is determined at the level of organellar DNA. *1*) Examinations have been made of the proteins eliminated in the cytoplasmic mutants affecting organellar function. Thus, in the case of yeast mitochondrial petites, cytochromes b, c_1, and aa_3 are almost uniformly absent. From this it has been concluded that the structure of these cytochromes may be encoded in mitochondrial DNA. However, it is not clear whether mitochondrial DNA contributes any structural information toward these cytochromes or whether their absence is the result of the dysfunction of a single regulator protein whose presence is essential for the integration of these cytochromes into the mitochondria. The latter possibility is supported by the findings referred to below, that heme a and cytochrome oxidase are synthesized in adapting yeast through the mediation of both mitochondrial and cytoplasmic protein synthesis (85). *2*) A petite-like phenotype has been observed in wild-type cells grown in the presence of chloramphenicol. Since cytoplasmic petites also lack the oligomycin sensitivity of mitochondrial ATPase, at least one of the proteins responsible for this functional attribute of the ATPase may be encoded by mitochondrial DNA (see discussion following). *3*) The converse approach, however, is much more useful. A number of characteristic mitochondrial enzymes have been shown to be retained in cytoplasmic petites (Table V), whose mitochondrial DNA is almost certainly so severely disordered that no single normal protein component could be manufactured off its transcripts. Since such petites are clearly able to continue the replication of their own mitochondrial DNA, DNA polymerase must be determined by nuclear DNA. As has been mentioned, some petites retain the capacity to make ribosomal RNA, albeit disordered, and functional tRNA molecules. They must therefore contain an RNA polymerase which is also likely to be determined at the level of nuclear genome. A more careful

Table V

Mitochondrial Components Made on Cytoplasmic Ribosomes

COMPONENT	COMMENT	REFERENCES
COMPONENTS PRESENT IN CYTOPLASMIC PETITE		
DNA polymerase	Most petites retain mitochondrial DNA	19, 29–31
RNA polymerase	Many petites retain ribosomal-type RNA and functional tRNA	32, 73, 75
Respiratory enzymes		
Succinic dehydrogenase		20, 86
Antimycin-sensitive NADH– cytochrome c reductase		
D-Lactate cytochrome c reductase		
L-Lactate cytochrome c reductase		
Aconitate hydratase		
Fumarate hydratase		
F_1-ATPase	Not sensitive to oligomycin inhibition	87
Elongation factors	Present even in absence of mitochondrial DNA	200, 206
COMPONENTS WHOSE SYNTHESIS IS NOT INHIBITED BY CHLORAMPHENICOL		
Most ribosomal proteins	Yeast mitochondrial ribosomes active after growth in chloramphenicol	15, 45
Cytochrome c		88
Outer mitochondrial membrane proteins		88
F_1-ATPase	Present in increased amounts in post- mitochondrial supernatant in glucose derepressing yeast in presence of chloramphenicol	89
Fc–Oligomycin-sensitivity-conferring protein		90
Fumarase, malate dehydrogenase, isocitric dehydrogenase, glutamate dehydrogenase, succinate dehydrogenase, succinate cytochrome c reductase, cytochromes c and c_1	Insensitive to chloramphenicol during aerobic adaptation and glucose derepression	91, 207
STRUCTURAL GENE PRESENT IN NUCLEAR DNA		
Cytochrome c		83

biochemical study of the properties of petites whose mitochondrial genome has been characterized would obviously be valuable.

Attempts have been made to study the nature of the proteins made by isolated mitochondria. It has been assumed that, since protein synthesis by isolated mitochondria can be inhibited by actinomycin D, those proteins which are made by the isolated organelles must derive genetically from the mitochondrial DNA. However, actinomycin D, even under conditions where inhibition is substantial, does not completely inhibit mitochondrial protein synthesis. Thus the possibility that the isolated organelle contains RNA transcripts of nuclear DNA or stable transcripts of mitochondrial DNA cannot be excluded. That mRNA may gain entry into mitochondria has been demonstrated in the organelles of frog liver, in which it has been shown that protein synthesis in intact mitochondria may be directed by the addition of polyuridylic acid to the incubation medium (68). Attempts to provide definitive identification of the products of isolated mitochondrial protein synthesis have so far had limited success. All that can be said at present is that radioactive amino acids are incorporated by isolated mitochondria into insoluble lipoprotein components of the inner mitochondrial membrane (64). Proteins of the outer mitochondrial membrane are not labeled in such incubations (92), nor are the specific mitochondrial proteins cytochrome c or malate dehydrogenase (64). The differential antibiotic sensitivity of protein synthesis within the mitochondria and within the extramitochondrial cytoplasm has proved useful in these investigations. Those proteins whose synthesis is inhibited by cycloheximide in vivo are presumed to be made by the cytoplasmic ribosomes, while in vivo inhibition of protein formation by the group of antibiotics specifically inhibiting mitochondrial ribosomal protein synthesis is presumed to account for those proteins produced within the mitochondria. While this approach has been quite fruitful, the results have not been as definitive as might have been anticipated. For example, many protein complexes, such as cytochrome oxidase, apparently require the cooperative functioning of both systems of protein synthesis (85). Furthermore, the designation of the site of synthesis of a protein as either the cytoplasmic ribosomes or the mitochondrial ribosomes does not necessarily designate the site of the genetic determination of the structure of such a protein. It is conceivable that mRNA molecules deriving from nuclear DNA might be translated on mitochondrial ribosomes, while mitochondrial mRNA might be translated on cytoplasmic ribosomes. With these limitations in mind, it is now possible to specify some of the proteins made by the cytoplasmic ribosomes (listed in Table V). However, care must be exercised in translating the findings for the mitochondria of primitive eukaryocytes such as yeast to mammalian mitochondria, in view of the substantial difference in the genome sizes in the two classes of mitochondria.

ORIGIN OF CHLOROPLAST PROTEINS

In the case of chloroplasts, the same difficulties are encountered in defining those proteins whose structure is determined by the chloroplast DNA as have been outlined above for mitochondria.

The use of the antibiotic, rifampicin, which preferentially inhibits the chloroplast RNA polymerase in *Chlamydomonas reinhardi* (62), has indicated that the plastid DNA polymerase is almost certainly made outside the chloroplast and from genetic information present in nuclear DNA. The antibiotic, spectinomycin, has provided further information on the nature of the proteins likely to be made within the chloroplast. This antibiotic preferentially inhibits chloroplast protein synthesis (63). After several generations of growth in the presence of spectinomycin, the plastids of *C. reinhardi* apparently retain normal levels of ribosomes, as indicated by ultrastructural studies. Certainly these algae continue to make the ribosomal RNA of plastid ribosomes under such a growth condition. This has been taken to suggest that most of the chloroplast ribosomal proteins as well as the plastid DNA–dependent RNA polymerase are not made by the chloroplast ribosomes themselves. Furthermore, in synchronous cultures of this alga, the doubling of the Hill reactivity (photosystem II) is prevented by spectinomycin, as is the increase in activity of ribulose-1,5-diphosphate carboxylase, while the doubling of chlorophyll and of the capacity to photoreduce NADP are not affected by this antibiotic. These results have been supported by studies (93) in yellow mutants of this alga inhibited during greening by chloramphenicol. These conclusions have been further supported by studies of a mutant of *Chlamydomonas*, ac-20, which is apparently unable to synthesize normal levels of chloroplast ribosomes (63) when grown heterotrophically. Many chloroplast components in these mutants are apparently unaffected by this marked reduction in chloroplast ribosome, suggesting that these components (listed in Table II of ref. 63) are manufactured independently of chloroplast protein synthesis. However, a few chloroplast components, such as ribulose diphosphate carboxylase and cytochromes 559 and q, are markedly reduced in the mutants. The absence is associated with a disorganization of chloroplast membranes and of pyrenoid formation. Similar results have been obtained with *Euglena gracilis* grown in the dark and illuminated in the presence of streptomycin, which selectively interferes with chloroplast development (66). The light-dependent increases in chlorophyll, carotenoid, photosynthetic CO_2 fixation, cytochrome 552, and ribulose diphosphate carboxylase are all but eliminated. It is, however, possible that all these defects exist because of the absence of one or a few proteins important for their organization and integration into the chloroplast membranes.

Biosynthesis of Nonprotein Components of Mitochondria and Chloroplasts

Apart from proteins, the other major constituents of mitochondrial and chloroplast membranes are lipids, heme compounds, and possibly glycolipids and glycoproteins.

MITOCHONDRIA

There is some dispute about the biosynthetic site of origin of the predominant mitochondrial phospholipids, lecithin and phosphatidylethanolamine, at least in rat liver mitochondria. Only in the liver has the origin of mitochondrial phospholipids been explored in detail. In this organ, the last steps in the synthesis of these two major phospholipids probably occur predominantly, if not exclusively, in the membrane of the endoplasmic reticulum, and the completed molecules are subsequently transferred through a non-energy-dependent mechanism to the membranes of mitochondria (94). The origin of the diglyceride precursors, and especially of the glycerol backbone of these phospholipids, is undetermined. Many investigators have observed the synthesis of phosphatidic acid in the mitochondrial membranes, apparently concentrated in the outer mitochondrial membrane (evidence summarized in ref. 94). The intramitochondrial localization of the synthesis of acyl dihydroxyacetone phosphate, an alternative precursor of phosphatidic acid (95), has not been determined. Although the synthesis of cytidine diphosphate (CDP)(diglyceride from phosphatidic acid and cytidine triphosphate (CTP) was first ascribed to the endoplasmic reticulum membranes (96), many workers have since found this activity concentrated in the mitochondria of several tissues (Table VI). The utilization of CDP diglyceride for the synthesis of phosphatidylglycerophosphate and phosphatidylglycerol appears largely confined to the mitochondrial membranes of the eukaryotic cell. Mitochondrial membranes, like bacterial membranes, are relatively rich in acidic phospholipids, of which cardiolipin is the most typically mitochondrial (94, 111). It has recently been shown that guinea pig liver mitochondria are able to manufacture their own cardiolipin (110). Present knowledge of phospholipid biosynthetic capacity of mitochondria is summarized in Table VI. From this it seems that the origin of mitochondrial phospholipids is different from one eukaryotic cell to another. Though the formation of lecithin from diglyceride and CDP choline appears to take place exclusively in the membranes of the endoplasmic reticulum in rat liver, this enzymatic activity is present in the mitochondrial membranes of *Tetrahymena pyriformis* and in the outer mitochondrial membranes of yeast mitochondria and probably beef heart mitochondria (97). Similar differences in enzyme localization probably also apply to CDP diglyceride synthetase (Table VI). The degree of development of the endoplasmic reticulum may well be an important factor in the determination of the intracellular localization of these enzymes. Though many of them are localized in the mitochondrial membranes, there is no reason to suspect that the lipid biosynthetic enzymes are made within the mitochondria or determined by the mitochondrial genome.

Mitochondria also participate in heme synthesis. The enzymes δ-amino levulinic acid synthetase and ferrochelatase are both largely associated with this organelle. The first enzyme appears to be present in the matrix and is likely manufactured on cytoplasmic ribosomes. The other enzyme is a component of the inner mitochondrial membrane (112). Coproporphyrin oxidase is also located in mitochondria. Recently Bosmann and his collaborators

Table VI

Phospholipid Biosynthesis in Mitochondria

ENZYME AND REACTION CATALYZED	SOURCE OF MITOCHONDRIA	COMMENTS	REFERENCES
1. Glycerophosphate acyl transferases Glycerol-3-phosphate + 2 Acyl CoA → phosphatidic acid Lysophosphatidic acid + Acyl CoA → phosphatidic acid	Rat liver	Enzyme located predominantly in outer membrane, also in endoplasmic reticulum	94, 97–100
2. CDP choline diglyceride phosphorylcholine transferase CDP choline + diglyceride → lecithin + CMP	*Tetrahymena,* *S. cerevisiae*	Claimed (97) to be present in outer mitochondrial membrane of rat liver but this unconfirmed (94, 99); in yeast in outer mitochondrial membrane	101, 102
3. Phosphatidyl inositol phosphokinase Phosphatidylinositol + ATP → phosphatidylinositol phosphate + ADP	Rat liver		103
4. CDP diglyceride synthetase Phosphatidic acid + CTP → CDP diglyceride + inorganic pyrophosphate	Beef heart Rat liver Chicken brain *S. cerevisiae*	In yeast, enzyme is present in both inner and outer mitochondrial membrane and probably also in rat liver, where it is also present in endoplasmic reticulum	104–106, 208
5. Phosphatidylglycerol "synthetase" CDP diglyceride + glycerol-3-phosphate → phosphatidylglycerophosphate + CMP Phosphatidylglycerophosphate → phosphatidylglycerol + inorganic P	Chicken liver Sheep brain Rat heart Rat liver	In rat liver mitochondria in inner membrane	107–109, 208
6. Cardiolipin "synthetase" Reactions 1 + 4 + 5 and phosphatidylglycerol + CDP diglyceride → cardiolipin + CMP	Guinea pig liver Rat liver	Exogenous CDP diglyceride not converted to cardiolipin in guinea pig, but it is in rat liver, where last enzyme is on inner membrane	110, 208, 209

(113, 114) have provided evidence that isolated mitochondria are capable of synthesizing both glycoproteins and glycolipids.

CHLOROPLASTS

The lipid composition of chloroplast membranes is in many respects unique. With the exception of the chromatophores, the photosynthetic organelle of photosynthetic bacteria, chloroplast membranes are distinguished by their high concentration of sulfoquinovosyl diglyceride, mono- and digalactosyl diglyceride, and phosphatidylglycerol (94). Furthermore, these chloroplast lipids are particularly rich in polyunsaturated fatty acids, especially linolenic acid, which is characteristic of plant tissues, and plastid phosphatidylglycerol, which contains the unique *trans*-Δ^3-hexadecenoic acid. Relatively few studies have been performed on light-induced stimulation of unsaturated fatty acid synthesis (94), which is apparently a function of the intact chloroplast (115). The biosynthesis of galactosyl diglycerides from diglyceride and uridine diphosphate (UDP) galactose has been noted in isolated spinach chloroplast and green pea leaf particles (116). This biosynthetic activity, however, does not appear to be confined to chloroplasts, since several non-chlorophyll-containing preparations have similar enzyme activity. Much further work is necessary for the complete elucidation of the biosynthetic origin of the many interesting lipids which are major components of chloroplast membranes. Kirk and Tilney-Bassett (61) have summarized the evidence suggesting, though by no means categorically proving, that the isolated chloroplasts are probably capable of synthesizing much of their own glycerides, phospholipid, glycolipids, and fatty acids. In no case, however, has it been shown that the chloroplast is the exclusive biosynthetic site of these chloroplast components. These authors have also summarized the evidence suggesting that chloroplasts may be capable of synthesizing the phytyl side chains of chlorophyll and plastoquinone, that they are probably responsible for the complete pathway of carotenoid synthesis, and that they may play a major part in the conversion of δ-amino levulinic acid to protoporphyrin and of the latter to protochlorophyllide, and in the photoconversion of protochlorophyllide to chlorophyllide A. It thus seems likely that the chloroplasts are capable of completing the synthesis of chlorophyll from δ-amino levulinic acid.

Reconstitution of Mitochondrial Membranes

Many studies, mainly from Racker's laboratory, on the resolution of mitochondrial membranes into their respective components have been reported. Though these studies do not provide direct information about the mode of assembly or biogenesis of these membranes in vivo, they nevertheless furnish valuable data on the interacting mechanisms that may be operative for such membrane assembly. They also indicate the specific structural requirements for the reconstitution and interaction of the appropriate membrane constituents. These studies have recently been reviewed (94, 117, 118) and will not be described in detail here. Suffice it to say that the reconstitution of functional membranes apparently requires the interaction of several proteins and phospholipids. Phospholipids are required for the reconstitution of respiration in lipid-depleted mitochondria. The mitochondrial lipid, cardiolipin, is particularly effective in this respect and for the activation of phospholipid-deficient cytochrome oxidase. As much as 70% of the lipid phosphorus of submitochondrial particles can be removed by phospholipase C treatment without substantial impairment in the capacity of the particles to carry out oxidative phosphorylation. The remaining phospholipid must be very important for oxidative phosphorylation, since phospholipids are obligatorily required for the reconstitution of succinoxidase activity and of oligomycin-sensitive ATPase (117). Studies of the accessibility of various reagents to components within various types of submitochondrial preparations have led to the proposal of an interesting topographic model of the inner mitochondrial membrane (117). In this model, the coupling factors and succinic dehydrogenase are placed on the inner face of the inner mitochondrial membrane. The outer face holds cytochrome c. Cytochromes b and c_1, and the hydrophobic membrane proteins which confer oligomycin sensitivity on F_1-ATPase, are placed in the body of the membrane, as is most of cytochromes a and a_3, which together span the full width of the membrane. However, a portion of cytochrome a abuts on the outer face and a portion of cytochrome a_3 on the inner face. This topography is thought to be essential for the coupling of phosphorylation with oxidation. The limited success in achieving coupled oxidative phosphorylation in reconstituted electron transport systems may be a reflection of the difficulty in regaining the natural topography of the membrane components (210). These studies indicate that the mechanisms they have uncovered are extremely relevant to the understanding of membrane biogenesis in vivo (see discussion of biogenesis of oligomycin-sensitive ATPase in yeast).

Assembly of Mitochondria and Chloroplasts

MITOCHONDRIAL ASSEMBLY

The genetic characteristics of mitochondrial DNA, the availability within mitochondria of the necessary components for the replication and transcription of its DNA, and the translation and synthesis of protein from the information coded within it, all suggest that mitochondria are formed by division and replication of preexisting organelles. The now classic experiments of Luck (119) have provided strong supportive evidence for this notion. Using a choline auxotroph of *Neurospora*, he grew this fungus in radioactive choline to mid-log phase, following which the radioactive choline was replaced by nonradioactive choline in the growth medium and growth was allowed to proceed for a further three generations. The choline labeled the mitochondrial lecithin. If lecithin is accepted as a stable marker of mito-chondrial membranes, at the end of this labeling experiment the radioactivity in mitochondria would be randomly distributed through the whole popula-tion of mitochondria if the origin of mitochondria is by growth and division of preexisting organelles. If mitochondria arise by de novo synthesis, those made when radioactive choline was present during growth would be labeled and those made during further growth in the absence of radioactive lecithin precursor would be essentially unlabeled. Quantitative autoradiographic studies by Luck provided strong support for the first proposition. Further support was obtained in a second series of experiments (5), making use of the fact that when this *Neurospora* mutant was grown in a low-choline medium its mitochondrial membrane was relatively poor in lecithin and high in buoyant density, while growth in a medium containing 10 times the amount of choline produced a population of mitochondria high in lecithin and low in buoyant density. Artificial mixtures of these two populations of mito-chondria were separated by density gradient centrifugation. When the choline auxotroph was grown in a low-choline medium and when midway through the growth the medium was supplemented with a large excess of choline, Luck observed a progressive decline in the buoyant density of the mito-chondria. There was no evidence for the presence of two populations of mitochondria in such cultures, suggesting that no clear distinction could be made between old and new mitochondria. While both these groups of experi-ments provide substantial support for the contention that mitochondria arise by the division and growth of preexisting organelles, they are never-theless subject to the criticism that exchange of lecithin between new and old mitochondria, or between new mitochondria and some other membrane component of the cells, cannot be completely eliminated. Luck's conclusion is further substantiated by an analogous series of experiments using living *Tetrahymena* cells labeled with radioactive thymidine and then grown through four cell division cycles in the absence of radioactive precursor (23). In these cells mitochondria were found to retain the label randomly distributed over the entire mitochondrial population. Although no categorical experiments of this sort have been performed in mammalian tissues, all the suggestive evidence available from electron microscopy of mammalian cells as well as from the pattern of labeling of rat liver mitochondria with tritiated thymidine (120) is compatible with the postulate that in mammalian cells, too, mito-chondria arise by division of preexisting mitochondria.

Despite the evidence cited above, the possibility that mitochondria are derived from nonmitochondrial precursor membranes has remained viable until relatively recently. The suggestion that mitochondria arise from non-mitochondrial precursor membranes has been extensively investigated in the yeast *Saccharomyces cerevisiae,* which offers a unique opportunity for the study of mitochondrial biogenesis. This organism is a facultative anaerobe having the ability to grow anaerobically in glucose or related carbohydrate carbon source and aerobically in medium containing nonfermentable carbon sources such as lactate, glycerol, or acetate. During anaerobic growth on hexose as carbon source, cytochromes aa_3, b, c_1, and c and characteristic respiratory function disappear, while reintroduction of oxygen stimulates the development of mitochondrial profiles and respiratory functions. Mitochondrial respiratory functions are also poorly developed during aerobic growth in the presence of large repressive concentrations of glucose. With this sugar as carbon source, mitochondrial function is fully expressed only when most of the glucose is exhausted. This situation affords an excellent system in which to study the regulation of mitochondrial biogenesis and function. The availability of both cytoplasmic and chromosomal mutants which affect mitochondrial function (see above under "Genetics of Mitochondria") and which are readily identifiable by their inability to grow on a nonfermentable carbon source such as glycerol or lactate furnish yet a further advantage for the study of mitochondrial formation in this organism.

During growth under anaerobic conditions, yeast requires the presence of ergosterol and unsaturated fatty acid as nutrient supplements, since both these components require oxygen for their biosynthesis. It has been shown that very small amounts of these two lipid supplements suffice to permit anaerobic growth (121). Largely because of difficulties with fixation and electron microscopy of yeast, the occurrence of mitochondrial precursors in anaerobic yeast has been the subject of controversy (94). It is now generally agreed that anaerobic yeast always contains mitochondrial precursors (15), to which Criddle and Schatz (122) have given the name promitochondria. The development and complexity of the promitochondria demonstrated by various workers in this field are probably functions of the procedures for electron microscopy as well as of the degree of lipid depletion of the anaerobic yeast cells studied (15). The promitochondria which persist under anaerobic conditions are bounded by a double membrane and contain amorphous electron-dense material in the matrix and quite distinct fibrillar DNA aggregates, with very little evidence of well-developed cristae (123). However, using freeze-etching techniques even these latter structures have been demonstrated in the mitochondria of anaerobic yeast (124). Upon aerobic incubation of anaerobically grown yeast containing these poorly developed promitochondria, cristae develop in these structures, forming typical aerobic mitochondria in association with the emergence of respiratory capacity (123). The biochemical verification of the persistence of incompletely functional mitochondrial entities in anaerobically grown yeast has been obtained in a number of studies. The unique mitochondrial DNA persists in concentrations similar to those encountered in aerobic yeast (125).

The phospholipid, cardiolipin, which is characteristic of mitochondrial membranes (111), persists in anaerobic yeast. Criddle and Schatz (122) have isolated promitochondria characterized by their morphology (124); by their characteristic lipid composition (126) containing high proportions of cardiolipin; and by their content of oligomycin-sensitive ATPase, mitochondrial DNA, and substantial specific activity of succinate dehydrogenase. That these promitochondria are indeed the precursors of fully functional aerobic mitochondria has been definitively demonstrated in an elegant experiment conducted by Schatz and his collaborators (127). The promitochondria of anaerobically grown yeast were labeled in vivo by incubating the cells with radioactive leucine and cycloheximide. Under these conditions only those proteins made by the mitochondrial ribosomes would be labeled. Such prelabeled cells were washed free of cycloheximide, and aerobic adaptation was performed in the presence of unlabeled leucine. All the protein radioactivity could be shown to be associated with the mitochondria. As a control, it was shown that the iso-1-cytochrome c, a protein formed de novo during adaptation, was essentially unlabeled. The adapted mitochondria were stained cytochemically for cytochrome oxidase activity, and subsequent analysis by quantitative electron microscope radioautography showed that cytochemically demonstrable oxidase and the protein radioactivity were present within the same particles.

The possession by anaerobic promitochondria of oligomycin-sensitive ATPase suggests that even under anaerobic circumstances mitochondrial protein synthesis proceeds to some extent. As discussed below, the existence of mitochondrial ATPase which retains oligomycin sensitivity implies the synthesis by the mitochondrial ribosomes of one or more proteins necessary for the expression of this function. Such mitochondrial protein synthesis has been demonstrated in anaerobic promitochondria (128). In contrast, petite yeast, in which no mitochondrial protein synthesis is demonstrable, contains an ATPase which is relatively resistant to oligomycin inhibition. Not only do anaerobic promitochondria retain the sensitivity of their ATPase to oligomycin, but their hydrolysis of exogenous ATP is inhibited by atractyloside, an agent which interferes with the transport of adenine nucleotides across the mitochondrial membrane. These observations indicate that the ATPase in anaerobic promitochondria is situated inside a semipermeable barrier capable of transporting adenine nucleotides (129), i.e., in a situation comparable to that in which it is found in fully aerobic mitochondria. Racker (117) has summarized the evidence indicating that the oligomycin-sensitive ATPase is an important component of the normal phosphorylative mechanism of intact mitochondria. It therefore seems that anaerobic yeast promitochondria retain a major portion of their phosphorylative capacity even in the absence of terminal respiratory activity; i.e., the two prime functions of mitochondria are not obligatorily coupled.

Anaerobic yeast exposed to oxygen rapidly develops respiratory capacity, which can be almost completely prevented by cycloheximide. Schatz and his colleagues (130) have been able to demonstrate that oxygen stimulates new synthesis by the mitochondrial protein-synthesizing machinery, even in the absence of cytoplasmic protein synthesis, of two distinct protein bands as examined by disc-gel electrophoresis. It therefore seems likely that the mitochondrial protein-synthesizing mechanism is responsible for the production of at least three proteins: the two whose synthesis is stimulated by exposure of anaerobic yeast to oxygen and one whose synthesis determines the integration of oligomycin-sensitive ATPase into the mitochondria, a protein apparently already made under anaerobic conditions. The interdependence of mitochondrial and cytoplasmic protein synthesis for the development of respiratory capacity has been demonstrated (Table VII) by the sequential use of chloramphenicol or cycloheximide as inhibitors of either mitochondrial or cytoplasmic protein synthesis during exposure to oxygen (131). In the absence of either inhibitor, the respiratory capacity, Q_{O_2}, increases from essentially zero to between 30 and 40 over a period of 4–7 hr. In the presence of cycloheximide throughout this period no respiratory adaptation occurs. In the presence of chloramphenicol alone, a small chloramphenicol-resistant adaptation occurs in the first hour, following which no further change takes place. If the chloramphenicol is removed at 3 hr of respiratory adaptation and replaced by cycloheximide, still no change occurs. If the incubation of adapting yeast in cycloheximide for 3 hr during which no

Table VII

Respiratory Adaptation in Saccharomyces cerevisiae

ADDITION	Q_{O_2} DURATION OF ADAPTATION	
	3 hr	7 hr
None	20	39
0–3 hr CHI 3–7 hr CAP	0	5
0–7 hr CAP	1.5	1.5
0–3 hr CAP + CHI 3–7 hr CAP	0	1.0
0–3 hr CAP 3–7 hr CHI	2.5	3.0
0–3 hr CAP + CHI 3–7 hr CHI	0	0

CHI, cycloheximide; CAP, chloramphenicol.
Results adapted from Rouslin, W., and Schatz, G. *Biochem. Biophys. Res. Commun.* **37**, 1002 (1969).
   Q_{O_2} values are presented to the nearest 0.5.

development of respiratory capacity takes place is followed by the removal of cycloheximide and its replacement by chloramphenicol, respiratory capacity develops to the extent of about 13% of the completely uninhibited situation. Thus, when anaerobically grown yeast cells are exposed to oxygen in the presence of cycloheximide, which inhibits cytoplasmic protein synthesis, they accumulate intermediates capable of being subsequently utilized for the development of respiring mitochondria. These intermediates are apparently formed by the mitochondrial protein-synthesizing mechanism. However, since a full complement of intermediates formed by the cytoplasmic protein-synthesizing mechanism do not accumulate unless simultaneous mitochondrial protein synthesis occurs, the initiation of the development of respiratory capacity upon exposure of anaerobic yeast to oxygen is not dependent on the global synthesis of all those components necessary for the formation of a fully functional mitochondrion and made by both the cytoplasmic and mitochondrial protein-synthesizing machinery. Some of the components necessary for organelle development are made in substantial amounts under anaerobic conditions, as evidenced by the existence of promitochondria in anaerobic yeast whose components are derived from the systems of cytoplasmic and mitochondrial protein synthesis. Indeed promitochondria contain proteins precipitable by antisera to cytochrome oxidase, suggesting the presence of some precursors of this enzyme complex in anaerobic promitochondria (130). As shown below, two of the protein constituents of the oligomycin-sensitive ATPase system, present in promitochondria, are almost certainly manufactured on the cytoplasmic ribosomes. Furthermore, evidence has been presented (132) and recently confirmed (133) that precursors of iso-2-cytochrome c, normally a minor component of the total yeast mitochondrial c cytochromes, accumulate as apoproteins in anaerobic yeast. This conclusion rests upon the demonstration that isolatable iso-2-cytochrome c develops upon exposure to oxygen in the absence of all cytoplasmic protein synthesis. It has also been shown that δ-amino levulinic acid is incorporated almost exclusively into iso-2-cytochrome c and not iso-1-cytochrome c, the more prominent c cytochrome, during respiratory adaptation in cycloheximide.

The experiments described above, which demonstrate the interdependence of mitochondrial and extramitochondrial protein synthesis for the development of fully functional mitochondria, have been confirmed with respect to a single functional constituent of mitochondria in the yeast *Saccharomyces carlsbergensis*. The effects of cycloheximide, chloramphenicol, and acriflavine added at different times during respiratory adaptation were measured upon the development of heme a and cytochrome oxidase activity (85). Anaerobically grown cells form heme a and cytochrome oxidase at a constant rate, reaching a maximum at 2 hr of adaptation for the former and 3 hr for the latter. The levels of these maxima were about a third of those present in fully aerobic cells and were substantially reduced by either cyclo-heximide or chloramphenicol. Cells induced by oxygen for periods up to 25 min accumulated sufficient precursor to form additional cytochrome oxidase activity upon further anaerobic incubation. This formation of intermediate or precursor was inhibited by cycloheximide but not by chloramphenicol or acriflavine. Cells developing cytochrome oxidase activity may within the first hour of oxygen induction, accumulate precursors apparently made by the mitochondrial protein-synthesizing mechanism (inhibited by chloram-phenicol) which permit the formation of further cytochrome oxidase in the presence of cycloheximide (inhibitor of cytoplasmic protein synthesis). Additional precursors or intermediates made by cytoplasmic protein syn-thesis accumulate only after 80 min of aerobic induction, so that further incubation in the presence of chloramphenicol permits the development of additional cytochrome oxidase activity. These precursors or intermediates are not made if the initial incubation is carried out in the presence of cyclo-heximide. It is clear from this investigation of even a limited aspect of res-piratory induction that the final formation of functional units is the result of a complex interaction of the two separate systems of protein synthesis in both space and time, and that oxygen induces both systems to manufacture mitochondrial proteins. Some insight is being gained into the molecular basis of this interaction from the study of preparations of the cytochrome oxidase complex of both *Neurospora* and yeast (130, 211). They have been separated by SDS gel electrophoresis into five or six protein bands, having estimated molecular weights of 30, 20, 13, 10, and 8×10^3 (for *Neurospora*) and 42, 34.5, 22, 15, 13.5, and 9.5×10^3 (for yeast), respectively. In *Neurospora* (211), only the 20,000 molecular weight component was labeled in the presence of cycloheximide, so that it is a candidate for synthesis by the mitochondrial system. Consistent with this is the observation by Schatz *et al* (130) that one or all of the three smaller proteins of the yeast cytochrome oxidase preparation contain bound heme a and are labeled in the presence of acriflavin and are therefore probably made on the cytoplasmic ribosomes.

Glucose derepression in yeast is an alternative model system which has been used for the study of mitochondrial biogenesis. Yeast cells grown aerobically in high glucose concentrations have relatively little respiratory capacity as long as large amounts of glucose remain in the growth medium. The synthesis of the proteins previously referred to which are made by promitochondria in response to oxygen stimulation during respiratory adaptation is also subject to glucose repression under aerobic conditions (130). The transfer of glucose-repressed cells to media containing little or no glucose results in the incremental emergence of characteristic mitochondrial functions. In this case too, the emergence of electron transport activity is dependent on the cooperative interaction of protein synthesized both within and outside the mitochondria (134, 135). However, the synthesis of cytochrome oxidase in these circumstances seems to be particularly under the control of mitochondrial protein synthesis (chloramphenicol inhibitable) (207). Tzagoloff (89, 90, 136, 212, 213) has studied the emergence, during the course of glucose derepression, of the oligomycin-sensitive ATPase characteristic of the inner mitochondrial membrane. This ATPase function has been resolved into a number of major components, F_1 or ATPase itself, F_c or OSCP (oligomycin-sensitivity-conferring protein), several membrane proteins and phospholipids, all of which are necessary for the expression of oligomycin (or rutamycin) sensitivity of mitochondrial ATPase (118). The resolved F_1-ATPase which apparently corresponds to the head pieces of the inner mitochondrial membrane and contains five polypeptide components (212) is both cold sensitive and insensitive to oligomycin or rutamycin inhibition. The restoration of cold resistance and oligomycin or rutamycin sensitivity is dependent upon the rebinding of the ATPase to the inner mitochondrial membrane (CFo + phospholipids) and involves the OSCP component. The identity of this last component among the nine polypeptides (i.e., four more than are present in the separated F_1-ATPase) of the total oligomycin-sensitive complex is not clear. During 6 hr of glucose derepression there is approximately a twofold increment in oxidative capacity of isolated derepressed mitochondria and a twofold increment in oligomycin-sensitive ATPase, which is inhibited by the presence of either chloramphenicol or cycloheximide. However in contradistinction to the situation in cycloheximide, the ATPase activity missing from the mitochondria following incubation in chloramphenicol, is recovered in the post-mitochondrial supernatant (see Table VIII). The identity of this activity with that normally found associated with the mitochondria is proved by its sensitivity to the agent dio-9 and by the fact that it is labile to cold, both properties of the resolved mitochondrial F_1-ATPase. An exactly similar situation is encountered for the case of the OSCP factor. Glucose-repressed mitochondria have little capacity to bind F_1-ATPase, but the capacity for binding and reconstitution increases during derepression. This increase is inhibited by either cycloheximide or chloramphenicol, but does emerge upon sequential incubation, during derepression in chloramphenicol followed by cycloheximide (136). This study suggests that the synthesis of proteins in the mitochondria, including the ATPase-binding protein(s), is stimulated by the products of cytoplasmic protein synthesis. However when yeast is labeled in the presence of cycloheximide, only the additional four peptides are labeled (i.e., those not seen in pure F_1) but these peptides newly synthesized in the mitochondrion do not associate with the whole ATPase complex in the absence of cytoplasmic protein synthesis. Of these four peptides that having the lowest molecular weight has interesting hydrophobic properties, being soluble in chloroform methanol (213). This component is also most heavily labeled in the presence of cycloheximide. Thus the F_1 and OSCP peptides which are essential for the oligomycin-sensitive ATPase are apparently made

Table VIII
Rutamycin-Sensitive ATPase in Glucose Derepressed Yeast

PERIOD OF DEREPRESSION (hr)	ADDITIONS	RUTAMYCIN-SENSITIVE ATPASE SPECIFIC ACTIVITY (μmoles P_i/min/mg protein)[a]	Total ATPase Activity (μmoles P_i/min)	
			MITOCHONDRIAL[a]	SUPERNATANT[b]
0		0.89	10.5	3.7
6		1.75	29.5	8.8
6	CAP 2 mg/ml	0.80	9.7	31.5
6	CHI 10^{-5}M	0.75	10.0	1.1

[a] This represents the difference between mitochondrial ATPase in the presence and absence of rutamycin (60 μg/mg protein).
[b] This represents the difference between supernatant ATPase in the presence and absence of Dio-9 (10 μg per mg protein) which inhibits F_1-ATPase which is resolved from the mitochondrial membrane.
CAP, chloramphenicol; CHI, cycloheximide.
Adapted from Tzagoloff, A. *J. Biol. Chem.* **244,** 5027 (1969).

on the cytoplasmic ribosomes, while several membrane peptides associated with the ATPase complex are made on the mitochondrial ribosomes.

This conclusion is supported by the finding that in cytoplasmic petite yeast which lacks the ability to make protein in the mitochondria, the mitochondrial ATPase is less firmly attached to the mitochondrial membrane and is much more resistant to inhibition by oligomycin.

All the experimental results so far discussed suggest that mitochondria are built upon a preexisting scaffold existing in the form of promitochondrial organelles into which appropriate mitochondrial elements may be inserted upon stimulation either by oxygen or by a fall in the glucose concentration.

Much less information is available on the mechanism of mitochondrial assembly and biogenesis in higher species. However, there is every suggestion that similar principles are operative. For example, cultured mammalian fibroblasts are responsive to the oxygen tension in the culture medium (137). These cells, when grown under diminished oxygen tensions in tissue culture, have lower total cell cytochrome oxidase activity and specific activity of mitochondrial cytochrome oxidase compared with those grown at normal oxygen tensions. Cells apparently contain the same numbers of mitochondria in these two situations, suggesting that upon stimulation by high oxygen tension the incompletely developed inner mitochondrial membranes of hypoxic cells become more fully respiratory-competent by the insertion of appropriate respiratory elements into preexisting mitochondrial membranes. During the growth and development of the rat liver in the immediate post-natal period, there occurs a rapid differentiation of the hepatic cells which includes the development of mitochondria. In this tissue, evidence has been provided for the asynchronous appearance of mitochondrial components during the course of development (138). Thus the rate of accumulation of cytochrome oxidase, succinate dehydrogenase, and succinate cytochrome c reductase is different for each enzyme. Furthermore, the fact that the specific activity of these enzymes increases over a three- to fourfold range when expressed against the concentration of cardiolipin, an element which has been shown to be exclusively associated with the inner mitochondrial membrane, suggests that in this system too it is possible that respiratory elements are either activated or newly synthesized and inserted into preexisting mitochondrial inner membranes. Also, skeletal muscle mitochondria of thyroidectomized rats exhibit lower respiratory activity than those from normal or thyroid-treated rats (139). Yet the mass of mitochondria within the hypothyroid skeletal muscle is at least as great as that in normal muscle. One speculation deriving from these observations is that thyroid hormone stimulates the insertion of new respiratory elements into preformed but incompletely developed mitochondria, as has already been suggested for adapting yeast. Some recent experiments of Gross (140) may be interpreted in this fashion also. He has shown that the treatment of hypothyroid rats with physiological doses of thyroid hormone results in an increase in the amount of cristal membrane in rat liver mitochondria associated with an increase of total hepatic respiratory activity.

ROLE OF LIPID IN MITOCHONDRIAL ASSEMBLY

The importance of the lipid component of mitochondrial inner membranes has been amply illustrated by the necessity of phospholipid for the reconstitution of all the respiratory and phosphorylative enzyme activities of the inner mitochondrial membrane (94). As has been mentioned previously, the phospholipid cardiolipin is largely localized to the inner mitochondrial membrane. Analysis of cardiolipin content of yeast under different physiological and genetic circumstances (111) further emphasizes the likelihood that mitochondria are assembled by sequential stepwise insertion of various components. Thus yeast which contains virtually no respiratory activity may nevertheless contain substantial concentrations of cardiolipin, albeit lower than those of fully respiratory-competent and active yeast cells. It appears that to some extent the cardiolipin content of yeast cells reflects the amount of mitochondrial "scaffold" membrane present in the cells under particular physiological circumstances. Thus anaerobic cells at stationary phase have about half the content of aerobic cells at stationary phase grown in glucose. There is, however, a manifold difference between the respiratory activity of these two yeast cells. Similarly, aerobic cells grown in glucose to mid-log phase, when mitochondrial activity is repressed, contain less than half the cardiolipin found in fully derepressed stationary-phase yeast cells.

When nonrepressive carbon sources are used, such as galactose or lactate, log-phase yeast cells contain as much cardiolipin and as much respiratory capacity as do stationary-phase glucose-grown cells. Cytoplasmic petites are known to have relatively poorly developed mitochondrial profiles

(141) and virtually no respiratory capacity. However, these mutants contain significant concentrations of cardiolipin, though less than that encountered in the corresponding wild-type cells. Despite the absence of respiratory capacity under any circumstances, these cytoplasmic petites respond to variations in the degree of glucose repression by the expected variations in the cardiolipin content. Chromosomal mutants in which mitochondrial function is abnormal, when grown under conditions calculated to encourage maximal mitochondrial development, contain cardiolipin concentrations equivalent to those of similarly grown wild-type cells. During glucose derepression, cardiolipin increases before or along with the increase in respiratory activity (142, 143). That the scaffold membrane of mitochondria may well be synthesized independently of the insertion of functional elements and of mitochondrial protein synthesis, is indicated by the observation that yeast grown in galactose has the same cardiolipin content whether or not chloramphenicol is present (143).

The need for either presynthesized promitochondrial membrane of appropriate lipid composition or for new lipid synthesis during respiratory adaptation has been demonstrated in a fatty acid auxotroph of yeast. This mutant requires unsaturated fatty acid for growth (144), and when it is grown anaerobically on severely limiting concentrations of unsaturated fatty acid, respiratory adaptation takes place upon exposure to oxygen only if further unsaturated fatty acid supplements are provided (145). Severe lipid depletion, in contrast to lipid supplementation, during anaerobic growth of wild-type yeast results in cells whose promitochondria have low amino acid incorporating ability and from which characteristic mitochondrial ribosomal RNA is not obtainable (15).

As is the case with the unsaturated fatty acid auxotrophs of *E. coli*, under certain circumstances *trans*-unsaturated fatty acids support the growth of this yeast mutant. The efficacy of various *trans*-unsaturated fatty acids is related to the temperature of growth (145). Thus, growth in linelaidic or palmitelaidic acid occurs at temperatures above 25° C but not at temperatures below 22° C. These observations have made it possible to demonstrate that the nature of the lipids associated with the mitochondrial membrane is an important element affecting the emergence of respiratory activity during derepression after growth on glucose. For example, cells grown at 30° C in palmitelaidic acid in high glucose concentration, harvested, and reincubated at 17° C in nongrowth medium containing a nonfermentable substrate, undergo respiratory derepression only if a *cis*-unsaturated fatty acid is provided. Therefore, preexisting scaffold membranes containing *trans*-unsaturated fatty acid cannot be utilized as acceptor membranes for new respiratory assemblies at 17° C. Alternatively, new lipid synthesis must take place, and for such new lipids to be effective in supporting the development of respiratory capacity, *cis*-unsaturated fatty acid must be incorporated into them. The emergence of respiratory activity can be correlated with the effect of these two classes of fatty acids upon protein synthesis during a period of derepression in the absence of growth. Clearly much further work is required on this system to elucidate the exact mechanism by which the nature of the lipid and its synthesis exert a controlling influence upon the development of functional mitochondrial assemblies. Unsaturated fatty acid depletion of the fatty acid mutant grown aerobically on glucose results in a specific loss of mitochondrial oxidative phosphorylation measured both in vivo and in vitro, without any effect on respiration (15). This defect in oxidative phosphorylation can be repaired in the absence of growth and protein synthesis (15).

Apart from what has already been discussed with respect to the origin of the phospholipids of mammalian mitochondria, almost nothing is known about the role of lipids in the assembly of mitochondria in higher animal cells. For the assembly of mitochondria, lipids are quite clearly required, and in the rat liver cell most of this phospholipid appears to be derived from molecules synthesized in the endoplasmic reticulum. This is supported by the kinetics of labeling the phospholipids, studied both biochemically and by electron microscope radioautography, of endoplasmic reticulum and mitochondria (146, 147). The incorporation of lipid molecules into the mitochondria probably occurs by an exchange reaction, involving a supernatant cell sap factor (94). That the described in vitro exchange reactions between phospholipids of the membranes of the endoplasmic reticulum and mitochondria are also significant in vivo has been suggested by a recent study of Wirtz and Zilversmit (148). These authors have shown that upon injection of radioactive phosphorus into rats treated with phenobarbital or carbon tetrachloride, the specific activities of lecithin and phosphatidylethanolamine of the microsomes and mitochondria were altered in parallel fashion. Their observations were interpreted as evidence for the origin of mitochondrial lecithin and phosphatidylethanolamine from the pool of molecules synthesized within the membranes of the endoplasmic reticulum.

CHLOROPLAST ASSEMBLY

A most useful system for the study of chloroplast formation arises from the observations made on etiolated plants exposed to the light. When seeds are germinated in the dark, the seedlings possess no chlorophyll, but manufacture this molecule when illuminated. This is known as the greening process during which the development of mature chloroplasts from proplastids may be studied. The proplastid contains a prolamellar body which, upon exposure to light, disperses with the formation initially of the primary chloroplast membranes or thylakoids which subsequently condense and form the mature chloroplast grana (61). Associated with these morphological changes is a synthesis of chlorophyll, chloroplast lipids (i.e., phosphatidylglycerol, sulfolipid, mono- and digalatosyl diglycerides—all containing polyunsaturated fatty acids), and the proteins of the photoreductive systems including those responsible for the Hill reaction. In higher plants the initial formation of chlorophyll after light exposure is the result of the conversion of proto-chlorophyllide, and any further production of chlorophyll is the result of de novo synthesis from small molecular precursors. Although the initial dispersal of the prolamellar body occurs at about the same time as the proto-chlorophyllide reduction to chlorophyllide, the two processes do not necessarily occur simultaneously. Thus it is possible to reduce about half of the protochlorophyllide before any detectable structural change occurs. It is not clear, however, whether a certain minimum percentage of protochloro-phyllide reduction is a prerequisite for subsequent prolamellar body dispersal and the initial ultrastructural changes which accompany chloroplast development. Formation of chloroplasts from proplastids in mature leaves apparently does not involve any necessary plastid division, as exemplified by observations made in *Phaseolus vulgaris* exposed to light. In the case of *Euglena gracilis* there is actually a reduced number of chloroplasts during the greening process, while *Chlamydomonas reinhardi* has only a single chloro-plast at all stages of its growth and differentiation. The detailed kinetic relations between the formation of membrane, of chloroplast lipids, and of photosynthetic capacity vary among cell species. In cells which accumulate a relatively large quantity of protochlorophyllide in the dark, there is an initial production of chlorophyll by the photoactivation of the conversion of the precursor chlorophyllide to mature chlorophyll. In this period there is no accompanying increment in photosynthetic capacity. Thus the development of the ability to liberate oxygen or fix CO_2 and to carry out the Hill reaction appears after an initial lag, and at this time the accumulation of this capacity in general parallels the de novo synthesis of chlorophyll.

The relation between the membrane changes, chlorophyll synthesis, and chloroplast lipid synthesis is by no means obligatory. Study of a series of barley mutants has revealed varying degrees of dissociation between these major transformations accompanying chloroplast development (149). For example, the xantha[11] mutant contains disorganized membrane structure and little chlorophyll; yet the chloroplast lipid synthesis accompanying exposure to light is apparently normal. The xantha m mutant has a chloro-plast containing only 25% of the usual amount of chlorophyll, makes no new membrane upon exposure to light, and yet incorporates the linolenic acid into monogalactosyl diglyceride at a greater-than-normal rate. What new chlorophyll is made upon illumination of this mutant is apparently incor-porated into preexisting prolamellar membrane or grana-like aggregates. The xantha b[12] mutant makes very little chloroplast lipid upon exposure to the light.

Most algae are capable of forming chlorophyll in the dark, but *Euglena gracilis* and *Ochromonas danica* as well as a group of yellow mutants of

Chlamydomonas reinhardi require light for chlorophyll formation. Consequently these cells have been most used for the study of chloroplast development. In dark-grown *Euglena*, the plastids are substantially dedifferentiated, so that they contain rather low concentrations of characteristic photosynthetic enzymes such as NADP-linked glyceraldehyde-3-phosphate dehydrogenase and photosynthetic cytochromes. Dark-grown *Euglena gracilis* contains barely detectable chloroplast ribosomal RNA, which does, however, increase to levels similar to those found in light-grown cells over 72 hr of greening (214). This induction by light seems to involve only the ribosomal RNA of the chloroplast, since other species of chloroplast RNA are present in approximately equal amounts in dark- and light-grown *Euglena* cells. In these cells, as well as in higher plants, the photosynthetic components are induced during chloroplast formation. In the yellow mutant (y-1) of *C. reinhardi* the plastid is not so severely dedifferentiated during dark growth, when the major loss is in chlorophyll concentration and capacity to carry out photosynthesis. Chlorophyll is lost by growth dilution in the dark (150). This process also seems to be responsible for the loss of photosynthetic capacity as measured by oxygen evolution, the Hill reaction, and the photoreduction of NADP. In contrast to *Euglena* and higher plants, however, enzymes like ribulose diphosphate carboxylase, NADP-linked glyceraldehyde-3-phosphate dehydrogenase, ferredoxin, and cytochrome continue to be present in high concentration in dark-grown cells of this mutant, which also retains the capacity for the manufacture of chloroplast lipids in the dark. During dark growth, the morphological dedifferentiation is represented by a gradual disorganization and decrement in the chloroplast lamellar system, though the pyrenoid is essentially unaffected. Upon illumination of such dark-grown cells, there is a parallel increment in membrane development, with redifferentiation of the chloroplast lamellar system, chlorophyll synthesis, and the emergence of photosynthetic activity measured by oxygen evolution, the Hill reaction, and NADP photoreduction (151). These changes are absolutely dependent upon the continued illumination of the greening cells, since return of the cells to the dark results in an abrupt cessation of chlorophyll synthesis. Unlike the changes which occur in these elements, there are relatively small alterations in those photosynthetic enzymes present in dark-grown *Chlamydomonas* (see above). These observations suggest that the photosynthetic membranes may be assembled during greening as a single-step process. However, subsequent work from this group (152) has suggested, on the basis of fractionation of the thylakoid membranes at various times during the greening process, that the proportions of cytochrome 554 and carotenoids relative to chlorophyll were changing continuously, and therefore that the thylakoid membranes are assembled by a multistep process. But during greening the buoyant density of the membranes increases from 1.12 to 1.14, a change which is also seen when greening *Chlamydomonas* cells, inhibited by chloramphenicol, are transferred, with continued illumination, to a medium lacking the antibiotic (see below and ref. 215). This density change is evident in membrane fragments sonicated to an average vesicle size of 700 Å. It was therefore suggested that at least at this level of resolution there is homogeneous growth of membrane without any evidence of active growing regions.

Labeling of chloroplast membrane proteins during the greening process in the presence of various antibiotics suggests that most of these proteins were derived from the extraplastid cytoplasm while some were made in the chloroplast. Thus as in the development of mitochondria from promito-chondria, so the development of a mature chloroplast from a dedifferentiated plastid in greening *C. reinhardi* involves the cooperative interaction of the plastid and extraplastid cytoplasmic protein-synthesizing mechanisms (93). Cycloheximide reduces chlorophyll synthesis, the development of photo-synthetic capacity, and the formation of photosynthetic membranes to about the same extent, but those membranes that do form are fused into grana. Low doses of chloramphenicol (25 μg/ml), while having no effect on chloro-phyll synthesis and photosynthetic membrane formation, inhibit the develop-ment of photosynthetic capacity. In its presence, the primary thylakoid membranes line up but do not fuse to form grana. The inhibition of grana formation can be repaired by removing the drug and continuing the incuba-tion in the light, even in the presence of cycloheximide. This repair is accom-panied by a full restoration of photosynthetic capacity on a chlorophyll basis. Higher doses of chloramphenicol (200 μg/ml) also inhibit chlorophyll syn-thesis and thylakoid membrane formation substantially. This too may be reversed by removal of the antibiotic (217). On disc gel electrophoresis of the thylakoid membrane proteins, two prominent bands, whose synthesis is inhibited by cycloheximide are evident. One of these (fraction c of ref. 217) appears in the extraplastid cytoplasm when chloroplast formation is in-hibited by high doses of chloramphenicol—a situation analogous to the behavior of F_1-ATPase in yeast during inhibition of glucose derepression by chloramphenicol (89). In these same experiments, the synthesis of peptide bands, which appeared to correspond to ribulose diphosphate carboxylase, was eliminated by chloramphenicol. Hoober (217) suggested that this enzyme might be made by chloroplast ribosomes, though coded for by a nuclear gene.

Eytan and Ohad (218) have used their own and the above observations to formulate a general and doubtless oversimplified model of chloroplast assembly. During greening in *C. reinhardi* light activates chlorophyll forma-tion, which in turn initiates, at the level of transcription off the nuclear genome, the synthesis of a series of cytoplasmically generated proteins (termed L proteins), which may continue to accumulate in the absence of chloroplast protein synthesis. These proteins, whose synthesis is inhibited by cycloheximide, may be incorporated into inactive photosynthetic assemblies, which can be rendered fully active by the addition of a series of "activation proteins" made in the chloroplast. The synthesis of the latter is inhibited by chloramphenicol but not cycloheximide and may continue in the dark provided the particles are rich in L proteins. This suggests that though the synthesis of cytoplasmic and chloroplast proteins necessary for the assembly of active photosynthetic lamellae are not obligatorily coupled, the cytoplasmic proteins do appear to exert some positive control on the formation of the chloroplast proteins, again reminiscent of the control of the generation of oligomycin-sensitive ATPase in yeast mitochondria during glucose derepression (136).

TURNOVER OF MITOCHONDRIA

The regulation of the mitochondrial complement of rapidly growing eukaryotic organisms such as yeast is achieved largely by the control of mitochondrial synthesis. In contrast, the regulation of organelle mass in mature mammalian cells is a function not only of the rate of synthesis but also of the rate of degradation of their mitochondria. That mitochondrial degradation is not necessarily a function of cell turnover has been known for many years. Rat liver cells are known to have a lifetime of 4 to 5 months (153). Yet within 4 weeks, the guanidino group of arginine of total liver proteins of rat livers continuously exposed to radioactive CO_2 has essentially the same specific activity as the excreted urea. Thus all the liver proteins had been renewed within this period. Since mitochondrial protein accounts for about 25% of total liver protein, this must also have accounted for the renewal of mitochondria contained within much longer lived liver cells (154). Verification of the turnover of mitochondria by the study of the incorporation of isotopic precursors into mitochondrial components was first done by Fletcher and Sanadi (155). They obtained a calculated half-life of about 10 days for total mitochondrial phospholipids, soluble and insoluble mitochondrial proteins, and cytochrome c labeled with ^{35}S-methionine or ^{14}C-acetate. The exponential decay of these components suggested random destruction. A similarity of the decay constants for each of these components led the authors to postulate a unitary breakdown of mitochondria. A later study with more specifically identified lipid and DNA components of the mitochondria appeared to provide further evidence for a unitary turnover of mitochondrial membranes and their associated constituents (156). Thus in both rat liver and rat kidney mitochondria similar turnover times were obtained for the mitochondrial DNA and for cardiolipin. These experiments, in which a half-life of approximately 10 days was obtained for the components studied, apparently confirmed the earlier studies referred to above as well as those of other investigators (summarized in 156). In most of these studies, however, the absolute half-life reported may well be in error for technical reasons, related primarily to the behavior of particular isotopic precursors employed.

The accurate evaluation of absolute rates of synthesis and degradation and of turnover is absolutely dependent upon the utilization of the appropriate methodology, which has recently been reviewed by Doyle and Schimke (157). The ideal method for the evaluation of the rate of synthesis of a tissue constituent and, if in the steady state, of its turnover rate, depends upon the ability to isolate at various time intervals the immediate precursor and its product in pure form for determination of accurate specific activities. Since these methods are frequently tedious, they have seldom been employed in turnover studies. The rate of renewal of a tissue constituent can be accurately determined by continuous labeling with the precursor and by examining the time course of approach of the product specific activity to that of the precursor. For the same reasons, this procedure has seldom been employed. In most turnover studies single isotope injections have been used and followed by the determination of the decremental changes in specific activity of the appropriate constituent with time. The validity of this approach depends upon a number of conditions. Steady-state conditions should obtain; the isotope injected should behave as a pulse; and the products resulting from the degradation of the molecular constituent under investigation should not be reutilized. ^{32}P, which has frequently been used to study the turnover of phosphate-containing macromolecules, may be particularly unsatisfactory since it does not behave in mammalian tissue as a pulse, as shown by the recent experiments on phospholipid turnover in rat liver by MacMurray and Dawson (146). Reutilization of protein precursors has frequently limited the accuracy or validity of half-lives determined with such precursors as leucine. At least in the case of liver, guanidino-labeled arginine functions as a precursor which is not extensively reutilized, since the guanidino group of free arginine is rapidly converted to urea, which is excreted. That reutilization is significant with other protein precursors is indicated by the fact that half-lives of several hepatic proteins determined with guanidino-labeled arginine are half the values obtained with other amino acid precursors such as leucine (158). An ideal precursor, limited to the study of heme proteins, is δ-amino levulinic acid, which is specifically incorporated into heme. Isotope from the heme breakdown is excreted quantitatively as bilirubin and related pigments.

Relative, but not necessarily absolute, turnovers may be easily determined by the double-isotope technique of Schimke *et al.* (158). This technique eliminates concern over reutilization and absolute precursor pool size provided certain precautions are observed. The turnover of the component must be exponential; animals must be in a steady state throughout the course of the experiment; differently labeled precursors, i.e., ^{3}H- and ^{14}C-labeled precursors, must be similarly metabolized and reutilized and should be administered under identical physiological circumstances. In the examination of organelle turnover studies, the above strictures should be continually borne in mind.

Mitochondria are made of three readily distinguishable compartments—the outer mitochondrial membrane (OMM), the inner mitochondrial membrane (IMM), and the matrix. Evidence has been presented above that the biosynthesis of the components of each of these mitochondrial compartments is probably under different control. It would therefore be surprising if each of these major mitochondrial compartments turned over as a unit. Indeed, using the above double-isotope technique, it has been shown that the proteins of OMM turn over more rapidly than those of IMM (159). Similar conclusions have been reached by following the specific activity decay curves of proteins in these two membranes labeled by a single injection of leucine (160) or methionine (161). Also the cytochrome b_5 of OMM has a more rapid turnover than cytochrome b of the IMM (162).

δ-Amino levulinic acid synthetase is localized in the mitochondrial matrix (112) and has an apparent half-life of 1.2 hr, while two other mitochondrial enzymes which are readily solubilized and perhaps also in a matrix location have half-lives less than 1 day—alanine and ornithine aminotransferase (163). It seems evident that, at least in the case of rat liver mitochondria, OMM and matrix components turn over more rapidly than those of the IMM.

It remains possible that IMM or a major unit of it turns over as an entity. Thus the hemes, labeled with δ-amino levulinic acid, of cytochromes a, b, and c, all of which are attached to the IMM and proteins labeled with the poorly reutilized precursor arginine all have similar half-lives (Table IX). This applies to both rat liver and heart mitochondria (164). In the case of mitochondrial DNA the half-life in heart (156) is close to that of the other mitochondrial components, while in liver it was somewhat longer, but reutilization of the tritium of thymidine has not been excluded in this situation. A unitary turnover rate for several components of the IMM suggests that lysosomes may be obligatorily involved in the catabolism of this membrane system. Evidence for the obligatory involvement of lysosomes in the degradation of mitochondria is seriously limited, though the lysosomes do have the biochemical capacity for such a process and indeed mitochondrial profiles have frequently been observed within lysosomal membrane in a variety of physiological and experimental situations (155).

Although in the case of both liver and heart, mitochondria are derived from a relatively homogeneous population of cells, there remains the possibility of microheterogeneity of mitochondrial populations (166) either within the same cells or within the aggregate of cells making up the studied tissue homogenate. Most studies have not taken account of this possibility for obvious technical reasons. The administration of thyroid hormone to thyroidectomized rats leads to an enhancement of mitochondrial enzyme activities (140). These biochemical changes are correlated with morphological evidence of increased mitochondrial mass. In the liver the increased functional

Table IX

Turnover of Rat Mitochondrial Components

COMPONENT	LABELED PRECURSOR	APPARENT HALF-LIFE (DAYS)	REFERENCES
		LIVER	
Total protein	^{3}H-Leucine	7.0	164
	^{14}C-Guanidino arginine	5.0	164
	Ca^{14}CO$_3$	5.4	163
Cytochrome c	Ca^{14}CO$_3$	5.0	163
	^{3}H-δ-Amino levulinic acid	5.5	164
Cytochrome b (heme b)	^{3}H-δ-Amino levulinic acid	5.5	162
Cytochrome a (heme a)	^{3}H-δ-Amino levulinic acid	5.5	164
		HEART	
Total protein	^{3}H-Leucine	6.1	164
	^{14}C-Guanidino arginine	6.2	164
Cytochrome c	^{3}H-δ-Amino levulinic acid	5.8	169
Cytochrome a (heme a)	^{3}H-δ-Amino levulinic acid	5.9	164
DNA	^{3}H-Thymidine	6.7	156

mitochondrial mass appears to be the result of both an increase in synthesis and a concomitant decline in degradation of mitochondria, while in the heart the latter process mainly accounts for the change in mitochondrial amount. Of particular interest is the evidence adduced for heterogeneity of degradation of mitochondrial DNA molecules in this model situation. Thus, on the basis of double-labeling experiments, after the thyroxine treatment of hypothyroid rats, mitochondrial DNA which has been labeled with thymidine prior to the initiation of treatment is more readily removed than that mitochondrial DNA synthesized after the attainment of high thyroid hormone levels (140). This is the first evidence of heterogeneity of mitochondrial turnover in a reasonably homogeneous cell population, although substantial functional and biogenetic heterogeneity has recently been claimed for rat liver inner mitochondrial membrane (219).

THE ENDOPLASMIC RETICULUM

While heterogeneity within the tissue or cell population of mitochondria is still problematic, it seems quite certain that the endoplasmic reticulum does not represent a homogeneous system of membranes but contains subclasses of differentiated membrane systems. The endoplasmic reticulum membrane can be readily separated into a rough endoplasmic reticulum system and at least two fractions of smooth membrane (4). Despite a substantial similarity in the lipid composition and enzymic constitution of these membrane fractions, some significant differences have been observed with respect both to their lipids (167) and their enzymic constituents (168). Tata (169) has demonstrated a heterogeneity within the rough endoplasmic reticulum based upon the increased synthesis of phospholipids in thyroid-stimulated hepatic growth.

Since the membranes of the endoplasmic reticulum are rich in phospholipid, the incorporation of phospholipid precursors into endoplasmic reticulum phospholipids has been employed as an indicator of this membrane's synthesis. Tata and his colleagues have suggested that in tissues whose growth has been stimulated by hormone administration there is coordinate synthesis of ribosomes and endoplasmic reticulum membranes. For example, in rats given either thyroxine or growth hormone, they observed the simultaneous appearance of RNA and phospholipid in the endoplasmic reticulum with a time course peculiar for each hormone. Following the administration of the two hormones together, an additive biphasic increase in the synthesis of microsomal phospholipids and RNA was evident. Coordination of microsomal phospholipids and RNA synthesis was also seen in the metamorphosing tadpole liver stimulated with triiodothyronine, in the thyrotropic-hormone-stimulated pig thyroid, and in the androgen-stimulated seminal vesicle. In all these situations a striking morphological picture of developing endoplasmic reticulum was evident.

Two model systems have been of particular value in studying the biosynthesis of endoplasmic reticulum membranes and particularly the smooth membranes. In the first, the administration of phenobarbital to rats results in a differential induction of certain of the hepatic endoplasmic reticulum enzymes, especially those involved in the metabolism of drugs and other hydrophobic substances. A concomitant proliferation of smooth endoplasmic reticulum membranes occurs (170). The accumulation, under these circumstances, of some endoplasmic reticulum enzymes is substantial, while the synthesis or accumulation of others is apparently unchanged. Thus, for example, NADPH–cytochrome c reductase accumulates very rapidly after phenobarbital administration, while cytochrome b_5 and ATPase show relatively little change. The mechanisms responsible for the differential hypertrophy of some functions of the liver endoplasmic reticulum are unclear. However, this model system does indicate quite clearly that individual components of the endoplasmic reticulum are capable of being independently inserted into or removed from the membrane substratum. The accumulation of certain classes of endoplasmic reticulum enzymes and of the membrane phospholipids seems to be ascribable partly to increased synthesis and partly to decreased degradation, the latter being particularly important in the case of the phospholipid components of the membrane (see 94). It is of interest, also, that upon cessation of phenobarbital treatment of rats, the induced enzymes returned to normal uninduced levels in a few days, apparently by an accelerated rate of degradation in the face of continuing synthesis (171), while the accumulated membranes remain in evidence for several weeks (172).

The developing rat liver provides the second model for the study of rapid endoplasmic reticulum membrane synthesis, which is most pronounced in the first few days following birth. The lipid composition of these membranes is similar to that of the adult membranes (173). Though the synthesis of lipid is accelerated in the developing endoplasmic reticulum membranes, the pattern of synthesis appears to be similar to that in the adult rat liver (174). However, the pattern of development of the various endoplasmic reticulum enzymes characteristically associated with this membrane is specific for each enzyme. The development of the two major electron-transport systems associated with the endoplasmic reticulum has been particularly chosen for study on the assumption that in this fashion possible micro-heterogeneity of the endoplasmic reticulum membrane would be eliminated (174, 175). If the membrane association of these systems has major functional meaning, it is most likely that the enzymes of a particular biochemical sequence are topographically closely related to one another, ensuring a maximal efficiency of biochemical transformation. Thus, if the membrane of the endoplasmic reticulum is assembled by the aggregation of individual functional membrane subunits, each of which is synthesized as a unit, all the enzymes of a particular biochemical sequence should appear in the developing membrane at once. Investigations have provided compelling evidence that this is not the case. Thus, in the NADPH electron-transport chain the first enzyme in the chain NADPH–cytochrome c reductase accumulated at a rapid rate after birth, while each of the subsequent enzymes and components in this sequence achieved adult specific activities progressively more slowly. A similar situation was found for the NADH electron-transport chain.

The heterogeneity of the patterns of development of endoplasmic reticulum enzymes affects many enzymes other than those involved in microsomal electron transport. Various nucleotide phosphatases develop at different rates. Glucose-6-phosphatase has been particularly studied. Early after birth, the enzyme activity rises briskly to levels exceeding those usually encountered in adult hepatic microsomes. The specific activity for this enzyme then gradually approaches that of the adult (175). Similar patterns of development have been seen for three enzymes involved in phospholipid synthesis—CDP choline diglyceride phosphorylcholine transferase, acyl coenzyme A 1-monoacyl glycerophosphorylcholine acyl transferase, and CDP diglyceride synthetase—although the detailed time course of the development of these enzymes is not identical (94, 176). In the case of most of these enzymes (glucose-6-phosphatase, NADPH–cytochrome c reductase, and the three phospholipid-synthesizing enzymes) activity in the perinatal period is much higher in the rough endoplasmic reticulum than in the smooth membranes (94, 175). Indeed the activity is barely detectable in the smooth endoplasmic reticulum just before and in some cases also soon after birth. These data suggest that the synthesis of many enzymes takes place on the ribosomes attached to the rough endoplasmic reticulum, with immediate insertion of the newly synthesized enzyme into the associated rough membrane. Glucose-6-phosphatase is apparently inserted uniformly and synchronously into the rough endoplasmic reticulum of a particular hepatocyte, though the developing hepatocyte population acquired the enzyme asynchronously (177). The enzyme appears in the smooth membranes as they develop. No evidence is obtainable from the pattern of development of glucose-6-phosphatase for the appearance of distinct foci of new membrane formation. Indeed, the techniques employed in these studies allowed no distinction between new and old membrane. The acquisition of functional enzyme in the smooth membrane seems then to be the consequence of one of three processes: membrane flow from rough to smooth endoplasmic reticulum, detachment of ribosomes from "matured" rough endoplasmic reticulum forming "matured" smooth membranes, or transport of newly synthesized enzyme through a nonmembranous phase. The first of these suggestions seems most likely, though the mechanism may be different for different enzyme proteins, as is suggested by the pattern of distribution of ATPase, an enzyme which has a higher specific activity in the smooth membranes at all stages of early perinatal development. It is possible that the smooth membranes studied in this latter case were contaminated by an ATPase-rich membrane not immediately derived from the endoplasmic reticulum. In any event, the role of the rough membrane as the site of membrane manufacture is supported by the fact that protein and lipid precursors are more rapidly incorporated into the rough endoplasmic reticulum of neonatal rats than into their smooth membranes (174). This is also true for protein precursors in the adult. Furthermore, the developmental pattern of various enzymes suggests that the synthesis of even closely related enzymes, functionally and presumably topographically, is independently regulated at least in time.

Such independent regulation of the individual components of the endoplasmic reticulum membranes is further substantiated by a comparison of their turnover times in the steady state or after phenobarbital treatment. Thus NADPH–cytochrome c reductase, cytochrome b_5, NAD glycohydrolase, total membrane proteins, and phospholipids all have different half-lives (162, 178–180; Table X). Using his double-isotope technique, Schimke and his colleagues have shown that the proteins of the smooth endoplasmic reticulum of adult rats have a heterogeneous turnover after phenobarbital treatment (179). Indeed, for many proteins of the endoplasmic reticulum, one determinant of the rate of turnover seems to be the size of the

Table X
Turnover of Components of Rat Liver Endoplasmic Reticulum

COMPONENT	LABELED PRECURSOR	APPARENT HALF-LIFE (DAYS)	REFERENCES
Total microsomal protein	^{3}H-Leucine	4.5	178
	^{14}C-Arginine (uniform)	5.2	179[a]
	^{14}C-Arginine (guanidino)	3.0	179
Washed microsomal[b] protein	^{3}H-Leucine	3.5	178
Rough endoplasmic reticulum protein	^{3}H-Leucine	4.6	178
	^{14}C-Arginine (uniform)	5.2	179
	^{14}C-Arginine (guanidino)	2.0	179
Smooth endoplasmic reticulum protein	^{3}H-Leucine	4.7	178
	^{14}C-Arginine (uniform)	5.8	179
	^{14}C-Arginine (guanidino)	2.1	179
NADPH–cytochrome c reductase[b]	^{3}H-Leucine	3.3	178
Cytochrome b_5[b]	^{3}H-Leucine	5.0	178
	^{14}C-Arginine (guanidino)	2.5	162
	^{3}H-δ-Amino levulinic acid	2.4	162
NAD glycohydrolase	^{14}C + ^{3}H-Leucine	18	180

[a] This study was performed using young rats, weighing only 80–100 g, which could influence the absolute values in comparison with those of other studies

[b] The washed microsomes should be regarded as the starting material for NADPH–cytochrome c reductase and cytochrome b_5.

protein (181). The rat liver ribosomal proteins also exhibit quite heterogeneous rates of turnover, which are nevertheless correlated with their molecular size. Larger proteins are more readily degraded, perhaps because their egress from lysosomes is not as rapid (182).

CONCLUSIONS

From the evidence reviewed above in relation to the biogenesis of the membranes of eukaroytic cellular organelles, virtually no indication can be derived of a completely de novo assembly of these biological membranes. In all cases new functional membrane formation appears to take place on or in relation to existing membranous organelles. This suggests that existing membrane may serve as a focus or scaffold for the orientation of newly synthesized membrane constituents. Irrespective of this constraint on the formation of new biological membranes within eukaroytic cells, all the evidence currently available seems to point to a multistep assembly process for these membranes. This appears to apply even to the biogenesis of mito-chondria, though some suggestion seems to be accumulating that a major element of the inner mitochondrial membrane may be degraded and turned over as an entity. The exact nature and identity of such a subunit must await further work.

The problem of membranous organelle biogenesis in eukaroytic organisms presents a number of unique challenges to the membrane biologist, all of which are still virtually virgin fields of endeavor. For example, it is known that all these membranes are composed largely of lipids and proteins, yet in few if any cases has it been possible to identify and study the actual interacting molecules involved in lipid-protein interactions. Knowledge of the mechanism and chemistry of the interplay between lipid and protein is vital for an understanding of membrane formation. Our current knowledge about the structural specificity and biosynthetic relation between the lipid and protein components of membranes is fragmentary indeed. The mechanisms responsible for maintaining the rather restricted composition of particular organellar membranes remain essentially undiscovered. A related problem concerns those mechanisms which might be responsible for the unique localization of particular components within particular membranes. This problem is seen in relation to the presence within a single cell of several isozymes which are differently located between the cytosol and the matrix of the mitochondrion. It is strongly suggested that many of these components are made outside the mitochondria and must somehow specifically gain access to the inner interstices of the organelle. Is this achieved by specific transport mechanisms for the characteristic mitochondrial components and isoenzymes? Are the binding properties of the mitochondrial matrix for specific proteins responsible for the localization of these components within this compartment? Some support is provided for such a suggestion by the recent report that the mitochondrial matrix is indeed an insoluble cellular network (183) which may well contain within it specific binding sites for components. Even more tantalizing perhaps is the possibility that mRNA molecules may be transported across the mitochondrial membrane, some of which may be translated within the mitochondria and some of which may be rejected as meaningless for the mitochondrial translation system. Indeed in relation to mitochondrial and chloroplast biogenesis, the communication mechanism which must exist between the nuclear and organellar genetic systems remains a challenge to those concerned with the genetic control of membrane biosynthesis. The appropriate control of organelle mass according to the physiological needs of the cell makes such a communication system a vital requirement.

Much remains to be done, but at least one prediction can be safely made: Advances in the understanding of membrane biogenesis will ultimately depend upon the development of new refined techniques of membrane separation and upon the application of the powerful tools of biochemical genetics to the manipulation of this problem. Most provocative among these will be the use of interspecific somatic cell hybrids for the study of the genetic control of organelle biogenesis (220, 221) and the nucleocyto-plasmic interaction in the expression if differentiated cell functions (222, 223) in mammalian cells.

REFERENCES

1. Korn, E. D. *Fed. Proc.* **28,** 6 (1969).
2. Wallach, D. H. F., and Gordon, A. *Fed. Proc.* **27,** 1263 (1968).
3. Rothfield, L., and Finkelstein, A. *Annu. Rev. Biochem.* **37,** 463 (1968).
4. Dallner, G., and Ernster, L. *J. Histochem. Cytochem.* **16,** 611 (1968).
5. Luck, D. J. L. *J. Cell Biol.* **24,** 445 (1964).
6. Mindich, L. *J. Mol. Biol.* **49,** 433 (1970).
7. Mindich, L. *Proc. Natl. Acad. Sci. U.S.A.* **68,** 420 (1971).
8. Chun, E. H. L., Vaughan, M. H., and Rich, A. *J. Mol. Biol.* **7,** 130 (1963).
9. Sager, R., and Ishida, M. R. *Proc. Natl. Acad. Sci. U.S.A.* **50,** 725 (1963).
10. Luck, D. J. L. and Reich, E. *Proc. Natl. Acad. Sci. U.S.A.* **52,** 931 (1964).
11. Wilkie, D. *The Cytoplasm in Heredity.* Methuen, London, 1964.
12. Coen, D., Deutsch, J., Netter, P., Petrochilo, E., and Slonimski, P. P. *Symp. Soc. Exp. Biol.* **24,** 449 (1970).
13. Williamson, D. H. *Symp. Soc. Exp. Biol.* **24,** 247 (1970).
14. Sherman, F., and Slonimski, P. P. *Biochim. Biophys. Acta* **90,** 1 (1964).
15. Linnane, A. W., and Haslam, J. M. *Current Topics in Cellular Regulation.* Ed. B. L. Horecker and E. Stadtman. Academic Press, New York, 1970, vol. 2, p. 101.
16. Correns, C. *Z. Induktive Abstammungs-Verbungslehre* **1,** 291 (1908).
17. Baur, E. *Z. Induktive Abstammungs-Verbungslehre* **1,** 330 (1908).
18. Sager, R., and Ramanis, Z. *Symp. Soc. Exp. Biol.* **24,** 401 (1970).
19. Rabinowitz, M., and Swift, H. H. *Physiol. Rev.* **50,** 376 (1970).
20. Borst, P., and Kroon, A. M. *Int. Rev. Cytol.* **26,** 108 (1969).
21. Schatz, G. *Membranes of Mitochondria and Chloroplasts.* Ed. E. Racker. Van Nostrand Reinhold, New York, 1970, p. 251.
22. Ashwell, M., and Work, T. S. *Annu. Rev. Biochem.* **39,** 251 (1970).
23. Parsons, J. A., and Rustad, R. C. *J. Cell Biol.* **37,** 683 (1968).
24. Borst, P., Ruttenberg, G. J. C. M., and Kroon, A. M. *Biochim. Biophys. Acta* **149,** 140 (1967).
25. Wolstenholme, D. R., and Gross, N. J. *Proc. Natl. Acad. Sci. U.S.A.* **61,** 245 (1968).
26. Wood, D. D., and Luck, D. J. L. *J. Mol. Biol.* **41,** 211 (1969).
27. Hollenberg, C. P., Borst, P., and Van Bruggen, E. J. F. *Biochim. Biophys. Acta* **209,** 1 (1970).
28. Suyama, Y., and Miura, K. *Proc. Natl. Acad. Sci. U.S.A.* **60,** 235 (1968).
29. Mounolou, J. C., Perrodin, G., and Slonimski, P. P. *Biochem. Biophys. Res. Commun.* **24,** 218 (1966).
30. Bernardi, G., Faures, M., Piperno, G., and Slonimski, P. P. *J. Mol. Biol.* **48,** 23 (1970).
31. Bernardi, G., Carnevali, I., Nicolaieff, A., Piperno, G., and Tecce, G. *J. Mol. Biol.* **37,** 493 (1968).

32. Fauman, M. *Biochemistry*. Submitted 1971.
33. Smith, D., Tauro, P., Schweizer, E., and Halvorson, H. O. *Proc. Natl. Acad. Sci. U.S.A.* **60,** 936 (1968).
34. Gross, N. J., and Rabinowitz, M. *J. Biol. Chem.* **244,** 1563 (1969).
35. Kirschner, R. H., Wolstenholme, D. R., and Gross, N. J. *Proc. Natl. Acad Sci. U.S.A.* **60,** 1466 (1968).
36. Levine, R. P., and Goodenough, U. W. *Annu. Rev. Genet.* **4,** 397 (1970).
37. Chiang, K. S. *Proc. Natl. Acad. Sci. U.S.A.* **60,** 194 (1968).
38. Chiang, K. S., and Sueoka, N. *Proc. Natl. Acad. Sci. U.S.A.* **57,** 1506 (1967).
39. Swift, H., and Wolstenholme, D. R. *Handbook of Molecular Cytology.* Ed. A. Lima-de-Faria. North Holland, Amsterdam, 1969, p. 972.
40. Tewari, K. K., and Wildman, S. G. *Symp. Soc. Exp. Biol.* **24,** 147 (1970).
41. Wells, R., and Birnstiel, M. *Biochem. J.* **112,** 777 (1969).
42. Stutz, E. *FEBS Letters* **8,** 25 (1970).
43. Bastia, D., Chiang, K. S., Swift, H., and Siersma, P. *Proc. Natl. Acad. Sci. U.S.A.* **68,** 1157 (1971).
44. Borst, P., and Grivell, L. A. *FEBS Letters* **13,** 73 (1971).
45. Kuntzel, H. *Nature* **222,** 142 (1969).
46. Smith, A. E., and Marcker, K. A. *J. Mol. Biol.* **38,** 241 (1968).
47. Galper, J. B., and Darnell, J. E. *Biochem. Biophys. Res. Commun.* **34,** 205 (1969).
48. Galper, J. B., and Darnell, J. E. *J. Mol. Biol.* **57,** 363 (1971).
49. Aloni, Y., and Attardi, G. *J. Mol. Biol.* **55,** 251 (1971).
50. Borst, P. *Symp. Soc. Exp. Biol.* **24,** 201 (1970).
51. Hoober, J. K., and Blobel, G. *J. Mol. Biol.* **41,** 121 (1969).
52. Scott, N. S., Munns, R., and Smillie, R. M. *FEBS Letters* **10,** 149 (1970).
53. Stutz, E., and Noll, H. *Proc. Natl. Acad. Sci. U.S.A.* **57,** 774 (1967).
54. Rawson, J. R., and Stutz, E. *Biochim. Biophys. Acta* **190.** 368 (1969).
55. Ingle, J., Possingham, J. V., Wells, R., Leaver, C. J., and Loening, U. E. *Symp. Soc. Exp. Biol.* **24,** 303 (1970).
56. Aliev, K. A., Filippovich, I. I., and Sisakayan, N. M. *Mol. Biol.* **1,** 206 (1967).
57. Reger, B. J., Fairfield, S. A., Epler, J. L., and Barnett, W. E. *Proc. Natl. Acad. Sci. U.S.A.* **67,** 1207 (1970).
58. Schwartz, J. H., Meyer, R., Eisenstadt, J. M., and Brawerman, G. *J. Mol. Biol.* **25,** 571 (1967).
59. Leis, J. P., and Keller, E. B. *Proc. Natl. Acad. Sci. U.S.A.* **67,** 1593 (1970).
60. Burkard, G., Eclancher, B., and Weil, J. H. *FEBS Letters* **4,** 285 (1969).
61. Kirk, J. T. O., and Tilney-Bassett, R. A. E. *The Plastids. Their Chemistry, Structure, Growth and Inheritance.* Freeman, London, 1967.
62. Surzycki, S. J. *Proc. Natl. Acad. Sci. U.S.A.* **63,** 1327 (1969).
63. Surzycki, S. J., Goodenough, U. W., Levine, R. P., and Armstrong, J. J. *Symp. Soc. Exp. Biol.* **24,** 13 (1970).
64. Roodyn, D. B., and Wilkie, D. *The Biogenesis of Mitochondria.* Methuen, London, 1968.
65. Kroon, A. M., and de Vries, H. *Symp. Soc. Exp. Biol.* **24,** 181 (1970).
66. Schiff, J. A. *Symp. Soc. Exp. Biol.* **24,** 277 (1970).
67. Wintersberger, E., and Viehhauser, G. *Nature* **220,** 699 (1968).

68. Dawid, I. B. *Symp. Soc. Exp. Biol.* **24,** 227 (1970).
69. Attardi, B., and Attardi, G. *Nature* **224,** 1079 (1969).
70. Attardi, G., Aloni, Y., Attardi, B., Ojala, D., Pica-Mattoccia, L., Robberson, D., and Storrie, B. *Cold Spring Harbor Symp. Quant. Biol.* **35,** 599 (1970).
71. Nass, M. M. K., and Buck, C. A. *Proc. Natl. Acad. Sci. U.S.A.* **62,** 506 (1969).
72. Nass, M. M. K., and Buck, C. A. *J. Mol. Biol.* **54,** 187 (1970).
73. Halbreich, A., and Rabinowitz, M. *Proc. Natl. Acad. Sci. U.S.A.* **68,** 294 (1971).
74. Cohen, M., and Rabinowitz, M. *J. Cell Biol.* **47,** 37A (1970).
75. Casey, J., Cohen, M., Rabinowitz, M., Fukuhara, H., and Getz, G. S. *J. Mol. Biol.* **63,** 431 (1972).
76. Cohen, M., Casey, J., Rabinowitz, M., and Getz, G. S. *J. Mol. Biol.* **63,** 441 (1972).
77. Bastia, D. Doctoral thesis. University of Chicago, 1971.
78. Stutz, E., and Rawson, J. *Biochim. Biophys. Acta* **209,** 16 (1970).
79. Scott, N. S., and Smillie, R. M. *Biochem. Biophys. Res. Commun.* **28,** 598 (1967).
80. Goldring, E. S., Grossman, L. I., Krupnick, D., Cryer, D. R., and Marmur, J. *J. Mol. Biol.* **52,** 323 (1970).
81. Weiss, S. B., Hsu, W. T., Foft, J. W., and Scherberg, N. H. *Proc. Natl. Acad. Sci. U.S.A.* **61,** 114 (1968).
82. Aloni, Y., and Attardi, G. *J. Mol. Biol.* **55,** 271 (1971).
83. Sherman, F., Stewart, J. W., Parker, J. H., Putterman, G. J., Agrawal, B. B. L., and Margoliash, E. *Symp. Soc. Exp. Biol.* **24,** 85 (1970).
84. Rifkin, M. R., and Luck, D. J. L. *Proc. Natl. Acad. Sci. U.S.A.* **68,** 287 (1970).
85. Chen, W. L., and Charalampous, F. C. *J. Biol. Chem.* **244,** 2767 (1969).
86. Slonimski, P. P. *La formation des enzymes respiratoires chez la levure.* Masson, Paris, 1953.
87. Schatz, G. *J. Biol. Chem.* **243,** 2192 (1968).
88. Clark-Walker, G. D., and Linnane, A. W. *J. Cell Biol.* **34,** 1 (1967).
89. Tzagoloff, A. *J. Biol. Chem.* **244,** 5027 (1969).
90. Tzagoloff, A. *J. Biol. Chem.* **245,** 1545 (1970).
91. Vary, M. J., Edwards, C. L., and Stewart, P. R. *Arch. Biochem. Biophys.* **130,** 235 (1969).
92. Neupert, W., Brdiczka, D., and Bücher, T. *Biochem. Biophys. Res. Commun.* **27,** 488 (1967).
93. Hoober, J. K., Siekevitz, P., and Palade, G. E. *J. Biol. Chem.* **244,** 2621 (1969).
94. Getz, G. S. *Adv. Lipid Res.* **8,** 175 (1970).
95. Hajra, A. K., and Agranoff, B. *J. Biol. Chem.* **243,** 1617 (1968).
96. Carter, J. R., and Kennedy, E. P. *J. Lipid Res.* **7,** 678 (1966).
97. Stoffel, W., and Schiefer, H. G. *Hoppe Seyler's Z. Physiol. Chem.* **349,** 1017 (1968).
98. Shephard, E. H., and Hubscher, G. *Biochem. J.* **113,** 429 (1969).
99. Sarzala, M. G., Van Golde, L. M. G., De Kruyff, B., and Van Deenen, L. L. M. *Biochim. Biophys. Acta* **202,** 106 (1970).
100. Zborowski, J., and Wojtczak, L. *Biochim. Biophys. Acta* **187,** 73 (1969).
101. Smith, J. D., and Law, J. H. *Biochim. Biophys. Acta* **202,** 141 (1970).
102. Ostrow, D. *Fed. Proc.* **30,** 1226 (1971).

103. Galliard, L., Michell, R. H., and Hawthorne, J. N. *Biochim. Biophys. Acta* **106,** 551 (1965).
104. Vorbeck, M. L., and Martin, A. P. *Biochem. Biophys. Res. Commun.* **40,** 901 (1970).
105. Petzold, C. L., and Agranoff, B. W. *J. Biol. Chem.* **242,** 1187 (1968).
106. Mangnall, D., and Getz, G. S. *Fed. Proc.* **30,** 1226 (1971).
107. Kiyasu, J. Y., Pieringer, R. A., Paulus, H., and Kennedy, E. P. *J. Biol. Chem.* **238,** 2293 (1963).
108. Davidson, J. B., and Stanacev, N. Z. *Can. J. Biochem.* **48,** 633 (1970).
109. Stanacev, N. Z., Stuhne-Sekalec, L., Brookes, K. B., and Davidson, J. B. *Biochim Biophys. Acta.* **176,** 650 (1969).
110. Davidson, J. B., and Stanacev, N. Z. *Biochem. Biophys. Res. Commun.* **42,** 1191 (1971).
111. Jakovcic, S., Getz, G. S., Rabinowitz, M., Jakob, H., and Swift, H. *J. Cell Biol.* **48,** 490 (1971).
112. McKay, R., Druyan, R., Getz, G. S., and Rabinowitz, M. *Biochem. J.* **114,** 455 (1969).
113. Bosmann, H. B., and Martin, S. S. *Science* **164,** 190 (1969).
114. Bosmann, H. B., and Case, K. R. *Biochem. Biophys. Res. Commun.* **36,** 830 (1969).
115. Appelqvist, L. A., Stumpf, P. K., and von Wettstein, D. *Plant Physiol.* **43,** 163 (1968).
116. Ongun, A., and Mudd, J. B. *J. Biol. Chem.* **243,** 1565 (1968).
117. Racker, E. *Essays Biochem.* **6,** 1 (1970).
118. Racker, E. *Membranes of Mitochondria and Chloroplasts.* Ed. E. Racker. Van Nostrand Reinhold, New York, 1970, p. 127.
119. Luck, D. J. L. *J. Cell Biol.* **16,** 483 (1963).
120. Droz, B., and Bergeron, M. *Compt. Rend.* **261,** 2757 (1965).
121. Jollow, D., Kellerman, G. M., and Linnane, A. W. *J. Cell Biol.* **37,** 22 (1968).
122. Criddle, R. S., and Schatz, G. *Biochemistry* **8,** 322 (1969).
123. Swift, H., Rabinowitz, M., and Getz, G. S. *Biochemical Aspects of the Biogenesis of Mitochondria.* Ed. E. C. Slater, J. M. Tager, S. Papa, and E. Quagliariello. Adiatrica Editrice, 1968, p. 71.
124. Plattner, H., and Schatz, G. *Biochemistry* **8,** 339 (1969).
125. Fukuhara, H. *Eur. J. Biochem.* **11,** 135 (1969).
126. Paltauf, F., and Schatz, G. *Biochemistry* **8,** 335 (1969).
127. Plattner, H., Salpeter, M. M., Saltzgaber, J., and Schatz, G. *Proc. Natl. Acad. Sci. U.S.A.* **66,** 1252 (1970).
128. Schatz, G., and Saltzgaber, J. *Biochem. Biophys. Res. Commun.* **37,** 996 (1969).
129. Groot, G. S. P., Kovac, L., and Schatz, G. *Proc. Natl. Acad. Sci. U.S.A.* **68,** 308 (1971).
130. Schatz, G., Groot, G. S. P., Mason, T., Rouslin, W., Wharton, D. C., and Saltzgaber, J. *Fed. Proc.* **31,** 21 (1972).
131. Rouslin, W., and Schatz, G. *Biochem. Biophys. Res. Commun.* **37,** 1002 (1969).
132. Fukuhara, H. *J. Mol. Biol.* **17,** 334 (1966).
133. Rabinowitz, M., Slonimski, P. P., and Hsu, H. Unpublished experiments.

134. Henson, C. P., Perlman, P., Weber, C., and Mahler, H. R. *Biochemistry* **7**, 4445 (1968).
135. Henson, C. P., Weber, C., and Mahler, H. R. *Biochemistry* **7**, 4431 (1968).
136. Tzagoloff, A. *J. Biol. Chem.* **246**, 3050 (1971).
137. Nakami, N., and Pious, D. A. *Nature* **216**, 1087 (1967).
138. Jakovcic, S., Haddock, J., Getz, G. S., Rabinowitz, M., and Swift, H. *Biochem. J.* **121**, 341 (1971).
139. Gustafsson, R., Tata, J. R., Lindberg, O., and Ernster, L. *J. Cell Biol.* **26**, 555 (1965).
140. Gross, N. J. *J. Cell Biol.* **48**, 29 (1971).
141. Yotsuyanagi, Y. *J. Ultrastruct. Res.* **7**, 141 (1962).
142. Gailey, F. B., and Lester, R. L. *Fed. Proc.* **27**, 485 (1968).
143. Mangnall, D. Unpublished experiments.
144. Resnick, M. R., and Mortimer, R. K. *J. Bacteriol.* **92**, 597 (1966).
145. Levin, B. Unpublished experiments.
146. McMurray, W. C., and Dawson, R. M. C. *Biochem. J.* **112**, 91 (1969).
147. Stein, O., and Stein, Y. *J. Cell Biol.* **40**, 461 (1969).
148. Wirtz, K. W. A., and Zilversmit, D. B. *Biochim. Biophys. Acta* **187**, 468 (1969).
149. Appelqvist, L. A., Boynton, J. E., Henningsen, K. W., Stumpf, P. K., and von Wettstein, D. *J. Lipid Res.* **9**, 513 (1968).
150. Ohad, I., Siekevitz, P., and Palade, G. E. *J. Cell Biol.* **35**, 521 (1967).
151. Ohad, I., Siekevitz, P., and Palade, G. E. *J. Cell Biol.* **35**, 553 (1967).
152. DePetrocellis, B., Siekevitz, P., and Palade, G. E. *J. Cell Biol.* **44**, 618 (1970).
153. Swick, R. W. *Arch. Biochem. Biophys.* **63**, 226 (1956).
154. Swick, R. W. *J. Biol. Chem.* **218**, 577 (1956).
155. Fletcher, M. J., and Sanadi, D. R. *Biochim. Biophys. Acta* **51**, 356 (1961).
156. Gross, N. J., Getz, G. S., and Rabinowitz, M. *J. Biol. Chem.* **244**, 1552 (1969).
157. Doyle, D., and Schimke, R. T. *Annu. Rev. Biochem.* **39**, 929 (1970).
158. Schimke, R. T., Ganschow, R., Doyle, D., and Arias, I. M. *Fed. Proc.* **27**, 1223 (1968).
159. DeBernard, B., Getz, G. S., and Rabinowitz, M. *Biochim. Biophys. Acta* **193**, 58 (1969).
160. Beattie, D. *Biochim. Biophys. Res. Commun.* **35**, 721 (1969).
161. Brunner, G., and Neupert, W. *FEBS Letters* **1**, 153 (1968).
162. Druyan, R., DeBernard, B., and Rabinowitz, M. *J. Biol. Chem.* **244**, 5874 (1969).
163. Swick, R. W., Rexroth, A. K., and Stange, J. L. *J. Biol. Chem.* **243**, 3581 (1968).
164. Aschenbrenner, V., Druyan, R., Albin, R., and Rabinowitz, M. *Biochem. J.* **119**, 157 (1970).
165. Swift, H., and Hruban, Z. *Fed. Proc.* **23**, 1026 (1964).
166. Swick, R. W., Stange, J. L., Nance, S. L., and Thomson, J. F. *Biochemistry* **6**, 737 (1967).
167. Glaumann, H., and Dallner, G. *J. Lipid Res.* **9**, 720 (1968).
168. Dallman, P., Dallner, G., Bergstrand, A., and Ernster, L. *J. Cell Biol.* **41**, 357 (1969).
169. Tata, J. R. *Biochem. J.* **116**, 617 (1970).
170. Orrenius, S., Ericsson, J. L. E., and Ernster, L. *J. Cell Biol.* **25**, 627 (1965).

171. Kuriyama, Y., Omura, T., Siekevitz, P., and Palade, G. E. *J. Biol. Chem.* **244,** 2017 (1969).
172. Remmer, H., and Merker, H. J. *Ann. N.Y. Acad. Sci.* **123,** 79 (1965).
173. Miller, J. E., and Cornatzer, W. E. *Biochim. Biophys. Acta* **125,** 534 (1966).
174. Dallner, G., Siekevitz, P., and Palade, G. E. *J. Cell Biol.* **30,** 73 (1966).
175. Dallner, G., Siekevitz, P., and Palade, G. E. *J. Cell Biol.* **30,** 97 (1966).
176. Mangnall, D. Unpublished experiments.
177. Leskes, A., Siekevitz, P., and Palade, G. E. *J. Cell Biol.* **49,** 264, 288 (1971).
178. Omura, T., Siekevitz, P., and Palade, G. E. *J. Biol. Chem.* **242,** 2389 (1967).
179. Arias, I. M., Doyle, D., and Schimke, R. T. *J. Biol. Chem.* **244,** 3303 (1969).
180. Bock, K. W., Siekevitz, P., and Palade, G. E. *J. Biol. Chem.* **246,** 188 (1971).
181. Dehlinger, P. J., and Schimke, R. T. *J. Biol. Chem.* **246,** 2574 (1971).
182. Segal, H. L., Matsuzawa, T., Haider, M., and Abraham, G. J. *Biochem. Biophys. Res. Commun.* **36,** 764 (1969).
183. Hackenbrock, C. R. *Proc. Natl. Acad. Sci. U.S.A.* **61,** 598 (1968).
184. Corneo, G., Zardi, L., and Polli, E. *FEBS Letters* **18,** 301 (1971).
185. Wolstenholme, D. R., and Dawid, I. B. *J. Cell Biol.* **39,** 222 (1968).
186. Nass, M. M. K. *J. Mol. Biol.* **42,** 529 (1969).
187. Stutz, E., and Vandrey, J. P. *FEBS Letters* **17,** 277 (1971).
188. Manning, J. E., Wolstenholme, D. R., Ryan, R. S., Hunter, J. A., and Richards, O. C. *Proc. Natl. Acad. Sci. U.S.A.* **68,** 1169 (1971).
189. Schäfer, K. P., Bugge, G., Grandi, M., and Küntzel, H. *Eur. J. Biochem.* **21,** 478 (1971).
190. Clayton, D. A., and Brambl, R. M. *Biochem. Biophys. Res. Commun.* **46,** 1477 (1972).
191. Piperno, G., Fonty, G., and Bernardi, G. *J. Mol. Biol.* **65,** 191 (1972).
192. Kasamatsu, H., Robberson, D. L., and Vinograd, J. *Proc. Natl. Acad. Sci. U.S.A.* **68,** 2252 (1971).
193. Robberson, D. L., Kasamatsu, H., and Vinograd, J. *Proc. Natl. Acad. Sci. U.S.A.* **69,** 737 (1972).
194. Ter Schegget, J., and Borst, P. *Biochim. Biophys. Acta* **246,** 249 (1971).
195. Arnberg, A., van Bruggen, E. F. J., Ter Schegget, J., and Borst, P. *Biochim. Biophys. Acta* **246,** 353 (1971).
196. O'Brien, T. W. *J. Biol. Chem.* **246,** 3409 (1971).
197. Attardi, G., and Ojala, D. *Nature New Biol.* **229,** 133 (1971).
198. Schmitt, H. *FEBS Letters* **15,** 186 (1971).
199. Kleinow, W., and Neupert, W. *FEBS Letters* **15,** 359 (1971).
200. Scragg, A. H. *FEBS Letters* **17,** 111 (1971).
201. Richter, D., and Lipmann, F. *Biochemistry* **9,** 5065 (1970).
202. Vasconcelos, A. C. L., and Bogorad, L. *Biochim. Biophys. Acta* **228,** 492 (1971).
203. Morimoto, H., Scragg, A. H., Nekhorocheff, J., Villa, V., and Halvorson, H. O. *Autonomy and Biogenesis of Mitochondria and Chloroplasts.* Ed. N. K. Boardman, A. W. Linnane, and R. M. Smillie. North Holland, Amsterdam, 1971, p. 194.
204. Schäfer, K. P., and Küntzel, H. *Biochem. Biophys. Res. Commun.* **46,** 1312 (1972).
205. Cohen, M., and Rabinowitz, M. Unpublished observations.

206. Richter, D. *Biochemistry* **10**, 4422 (1971).
207. Mahler, H. R., and Perlman, P. S. *Biochemistry* **10**, 2979 (1971).
208. Hostetler, K. Y., and van den Bosch, H. *Biochim. Biophys. Acta* **260**, 380 (1972).
209. Hostetler, K. Y., van den Bosch, H., and van Deenen, L. L. M. *Biochim. Biophys. Acta* **260**, 507 (1972).
210. Racker, E., and Kandrach, A. *J Biol. Chem.* **246**, 7069 (1971).
211. Weiss, H., Sebald, W., and Bücher, T. *Eur. J. Biochem.* **22**, 19 (1971).
212. Tzagoloff, A., and Meagher, P. *J. Biol. Chem.* **246**, 7328 (1971).
213. Tzagoloff, A., and Meagher, P. *J. Biol. Chem.* **247**, 594 (1972).
214. Brown, R. D., and Haselkorn, R. *Proc. Natl. Acad. Sci. U.S.A.* **68**, 2536 (1971).
215. Eytan, G., and Ohad, I. *J. Biol. Chem.* **247**, 112 (1972).
216. Hoober, J. K. *J. Biol. Chem.* **245**, 4327 (1970).
217. Hoober, J. K. *J. Cell Biol.* **52**, 84 (1972).
218. Eytan, G., and Ohad, I. *J. Biol. Chem.* **247**, 122 (1972).
219. Werner, S., and Neupert, W. *Eur. J. Biochem.* **25**, 379 (1972).
220. Clayton, D. A., Teplitz, R. L., Nabholz, M., Dovey, H., and Bodmer, W. *Nature* **234**, 560 (1971).
221. Attardi, B., and Attardi, G. *Proc. Natl. Acad. Sci. U.S.A.* **69**, 129 (1972).
222. Schneider, J. A., and Weiss, M. C. *Proc. Natl. Acad. Sci. U.S.A.* **68**, 127 (1971).
223. Peterson, J. A., and Weiss, M. C. *Proc. Natl. Acad. Sci. U.S.A.* **69**, 571 (1972).

14

Mark E. Tourtellotte

Mycoplasma Membranes: Structure and Function

In the course of evolution leading to what we define as free-living cells, one of the most important events was, without doubt, the development of the cell membrane. The full significance of membrane systems is only now beginning to be appreciated. In addition to providing the physical boundary which separates the living cell from its environment, membranes are intimately involved in essential life processes, including transport of metabolites, nerve conduction and brain function, muscle contraction, cell contact recognition, protein synthesis, oxidative phosphorylation, and photosynthesis (1, 3, 6). Understanding membrane-related functions on a molecular level requires a knowledge of membrane structure. Despite extensive investigation, a definitive answer to the question of molecular architecture of membranes has proved decidedly elusive (1, 2, 6). All membranes, both cytoplasmic cell membranes and the internal membranes of higher cells, are composed principally of protein and lipid (3). An understanding of membrane structure requires a knowledge of the conformation of lipid and protein and of the interrelation between these components.

Various models have been proposed for the arrangement of protein and lipid in membranes. The best known is the Danielli-Davson bilayer model (11), which was proposed in the 1930s and refined by Robertson (36) as the "unit membrane" hypothesis. The main feature (Fig. 1) is a bimolecular layer of lipid; the hydrocarbon chains of the lipid bilayer form a continuous hydrophobic phase, and the protein is located on the hydrophilic surface of the lipid. In the 1960s, alternative models were proposed (8, 18) in which the protein is intimately associated with the apolar portion of the lipid molecules. The Benson (8) model (Fig. 1) is an example. The fundamental structural feature is hydrophobic lipid-protein association. This is also the basic feature of the Green model (18), where the membrane is viewed as a network of lipoprotein subunits. Other membrane models (42, 58) combine features of both the bilayer and the Benson models.

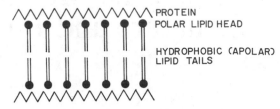

PROTEIN
POLAR LIPID HEAD

HYDROPHOBIC (APOLAR)
LIPID TAILS

DANIELLI–DAVSON–ROBERTSON BILAYER MODEL

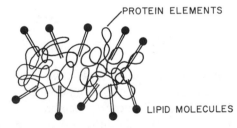

PROTEIN ELEMENTS

LIPID MOLECULES

BENSON MODEL

Figure 1

Models for membrane structure, viewed in cross section. (Adapted from Benson, A. A., *J. Am. Oil Chem. Soc. 43*, 265, 1966; Danielli, J. F., and Davson, H., *J. Cell. Comp. Physiol. 5*, 495, 1935; and Robertson, J. D., *Progr. Biophys. Biophys. Chem. 10*, 344, 1960.)

In the Danielli-Davson model (11) the proteins on the surface of the lipid bilayer were postulated to have little secondary structure. Recent spectroscopic studies have shown this to be incorrect. Infrared studies (23) indicate the absence of significant amounts of β-structure, and optical rotatory dispersion and circular dichroism studies of a wide range of membranes show that membrane proteins contain significant amounts of α-helix (6).

Extensive morphological and biophysical investigations of many different membranes have failed to show definitively that a particular model is correct (1, 3, 6). The association of protein and lipid, whether primarily polar or nonpolar, is uncertain.

The mycoplasmas have no cell wall and no endoplasmic reticulum. They are enclosed only by a homogeneous cytoplasmic membrane which, by means of simple techniques, can be separated from other cellular material. In addition, the fatty acid composition of the membrane can be drastically altered. Thus, the mycoplasmas provide an ideal system for studying membrane structure and associated functions (5).

Definition of Mycoplasmas

The mycoplasmas, like the bacteria, are simple (prokaryotic) cells in that they contain no intracellular inclusions such as mitochondria and endoplasmic reticulum. Unlike the bacteria, however, the mycoplasmas lack a cell wall, and this accounts for the marked pleomorphism of mycoplasma cells observed under the microscope (Fig. 2).

The first mycoplasma isolated was *Mycoplasma mycoides*, which was grown on artificial medium in 1898 by Nocard and Roux. Previous to this, Pasteur had recognized a disease, bovine pleuropneumonia, caused by this agent, but was unable to grow it in broth. Because, like the viruses, mycoplasma was filtrable through porcelain filters, until Nocard and Roux successfully propagated the agent on a complex but cell-free medium, it was, in fact, thought to be a virus. In 1931, Elford of the National Institute for Medical Research in London, who developed the first filters in which pore size could be precisely determined, showed that the pleuropneumonia agent contained viable cells 125–150 mμ in diameter. These particles were smaller than many viruses, and yet were free-living (5). Subsequent studies have shown that the mycoplasmas will squeeze through small filter pores and that the smallest viable cells are probably about 200 mμ in diameter. Still, the mycoplasmas are the smallest free-living cells, and it has been shown by Morowitz and others that the DNA in *Mycoplasma hominis* is approximately 500×10^6 daltons—smaller by a factor of 5 than the DNA of *Escherichia coli*. Since the coding ratio of DNA to protein is approximately 20:1, if we assume an average protein molecular weight of 40,000, one cistron of DNA would be 800,000 daltons; thus this mycoplasma would contain approximately 650 cistrons, a much smaller number than that observed in most bacteria (27).

As might be predicted from the limited number of cistrons, mycoplasmas are quite limited in their biosynthetic abilities. Although they synthesize DNA and RNA, and contain ribosomes and other essentials for protein synthesis, even the simplest strain is incapable of synthesizing amino acids and nucleotides, which are therefore required in growth media.

Since most strains require cholesterol as a growth factor, which bacteria and plants do not, it is tempting to think of these cells as "animals."

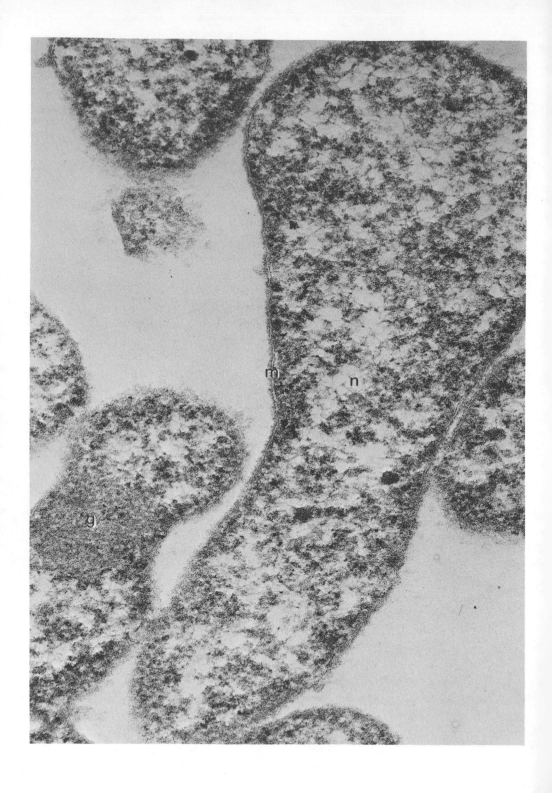

However, the evidence that mycoplasmas are prokaryotic cells and in no other way resemble animal cells is most convincing. The requirement for cholesterol is most likely due to evolutionary convergence, especially since most mycoplasmas are animal parasites and thus have access to an abundance of this sterol (52).

ISOLATION OF MYCOPLASMA MEMBRANES

Any study of cell membranes must seriously consider whether the process of isolation disturbs the structural integrity of the membrane. "Membranes" can be readily isolated from mycoplasma by mechanical means such as the French press, X-press, Hughes press, and sonic and ultrasonic disintegrators, but the use of a press releases proteins from the membrane to a considerable degree and disrupts parts of the membrane into very small pieces. The use of sonic and ultrasonic disintegrators has the same disadvantages plus the additional effect of reorienting lipid-protein interactions.

Fortunately, the membranes of at least one mycoplasma are subject to osmotic shock. Washed cells are placed in a hypotonic salt solution and allowed to autolyze. As judged by localization of ATPase, DPNH oxidase, and lipid, all of which are membrane-associated, the membranes under these conditions of lysis appear to remain reasonably intact. Interestingly, cells do not significantly lyse in the presence of divalent cations such as Mg^{++} even in distilled water. It is obvious that divalent cations play a large role in maintaining the integrity of the membrane (32).

Membranes are then further purified by washing in buffer and centrifugation on a sucrose density gradient. *Mycoplasma laidlawii* membranes have a density of approximately 1.17–1.18. As viewed by electron microscopy, both in thin section and after freeze-etching, the isolated membranes are indistinguishable from membranes of whole cells (15, 54).

CHEMICAL COMPOSITION OF MEMBRANES

Like other biologic membranes, *M. laidlawii* membranes consist essentially of protein (55–60%) and lipid (36–40%). They also contain significant amounts of the amino sugars, glucosamine, and galactosamine, which do not appear to be covalently linked to protein. Trace amounts of DNA and RNA are present and are readily removed by DNAase and RNAase.

Figure 2

Electron micrograph of *M. laidlawii* showing cell membrane (*m*), nuclear material (*n*), and granular region (*g*). Note the pleomorphism and absence of mitochondria, inner membrane, and cell wall. (From Maniloff, J. *J. Bacteriol. 102*, 561, 1970.)

Membrane Proteins

The separation of mycoplasma membrane protein and lipid was first achieved in 1967 using the detergent, sodium dodecyl sulfate (SDS), to solubilize the membranes and sucrose density gradients to separate the protein from the lipid (15). Rodwell *et al.* (37), using the detergents SDS and deoxycholate along with ammonium sulfate precipitation, also separated *M. laidlawii* membrane protein from lipid. Since this protein (80% of total membrane protein) was insoluble in aqueous salt solutions, it was at first thought to be similar to "mitochondrial structural protein" and to contain high levels of nonpolar amino acids. However, amino acid analyses of membrane protein by Engelman and Morowitz (14) failed to substantiate this hypothesis: Mycoplasma membranes contain substantial amounts of polar amino acids; the only unusual finding was the low content of cysteine (0.2 mole %).

Optical rotatory dispersion (ORD), circular dichroism (CD), and infrared spectroscopy spectra indicate a significant amount of α-helicity with little or no β-structure detectable (48). Disc-gel electrophoresis on polyacrylamide gels in the presence of SDS reveals the presence of between 20 and 30 separate bands ranging in molecular weight from approximately 15,000 to 100,000 (28).

It is interesting that between 5 and 10% of the membrane "protein" is extracted by chloroform-methanol (2:1) with the lipid fraction. However, this material has not as yet been characterized. It is apparently *not* covalently linked to lipid.

Membrane Lipids

Most of the lipids of *M. laidlawii* are polar (97+%); the nonpolar lipids include a trace of diglycerides and approximately 1% carotenoids (5).

Neutral lipids. The carotenoids, which are contained in the non-sterol-requiring strains (i.e., *M. laidlawii*), consist of four major and three minor polyterpenes. Neurosporene is the most polar of the major pigments and is identical to the all-*trans*-compound found in tomatoes. Because of its high molar extinction coefficient, this carotenoid is primarily responsible for the yellow color of the organism (43).

Sterols. Although *M. laidlawii*, the strain mainly discussed in this chapter, does not require sterols for growth, it is perhaps desirable to discuss briefly the sterol requirements of other mycoplasmas, for as explained later, cholesterol has a profound effect on thermal transitions of polar lipids and a strong influence on membrane structure.

Depending on the strain, between 10% and 30% of the total lipid is cholesterol. Although cholesterol esters are also incorporated by these strains (2–3%), there is no evidence that the free cholesterol molecule is chemically altered by the cell. Furthermore, cholesterol esters do not replace free cholesterol as a growth requirement. Other sterols which do replace cholesterol are cholestanol, lathosterol, β-sitosterol, stigmasterol, and ergosterol. Steroids which do not replace cholesterol are coprostanol, 7-dehydrocholesterol, epicholesterol, and cholestane-3-one. Thus, to support mycoplasma growth *1*) the sterol must be planar, i.e., the A and B rings must be *trans*-fused, *2*) it must have an equatorial 3-β-hydroxyl group, and *3*) there must be a hydrocarbon tail (43).

Polar lipids. The polar lipids of *M. laidlawii* can be divided into five fractions: *1*) phosphatidylglycerol (36%), *2*) O-amino acid ester of phosphatidylglycerol (6%), *3*) monoglucosyl diglyceride (14%), *4*) diglucosyl diglyceride (32%), and *5*) a phosphorylated derivative of monoglucosyl diglyceride, where a phosphate is esterified on one of the hydroxyls of the glucose ring structure (12%). The structure of these lipids is shown in Figure 3. It is interesting that most of the polar lipids are glycolipids rather than phospholipids (43).

Once synthesized, none of the polar lipids show any turnover and are metabolically stable in actively growing cells. When ^{14}C-labeled fatty acids, glucose, glycerol, and ^{32}P were incorporated into a particular fraction, no turnover was observed over a period of 4 hr in which protein and cell mass more than doubled (26).

These results suggest a structural function for the polar lipids in this organism. However, as discussed later, the absence of turnover does not imply that polar lipids do not play a dynamic role in membrane function.

MOLECULAR INTERACTIONS IN BIOLOGICAL MEMBRANES

Before proceeding in the discussion of biological and physical effects of fatty acids in the polar lipids of mycoplasma membranes, it is first necessary to consider the types of molecular interactions involved in membranes.

Because of the ease of solubilizing membranes by detergent or sonic oscillation and the ease of extracting lipids with organic solvents, it is obvious that covalent bonds ($\Delta H = -50$ to -100 kcal/mole) are of little or no importance in the interaction of membrane components. Examination of the structures of the lipids and proteins suggests three possible types of interactions, all of which are weak, having energies between 1 and 10 kcal/mole. *1*) Electrostatic interactions could occur in several forms, including binding between oppositely charged groups or divalent-cation mediated interactions between two negatively charged groups. These interactions could be direct (salt linkage) or water-mediated. *2*) Hydrogen bonding could occur between

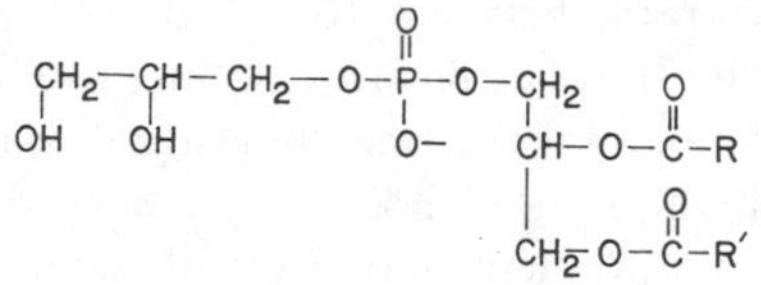

PHOSPHATIDYL GLYCEROL

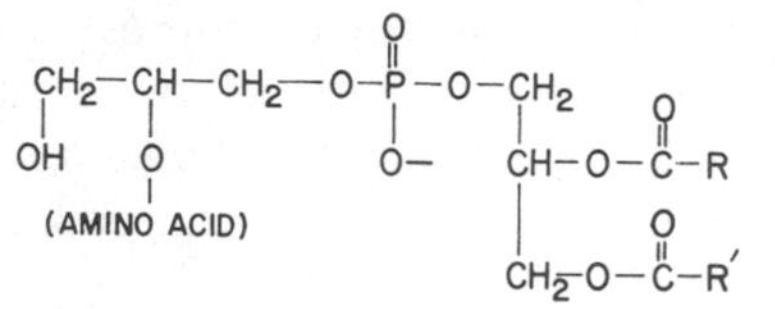

O-AMINO ACID ESTER OF PHOSPHATIDYL GLYCEROL

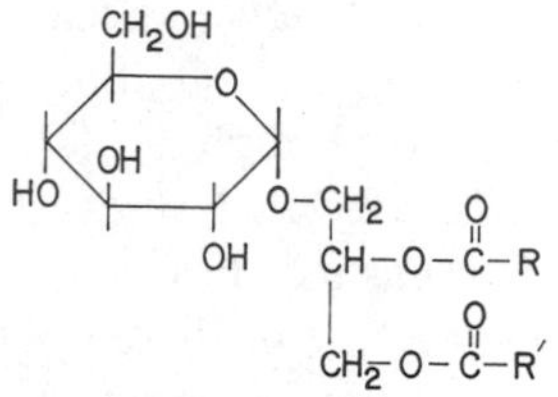

MONOGLUCOSYL DIGLYCERIDE

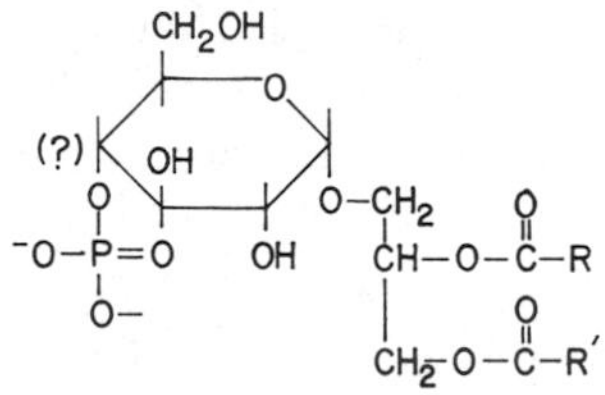

PHOSPHORYLATED MONOGLUCOSYL DIGLYCERIDE (?)

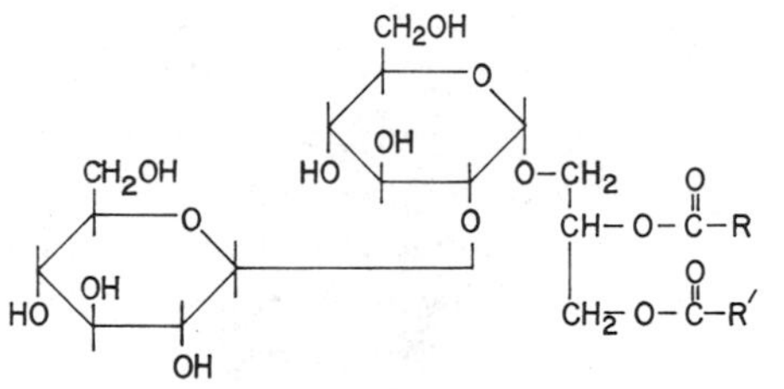

DIGLUCOSYL DIGLYCERIDE

Figure 3
Molecular structure of the polar lipids of *M. laidlawii*.

a variety of groups. *3*) London–van der Waals dispersion forces could occur between nonpolar groups. In addition to van der Waals forces, hydrophobic associations would result from an increase in entropy derived from transfer of hydrophobic groups from an aqueous to a nonaqueous phase. Since both lipids and proteins contain polar and nonpolar groups, electrostatic and hydrophobic interactions could occur between lipid-lipid, lipid-protein, and protein-protein (57).

Electrostatic Forces

Electrostatic forces involving cations are known to be involved in maintaining the integrity of membranes. As long ago as 1936, it was demonstrated that in the absence of Ca^{++} ions, punctured cells do not reseal. With *M. laidlawii*, lysis in hypotonic solution is prevented in the presence of Mg^{++} (5), and reaggregation of membranes solubilized in detergent is dependent on the presence of divalent cations (15).

Evidence is available that some lipid-protein interactions are of a polar character. The inability of nonpolar solvents such as chloroform to extract most of the lipid from many membranes is thought to imply that nonpolar or hydrophobic interactions between lipids and proteins are not extensive. The addition of a polar solvent, as in the normal chloroform-methanol extraction procedures, liberates nearly all the lipid. These results have been interpreted as favoring a polar interaction between the charged amino acid residues and phospholipid polar groups, probably mediated by water (39). The fact that relatively large amounts of cholesterol can be extracted by nonpolar solvents alone is consistent with this interpretation, since cholesterol is thought to nestle among the hydrocarbon tails of phospholipids containing a monounsaturated fatty acid, being bound in the membrane primarily by nonpolar (London–van der Waals) forces (56).

Another study demonstrating the existence of polar lipid-protein interactions is that of Brown (9) on a species of halophilic bacteria. This study essentially consisted of electrometrical titrations with acids and bases of the membrane as it underwent spontaneous disaggregation at decreasing ionic strength. Brown concluded that hydrogen bonds (water-mediated electrostatic interactions) between the α-amino groups of membrane protein and the acidic phosphate groups of the phospholipids are important forces in maintaining membrane integrity. He suggested that hydrophobic interactions between nonpolar amino acid residues and the hydrocarbon tails of phospholipid molecules might exist in this membrane.

Hydrophobic Interactions

London–van der Waals forces between individual atoms are only slightly greater than random thermal energy at room temperature, but the forces between molecules are additive, and London–van der Waals interactions are profoundly affected by the distance between the interacting atoms. The interaction of two lecithin molecules containing saturated fatty acids 18 carbons long in closest possible contact results in van der Waals dispersion forces of -18 kcal; the electrostatic interaction between the polar headgroups of these molecules is approximately -10 kcal (40). It is apparent that hydrocarbon tails can develop interactions of considerable strength, especially when the increase in entropy derived from transfer of hydrophobic groups from a polar to a nonpolar phase is included.

INTERACTIONS BETWEEN LIPIDS

The lipid bilayer model for membrane structure has prompted many monolayer and bilayer experiments to establish the effect of saturation, unsaturation, branching, and length of acyl chain on the binding forces between these chains. As shown in Table I, as the number of carbon atoms in a saturated fatty acid is increased, the monolayer becomes more condensed (attractive force increases); conversely, unsaturated or branched-chain acids result in expanded films because of steric hindrance. This is because the branched chains contain a methyl group and the unsaturated acids have a "kink" in their chains imposed by changing the normal bond angle from 109° in saturated hydrocarbon chains to 125° due to the "*cis*"-double bond. Other biologically important components such as phytol, the isoprenoid chain of chlorophyll which contains four branched methyl groups, and carotenoids, which also contain branched groups, give highly expanded films. Conversely, the addition of cholesterol to monolayers containing mono-unsaturated fatty acids results in a condensation of the film, i.e., the surface

Table I

Effects of Chain Length and Branching on Packing of Fatty Acids

COMPOUND	STRUCTURE	SURFACE AREA (A^2/MOLECULE)
Myristic acid	14:0[a]	37[b]
Pentadecylic acid	15:0	37
Palmitic acid	16:0	24
Stearic acid	18:0	23.5
Arachidic acid	20:0	23
Oleic acid	18:1	48
Erucic acid	22:1	40
2-Methyl-octadecanoic acid	18:branched	30
16-Methyl-heptadecanoic acid	17:branched	32

[a] Number left of colon refers to number of carbon atoms; number right of colon refers to number of double bonds.
[b] All measurements made at constant surface pressure of 5 dyn/cm.
All values taken from O'Brien, J. S. *Theor. Biol.* **15**, 307 (1967).

area of a mixture of cholesterol and fatty acid is less than the sum of the areas occupied by each component when alone (29).

If it is assumed that the nonpolar interactions of the hydrocarbon tails of lipids are important in vivo, differences should exist in the membranes which could be correlated with variations in the lipid composition. Examining the rate of hemolysis of red blood cells from ox, sheep, rabbit, man, and rat, de Gier *et al.* (12) observed that leakage of hemoglobin is lowest with ox and is increased in the above order to rat. It has been shown that erythrocytes from these animals differ in two significant ways: *1*) the content of unsaturated lipids and *2*) the amount of sphingolipids. Significantly, the erythrocytes containing the *lowest* concentrations of unsaturated lipids and the highest amounts of sphingomyelin were the least permeable to hemoglobin; as the rate of hemolysis increased, so did the amount of unsaturation, suggesting a direct correlation between strength of hydrophobic lipid-lipid bonding and membrane permeability.

In a study of glycerol permeation of *M. laidlawii*, McElhaney *et al.* (25) showed that for both intact cells and liposomes formed from extracted lipids grown in medium containing three different fatty acids, the permeability was dependent on the geometric configuration (*cis* and *trans*) and number of double bonds present in the fatty acyl chains. The ratio of glycerol permeation increased in the order elaidic (18:1 *trans*) > oleic (18:1 *cis*) > linoleic (18:2).

VARIATIONS IN FATTY ACID COMPOSITION OF MEMBRANE LIPIDS

In most organisms, including bacteria and higher forms, the fatty acid composition of cell lipids remains relatively constant, and it has been suggested, because of this consistent composition, that the fatty acids involved have a high degree of specific interactions with membrane proteins (8). As shown in Table II, the fatty acid composition can be drastically altered when *M. laidlawii* cells are grown in lipid-extracted medium supplemented with specific fatty acids. This organism is incapable of synthesizing unsaturated fatty acid but does, under certain conditions, synthesize saturated, even-carbon-numbered acids; this accounts for the presence of lauric, myristic, and palmitic acids in all experiments. When the organism is grown on pentadecanoic acid (15:0), 83% of the total acids are 15:0 and 97.3% of the total are saturated acids. At the other extreme, elaidic acid (18:1 *trans*) accounts for 74% of the total, and isopalmitate, a branched-chain acid, for approximately 80% (26).

As demonstrated by gas chromatography, in addition to wide quantitative variations in fatty acid composition, *M. laidlawii* readily incorporates odd-chain, branched-chain, and cyclopropane fatty acids. The rates of growth at 37°C in the presence of any of these fatty acids are the same.

Table II

Fatty Acid Composition of *M. laidlawii* B Membrane Polar Lipids Grown in Medium Containing Fatty Acid Indicated

ADDED FATTY ACID	Incorporation of Fatty Acids into Polar Lipids (Moles %)								
	12:0	14:0	15:0	16:0	18:0	18:1	18:2	16:0 *i*	19:0 *cp*
None	7.5	24.8	0	53.5	3.0	6.6	4.4	0	0
Pentadecanoic (15:0)	2.3	3.3	83.0	3.8	3.9	1.6	1.1	0	0
Palmitic (16:0)	9.5	9.2	0	74.3	2.4	1.6	1.0	0	0
Stearic (18:0)	5.0	5.0	0	8.0	65.0	10.3	6.7	0	0
Oleic (18:1 *c*)	3.5	6.4	0	20.2	0.9	68.9	1.0	0	0
Elaidic (18:1 *t*)	3.9	4.7	0	16.5	0.7	73.7	0.5	0	0
Isopalmitic (16:0 *i*)	6.3	3.1	0	3.3	0.5	5.6	2.5	78.8	0
Dihydrosterculic (19:0 *cp*)	1.1	5.5	0	15.4	1.8	15.3	0.2	0	60.0

The fatty acids are designated by the number of carbon atoms, followed by the number of double bonds; *c* and *t* indicate *cis*- and *trans*-configuration of the double bond; *i* indicates a methyl branch on the penultimate carbon atom; *cp* indicates a cyclopropene ring.

From McElhaney, R., and Tourtellotte, M. E. *Science* **164**, 434 (1969). Copyright 1969 by the American Association for the Advancement of Science.

When cells are grown in the presence of long-chain saturated fatty acids (17:0, 18:0, 19:0, and 20:0), they become swollen and eventually lyse. That the swelling and lysis are osmotic phenomena is demonstrated by the fact that lysis can be prevented by increasing the osmolarity of the growth medium with NaCl, KCl, and sucrose. Cells grown in medium containing short-chain, branched-chain, cyclopropane, and unsaturated fatty acids grow as short to long filaments and show no evidence of swelling (26).

Although the precise mechanism of this osmotic lysis is presently unknown, results suggest that it occurs when membrane lipids become too "crystalline" due to hydrophobic interactions of the hydrocarbon tails. Growing cells on various fatty acids has little effect on the relative amounts of polar lipid fractions, so the effects observed cannot be due to changes in polar headgroups. In the series of saturated fatty acids from 12 to 20 carbons, at 37°C no lysis occurs with 12–16 carbon acids, the 17-carbon acid shows some lysis, while 18-, 19-, and 20-carbon acids show complete lysis and cell death. If these hydrophobic interactions are important, it would be predicted that a temperature effect would also occur, since at lower temperatures the shorter-chain acids would be more "crystalline." As the growth temperature is lowered from 37° to 25°C, the critical chain length causing lysis decreases from 17 at 37°C to 16 at 30°C and to 15 at 25°C.

Further evidence that the hydrophobic interactions of the fatty acid tails are important to cell integrity is shown in Table III (55). This table demonstrates the effect of adding small amounts of various fatty acids of differing structures on lysis caused by growth in medium containing stearic

Table III

Effect of Fatty Acids of Various Structures on Cell Lysis and Death Caused by Growth in Medium Containing Stearic Acid

STEARIC ACID (0.16 μmole/ml) PLUS: (0.04 μmole/ml)	OD 420 mμ	STEARIC ACID (0.16 μmole/ml) PLUS: (0.04 μmole/ml)	OD 420 mμ
18:0 Alone	0.05	+18:1 (9c)	0.42
+8:0	0.05	+18:1 (9t)	0.29
+10:0	0.10	+18:2 (9c, 12c)	0.45
+12:0	0.38	+18:2 (9t, 12t)	0.39
+13:0	0.38	+18:3 (9, 12, 15c)	0.44
+14:0	0.39	+18:3 (6, 9, 12c)	0.43
+14:1 (9c)	0.46	+19:0	0.05
+15:0	0.09	+19:0 (cp)	0.43
+16:0	0.11	+20:0	0.05
+16:1 (9c)	0.45	+20:1 (11c)	0.45
+16:1 (9t)	0.35	+20:2 (11, 14c)	0.44
+16:0 i	0.46	+20:4 (5, 8, 11, 14c)	0.43
+17:0	0.05	+20:5 (5, 8, 11, 14, 17c)	0.46
+18:1 (9c)	0.45	+22:6 (4, 7, 10, 13, 16, 19c)	0.45
+18:1 (9t)	0.45		
+18:1 (11c)	0.44		
+18:1 (11t)	0.40		
+18:1 (6c)	0.42		

M. laidlawii cells grown in lipid-extracted tryptose containing lipid-poor bovine serum albumin + glucose + 0.16 μmole stearic + 0.04 μmole of the fatty acid indicated. Optical density of cells grown in stearic acid alone before lysis and cell death was 0.40; *c* and *t* indicate *cis*- and *trans*-configuration of double bond; *i* indicates methyl branched; *cp* indicates cyclopropane ring.

acid. It will be noted that any acid which causes a looser packing of the hydrocarbon chains prevents lysis and cell death caused by growth in medium containing stearic acid alone.

From all this evidence, it appears that for proper membrane function, the lipids must be in a quasiliquid state, "crystallization" of the lipids being fatal to the cell.

ANALYSIS OF LIPIDS AND MEMBRANES BY CALORIMETRY

If, as suggested by the previous data, the lipids in a membrane exist in a liquid-crystalline state, a "melt" or change in physical structure should be detectable by physical means.

It has been known for some time that phospholipids in aqueous dispersions undergo a reversible thermotropic gel-liquid crystal phase transition. The transition has been studied by differential scanning calorimetry (DSC), x-ray diffraction, and light microscopy, and has been found to arise from the melting of the hydrocarbon chains in the interior of the lipid bilayer (21). Unlike transitions between liquid- and crystalline mesophases, the melt does not result in a molecular rearrangement, and the lipids exist in the lamellar phase both above and below the transition temperature.

Thus, if a substantial amount of the lipid in membranes exists in a bilayer, DSC should provide a convenient means for its detection.

Thermal analysis (22) is usually performed in two ways: differential thermal analysis (DTA) and DSC.

Differential Thermal Analysis

In DTA the sample and an inert reference material are heated at the same rate, and the difference in temperature between them is recorded. The differential temperature remains constant or zero until a thermal reaction occurs in the sample, at which time the differential temperature increases until the transition is completed and then decreases again to zero. Thus, a peak is obtained on the curve, the direction of the peak indicating whether the transition is exothermic or endothermic. Unfortunately, it has proved difficult with present instrumentation to accurately relate the area under the DTA curve to the heat of transition. A better method for this is DSC.

Differential Scanning Calorimetry

In DSC the temperatures of sample and reference material are maintained at an equal level. As the sample and reference material are heated or cooled, when a thermal transition occurs, the differential power required to maintain the sample temperature equal to the reference material temperature is measured. This differential power is directly related to the heat absorbed (endothermic) or evolved (exothermic) by the sample during the transition. Since the systems used are nonlinear, calibration of the instrument at several points using known standards is required.

For pure compounds with sharp, well-defined transitions, the transition temperature (T_m) corresponds to the point of departure from the base line. Reproducibility is often better than $\pm$ 0.5°C. With nonhomogeneous systems such as biological lipids, where the thermal transition occurs over a broad temperature range, it is somewhat more difficult to establish a transition temperature, since there is usually no sharp break from the base line. However, even under these conditions, DSC is most dependable in relating the area under the curve to the heat of transition (ΔH). To establish both ΔH and T_m with a high degree of accuracy, both DSC and DTA should be employed.

The melt of the hydrocarbon tails of phospholipids is a cooperative phenomenon. In DSC curves of 50 weight per cent dispersions in water of dipalmitoyl phosphatidylcholine/cholesterol mixtures, as the cholesterol concentration is increased from 0 to 50 mole %, the thermal transition peak broadens and decreases in area until at a 1:1 molar ratio of lecithin to cholesterol, no transition is observed. Complementary x-ray diffraction and nuclear magnetic resonance studies suggest that the cholesterol controls the fluidity of the hydrocarbon chains by restricting the motion of the fluid chains and preventing close interactions of the chains which would result in crystallization (4).

Studies of synthetic phospholipids have shown that fatty acyl chain lengths, degree of unsaturation, and branching profoundly affect the temperature of transition. As expected, the longer the saturated carbon chain, the higher the transition temperature; unsaturation and branched chains result in a lower transition temperature (4).

In addition to the hydrocarbon tails, the polar headgroups also affect the T_m; for example, dimyristoyl ethanolamine melts at approximately 45° C, while dimyristoyl lecithin melts at about 25° C (34). The lower temperature of transition for lecithin may reflect decreased stability due to the more bulky headgroup. Transition temperatures appear unaffected by high concentrations of either mono- or divalent cations when compared with transitions in distilled water. Ethylene glycol–water (1:1 v/v) also has no effect. This is important because transitions in some membranes occur below the ice point, and a suspending medium other than water is therefore necessary (34).

DIFFERENTIAL SCANNING CALORIMETRY OF MYCOPLASMA LIPIDS AND MEMBRANES (48)

Figure 4 shows representative endothermic transitions of lipids, membranes, and whole cells of *M. laidlawii*. For the lipids, heats of transition are 3–4 cal/g. Lipids from cells grown in tryptose supplemented with stearate are shown in Figure 4, curve A, while curve B is a scan of intact membranes from the same cells. Curves C and D are from lipids and membranes of cells grown in unsupplemented tryptose. Transitions of lipids and membranes rich in oleate are shown in curves E and F. The transition for whole cells, shown in curve G, indicates that the effects observed are not artifacts introduced during the preparation of membranes. For both membranes and lipids with transitions above the ice point, the transitions are the same in distilled water, 0.25 M NaCl, and 0.05 M $MgSO_4$. In 50% ethylene glycol, transition temperatures are depressed 3–5° C, but the heats of transition are unaffected within the limits of experimental error.

Lipid transitions in membranes, as in protein-free membrane lipids, are reversible. Transition temperatures and peak shapes are independent of scan rate. In Figure 4, curve D, the smaller transition at higher temperature arises from protein denaturation (44) because it is irreversible, occurs at approximately the same temperature for all membranes regardless of lipid composition, and is greatly decreased by treatment of the membranes by pronase. In

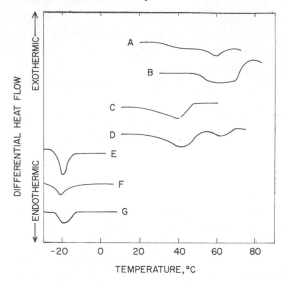

Figure 4

Calorimeter scans of *M. laidlawii* lipids, membranes, and whole cells. *A*. Total membrane lipids from cells grown in tryptose with added stearate. *B*. Membranes from stearate-supplemented tryptose. *C*. Total membrane lipids from cells in unsupplemented tryptose. *D*. Membranes from unsupplemented tryptose. *E*. Total membrane lipids from cells grown in tryptose with added oleate. *F*. Membranes from oleate-supplemented tryptose. *G*. Whole cells from oleate-supplemented tryptose. The first four preparations were suspended in water; for the latter three scans, the solvent was 50% ethylene glycol containing 0.15 M NaCl. (From Steim, J.M., *et al. Proc. Natl. Acad. Sci. U.S.A. 63*, 106, 1969.)

Figure 4, curve B, the protein transition is obscured by the lipid transition. In such cases, where protein and lipid transitions overlap, the peak area on the initial scan is always greater than on subsequent scans. This effect would be expected from irreversible protein denaturation.

As can be seen, transitions of *M. laidlawii* lipids in water show the same dependence upon fatty acid composition exhibited by synthetic lipids in the lamellar phase (21). When negatively stained with phosphotungstic acid (7), preparations of *M. laidlawii* lipids with transitions both above room temperature (stearate-enriched) and below room temperature (oleate-enriched) appear lamellar in the electron microscope. Similar smectic bilayer arrays are well established for other phospholipids in water at the concentrations used in these experiments. Thus the transitions observed with *M. laidlawii* lipids, like those of synthetic phospholipids, appear to arise from melting of the fatty acid chains within the interior of the bilayers.

The membrane transitions occur at the same temperatures as the lipid transitions. They are reversible and remain after thermal protein denaturation or treatment with pronase. There is little doubt that these reversible membrane transitions arise from a change in state of the lipids in the membranes which is very similar to that observed with the protein-free membrane lipids dispersed in water. Such melting of hydrocarbon chains is a cooperative phenomenon and should be greatly affected by any factors affecting the hydrocarbon chains. For example, oleate-rich lipids from *M. laidlawii* melt 80° below the stearate-rich system, and synthetic dioleoyl lecithin melts 50° below dipalmitoyl lecithin. Substitution of isopalmitic acid, with two terminal methyl groups, for palmitic acid lowers the transition temperature of *M. laidlawii* lipids by 20° (33). Growth in medium containing different fatty acids appears to have no significant effect on the relative amounts of polar lipid classes (headgroups) (33). It has been suggested, therefore, that the transitions observed in membranes result from a melt of hydrocarbon chains within bilayers, schematically depicted in Figure 5 (48).

By comparing areas under the curves of extracted lipids and whole membranes, the heats of transition were calculated based on the amount of total lipid in each preparation. The heat of transition in whole membranes was 3.6 ± 0.5 cal/g, while in aqueous dispersions of protein-free lipid it was 3.9 ± 0.1 cal/g. These values are less than that of synthetic lipids (approximately 10 cal/g) but close to that reported for cholesterol-free lipid extracts of myelin (34).

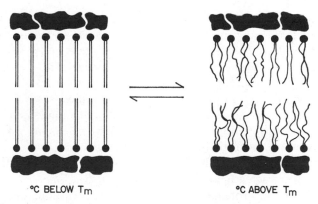

°C BELOW T_m °C ABOVE T_m

Figure 5

Schematic representation of the reversible thermal transition postulated for *M. laidlawii* membranes, showing the crystalline state of hydrocarbon chains below the transition temperature, T_m, and the liquid-like state above the transition. Protein is shown covering the outside surfaces of the membrane. The polar ends of the lipids are shown as black circles. (From Steim, J. M., *et al.* Proc. Natl. Acad. Sci. U.S.A. 63, 108, 1969).

The heat of transition of the membranes is approximately 90% of that of the aqueous dispersions of lipid, which is known to be in a bilayer array. Since it is known that cholesterol reduces and eventually destroys the co-operative chain melt, extensive hydrophobic association with protein would result in the same effect. Although these results do not necessarily rule out that 5–10% of the hydrocarbon chains could be interacting with protein, they clearly demonstrate that the fundamental structure of the mycoplasma membrane cannot consist primarily of hydrophobic lipid-protein interactions.

X-RAY DIFFRACTION

Results of DSC have been confirmed by low-angle x-ray diffraction (13), which shows that mycoplasma membrane fatty acyl chains undergo a thermal phase transition dependent on the fatty acid composition. Below the transition the chains are in a hexagonal array characteristic of long-chain paraffins in a crystalline state and give a sharp diffraction maximum near 4.2 Å. Above the transition, which occurs over a range of 6–10°, the chains are in a more fluid state, and a broad maximum at 4.6 Å is observed which is characteristic of a hydrocarbon melt. These data strongly suggest that the lipids exist in a bilayer configuration both below and above the transition temperature. In addition, a grain size of 400 Å was calculated, demonstrating that lipid bilayer structure in the cell membrane is extensive. This material is described in detail in Chaper 4.

ELECTRON SPIN RESONANCE (ESR)

Since the theory and interpretation of ESR are covered in Chapter 8, only the results of using nitroxide-labeled fatty acids in mycoplasma membranes are considered here. In studies by Tourtellotte *et al.* (54), the nitroxide probe, an N-oxyl-4′,4′-dimethyloxazolidine derivative of 12-ketostearic acid (12NS), was incorporated by ester linkage into the polar lipids (Fig. 6). The ESR spectrum of 12NS in intact cells and membranes was quite similar to that in aqueous dispersions of extracted lipid; the mobility of 12NS was greater in oleate-grown cells than in stearate-grown cells, as would be predicted if extensive hydrophobic lipid-lipid interaction was occurring. In addition, mobility was slightly, but significantly, more restricted in membranes than in extracted lipids (Fig. 7).

Denaturation of protein to 95°, glutaraldehyde fixation, and Mg^{++} had no effect on the correlation time (mobility), suggesting again that the bulk of lipid is not interacting with protein. A study by Rottem *et al.* (38) demonstrated that as the nitroxide group is moved further from the carboxyl group of the fatty acid, mobility of the probe increases, suggesting that the middle of the bilayer is in a more liquid state, and as the probe gets closer to the carboxyl group, its mobility becomes more restricted. This finding gives some insight into what is occurring, in molecular terms, within the bilayer. Since

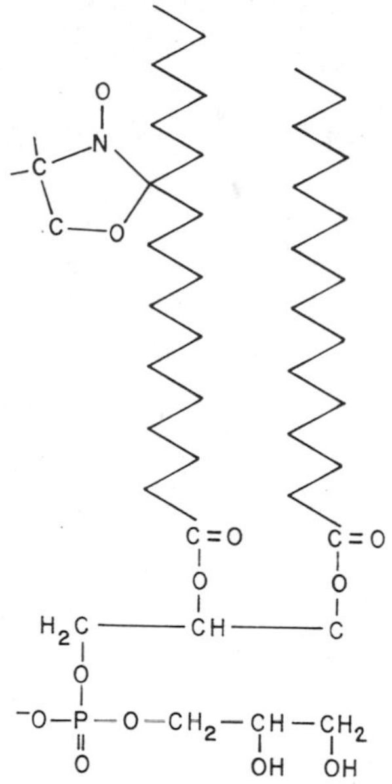

Figure 6
Structure of nitroxide-labeled fatty acid incorporated into phosphatidylglycerol.

the polar heads of the lipids in a bilayer are restricted by mutual interaction or ionic interaction with protein, this result is not unexpected.

In an Arrhenius plot of correlation time of 12NS mobility against temperature from cells grown in stearic acid, a discontinuity in the slope at approximately 45°C suggests a thermal transition in these membranes (54), lending further support to DSC and x-ray diffraction results.

The ESR spectra of 12NS bound to a wide variety of proteins (including bovine serum albumin, mitochondrial structural protein, and the polypeptide gramicidin D) is highly immobilized (Fig. 8B), suggesting strong interaction between 12NS and protein. In no case was this type of interaction observed in membranes. Within the sensitivity range of the instrumentation, if 1% of the 12NS probe were interacting *strongly* with protein, it would be detectable; and, in fact, addition of 10^{-4} M gramicidin to mycoplasma membranes containing 12NS in the polar lipids produces an immobilized spectrum (Fig. 8A). From these data it is reasonable to conclude that lipid does not hydrophobically interact with protein in mycoplasma membranes. Unfortunately, ignorance of how 12NS interacts with protein prevents final acceptance of this conclusion; it is obvious that proteins can exist in an infinite number of configurations, and membrane proteins could exist in a state not allowing tight bonding as observed with bovine serum albumin and gramicidin.

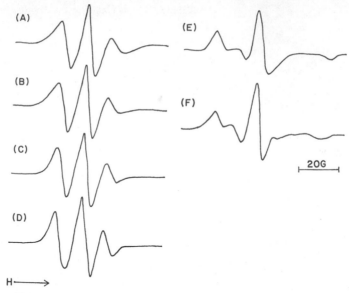

Figure 7

A–D. Electron paramagnetic resonance spectra of mycoplasma cells and lipids with incorporated 12NS. *A.* Stearate-enriched membranes. *B.* Aqueous dispersion of isolated lipids from *A.* *C.* Oleate-enriched membranes. *D.* Aqueous dispersion of lipids from *C.* *E–F.* 12NS bound to bovine serum albumin in aqueous medium: *E*, at 30°C; *F*, at 60°C. (From Tourtellotte, M. E., *et al. Proc. Natl. Acad. Sci. U.S.A. 66*, 911, 1970.)

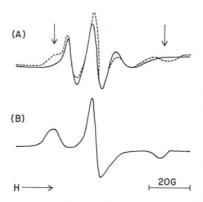

Figure 8

Electron paramagnetic resonance spectra of 12NS at 48°C. *A.* Incorporated into mycoplasma membranes before (solid line) and after (broken line) addition of 10^{-4} gramicidin D. *B.* Aggregated with 10^{-4} gramicidin D in aqueous solution. (From Tourtellotte, M. E., *et al. Proc. Natl. Acad. Sci. U.S.A. 66*, 913, 1970.)

Nevertheless, one inescapable conclusion emerges from all these studies: The bulk of the lipid in *M. laidlawii* membranes exists as a bilayer structure with extensive hydrophobic lipid-lipid interaction.

Since the techniques used in DSC, x-ray diffraction, and ESR are essentially averaging techniques, they demonstrate the state of the major portion of the membrane structure; they do not rule out the possibility that a small portion of the lipid is interacting hydrophobically with protein.

Although it has been shown by both DSC and ESR that denaturation of protein has little effect on the lipid phase, the reverse situation may not be true; protein conformation may in fact be dependent on the state of the lipids.

CIRCULAR DICHROISM AND FLUORESCENCE

Reinert and Steim (35) examined protein structure by circular dichroism (CD) and tryptophan fluorescence as the lipids in the membrane underwent a phase transition. Using *M. laidlawii* membranes in which thermotropic lipid transitions began at 20°C and ended at approximately 45°C, they observed no change in CD spectra in this region (Fig. 9).

The general features are similar to those reported for other membranes (17, 45). An appreciable helical contribution appears to be present, but the optically active bands are bathochromically shifted (47, 59) with respect to those observed for most synthetic polypeptides and proteins in solution.

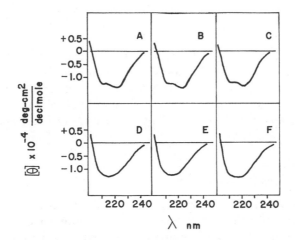

Figure 9

Circular dichroism spectra at various temperatures of membranes from cells grown in tryptose medium with serum fraction: *A*, 9°C; *B*, 30°C; *C*, 45°C; *D*, 55°C; *E*, 63°C; *F*, 30°C. Spectra *A–E* were taken in sequence; after protein denaturation the sample was cooled to 30°C and spectrum *F* was recorded.

The trough and shoulder appear at about 223 and 211 nm, respectively, with a crossover at 204 nm. In different preparations the ellipticity of the shoulder at 211 nm varies, but the response to temperature is the same in all preparations studied. Note that essentially the same spectra are obtained from 9–45°C (Fig. 9A–C), the temperature range in which the lipid phase change occurs. Above the temperature of the lipid transition, spectral changes occur (Fig. 9D, E) which correlate with higher temperature protein denaturation. Fine structure is lost, and a broad trough appears at 220 nm. The protein denaturation is irreversible (in agreement with the calorimetry studies), since after the sample is cooled to 30°C the original conformation is not restored (Fig. 9F).

In agreement with the CD results, no changes in fluorescence emission spectra were observed in the region of the lipid transition (35). From these results the authors concluded that protein conformation is independent of the physical state of the lipids. They emphasized, however, that this conclusion is valid only within the limits of the experimental techniques employed, and the possibility still exists that a limited amount of the lipid bilayer may be occupied by protein molecules. Thus, like DSC, x-ray diffraction, and ESR, CD and fluorescence are averaging techniques and may be unable to detect minor or subtle interactions between membrane components.

ELECTRON MICROSCOPY OF MYCOPLASMA MEMBRANES

Thin Sectioning

As observed in other membranes, electron micrographs of thin-sectioned *M. laidlawii* cells and membranes show a characteristic unit membrane structure (24, 28), supporting the hypothesis of Danielli and Davson (11) and Robertson (36) that the membrane consists of a lipid bilayer. This also is in substantial agreement with results of other methods of analysis already discussed in this chapter.

Freeze-Fracture and Freeze-Etching

Unlike most techniques for preparation of material for electron microscopy, the freeze-fracture and etching techniques do not require fixatives or dehydrating agents. Cells are rapidly frozen at liquid nitrogen temperature, fractured under vacuum, and shadowed with carbon and platinum, thus leaving a replica of the original material. Deep etching is accomplished by allowing ice to sublime away for 10–15 sec, exposing surface area previously under the ice layer (Fig. 10). Shadowing is then done in the usual manner. All evidence suggests that morphological integrity of the original material is retained (1). This technique has revealed features of membrane ultrastructure not previously observed. Fracture of the cell results in the appearance of globular units approximately 50–85 Å in diameter on the exposed membrane face. There has been considerable controversy about whether these particles are located on the surface or within the matrix of the membrane (1).

According to Branton the cleavage of the membrane is most often down the middle of the hydrophobic lipid bilayer. The 85-Å particles observed lie

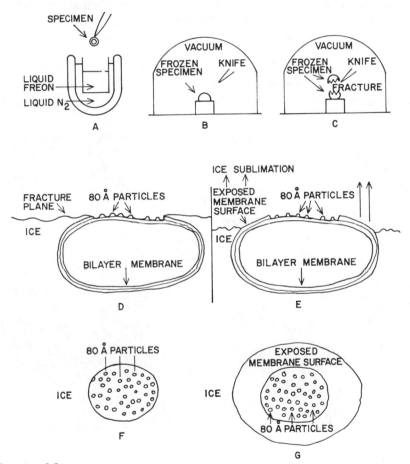

Figure 10

Diagram of freeze-fracture and freeze-etch procedure. Cells on grid are rapidly frozen at liquid nitrogen temperature (*A*), placed in a vacuum at −110°C (*B*), and fractured with a cold knife and shadowed with carbon and platinum (*C*). *D*. Fracture of cell without etching. *E*. Fracture with deep etching exposing outer surface of membrane. *F*. Surface view of fracture face. *G*. Surface view after deep etching.

within this region of the membrane and not on the outer surface. That this is in fact the case has been verified by Tillack and Marchesi (51), who labeled the surface of red blood cell membranes with fibrinous actin, and Pinto da Silva and Branton (31), who labeled it with covalently bound ferritin. In both cases, when cells were fractured and deep etched, actin and ferritin were observed only on the surfaces exposed by deep etching; the labels were never observed in the region of 85-Å particles.

As shown in Figure 11, *M. laidlawii* membranes, when freeze-fractured, reveal these same 50- to 85-Å particles; it is interesting to note that there are many more particles and they are more regularly arrayed in membranes from oleate-grown cells than in membranes from stearate-grown cells (54).

In red blood cells, evidence suggests that the particles are proteins or glycoproteins. Engstrom and Branton (16) working with red blood cells and Tourtellotte and Branton (53) working with *M. laidlawii*, have shown that

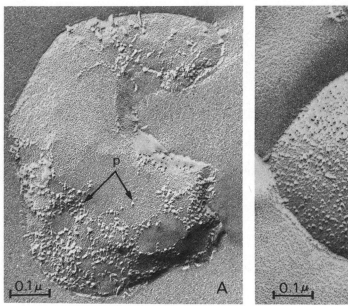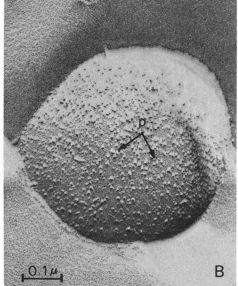

Figure 11

Fracture faces of mycoplasma cells. The stearate-supplemented cells (*A*) are larger and less regular in shape, and bear fewer particles (*p*) than the smaller, more coccoid oleate-supplemented cells (*B*). Neither the particles nor the smooth areas between them are membrane surface features. Both are interpreted as morphological features of a hydrocarbon region within the plane of the membrane. ×100,000. (From Tourtellotte, M. E., *et al. Proc. Natl. Acad. Sci. U.S.A. 66,* 913, 1970.)

the particles can be partially removed by digestion of membranes with the proteolytic enzyme pronase; this strongly indicates that the particles are protein. Although some pronase preparations reportedly contain lipase activity, Tourtellotte *et al.* were unable to detect any degradation of *M. laidlawii* lipids over a 4-hr period of exposure of membranes to pronase, in which time 80% of the protein and a significant number of the particles were removed.

It is interesting to note that as red blood cell membranes are being digested by pronase, in approximately 30 min the particles aggregate in large clusters and then gradually reduce in numbers until, after 8 hr, they are at least 90% gone.

The results of freeze-etching (which suggest 10–20% protein in the bilayer) and of DSC, x-ray diffraction, ESR, CD, and fluorescence (all of which suggest little or no hydrophobic lipid-protein interaction) appear paradoxical. However, assuming that 20% of the bilayer area is protein, that the protein particles are 80 Å in diameter, and that on the average, each lipid molecule occupies an area of 65 $Å^2$, for each protein there would be 310 polar lipid molecules. Of the 310 lipid molecules, 30 would interact directly with the protein (9.6% lipid-protein interaction). If 80-Å proteins occupy 10% of the total surface, then there is 4.3% lipid-protein interaction. Because the observed particles are shadowed with carbon and platinum, 80 Å probably represents an upper limit on protein particle size. If these particles were on the average 50 Å in diameter or smaller, the amount of lipid-protein interaction would be correspondingly less. It thus may not be coincidental that *1*) the heat of transition of membranes is only 90% that of extracted lipids, *2*) the mobility of nitroxide-labeled fatty acids is greater (5–10%) in aqueous dispersions of lipids than in intact membranes, and *3*) in membranes solubilized in sodium dodecyl sulfate (SDS) which are allowed to re-form by dialyzing out the SDS, the higher protein/lipid ratios result in decreased mobility of the nitroxide-labeled fatty acid suggesting an increase in hydrophobic lipid-protein interaction.

Although these data appear to reconcile the seeming paradox, the fact remains that ESR is sensitive enough to detect strong lipid-protein interactions at the 1% level. If this interpretation is correct, one might assume that *1*) the hydrophobic lipid-protein interactions are weaker than or different from those displayed between bovine serum albumin and 12NS or *2*) that the interaction between the 80-Å proteins and lipid is of a polar nature (Fig. 13B). The latter, however, seems most unlikely since polar interactions of 80-Å protein particles occupying 10–20% of the membrane should be readily detectable by thin-section electron microscopy using conventional procedures which stain polar groups in the membrane. Admittedly, this reasoning is not totally convincing. Other data obtained from sugar transport kinetics are, however, far more compelling.

ACTIVE TRANSPORT

In most organisms, including *M. laidlawii*, the evidence that active transport is protein-mediated is overwhelming (30).

Accumulation of both potassium and 2-deoxyglucose against a gradient has been demonstrated in *M. laidlawii*. Transport was energy-dependent, was inhibited by sulfhydryl group inhibitors, showed saturation kinetics, and, in the case of 2-deoxyglucose, showed a high degree of structural specificity (10, 61).

Thus, if lipids are interacting by hydrophobic bonding to proteins (either transport protein or "energizing" protein), the transport process should be markedly affected by the fatty acid composition of the lipids. The rate of transport when lipids are in a "liquid" phase should be substantially greater than that when lipids are in a "crystalline" phase where interaction with protein would be stronger.

Substantiation of this prediction comes from an Arrhenius plot of the rates of 2-deoxyglucose transport at various temperatures above and below the thermal phase transition of the bulk lipid phase (Fig. 12). Examination of Figure 12 shows, as Arrhenius kinetics would predict, a straight line for oleate-grown cells between 10° and 37°C. The lipid melt for oleate membranes occurs at approximately $-10°$ to $-15°$C, and since membranes

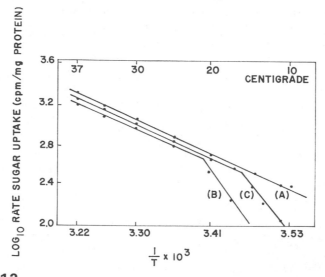

Figure 12
Arrhenius plot of rate of 2-deoxyglucose transport in *M. laidlawii* grown in medium containing oleic acid and in unsupplemented media. *A*. Oleate cells. *B*. Unsupplemented cells containing *more* saturated lipids. *C*. Unsupplemented cells containing *less* saturated lipids.

would be liquid no discontinuity would occur between 10° and 37°C. As would also be predicted, in the case of unsupplemented cells where the liquid transition occurs between 15° and 30°C, a discontinuity in transport kinetics was observed.

Two different batches of unsupplemented cells were used: one harvested at 16 hr and the second at 19 hr. As measured by gas chromatography, the polar lipids of the older culture had a higher proportion of long-chain saturated acids than did the younger. Because of the higher degree of long-chain saturates, the thermal transition of the lipids would be at a higher temperature. As shown in Figure 12, the discontinuity of the rate of transport also occurred at a higher temperature, demonstrating rather conclusively that transport kinetics is dependent on the physical state of the hydrophobic region of the bilayer (61).

Two observations should be made concerning this study: *1*) although the lipid transition is broad, the discontinuity appears to be *sharp*, and *2*) the discontinuity occurs at or near the start of the broad lipid melt and *not* near the peak of the transition. The sharp break in the Arrhenius plot may be more imagined than real, since transport rates were measured at 4° intervals; thus the sharp break may in fact be a more gradual curve. In regard to the discontinuity occurring near or at the start of the lipid melt, it may be that a specific polar lipid whose fatty acid composition is slightly more liquid than the total is interacting with the transport system. Another possibility also exists. ESR has demonstrated that strong hydrophobic lipid-protein interactions do not occur (38, 54). In a lipid bilayer, close packing of long-chain saturated fatty acids can generate a considerable energy of interaction. A cursory examination of the lengths of hydrocarbon groups of amino acids and the possibility of a hydrophobic region of protein presenting an uninterrupted surface of 25 Å, the length of a stearic acid chain, is remote. Thus hydrophobic interactions between lipid and protein particles might be expected to be *less* than those between lipid and lipid. This would also explain the discontinuity in transport kinetics occurring at the start of the lipid transition.

CONCLUSIONS, SPECULATIONS, AND POSSIBLE MEMBRANE MODELS

Based on the total information on *M. laidlawii* membranes available from DSC, x-ray diffraction, CD, fluorescence, ESR, electron microscopy (both thin-section and freeze-etch), and transport kinetics, the following picture emerges:

1. The major portion of the membrane lipid exists in a bilayer array.
2. The lipid bilayer is interrupted with 50- to 85-Å protein particles which traverse the bilayer and occupy between 10 and 20% of the area.

3. The amino acid composition of membrane proteins includes "normal" amounts of polar amino acids; thus, it is unlikely that these protein particles consist predominantly of nonpolar amino acids. It is known that a large number of polar groups of myoglobin are on the exterior surface, and hydrophobic interactions between nonpolar amino acid residues inside myoglobin contribute substantially to the structural stability of this protein (19). It is possible in cell membranes, in the region of the bilayer, that proteins may resemble "inside-out" myoglobin with hydrophobic amino acid residues exposed to the exterior and the polar groups on the interior of the molecule (Fig. 13A).

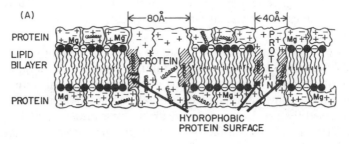

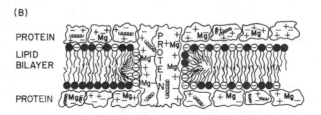

Figure 13

Schematic representation of possible structure of *M. laidlawii* membrane, showing (*A*) hydrophobic interaction between lipids and protein particles and (*B*) ionic interaction between lipids and protein particles. Structure *B* is considered most unlikely. See text.

4. If these protein particles were spheres whose outer surfaces consisted solely of exposed hydrophobic groups and were located entirely in the lipid bilayer, based on thermodynamic considerations they would aggregate when heated above the lipid transition temperature and certainly at 95°C; however, they do not. This suggests that at least part of these particles have polar groups exposed to the exterior which probably interact ionically with other

proteins or lipid headgroups present on the outer and inner surfaces of the membrane (Fig. 13A).

5. The protein particles could have ionic surfaces exposed in the bilayer which interact with polar headgroups of lipids (Fig. 13B). However, this seems unlikely for two reasons: *1*) electron microscopic examination after thin sectioning and staining with materials that interact with ionic groups has failed to detect such particles, which because of their size and frequency, should be visible, and *2*) if these particles are involved in transport, the effect of hydrophobic fatty acyl chains on transport kinetics is difficult to explain.

6. A more plausible explanation of the effect of hydrophobic fatty acyl chains on transport kinetics involves their effect on the ability of the transport proteins to undergo conformational changes in the process of substrate transport. Conformational changes of a protein would be more readily accomplished in a liquid than in a crystalline environment. For the transport process two types of "mobile carrier" can be assumed: *1*) an ionic or "charge" carrier where substrate is propagated across the membrane through a polar "pore," in which case no conformational change in the transport protein would be expected, and *2*) a carrier system in which, when substrate is bound to the binding site of the transport protein, the protein, or parts of it, undergo conformational changes in which the binding site–substrate complex is transported from the outside to the inside of the membrane. In the case of an "ionic carrier," where no conformational changes in protein occur, the effect of hydrophobic fatty acyl groups on the rate of transport is difficult to explain, since the transport process would occur entirely within the ionic environment of the protein. The effect of the lipids on transport is more consistent with the second type of "mobile carrier," which involves conformational changes in protein structure.

7. If the protein particles are involved in transport, they are probably dimers, tetramers, or larger. From their size (80 Å), their calculated molecular weight, assuming they are all protein, is 210,000. Since the largest membrane protein in *M. laidlawii* is no larger than 80,000–100,000 (28, 49), and many are much smaller, the particles are at least dimers. It is tempting to speculate, since transport requires energy, that these particles are "transport units" involving perhaps binding protein, transport protein, and "energy-coupling" protein. Close examination of freeze-etched specimens also reveals particles approximately 50 Å in diameter; if shadowing overestimates the size of the particles to a considerable extent, they might be as small as 40 Å in diameter. A globular protein of this size has a molecular weight of approximately 30,000, which is surprisingly close to molecular weight values for a number of transport proteins isolated from *Salmonella typhimurium* and *E. coli* (30).

Isolation of specific membrane proteins and reaggregation of these proteins with lipid bilayers are probably necessary for the understanding of membrane-related phenomena. Mycoplasma membranes solubilized in detergent have been shown to reaggregate into pieces, which on thin-section electron microscopy show what appears to be unit membrane structure. In sucrose density gradients, the re-formed membranes have the same density as the original (5, 14, 28, 50). It is too early to state how successful these techniques will be in regard to obtaining functionally active membranes.

It is tempting to extrapolate the present interpretation of the structure of *M. laidlawii* membranes to the structure of other cells as well. Evidence for such an extension includes: *1*) electron microscopy following freeze-fracture of many cells including bacteria, plants, and animals, with the exception of myelin, has demonstrated the presence of 85-Å particles in the hydrophobic region of the membrane (1); *2*) discontinuities in Arrhenius plots of transport kinetics, dependent on fatty acyl groups, have been demonstrated in bacteria (41, 60); *3*) differential scanning calorimetry has demonstrated thermotrophic lipid phase transitions in *E. coli* and *Micrococcus lysodeikticus* (46); and *4*) ESR studies of *Neurospora* mitochondria suggest the lipids exist in bilayer arrays (20).

REFERENCES

Review Articles

1. Branton, D. *Annu. Rev. Plant Physiol.* **20,** 209 (1969).
2. Chapman, D., and Wallach, D. F. H. *Biological Membranes.* Ed. D. Chapman. Academic Press, New York, 1968, p. 125.
3. Korn, E. D. *Annu. Rev. Biochem.* **38,** 263 (1969).
4. Ladbrooke, B. D., and Chapman, D. *Chem. Phys. Lipids* **3,** 304 (1969).
5. Razin, S. *Annu. Rev. Microbiol.* **23,** 317 (1969).
6. Stoeckinius, W., and Engelman, D. *J. Cell Biol.* **42,** 613 (1969).

Specific References

7. Bangham, A. D., and Horne, R. W. *J. Mol. Biol.* **8,** 660 (1964).
8. Benson, A. A. *J. Am. Oil Chem. Soc.* **43,** 265 (1966).
9. Brown, A. D. *J. Mol. Biol.* **12,** 491 (1965).
10. Cho, H. W., and Morowitz, H. J. *Biochim. Biophys. Acta* **183,** 295 (1969).
11. Danielli, J. F., and Davson, H. *J. Cell Comp. Physiol.* **5,** 495 (1935).
12. de Gier, J., Van Deenen, L. M., and Van Senden, M. *Experimentia* **22,** 40 (1966).
13. Engelman, D. *J. Mol. Biol.* **47,** 115 (1970).
14. Engelman, D., and Morowitz, H. J. *Biochim. Biophys. Acta* **150,** 385 (1968).
15. Engelman, D., Terry, T., and Morowitz, H. J. *Biochim. Biophys. Acta* **135,** 381 (1967).
16. Engstrom, L., and Branton, D. In preparation.
17. Gordon, A. S., Wallach, D. F. H., and Straus, J. H. *Biochim. Biophys. Acta* **183,** 405 (1969).
18. Green, D. E., and Perdue, J. *Proc. Natl. Acad. Sci. U.S.A.* **55,** 1295 (1966).
19. Herskovits, T. T., and Jaillet, H. *Science* **163,** 282 (1969).

20. Keith, A., Bulfield, G., and Snipes, W. *Biophys. J.* **10**, 618 (1970).
21. Ladbrooke, B. D., Jenkinson, T. J., Kamat, V. B., and Chapman, D. *Biochim. Biophys. Acta* **164**, 101 (1968).
22. Ladbrooke, B. D., Williams, R. M., and Chapman, D. *Biochim. Biophys. Acta* **150**, 333 (1968).
23. Maddy, A. H., and Malcom, B. H. *Science* **150**, 1616 (1965).
24. Maniloff, J. *J. Bacteriol.* **102**, 561 (1970).
25. McElhaney, R. N., de Gier, J., and Van Deenen, L. M. *Biochim. Biophys. Acta* **219**, 245 (1970).
26. McElhaney, R., and Tourtellotte, M. E. *Science* **164**, 433 (1969).
27. Morowitz, H. J. *The Mycoplasmatales and the L-Phase of Bacteria.* Ed. L. Hayflick. Appleton, New York, 1969, p. 405.
28. Morowitz, H. J., and Terry, T. *Biochim. Biophys. Acta* **183**, 276 (1969).
29. O'Brien, J. S. *J. Theor. Biol.* **15**, 307 (1967).
30. Pardee, A. B. *Science* **162**, 632 (1968).
31. Pinto da Silva, P., and Branton, D. *J. Cell Biol.* **45**, 598 (1970).
32. Pollack, J. D., Razin, S., and Cleverdon, R. C. *J. Bacteriol.* **90**, 617, (1965).
33. Rader, R., Tourtellotte, M. E., Reinert, J., and Steim, J. M. In preparation.
34. Reinert, J., and Steim, J. M. *Science* **168**, 1580 (1970).
35. Reinert, J., and Steim, J. M. In press (1971).
36. Robertson, J. D. *Progr. Biophys. Biophys. Chem.* **10**, 344 (1960).
37. Rodwell, A. W., Razin, S., Rottem, A., and Argaman, M. *Arch. Biochem. Biophys.* **122**, 621 (1967).
38. Rottem, S., Hubbell, W., Hayflick, L., and McConnell, H. *Biochim. Biophys. Acta* **219**, 104, (1970).
39. Rumsby, M. G., and Finean, J. B. *J. Neurochem.* **13**, 1501 (1966).
40. Salem, L. *Can. J. Biochem. Physiol.* **40**, 1287 (1962).
41. Schairer, H. V., and Overpath, P. *J. Mol. Biol.* **44**, 209 (1969).
42. Sjöstrand, F., and Barajas, L. *J. Ultrastruct. Res.* **32**, 293 (1970).
43. Smith, P. F. *The Mycoplasmatales and the L-Phase of Bacteria.* Ed. L. Hayflick. Appleton, New York, 1969, p. 469.
44. Steim, J. M. *Arch. Biochem. Biophys.* **112**, 599 (1965).
45. Steim, J. M. *Adv. Chem.* **84**, 259 (1968).
46. Steim, J. M. In press (1971).
47. Steim, J. M., and Fleischer, S. *Proc. Natl. Acad. Sci. U.S.A.* **58**, 1292 (1967).
48. Steim, J. M., Tourtellotte, M. E., Reinert, J., McElhaney, R., and Rader, R. *Proc. Natl. Acad. Sci. U.S.A.* **63**, 104 (1969).
49. Terry, T. In preparation.
50. Terry, T., Engelman, D. M., and Morowitz, H. J. *Biochim. Biophys. Acta* **135**, 391 (1967).
51. Tillack, T. W., and Marchesi, V. T. *J. Cell Biol.* **45**, 649 (1970).
52. Tourtellotte, M. E. *The Mycoplasmatales and the L-Phase of Bacteria.* Ed. L. Hayflick. Appleton, New York, 1969, p. 451.
53. Tourtellotte, M. E., and Branton, D. In preparation.
54. Tourtellotte, M. E., Branton, D., and Keith, A. *Proc. Natl. Acad. Sci. U.S.A.* **66**, 909 (1970).
55. Tourtellotte, M. E., and McElhaney, R. N. Unpublished data.

56. Van Deenen, L. L. M. *Progress in the Chemistry of Fats and Other Lipids*. Ed. R. T. Holman, Pergamon Press, New York, 1965, vol. 8, part 1.
57. Vanderheuval, F. A. *J. Am. Oil Chem. Soc.* **42,** 481 (1965).
58. Vanderkooi, G., and Green, D. E., *Proc. Natl. Acad. Sci. U.S.A.* **66,** 615 (1970).
59. Wallach, D. F. H., and Zahler, P. H. *Proc. Natl. Acad. Sci. U.S.A.* **56,** 1552 (1966).
60. Wilson, G., Rose, S., and Fox, C. F. *Biochem. Biophys. Res. Commun.* **38,** 617 (1970).
61. Zupnik, J., and Tourtellotte, M. E. In preparation.

15

Harvey R. Herschman

Alterations in Membranes of Cultured Cells as a Result of Transformation by DNA-Containing Viruses

Relation Between Neoplasia and Transformation

Evidence for viral induction of tumors in experimental animals has accumulated over the past half century. Although these experimental observations have been in the literature for a relatively lengthy period, the bulk of our knowledge of the basic cellular biochemical alterations that occur as a result of viral oncogenesis has only recently emerged. It has been difficult for the biochemist to deal with the basis of tumor induction because of the heterogeneous nature of the neoplastic process. The biochemist or molecular biologist who hopes to provide a chemical description of the oncogenic process wishes to observe both the interaction between the virus and the infected cell, and the properties of progeny of the initial cells which have been converted to a neoplastic state. These initial processes are difficult if not impossible to control or to observe in experimental animals; consequently, the study of established neoplasms remained for many years the only alternative. In this study, however, the biochemist is confronted *not* with the immediate product of a virally induced oncogenic event, but the interaction of that event with the forces that both restrict and select the cells which eventually populate the neoplasm. Because of the alterations in the cellular population due to mutation, immunological selection, hormonal variation, etc., the final neoplastic population may vary both from animal to animal and from the initial converted cell. Consequently, the study of established virally induced neoplasms has revealed little about the cellular alterations that accompany the initial oncogenic process.

The major advances that have contributed so greatly to our ability to analyze the process and results of viral carcinogenesis have been the utilization of cell culture techniques and the analysis of interactions of viral populations with cultured cells. Normal cells growing in culture do not give rise to tumors when injected into syngenic animals. Many cultured lines of normal cells, when infected with certain oncogenic viruses, undergo a series of genetically stable changes which culminate in the emergence of a cell type that contains no detectable viral particles, but will produce tumors upon inoculation into a host animal. This series of events, which gives rise to a stable population of culturable cells possessing oncogenic potential is termed "viral transformation." In this experimental system many of the disadvantages of the in vivo system are eliminated. Cell populations may be easily controlled. Alterations in environmental factors may be carried out easily. The immediate progeny of a virally transformed cell can easily be examined. The relevant question for the researcher interested in virally induced tumors then becomes, "Are there significant differences between virally transformed cultured cells and virally induced neoplasms in experimental animals?" There are undoubtedly differences between transformed cultured cells and neoplastic cells, due to the selective forces described previously which operate in vivo. In his excellent review of viral transformation, to which the reader is referred for a concise discussion of many of the molecular aspects of the transformation process, Dulbecco (19) suggests that a "reasonable view of the relationship of transformed cells to neoplastic cells" would equate transformed cells with those *initially present* in the in vivo neoplasm when it first occurs, as opposed to those cells present after selection pressures have had an opportunity to operate.

Oncogenic Viruses

This discussion is limited to the small DNA-containing papova viruses, SV40 and polyoma (PY) virus. Both these viruses contain circular DNA molecules of molecular weight approximately 3×10^6. This amount of DNA is sufficient to code for 6–10 genes. When cultured cells are infected with these DNA viruses, two different phenomena may occur. In the first instance, viral infection may cause a lytic response resulting in the multiplication of the virus and the death of the host cells. Alternatively, the process of transformation of the host cell may occur, with the resulting emergence of an altered cell with oncogenic potential. Which of these two alternative pathways the infection will follow depends to a great degree on the *characteristics of the host cell.*

The response of a given cell to viral infection is determined in part by the genetic characteristics of that cell.

As a result of viral transformation, the second of the two possible pathways, several changes occur in the host cell: *1*) the viral DNA is integrated into the DNA of the transformed cell (42); *2*) a new protein, termed the T antigen, is found in the nucleus (38, 40); *3*) new surface antigens are present on the transformed cells (18, 29); *4*) the plating efficiency of the cells is increased, and the growth characteristics in culture of the cells are drastically altered; and *5*) such transformed cells give rise to tumors when injected into syngenic hosts. For a review of the effects of oncogenic viruses, the reader is referred to Svoboda (50).

Cells in Culture

Rodent embryo cells have been extensively employed in the study of cells in culture. They require a complex medium, which includes calf serum. When cells such as these are inoculated into a culture plate, they attach firmly to the substrate and, after a lag period, they begin to divide. Eventually they cover the surface of the culture dish with a monolayer of cells. Various cell populations stop dividing and remain at a constant "saturation density," which is an heritable property of the cell line in question. The mechanism whereby cell division is stopped as a result of cell-to-cell contact is termed "contact inhibition of cell division" (47) or "density-dependent inhibition."

The two most common cell culture lines used for the study of viral transformation are a line of hamster embryo kidney fibroblasts known as BHK cells and a line of mouse embryo fibroblasts known as 3T3 cells (46, 51). 3T3 cells have been grown under transfer and feeding conditions that limit cell-to-cell contact. If cells are allowed to engage in extensive contact, then a cell which has lost the ability to respond to contact by ceasing division, but instead continues to divide, soon gives rise to a *population* of cells which no longer exhibits contact-dependent inhibition of growth. Selection pressure results in the emergence of a population that continues to divide even though the cells are confluent, and which thus has a higher saturation density. Viral transformation, by SV40 or polyoma virus, of 3T3 cells also results in the establishment of cell lines that do not show contact inhibition of division, but instead grow to extremely high saturation densities. Indeed, this is one of the most characteristic in vitro traits of transformed cells.

Recently Aaronson and Todaro (1) have demonstrated that loss of contact inhibition of division is one of the properties of transformed cells which most closely correlates with tumorigenicity. The relation derived by these investigators between saturation density and tumorigenicity is illustrated in Figure 1. It is clear from these data that loss of contact inhibition of division (i.e., growth to a higher saturation density) correlates well with an increased

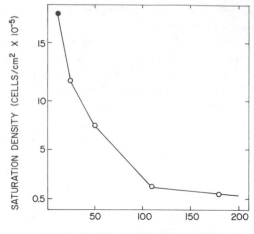

Figure 1

Relation between saturation density and tumor-forming ability. Animals were injected with 10⁷ cells. O, Balb/3T12 sublines; ●, SV-Balb/3T3 cells. (Data from Aaronson, S. A., and Todaro, G. J. *Science 162,* 1024, 1968.)

ability to promote tumor induction in vivo. Pollack *et al.* (37) have demonstrated this relation in an essentially reciprocal experiment. These authors selected against transformed cells that continued to divide following the establishment of cell-to-cell contact by adding 5-fluorodeoxyuridine to confluent cultures of transformed cells. The only cells surviving such selection are those that have ceased to divide prior to the application of the drug. The variant transformed cell lines established in this manner had a lower saturation density (i.e., a greater response to contact inhibition of division) and a correspondingly reduced malignant capability, despite the continued presence of the integrated viral genome. It is important to point out that viral genomes rescued from such variant cells, when used to transform normal fibroblasts result in a transformed cell line which grows to the saturation density characteristic of normally transformed cells and which has a high degree of malignant potential. The genetic alteration which leads to the establishment of the low-density relatively nonmalignant cell line is thus evidently an alteration not in the integrated viral genome, but in the cellular genome.

There is clearly a strong correlation between the ability of cells to continue to divide in cell culture despite extensive cell contact and their ability to give rise to tumors in syngenic hosts. The remainder of this chapter is devoted to consideration of the alterations in the surface properties that occur in cultured cells as a result of viral transformation and the effects of such alterations on saturation density and, therefore, on tumorigenicity.

PLANT AGGLUTININS

In 1888 Stillmark (45) reported that extracts of the castor bean (*Ricinus communis*) were able to cause the agglutination of erythrocytes. Subsequently it was shown that extracts of a wide variety of plant seeds were capable of causing a similar reaction. [For a review of the early literature on plant agglutinins the reader is referred to Bird (4).] The first of these hemagglutinins were relatively nonspecific, i.e., they agglutinated the red cells of a variety of species. Somewhat later, extracts of other seeds were shown to agglutinate the red cells of different species with varying degrees of specificity.

The chemical specificity of this group of compounds has most clearly been demonstrated, however, in the ability of some of the plant agglutinins to react with specific blood group antigens. Boyd (8) described the first of these reactions, that of a lima bean agglutinin specific for the blood group A antigen. Subsequently group-specific plant agglutinins were discovered for a variety of blood groups. Many of the hemagglutinins have been shown to cause the clumping of nucleated cells as well as red cells, although until recently the emphasis has been on studies with red cells. In 1964 Tunis (52) demonstrated that kidney beans contained, in addition to an activity that causes red cell clumping, a nonhemagglutinating cytoagglutinin, i.e., a fraction that would *not* agglutinate red cells, but would cause the agglutination of nucleated cells.

The interaction of the plant agglutinins with erthryocytes and with the purified antigens of erythrocytes resembles closely the interactions of blood-group-specific antibodies (9). Like antibody molecules, those plant agglutinins that have been studied physically are divalent, i.e., they have two combining sites. Since cells are polyvalent antigens, a network interaction occurs with a resulting agglutination. As in the case of the antigen-antibody reaction, if a monovalent antigen (termed a hapten) is prepared, the agglutinins will react with the hapten and form a nonprecipitating complex. Haptens can therefore be used to inhibit or reverse the agglutination reaction, as shown in Figure 2. By using potential haptens of defined chemical structure, it is therefore possible to characterize the chemical nature of the membrane determinants responsible for the cellular agglutination reaction.

Agglutinins for Malignant Cells

While attempting to study the differences in the surfaces of normal and tumor cells, Aub and his coworkers (3) began to investigate the effects of a variety of enzymes on such cells. One of the enzymes employed in this study was a commercial preparation of lipase from wheat germ. Aub *et al.* observed that tumor cells treated with this preparation clumped together, while non-neoplastic cells did not appear to be agglutinated nearly so readily. This

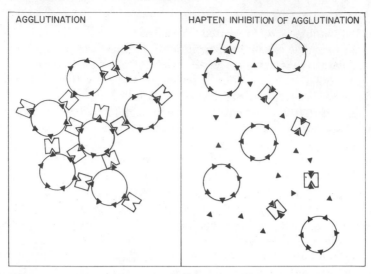

Figure 2

Hapten inhibition of a cellular agglutination reaction. Polyvalent agglutinable cells are represented by spheres with solid triangles. Monovalent hapten inhibitors of agglutination are represented by free solid triangles. As shown in the illustration, interaction of the hapten with an agglutinin prevents cellular agglutination.

observation was the first report of a phytoagglutinin with a specific affinity for tumor cells.

WHEAT GERM AGGLUTININ

Aub and his associates also found that the agglutination activity present in the wheat germ preparation was stable to heating or to exposure to *p*-chloromercuribenzoate, although the lipase activity was destroyed as a result of either treatment. The enzyme activity, on the other hand, was insensitive to periodate treatment, while the agglutinating activity of the material was inactivated. These observations led these experimenters to suggest that the agglutination activity of the preparation resided in a macromolecule distinct from the lipase. Because of the periodate sensitivity, they proposed that the agglutinin activity was the result of a mucopolysaccharide impurity in the lipase preparation.

Burger and Goldberg (15) have purified the wheat germ agglutinin (WGA). Extracts of commercial wheat germ lipase preparations were heat-inactivated, fractionated by ammonium sulfate, and finally chromatographed on Sephadex G-75. The agglutinin purified by this procedure is a glycoprotein which runs as a single peak in the analytical ultracentrifuge, with a sedimentation value of 2.74S and a calculated molecular weight of 26,000. Acrylamide-gel electrophoresis of the purified preparation resulted in two bands of about equal intensity. Agglutination activity was destroyed by proteolytic enzymes, but not by RNAse, DNAse, or neuraminidase. The agglutinating activity of the purified preparation is sensitive to periodate. Determination of reducing sugars indicated that about 4.5% of the agglutinin was carbohydrate.

The purified glycoprotein agglutinates L1210 leukemia cells and poly-oma-transformed BHK cells (both neoplastic cells) but not BHK cells. By utilizing the hapten-inhibition technique described previously, Burger and Goldberg demonstrated specific inhibition of the agglutination reaction with the monosaccharide N-acetyl-glucosamine and its disaccharide, chitobiose. This hapten inhibition was quite specific over a wide range of mono- and disaccharides. Agglutinated cells could be disaggregated by the addition of N-acetyl-glucosamine to the clumps. It is of interest to note that while sialic acid was unable to act as a hapten, this compound plays some role in the agglutination of transformed cells by WGA, since its specific removal from the surface of such cells by neuraminidase prevented agglutination by WGA.

Burger subsequently described a method of purification for WGA which exploited its biochemical specificity. Ovomucoid, a glycoprotein containing large quantities of N-acetyl-glucosamine, was immobilized in a column with kieselguhr. Wheat germ lipase was applied to the column, and elution was begun with 0.1 M glucose to remove nonspecifically bound protein. Elution was then continued with an N-acetyl-glucosamine gradient, and finally with chitobiose. This specific hapten-elution technique resulted in a one-step purification of WGA (13).

CONCONAVALIN A

Conconavalin A (ConA), a hemagglutinin found in the jack bean (2), was the first of the plant agglutinins to be isolated in crystalline form (48). It agglutinates dextran, glycogen, and some strains of yeast and bacteria in addition to erythrocytes of some, but not all, vertebrate species. ConA appears to have two agglutinating sites per molecule (28, 54). Inbar and Sachs (27) examined the ability of ConA to agglutinate various transformed and un-transformed cell lines from several rodent species. Normal cells from hamster, rat, and mouse embryo secondary cultures, as well as mouse spleen cells, rat bone marrow cells, and the untransformed 3T3 cell line were not agglutinated by 500 μg/ml ConA after 30 min of incubation. In contrast, polyoma-transformed lines of rat, mouse, and hamster embryo cultures and polyoma-transformed 3T3 cells were agglutinated, as were SV40-transformed embryo cultures and 3T3 cells. Leukemic cell lines derived from mouse erythroid-, myeloid-, and lymphoid-induced tumors were also agglutinated, as were transformed hamster cells produced by chemicals or x-irradiation.

Hapten-inhibition tests performed with the ConA system have demon-strated the ability of α-methyl-D-glucopyranoside (α-MG), to prevent the interaction of ConA with dextran (20). Inbar and Sachs (27) demonstrated that α-MG could reverse the ConA-dependent agglutination of virally and chemically transformed cell lines. This reversibility was specific; in particular N-acetyl-glucosamine, the hapten inhibitor of WGA, could not reverse ConA agglutination of transformed cells. Recent preliminary experiments (33) have shown that oncogenic transformation by RNA viruses does not result in an increased agglutinability by either WGA or ConA. This correlation between transformation and agglutination is evidently somewhat restricted and has most convincingly been illustrated with the papova viruses.

SOYBEAN AGGLUTININ

A hemagglutinin from untoasted soybean flour (SBA) has been purified to chromatographic homogeneity by Lis *et al.* (31). The simple sugar, N-acetyl-D-galactosamine, was subsequently shown to inhibit the interaction between this agglutinin and animal cell surfaces (30). Normal mouse, rat, hamster, and human cells could not be agglutinated by this glycoprotein, while virally transformed mouse, rat, and human cell lines were susceptible to agglutination. In contrast to the results of studies with the other species, virally and chemically transformed hamster cells were *not* rendered agglutinable by SBA, although they were agglutinable by ConA. These results hint at the complexity of the membrane matrix and the variations in the exposure of different membrane structures as a result of viral transformation. A summary of the specificity studies with the agglutinins discussed is presented in Table I.

Table I
Specificity of Plant Agglutinins

AGGLUTININ	MONOSACCHARIDE INHIBITOR OF AGGLUTINATION
Wheat germ agglutinin	N-acetyl-glucosamine
Conconavalin A (jack bean)	α-Methyl-D-glucopyranoside
Soybean agglutinin	N-acetyl-galactosamine

Cryptic Agglutination Sites on Untransformed Cells

The simplest and perhaps most attractive hypothesis to account for the selective agglutinability of virally transformed cells proposes that the viral genes integrated into the host chromosome either *1*) direct the synthesis of unique macromolecules which become an integral part of the cellular membranes or *2*) direct the genome of the cell to synthesize de novo such new macromolecules by a process of derepression or activation of a dormant cellular gene. At least two other possibilities exist, however. The untransformed cell may, for example, possess only a few agglutinin receptor molecules. The number of sites on the surface of the untransformed cell would not be sufficient to permit agglutination; the transformation process would modulate the regulatory mechanism responsible for the synthesis of such sites and increase their surface density. Alternatively, the untransformed cells might possess all the agglutination sites in a "cryptic" form. Viral transformation would result in a conversion of the membrane to a form in which the previously existing but unavailable agglutination sites become exposed at the cell surface.

Burger (12) demonstrated that hypotonic shock of either L1210 mouse leukemia cells or polyoma-transformed BHK cells releases material from the surface of these cells which can inhibit the WGA agglutination reaction or reverse the agglutination of transformed cells previously clumped with WGA. The shocked cells, which are still viable, can no longer be agglutinated with WGA. The conclusion that the virus-transformation-specific WGA agglutination site is easily solubilized from the surface of the transformed cell seems clearly justified from these results. Similar experiments performed with untransformed, nonagglutinable cells, however, also released a fraction that combines with WGA and inhibits the agglutination reaction of transformed cells. The latter observation suggested that the agglutination site for WGA might be present on untransformed cells in a cryptic fashion, rendering it unavailable to the agglutinin.

Previous studies on hemagglutinins have shown that agglutination sites are present in a cryptic fashion in some cases and can be unmasked by a variety of enzyme treatments of the erythrocytes in question. Burger (13) treated BHK, 3T3, rat liver, and primary chick embryo cells with a variety of proteases, and found that these normally nonagglutinable cells were rendered agglutinable by purified WGA after such treatments. Minimal concentrations for the conversion of 3T3 cells to an agglutinable state after 6 min exposure are: trypsin, 0.05 mg/ml; chymotrypsin, 0.08 mg/ml; ficin, 0.005 mg/ml; papain, 0.008 mg/ml. In a similar series of experiments Inbar and Sachs (27) demonstrated that hamster, mouse, human, and rat untransformed cells, which could not be agglutinated by ConA, were converted to a ConA-agglutinable form by mild proteolytic (trypsin or pronase) treatment. Like the WGA agglutination site, the ConA site is also apparently present, but in an unavailable form, in the membranes of untransformed cells of a variety of species. Similarly, normally nonagglutinable untransformed mouse, rat, and human cell lines could be agglutinated by SBA after brief treatments with pronase or trypsin. Transformed hamster cells, which were *not* agglutinable with SBA (but could be agglutinated with either WGA or ConA) could be agglutinated by SBA only after prolonged pronase incubation. In each of these cases, the agglutination of normal cells resulting from proteolytic treatment could be reversed specifically by the proper monosaccharide hapten.

These results with three agglutinins on normal and transformed cell lines of a variety of species suggest that the various agglutination sites, which may be indicative of three qualitatively different saccharide-containing membrane constituents, are present in untransformed cells, but in a cryptic fashion. They have a "cover" which prevents their interaction with the agglutinins. This "cover" is at least partially protein in nature, since proteolytic digestion exposes the agglutination sites. The integration of the viral genome, and the adoption of the transformed state, either eliminates the "cover" from the surface of the transformed cell or causes a configurational change in the surface of the cell which renders the previously cryptic agglutination sites available at the cell surface.

Correlations Between Saturation Density, Tumorigenicity, and Agglutinability

As described earlier, Pollack *et al.* (37) have isolated "flat" variants of mouse and hamster SV40-transformed cell lines which have saturation densities much lower than those of the parent transformed lines. In addition, the reduced tumorigenicity of the variant cell lines correlates well with the re-establishment of contact inhibition of growth. Because contact inhibition of growth is a difficult parameter to measure biochemically, Pollack and Burger (36) utilized agglutinability to examine in more detail the relation between surface architecture and ability to give rise to tumors in these variant cell lines. Five cell lines, 3T3 and four transformed 3T3 lines of varying saturation density, were examined for their agglutinability with WGA. The cell lines were also characterized with respect to saturation density, which varied over a 10-fold range. *The relative agglutinability by WGA of the five cell lines was in the same order as their saturation densities, i.e., as their ability to promote the formation of tumors in vivo.* Trypsinization of the variants with low saturation density rendered these cells agglutinable by WGA, as had previously been shown for untransformed 3T3. These experiments clearly show that it is the degree of contact inhibition and therefore the relative malignant potential, and not the presence or absence of the viral genome, that dictates the degree of agglutinability of transformed cell lines. As in the case of normal cells, the variant transformed cell lines possess the agglutination sites for WGA, but in the cryptic form. It is important to realize that the genetic alteration responsible for the establishment of the unagglutinable, density-inhibited, transformed variant has occurred in the *cellular* genome and not in the viral genome.

Exposure of Agglutinin Sites and Contact Inhibition of Division

The agglutination sites for WGA, ConA, and SBA are all present in the membrane of normal cells, but unavailable to the various agglutinins. Transformation eliminates contact inhibition of division, and concomitantly exposes these agglutinin sites. Is it possible to either *1*) reduce contact inhibition and expose the agglutinin sites by some means other than viral transformation, or *2*) restore contact inhibition by artificially covering the agglutination sites on transformed cells? Success at either one of these experimental approaches would provide strong correlative evidence for the relation between these membrane components and the contact-mediated control of growth (and therefore malignant potential).

In order to examine this question, Burger (14) treated confluent monolayers of 3T3 mouse fibroblasts with low levels of various proteolytic enzymes. Treatment with pronase, ficin, or trypsin, which exposed the cryptic WGA agglutination sites, was able to overcome density-dependent inhibition of growth and initiate a subsequent round of cell division (Fig. 3). While the molecular nature of the membrane-associated events responsible for this phenomenon is still unknown, it is clear from this experiment that a treatment known to expose normally cryptic membrane components found on the surface of the transformed cells can bring about a reversal of contact inhibition

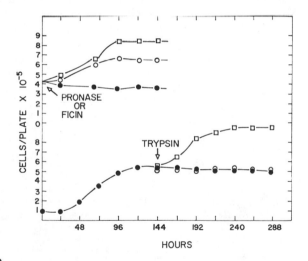

Figure 3

Initiation of cell division in growth-inhibited fibroblast cultures by proteolytic enzymes. 3T3 cells were grown to saturation density. Cells were then treated for 45 min with the proteolytic enzymes indicated. After treatment, conditioned medium was added to the culture plates. Daily thereafter cell number was determined. Upper graph: ●, control cells; ○, cells treated with 0.0005% ficin; □, cells treated with 0.005% pronase. Lower graph: ●, control cells; □, cells treated with 0.007% trypsin; ○, cells treated with trypsin inactivated with diisopropylfluorophosphate. (Data from Burger, M. M. *Nature* 227, 170, 1970.)

of division. Burger and Noonan (16) have also reported an experiment complementary to the one just presented. In these studies attempts were made to recover growth control in virally transformed cells by artificially covering the exposed agglutination sites. This experiment attempts to create a surface more closely resembling that of the untransformed cells. Burger and Noonan chose to use ConA for these studies, since WGA agglutination can be overcome by serum glycoproteins necessary for cell growth. In order to avoid the criticism that inhibition of growth of transformed cells by ConA might be the result of immobilization or toxicity rather than a true contact-dependent phenomenon, they used monovalent ConA. ConA was "split" by controlled proteolytic digestion. The resultant product was shown to inhibit ConA agglutination of PY-3T3 cells by untreated ConA, without itself causing the agglutination of PY-3T3 cells, indicating it had been converted to a monovalent form. Figure 4 shows the growth response of transformed cells to the presence of varying concentrations of monovalent ConA. Clearly, increased concentrations of "split" ConA are able to restore successively greater degrees of contact inhibition of division to virally transformed cells. In the presence of 75 μg/ml of split ConA the transformed cells have phenotypic growth control characteristics almost identical to those of the untransformed parent cell line. These data, along with those previously dis-

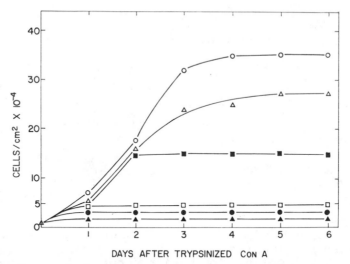

DAYS AFTER TRYPSINIZED Con A

Figure 4

Growth response of PY-3T3 cells to varying concentrations of trypsinized ConA. ○, PY-3T3 cells; ▲, 3T3 cells; △, PY-3T3 cells + 10 μg/ml trypsinized ConA; ■, PY-3T3 cells + 25 μg/ml trypsinized ConA; □, PY-3T3 cells + 50 μg/ml trypsinized ConA; ●, PY-3T3 cells + 74 μg/ml trypsinized ConA. (Data from Burger, M. M., and Noonan, K. D. *Nature 228***, 512, 1970.)**

cussed for the relative agglutinability of the "flat" variants, suggest a quantitative correlation between exposure of agglutination sites and the degree of density-dependent inhibition of division. As Burger and Noonan point out, however, proof of this assertion requires a method capable of counting the density of exposed and covered agglutination sites. If this inhibition of cell division is in fact due to the blocking of the exposed agglutination sites and not to some nonspecific toxic effect resulting from irreversible damage to the cell surface, then growth of the ConA-inhibited transformed cells might be expected to resume when the agglutinin is removed. Removal of ConA can easily and specifically be carried out by addition of the hapten α-MG. As shown in Figure 5, the addition of α-MG to PY-3T3 cells in the presence of split ConA reverses the phenotypic density-dependent inhibition of division and leads to resumption of cell division.

The effects of proteolytic enzymes on normal cells and of "split" ConA on transformed cells emphasize the importance of the role of carbohydrate-containing macromolecules of the cell surface in the social regulation of cell division. It seems clear that there exists a covering "layer" or structure on the surface of normal cells which modulates growth control and is altered or

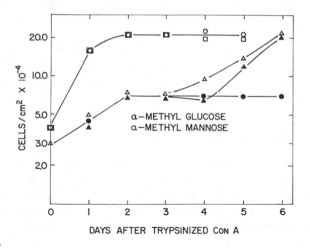

DAYS AFTER TRYPSINIZED CON A

Figure 5

Haptenic reversal of the effect of trypsinized ConA on the growth of PY-3T3 cells. O, PY-3T3 cells; ●, PY-3T3 cells + 50 μg/ml trypsinized ConA; □, PY-3T3 cells + 50 μg/ml trypsinized ConA incubated with 10^{-2} M α-methyl-D-glucoside for 12 hr; △, PY-3T3 cells + 50 μg/ml trypsinized ConA + 10^{-2} M α-methyl-D-glucoside or 10^{-2} M α-methyl-D-mannoside added to the stationary culture at day 3; ▲, PY-3T3 cells + 50 μg/ml trypsinized ConA + 10^{-3} M α-methyl-D-glucoside or 10^{-3} M α-methyl-D-mannoside added to the stationary culture at day 3. (Data from Burger, M. M., and Noonan, K. D. *Nature 228*, 512, 1970.)

removed by viral transformation. "Beneath" this layer lie the agglutination sites, whose primary function is unknown. It is remarkable that covering the agglutination sites and restoring an artificial, and probably a chemically very different, covering layer can restore growth control to transformed cells.

GLYCOLIPID ALTERATIONS AS A RESULT OF VIRAL TRANSFORMATION

The study of alteration in the glycolipids of normal and virally transformed cultured cells began, as did the agglutination studies, as a result of earlier observations with neoplastic tissue. Increases in a particular "tumor glycolipid" and a concomitant decrease in other glycolipids had previously been described for human cancer tissue (24). Arguing by analogy, Hakomori and Murakami (25) predicted that such alterations in cellular glycolipid components might accompany viral transformation of mammalian cells.

While a complete description of the structure and chemistry of the gangliosides or sphingoglycolipids of mammalian cells is beyond the scope of this chapter, a summary of the relevant nomenclature and structures is appropriate. Gangliosides contain sialic acid, specific carbohydrate sequences in oligosaccharide chains, a long-chain dihydroxyamine, and a fatty acid. The lipid portion of these compounds is termed a ceramide, an amide of the long-chain dihydroxyamine (a sphingosine) and a long-chain fatty acid. The structure of ceramide is shown in Figure 6. The various oligosaccharides found in the gangliosides are attached to ceramide through the C_1 hydroxyl. The structure of the various relevant gangliosides, their common names, and designations (49) are summarized in Table II.

Hematoside Concentration in Transformed Cells

The first study of glycolipids of normal and virally transformed cultured cells concentrated on the levels of hematoside and its desialated precursor, lactosyl ceramide (25). Hakomori and Murakami used BHK cells, a spontaneously transformed clone of BHK, and polyoma-transformed BHK cells for their initial analyses. These workers demonstrated that the transformed

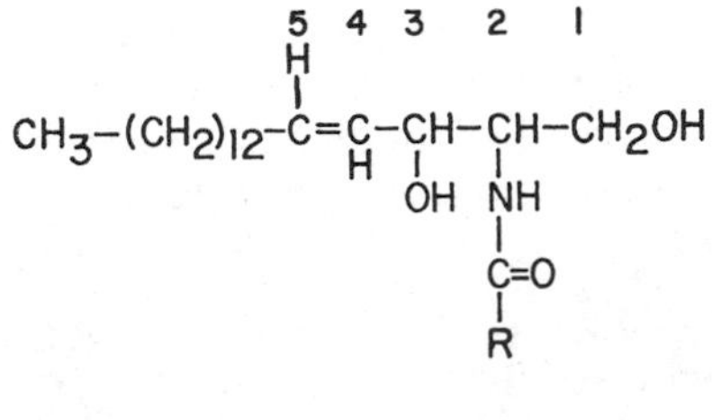

Figure 6
Structure of ceramide.

Table II

Structure of Gangliosides

COMMON NAME	STRUCTURE	DESIGNATION
Hematoside	N-acetyl-neuraminyl-galactosyl-glucosyl ceramide	GM_3
Tay-Sachs ganglioside	N-acetyl-galactosaminyl-(N-acetyl-neuraminyl)-galactosyl-glucosyl ceramide	GM_2
Monosialoganglioside	Galactosyl-N-acetyl-galactosaminyl-(N-acetyl neuraminyl)-galacosyl-glucosyl ceramide	GM_1
Disialoganglioside	N-acetyl-neuraminyl-galactosyl-N-acetyl-galactosaminyl-(N-acetyl-neuraminyl)-galactosyl-glucosyl ceramide	GD_{1a}

cells had indeed lost their density-dependent control of saturation density. They also showed that transformed cells had reduced levels of hematoside when compared with normal cells and a concomitant increase in the concentration of lactosyl ceramide. Their data are summarized in Table III. These observations led Hakomori and Murakami to postulate a generalization they termed the "incompleteness of the carbohydrate chain" proposal, in which they suggest that one of the major metabolic events accompanying malignant transformation is the incomplete synthesis of the carbohydrate chains of cellular glycolipids. Two important reservations concerning the significance of this observation to the events controlling saturation density and oncogenic potential are relevant here. The incomplete "synthesis" of hematoside observed, more properly described as a reduced concentration of hematoside, could be the result of either a reduced sialyl transfer activity or an enhanced neuramidase activity: the measurements summarized in Table III are, of course, unable to distinguish between these two alternatives. It is also important to point out that at this time Hakomori and Murakami did not present any evidence demonstrating that the hematoside they measured was located on the surface of the cell. Both these reservations are discussed more fully in the context of experiments to be described subsequently.

In their initial paper on the subject of glycolipid concentrations in virally transformed cells Hakomori and Murakami presented one other significant experimental result which brings together these observations with previously described ones. They were able to repeat the specific cytoagglutination of transformed cells by WGA described by Burger and Goldberg. They also repeated the monosaccharide hapten-inhibition experiments and confirmed the observation that N-acetyl-glucosamine was the only *monosaccharide*

Table III

Hematoside and Galactosyl-Glucosyl Ceramide in Normal and Transformed BHK Cells

CELL LINE	HEMATOSIDE (GM$_3$) (μg/100 mg PROTEIN)	GALACTOSYL-GLUCOSYL CERAMIDE (μg/100 mg PROTEIN)
BHK/C$_{13}$ fibroblasts	475	13
Spontaneously transformed BHK/C$_{13}$ fibroblasts	310	117
PY-BHK/C$_{13}$	105	125

Data from Hakomori, S., and Murakami, W. T. *Proc. Natl. Acad. Sci. U.S.A.* **59**, 254 (1968).

inhibitor of the WGA agglutination reaction. They then turned their attention to the ability of various glycolipid fractions of normal and transformed cells to inhibit this reaction. Both hematoside and lactosyl ceramide were unable to act as hapten inhibitors of the agglutination reaction. The ganglioside fraction of the transformed cell line, however, was able to inhibit agglutination. This fraction is composed primarily of hematoside and lactosyl ceramide, but does contain "a small number of gangliosides or higher glycolipids in much smaller quantities." In contrast, the corresponding ganglioside fraction of the untransformed cell line did *not* have hapten activity. However, exposure of this ganglioside fraction of the normal cells to the Smith degradation (44) resulted in a product that is able to inhibit the agglutination of transformed cells. The authors suggest that this procedure, which destroys periodate-susceptible groups, removes a masking group (either galactose or fucose) from the normal ganglioside and transforms it to one capable of acting as a hapten. Digestion of the same fractions with 0.1 M H$_2$SO$_4$ at 80°, which specifically removes sialic acid, does not result in the production of hapten activity. We are left, therefore, with the provocative result that the concentrations of hematoside, which composes the bulk of the glycolipid fraction of normal cells, are significantly reduced as a result of transformation; however, a minor glycolipid is unmasked as a result of transformation and is able to block the agglutination potential associated with transformation. Unfortunately the latter observation has not yet been elaborated upon in the literature, despite its obvious potential significance. The relevance of the study of alterations of cellular glycolipids to viral oncogenesis, however, is clearly illustrated by these initial studies.

In an extension of this work, Hakomori *et al.* (26) measured the concentrations of glycolipids of polyoma- and SV40-transformed 3T3 cells. While BHK cells had been found to contain "N-acetyl-hematoside" [N-acetylneuraminyl (2 → 3) galactosyl (1 → 4) glucosyl ceramide] as the principal glycolipid, 3T3 cells contained both "N-acetyl-hematoside" and

"N-glycolyl hematoside" [N-glycolyl neuraminyl (2 → 3) galactosyl (1 → 4) glucosyl ceramide]. Although viral transformation caused a reduction in both hematosides of 3T3 cells, as previously observed for transformed BHK cells, no concomitant increase in lactosyl ceramide was observed in SV-3T3 or PY-3T3 cells.

In this paper Hakomori and his coworkers also approached the question of the cellular localization of glycolipids, and the relation of such localization to oncogenic capacity. Antisera were prepared against the two hematosides described. These antisera were used to determine whether hematoside is present on the surface of the cell. To answer this question, Hakomori *et al.* employed the technique of complement-dependent cytolysis of cells labeled with ^{51}Cr. If an antigen is present on the surface of a cell, the homologous antibody reacts at this site. Subsequent addition of complement results in cell lysis and liberation of the radioisotope. Thus surface antigenicity can be distinguished from antigens present within the cell or present on the cell membrane in a cryptic fashion, and not available to antibody. The results of these experiments are shown in Figure 7. Although the *total* amount of hematoside in virally transformed cells is lower than that in normal cells, the

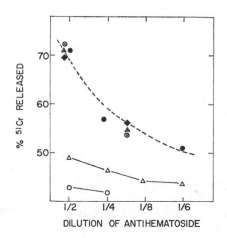

Figure 7

Complement-dependent immune cytolysis of transformed and untransformed fibroblasts with antihematoside antisera. Control values with preimmune sera were always less than 40%. Reactions were carried out with 0.1 ml of antisera, 0.1 ml of complement (1:5 dilution of guinea pig sera), and 0.1 ml of a cell suspension which contained 0.8 × 10⁶ cells. Values for total radioactivity were determined by measuring the radioactivity of the supernatant when cells were lysed in 0.5% acetic acid. ○, BHK cells; △, 3T3 cells; ●, SV 3T3 cells; ◆, PY-BHK cells; ⊗, trypsinized BHK cells; ▲, trypsinized 3T3 cells. (Data from Hakomori, S., et al. Biochem. Biophys. Res. Commun. 33, 563, 1968.)

cytolysis due to the presence of antihematoside serum is much greater with transformed than with normal cells. This observation indicates that more hematoside is exposed at the cell surface of transformed cells despite the reduced over-all concentration of this compound. The specificity of the hematoside antisera was shown by the inhibition of lysis of the transformed cells in the presence of added hematoside as hapten. How can the higher levels of hematoside known to be present in normal cells be reconciled with the immunological experiments which indicate that transformed cells have more of this compound present on the cell surface? One possible explanation for this observation proposes that hematosides of the surface membrane of normal cells are present in a "cryptic" fashion, which cannot interact with the antibody. Viral transformation results in the exposure of the cryptic antigen. This proposal is precisely the same as that for the cryptic existence of masked agglutination sites on normal cells, demonstrated by Burger (13). To test this proposal, Hakomori *et al.* treated normal cells with trypsin, in analogy to the previous experiments with the agglutinins, and then tested the cytolytic susceptibility of treated cells to antihematoside antisera. After trypsin treatment (the concentration and duration of which are not described), the untransformed cells were as reactive with antihematoside antisera as the transformed lines (Fig. 7). These results suggest that, like the agglutination sites previously discussed, the hematosides of normal cells are masked by a trypsin-sensitive "cover" which is removed as a result of viral transformation. As a result of this study it becomes clear that both the total content *and the spatial arrangement* of cellular glycolipids of mammalian cells undergo significant alteration as a result of viral transformation. Which, if either, of these considerations is functionally correlated with oncogenic potential, however, remains an open question.

Cell-Density-Dependent Changes in Glycolipid Concentrations

The previously described studies on agglutinability and glycolipid com-components of transformed and normal cells have not taken into account the possible existence of a systematic chemical alteration in the surfaces of cells *as a result* of cell-to-cell contact. They appear to have instead made the tacit assumption that a chemical difference between normal and transformed cells existing in preconfluent cultures determines the growth response of such cells upon reaching confluency. In a subsequent paper, Hakomori (23) describes the results of quantitative measurements of the gangliosides of growing and confluent cultures of normal, spontaneously transformed, and polyoma-transformed BHK cells. His results, summarized in Table IV, demonstrate that *1*) hematoside is reduced in transformed cells; *2*) lactosyl ceramide is increased in transformed cells; *3*) hematoside concentrations do not change significantly in growing compared with confluent cultures of normal or transformed BHK cells; 4) there appear to be either low or nonexistent levels of GD_{1a}, GM_1, and GM_2 in both normal and transformed BHK cells; and 5) lactosyl ceramide levels do not seem to change appreciably as a result of cell-density variations. The most interesting observation, however, concerns a ceramide trihexoside (CTH; galactosyl-galactosyl-glucosyl ceramide). This sphingolipid is present in appreciable quantities in normal cells, but is absent or present in very small quantities in polyoma-transformed cells. Its concentration in untransformed cells is elevated when the cells have reached saturation density and reduced when the cells are in the log phase of growth. Once again the causal relation between the phenomena is not known, but it is clear that 1) proliferating BHK cells have reduced levels of CTH, 2) BHK cells at confluency have elevated CTH levels, and 3) polyoma-transformed BHK cells, which do not demonstrate contact-dependent regulation of cell division, do not contain CTH. Radioactive precursor studies demonstrated an increased rate of labeling of CTH at high cell density. At present, the levels of CTH have not been reported either 1) in 3T3 cells or 2) as a result of SV40 transformation. Such studies will enable us to determine the generality of this observation. These results are also subject to the same reservations described for the earlier data presented on hematoside; CTH concentrations are described *for the whole cell.* We currently do not know if this compound is present in an exposed or a cryptic form, or if it is in fact, a component of the cell surface membrane.

Changes in Higher Sialogangliosides as a Result of Viral Transformation

Hakomori did not observe significant levels of GD_{1a}, GM_1, or GM_2 in either BHK cells or PY-BHK cells. He did, however, observe significant concentrations of GM_1 and GD_{1a} in human diploid fibroblasts growing in cell culture (Table IV). When such cultures were transformed by SV40, the levels of hematoside, GD_{1a}, and GM_2 were dramatically reduced. Concomitantly, high levels of GM_2 (absent in the untransformed cell line) were present in the transformed line. GD_{1a} and GM_1 levels appeared to be a function of growth stage; both gangliosides were more apparent in confluent cultures than in growing cultures.

Recently Mora *et al.* (34) have analyzed sialogangliosides from 3T3 and doubly transformed SVPY-3T3 cells. Like Hakomori, they observed a reduction in the total amount of gangliosides (Table V). They did not, however, observe a decrease in the hematoside concentration of transformed 3T3 cells, in contrast to the results of Hakomori *et al.* These workers also were unable to find a decrease in hematoside in several other transformed cell lines of established fibroblast cultures from inbred mouse strains. They did observe

Table IV

Cellular Gangliosides in Cultured Hamster and Human Fibroblasts at Various Stages of Growth and Varying Cell Densities

CELL LINE	TUMORI-GENICITY	CONTACT INHIBITION OF DIVISION	CELL DENSITY OR STAGE OF GROWTH	GLUCOSYL-CERAMIDE
BHK/C$_{13}$	−	+ +	Growing	6
			Confluent	4
Spontaneously transformed BHK/C$_{13}$	+	+	Growing	7
			Confluent	9
PY-BHK/C$_{13}$	+ +	−	Low	8
			High	8
Contact-inhibited human fibroblasts		+ +	Growing	7
			Confluent	10
SV40-transformed human fibroblasts		−		28

[a] Average of three experiments (range in parentheses), values without symbols are the average of two experiments. Data from Hakomori, S. *Proc. Natl. Acad. Sci. U.S.A.* **67,** 1741 (1970).

significant reductions in GD$_{1a}$, GM$_1$, and GM$_2$, as a result of viral transformation. It is of interest to note that these higher gangliosides are absent from BHK cells (Table IV). The reader will recall that transformed hamster cells specifically were *not* agglutinable by SBA, in contrast to transformed rat, mouse, and human fibroblasts.

A subsequent elaboration of these experiments (10) has extended the observations concerning higher gangliosides to both SV40- and polyoma-transformed 3T3 lines and other lines of mouse embryo origin. These results were essentially the same as their previous observations: no change in hematoside (GM$_3$) and a decrease in GD$_{1a}$ and GM$_1$. Brady and Mora also analyzed the ganglioside patterns of several "spontaneous" transformants of 3T3 and other untransformed established lines. These cell lines appear to have "nearly normal" ganglioside patterns, despite their high cell densities and oncogenic potential, in contrast to the results observed by Hakomori and Murakami (25) and Hakomori (23) for spontaneously transformed BHK cells.

While the conflicts in the experimental results of these two groups have not been resolved at this date, it is clear that viral transformation of cultured mammalian cells results in significant alterations in their glycolipid components. The previously described experiments have been primarily confined to the study of whole cells, however. Only one of the reports has attempted

GLUCOSYL-GALACTOSYL-CERAMIDE	GALACTOSYL-GALACTOSYL-GLUCOSYL CERAMIDE	Gangliosides (μg/10 mg protein) HEMATOSIDE (GM$_3$)	GD$_{1a}$	GM$_2$	GM$_1$
8(7–9)[a]	12(8–16)[a]	46(44–48)[a]	<3[a]	0	<3[a]
5(3–7)[a]	30(29–31)[a]	52(50–54)[a]	<5[a]	0	<3[a]
8	0	45	0	0	0
9	0	56	0	0	0
18	<1	20	0	0	0
22	<1	18	0	0	0
0	0	42(40–44)[a]	15(14–16)[a]	0	10 (9–11)[a]
0	0	55(52–58)[a]	38(35–41)[a]	0	25(23–27)[a]
0	0	15	8	35(32–38)[a]	<3

to deal with the *availability* of the glycolipid components of the cell at its surface, and none has dealt with the glycolipid concentrations of subcellular fractions of transformed and normal cells. Progress in this important area will certainly depend on quantitative characterization of changes in the glycolipid fractions of subcellular components as a result of viral transformation and on a more extensive characterization of the glycolipid concentrations and availability at the surface of the cell membrane.

GLYCOPROTEINS OF NORMAL AND TRANSFORMED CELLS

The agglutinability changes observed for transformed cells have prompted the investigation into the glycopeptide constituents of the cell surface of normal and transformed cells, as well as their glycolipid components. The primary approaches to this problem have utilized *1*) preliminary subcellular fractionation, solubilization by detergent, and chromatographic separation of the components of the solubilized subcellular fraction, or *2*) enzymatic degradation of the surface of the cell, followed by a chromatographic separation of the solubilized products.

The first procedure has been employed to analyze the differences in the carbohydrate-containing macromolecules of the mitochondrial fractions of SV-3T3 and 3T3 cells (32). Cultures were extensively labeled with

Table V

Cellular Gangliosides in Transformed and Untransformed Cultured Mouse Cells

CELL LINE	TUMORI-GENICITY	CONTACT INHIBITION DIVISION	Gangliosides (μmoles/mg of protein)				
			GD_{1a}	GM_1	GM_2	GM_3 (HEMA-TOSIDE)	TOTAL
3T3 (harvested in saline)			2.4	2.6	1.8	4.0	10.8
	−	+ +					
3T3 (harvested in saline + EDTA)			2.8	2.8	2.4	4.8	12.8
SVPY-3T3 (harvested in saline)			0.6	0.8	0.4	4.8	6.6
	+ +	−					
SVPY-3T3 (harvested in saline + EDTA)			0.6	0.6	0.8	5.3	7.3

Data from Mora, P. T., *et al. Proc. Natl. Acad. Sci. U.S.A.* **63**, 1290 (1969).

^{3}H-glucosamine and ^{14}C-glucosamine, respectively. The cells were then harvested and pooled, and the mitochondrial fraction was isolated. The membranes were solubilized in 2% sodium dodecyl sulfate and fractionated on Sephadex G-150 columns. The results of this analysis are shown in Figure 8. There are clearly significant differences in the glycoprotein profiles of the mitochondrial fractions of transformed and untransformed cells. Five distinct peaks are shown in this analysis. When labeled nuclei and endoplasmic reticlum were also solubilized and chromatographed in this manner, two consistent results were observed: *1*) all subcellular fractions analyzed (nuclei, mitochondria, endoplasmic reticulum) had qualitatively similar elution profiles, like the one shown in Figure 8; and *2*) the relative labeling in the various peaks (larger amount of label in peaks 2 and 3 in transformed cells, the reverse for peaks 4 and 5) was also true for all three subcellular fractions. The glycoprotein patterns for surface membrane fractions solubilized in this manner were not reported, however, although preparation of a surface membrane fraction is described in the scheme of subcellular fractionation (53).

In order to obtain a more extensive resolution of the glycopeptides of transformed and normal cells, several groups of workers have first digested these subcellular fractions with proteolytic enzymes such as pronase, and then chromatographed the resulting doubly labeled fractions on Sephadex G-50. Nuclei, mitochondria, surface membrane, endoplasmic reticulum, and soluble glycoprotein have been investigated in this fashion (32). The elution

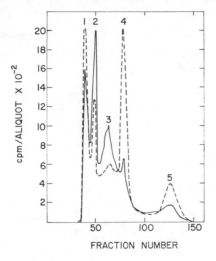

FRACTION NUMBER

Figure 8

Fractionation of mitochondrial membrane glycoproteins from transformed and untransformed fibroblasts on Sephadex G-150. 3T3 and SV-3T3 cells were labeled with ^{3}H-glucosamine and ^{14}C-glucosamine, respectively. Mitochondria were isolated by differential centrifugation. Membrane glycoproteins were solubilized in 2% sodium dodecyl sulfate and fractionated on a Sephadex G-150 column in the presence of 1% sodium dodecyl sulfate. Broken line indicates ^{3}H-containing glycoproteins (3T3 cells); solid line indicates ^{14}C-containing glycoproteins (SV-3T3 cells). (Data from Meezan, E., *et al. Biochemistry 8*, 2518, 1969.)

patterns obtained are similar for all fractions and consist of four fractions (Fig. 9). Labeling is more extensive in peaks 1 and 4 of the normal cells than of transformed cells in all the particulate subcellular fractions. Peak 2 is more extensively labeled in the transformed line. The most interesting qualitative shift occurs in peak 3 of the nuclear and mitochondrial fractions. In this case the two peaks do not overlap, but are offset to some degree (Fig. 9). This qualitative difference was *not* observed with surface membrane or endoplasmic reticulum elution profiles, however. The similar elution profiles of the various preparations suggest that different subcellular particulate fractions within a given cell line may have similar glycopeptides, and consequently similar monosaccharide transferases, involved in the synthesis of many cellular membrane structures. The authors themselves, however, raise the specter of significant cross-contamination among their various fractions as a result of the simplified and relatively uncharacterized subcellular fractionation scheme employed in these studies.

In a similar set of experiments the glycopeptides of the surfaces of BHK and PY-BHK cells have also been examined. Cultures were labeled with

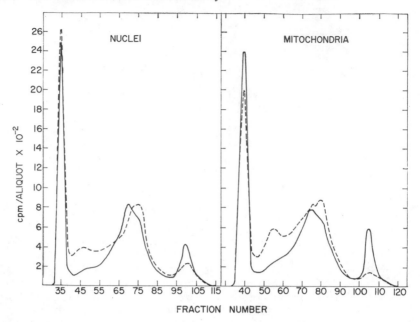

Figure 9

Fractionation of membrane glycopeptides from normal and SV40-transformed 3T3 fibroblasts on Sephadex G-50. ³H-glucosamine labeled SV-3T3 cells and ¹⁴C-glucosamine labeled 3T3 cells were mixed and fractionated into nuclei, mitochondria, surface membrane, endoplasmic reticulum, and soluble fractions (53). Each subcellular fraction containing doubly labeled glycoproteins was digested with pronase. The resulting mixture of glycopeptides was fractionated on a Sephadex G-50 column. Solid line indicates ³H-containing glycopeptides (SV-3T3 cells); broken line indicates ¹⁴C-containing glycopeptides (3T3 cells). (Data from Meezan, E., *et al. Biochemistry 8*, 2518, 1969.)

either ^{3}H-fucose or ^{14}C-fucose for 3 days. Glycoproteins were removed from the surface of the monolayers by digestion with trypsin. After dialysis, the material released by trypsin was subjected to pronase digestion, then chromatographed on G-50 Sephadex (11). The elution profiles of such pronase-digested fractions of trypsin-releasable surface membrane (Fig. 10) are quite different from the previously described results with pronase-digested isolated membranes of SV-3T3 and 3T3. There is a significant difference in the elution patterns of the glycopeptides of normal and transformed cells. The membranes of transformed cells appear to yield glycopeptides of higher molecular weight than do those of their parent strain.

These studies do not yet reveal any *specific* alterations in unique glycoprotein species which accompany the phenomenon of transformation. They

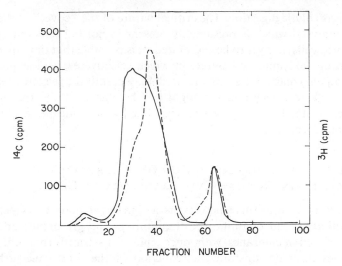

Figure 10

Fractionation of surface membrane glycopeptides from normal and polyoma-transformed BHK fibroblasts on Sephadex G-50. PY-BHK cells and BHK cells were grown as monolayers for 72 hr in the presence of ³H-fucose or ¹⁴C-fucose, respectively. After being washed, the cells were incubated with trypsin (1 mg/ml). The tryptic digests were pooled and digested with pronase. The glycopeptides were then fractionated on a Sephadex G-50 column. Solid line indicates ³H-containing glycopeptides (PY-BHK cells); broken line indicates ¹⁴C-containing glycopeptides (BHK cells). (Data from Buck, C., *et al. Science 172*, 169, 1971.)

do, however, demonstrate the existence of qualitative and quantitative changes in the glycoprotein component of the cell surface and hold out the possibility that they will, in conjunction with other procedures designed for the separation of such macromolecules, permit a more complete analysis of this aspect of the surfaces of normal and transformed cells.

Recently Sakiyama and Burge (41) have reported, in a preliminary communication, that 3T3 cells transformed by SV40 lack a surface glycoprotein that is present in the 3T3 parent line. Cells were labeled with either ³H- or ¹⁴C-glucosamine and extracted with isotonic saline containing 10 mM ethylenediamine tetraacetate (EDTA). Acrylamide-gel electrophoretic analysis of the EDTA extracts demonstrated that a protein of molecular weight approximately 100,000 was present in 3T3 extracts, but not in the SV40-transformed line.

Once again the reader's attention is drawn to the earlier "unmasking" experiments of Burger (13) and Hakomori (26). The exposure of cryptic agglutination and antigenic sites on untransformed cells was brought about by mild proteolytic digestion. The critical nature of the "cover" of the cryptic site, the material which is presumably present in normal cells but absent in transformed cells, has yet to be elaborated. It is possible that the experimental approach of Sakiyama and Burge, in which selective extraction procedures rather than enzymatic digestions are used, may permit the selective solubilization of this factor. Such a factor might well be able to inhibit the agglutination reaction not by acting as a hapten, but by blocking the agglutination site on transformed cells.

Synthesis and Degradation of Glycoproteins and Glycolipids in Normal and Transformed Cells

As we have seen, there now exists a significant body of evidence that links qualitative and quantitative alterations in the glycolipids and glycoproteins of cell membranes with oncogenic virus transformation. Several laboratories have demonstrated a reduction in the total concentration of gangliosides in transformed cells (23, 34). Similarly, the total amounts of sialic acid in SV40-transformed 3T3 cells and polyoma-transformed BHK cells are reduced when compared with their respective parent cell strains (23, 34, 35). N-acetyl-galactosamine is also reduced in SV-3T3 cells compared with 3T3 cells (32, 53), while a reciprocal elevation of N-acetyl-glucosamine occurs. Two alternative explanations provide equally meager means of explaining these observations. The total content of carbohydrate-containing protein or lipid constituents might be reduced in transformed cells by *1*) a reduction in the levels of the relevant biosynthetic enzymes, the glycosyl transferases, or *2*) by an elevated level of the appropriate catabolic enzymes, the glycosidases. The mechanism of synthesis of glycosyl polymers is not a well-understood area of biochemistry. Glycosyl transferases are thought to exhibit a unique specificity for their receptors. Thus transfer of a single monosaccharide might be carried out by differing transferases depending on the specific glycoprotein or glycolipid acceptor in question (39). Despite these difficulties, several investigators have measured the activities of glycosyl transferases and glycosidases in transformed and normal cells in order to determine the mechanism resulting in lowered cellular concentrations of carbohydrate-containing macromolecules.

Glycosyl Transferases

Bosmann *et al.* (6) have measured the levels of two fucosyl transferases, an N-acetyl-galactosaminyl transferase and a galactosyl transferase, from both SV40- and polyoma-transformed 3T3 cells. These enzymes are part of a particulate complex and are extracted with a nonionic detergent (22). The specific activities of each of these enzymes were *greater* in the transformed cells than in 3T3 cells. These workers conclude that increased glycosyl transferase activity is a concomitant of viral transformation, despite the reduced concentration of N-acetyl-galactosamine present in transformed cells. Grimes has reexamined the level of fucosyl transferase and found results contradictory to those reported earlier. In this case the fucosyl transferase activity of 3T3 cells was nearly twice that of SV40-transformed cells (21). Several possible explanations exist for these disparate results: *1*) fucosidase activity is apparently more extensive in the latter report, *2*) the method of preparation of the glycosyl transferases differs, and *3*) the always real possibility exists that these activities may vary from one clone to another of the same cell strain. While these disclaimers may make it possible to reconcile the data to some small degree, they do not permit us to draw any conclusions concerning the relation between oncogenic viral transformation and glycosyl transferase activity.

Probably the most carefully characterized monosaccharide of the cell surface is sialic acid. Studies performed in several laboratories agree that virally transformed cells have only approximately 60% the sialic acid concentration found in their respective parent lines (21, 32, 35). Sialic acid transferase activity of particulate preparations from 3T3 and SV-3T3 cells has recently been measured (21). The characteristics of the transferase systems in untransformed and SV40-transformed 3T3 cells appear to be identical. The specific activity of the sialic acid transferase of SV-3T3 cells was reduced to approximately 60% of that of 3T3 cells, however, regardless of whether growing or confluent cells were assayed. In this case there appears to be a direct correlation between the specific activity of sialic acid transferase and the concentration of cell-bound sialic acid.

Studies of the chemical composition of transformed cells have indicated that all the higher sialogangliosides are reduced in concentration as a result of viral transformation. Cumar *et al.* (17) have measured the levels of a key enzyme in the biosynthesis of higher gangliosides, uridine diphosphate N-acetyl-galactosamine:hematoside N-acetyl-galactosaminyl transferase. This enzyme adds the first of several monosaccharides to hematoside in the course of synthesizing many of the higher gangliosides. Their results show that microsomal preparations of either polyoma- or SV40-transformed cells have extensively reduced capacity for the transfer of N-acetyl-galactosamine from uridine diphosphate N-acetyl-galactosamine to hematoside. The integration of either viral genome results in a profoundly reduced specific activity for this enzyme.

Glycosidases

A wide variety of glycosidases have been investigated in virally transformed cells in order to determine whether the reduced levels of cellular monosaccharides might be the result of increased glycosidase activity. Seven specific glycosidases were examined in 3T3, SV-3T3, PY-3T3, and SVPY-3T3 (doubly transformed) cells. For all the glycosidases examined, transformed cells had greater specific activities than 3T3 cells. The relative increases in specific activities of the various glycosidases ranged between 1.4 and 4.8 times that of the corresponding glycosidase activity of 3T3 cells (5). These analyses have recently been extended to the BHK system, with essentially similar results. The levels of 10 glycosidases were found to be elevated in PY-BHK cells when compared with BHK cells. In the same study the transformed clone also possessed a greater specific activity for glycoprotein N-acetyl-galactosamine transferase (7).

It is not known whether these cellular glycosidases act extracellularly. If this were the case, the increased levels of glycosidases present in transformed cells might well help to account for the reduced level of surface glycoproteins and glycolipids in transformed cells. The data on the synthesis, content, and degradation of cellular glycoproteins and glycolipids are both sparse and contradictory at this point. It is clear, however, that significant differences exist when comparisons are drawn between normal and virally transformed cells in the ability to transfer monosaccharides into macromolecular complexes, in the qualitative and quantitative constitution of such complexes, and in the potential for the degradation of such complexes. The nature of the relation between these synthetic and catabolic capabilities, which appear to be altered so significantly as a result of viral transformation, and the oncogenic potential of the transformed cell is one of the most significant problems facing the cell biologist today.

Conclusions

The study of the molecular alterations of mammalian cell surfaces as a result of viral transformation is clearly an embryonic, but burgeoning, field of research. The development of cell culture techniques, clonal cell populations, and oncogenic transformation under rigidly controlled conditions has essentially created a new discipline of molecular oncology. The fortuitous discovery that a group of proteins, the plant agglutinins, possesses the ability to discriminate between "normal" cells and those transformed by papova viruses has provided biochemists with a series of probes with which to characterize the architecture of the surface of these cells. The use of these agglutinins and specific antisera to cellular membrane constituents should provide clues to the spatial arrangements of the macromolecular components of the cell surface of transformed and normal cells. Competition experiments between various agglutinins and antisera with normal cells, trypsinized cells, and virally transformed cells will be able to provide information on the relative proximity and density of the various receptors. Fluorescence-labeled agglutinins and antisera will provide information on localization of such determinants; electron microscopic studies with ferritin- or peroxidase-labeled molecules will provide ultrastructural correlates.

Viral transformation results in a decreased content of higher sialogangliosides and an altered pattern of glycolipid and glycoprotein synthesis and degradation. The specific alterations that occur in the metabolism of these macromolecules are not yet clear; the literature is current and consequently contradictory. Changes both in the chemical constitution of the surface of transformed cells and in the regulation of synthesis of the components of cell membranes are, however, certain. While the current data appear to depend on cell line, virus, and laboratory of choice, a consensus should soon emerge on these questions.

The importance of isolation of "flat" revertants of virally transformed cells cannot be overemphasized. These cells, which exhibit contact inhibition of division, reduced oncogenic potential, and revertant surface properties, contain the viral genome. Rescue of the genome by cell fusion, and transformation of a normal cell with the rescued virus, result in a transformed cell that does not exhibit contact-dependent inhibition of division, which has exposed agglutination sites, and which gives rise to tumors. *The stable integration of the viral genome into the genetic apparatus of the host cell is therefore a necessary but not a sufficient condition to bring about the increase in tumorigenicity, loss of contact inhibition of division, and change in surface membrane properties associated with stably transformed cells.* Some interaction between the integrated viral genome and the cellular genome is required for these manifestations of transformation to be expressed. This interaction may be interfered with by a genetic event *associated with the cellular genome.* Great progress in the understanding of the nature and control of the events leading to the permanent integration into and maintenance of the viral genome in the genetic apparatus of the cell has been achieved. This chapter has described recent advances in the description of alterations of surface properties of cells which accompany and appear to be necessary for the expression of malignant potential and loss of growth regulation that are associated with the transformed state. An explanation of the molecular mechanism by which the integrated viral genome alters the genetic control of the expression of surface properties of its host cell, however, has not yet begun to emerge.

The observations regarding altered agglutinability or chemical composition of transformed clones is to some degree subject to the same criticism as the discussion of neoplasias presented earlier in this chapter. The objectives and limitations of such studies on these cells need to be clearly understood. From such studies we can hope to learn *1*) the effect of the presence of the viral genome on the structure of the cell surface, *2*) the correlative relation between virally induced alterations in the surface structure and composition of cells and their relation to malignant potential, and *3)* possibly the relation between manipulation of the surface of a potentially malignant cell and the effect of such manipulation on the ability to promote tumor formation. Such studies will not, however, determine the mechanism of viral oncogenesis, i.e., the manner in which the viral genome influences the cellular genome's regulation of its surface properties, growth characteristics, and malignant potential. Answers to these questions will result from comparative metabolic and enzymatic studies of the sorts alluded to earlier in this chapter. The study of the correlates of surface structure, saturation density, and malignant potential of clonal transformed cells holds as its outstanding and elusive promise, however, one of the keys to the phenotypic reversal of the oncogenic state.

NOTE ADDED IN PROOF

This chapter covered literature through May 1971 at the time it was sent to the publisher. Since then a number of publications have appeared which have been concerned with the relation of agglutinin binding and transformation. Several groups have labeled purified WGA and/or ConA with tritium or ^{125}I and demonstrated that normal and virally transformed fibroblasts bind labeled lectins (Cline, M. J., and Livingston, D. *Nature* **232,** 155, 1971; Ozanne, B., and Sambrook, J. *Nature* **232,** 156, 1971; Arndt-Jovin, D., and Burg, P. *J. Virol.* **8,** 716, 1971). Cells which are nonagglutinable (i.e., nontransformed) and virally transformed cells appear to have lectin-binding sites. A recent study utilizing ferritin-conjugated ConA has proposed a rearrangement of lectin-binding sites leading to agglutinability of transformed cells (Nicolson, G. L. *Nature* **233,** 244, 1971). Recently Inbar, Ben-Bassat, and Sachs (*Proc. Natl. Acad. Sci. U.S.A.* **68,** 2748, 1971) have described results suggesting a specific membrane-associated metabolic activity in transformed cells which is responsible for ConA agglutinability. These authors have extended their experiments and elaborated on their model of the structural and metabolic membrane changes which accompany malignancy and agglutinability (Inbar, M., Ben-Bassat, H., and Sachs, L. *Nature* **236,** 3, 1972).

The nature of the changes of transformed cell membranes leading to altered agglutinability will be more easily understood when the receptors for lectins have been characterized. Allan, Auger, and Crumpton's use of sepharose-bound ConA has led to the isolation of several ConA-binding glycoproteins from pig lymphocytes (*Nature* **236,** 13, 1972). The application of similar methodology to the membrane of normal and transformed cells should yield valuable information on the relation between lectin receptors, agglutinability, and malignancy.

REFERENCES

1. Aaronson, S. A., and Todaro, G. J. *Science* **162**, 1024 (1968).
2. Assman, F. *Arch. Gesamte Physiol.* **137**, 489 (1911).
3. Aub, J. C., Tieslau, C., and Lankester, A. *Proc. Natl. Acad. Sci. U.S.A.* **50**, 613 (1963).
4. Bird, G. W. C. *Br. Med. Bull.* **15**, 165 (1959).
5. Bosmann, H. B. *Exp. Cell Res.* **54**, 217 (1969).
6. Bosmann, H. B., Hagopian, A., and Eylar, E. H. *J. Cell Physiol.* **72**, 81 (1968).
7. Bosmann, H. B., and Pike, G. Z. *Life Sci.* **9**, 1433 (1970).
8. Boyd, W. C. *Fundamentals of Immunology*, 2d ed. Interscience, New York, 1947.
9. Boyd, W. C. *Introduction to Immunochemical Specificity*. Interscience, New York, 1962.
10. Brady, R. O., and Mora, P. T. *Biochim. Biophys. Acta* **218**, 308 (1970).
11. Buck, C., Glick, M. C., and Warren, L. *Science* **172**, 169 (1971).
12. Burger, M. M. *Nature* **219**, 499 (1968).
13. Burger, M. M. *Proc. Natl. Acad. Sci. U.S.A.* **62**, 994 (1969).
14. Burger, M. M. *Nature* **227**, 170 (1970).
15. Burger, M. M., and Goldberg, A. *Proc. Natl. Acad. Sci. U.S.A.* **57**, 359 (1967).
16. Burger, M. M., and Noonan, K. D. *Nature* **228**, 512 (1970).
17. Cumar, F. A., Brady, R. O., Kolodny, E. H., McFarland, W., and Mora, P. T. *Proc. Natl. Acad. Sci. U.S.A.* **67**, 757 (1970).
18. Defendi, V. *Proc. Soc. Exp. Biol. Med.* **113**, 12 (1963).
19. Dulbecco, R. *Science* **166**, 962 (1969).
20. Goldstein, I. J., Hollerman, C. E., and Smith, E. E. *Biochemistry* **4**, 876 (1965).
21. Grimes, W. J. *Biochemistry* **9**, 5083 (1970).
22. Hagopian, A., Bosmann, H. B., and Eylar, E. H. *Arch. Biochem. Biophys.* **128**, 387 (1968).
23. Hakomori, S. *Proc. Natl. Acad. Sci. U.S.A.* **67**, 1741 (1970).
24. Hakomori, S., Koscielak, J., Bloch, K. J., and Jeanloz, R. W. *J. Immunol.* **98**, 31 (1967).
25. Hakomori, S., and Murakami, W. T. *Proc. Natl. Acad. Sci. U.S.A.* **59**, 254 (1968).
26. Hakomori, S., Teather, C., and Andrews, H. *Biochem. Biophys. Res. Commun.* **33**, 563 (1968).
27. Inbar, M., and Sachs, L. *Proc. Natl. Acad. Sci. U.S.A.* **63**, 1418 (1969).
28. Kalb, A. J., and Levitzki, A. *Biochem. J.* **109**, 669 (1968).
29. Koch, P. A., and Sabin, A. B. *Proc. Soc. Exp. Biol. Med.* **113**, 4 (1963).
30. Lis, H., Sela, B., Sachs, L., and Sharon, N. *Biochim. Biophys. Acta* **211**, 582 (1970).
31. Lis, H., Sharon, N., and Katchalski, E. *J. Biol. Chem.* **241**, 684 (1966).
32. Meezan, E., Wu, H. C., Black, P. H., and Robbins, P. W. *Biochemistry* **8**, 2518 (1969).
33. Moore, E. G., and Temin, H. M. *Nature* **231**, 117 (1971).
34. Mora, P. T., Brady, R. O., Bradley, R. M., and McFarland, V. W. *Proc. Natl. Acad. Sci. U.S.A.* **63**, 1290 (1969).

35. Ohta, N., Pardee, A. B., McAuslan, B. R., and Burger, M. M. *Biochim. Biophys. Acta* **158,** 98 (1968).

36. Pollack, R., and Burger, M. M. *Proc. Natl. Acad. Sci. U.S.A.* **62,** 1074 (1969).

37. Pollack, R., Green, H., and Todaro, G. *Proc. Natl. Acad. Sci. U.S.A.* **60,** 126 (1968).

38. Pope, J. H., and Rowe, W. P. *J. Exp. Med.* **120,** 121 (1964).

39. Roseman, S. *Biochemistry of Glycoproteins and Related Substances.* Ed. E. Rossi and E. Stoll. Karger, Basel, 1966, p. 244.

40. Sabin, A. B., and Koch, M. A. *Proc. Natl. Acad. Sci. U.S.A.* **52,** 1131 (1964).

41. Sakiyama, H., and Burge, B. W. *Abstracts; 1970 Tumor Virus Meeting.* Cold Spring Harbor Laboratory, Cold Spring Harbor, 1970.

42. Sambrook, J., Westphal, H., Srinivasan, P. R., and Dulbecco, R. *Proc. Natl. Acad. Sci. U.S.A.* **60,** 1288 (1968).

43. Sela, B., Lis, H., Sharon, N., and Sachs, L. *J. Membrane Biol.* **3,** 267 (1970).

44. Smith, F., and Unrau, A. M. *Chem. Ind.* 881 (1959).

45. Stillmark, H. Der giftige Eiweisskorper Ricin: Seine Wirkung auf das Blut. Inaugural Dissertation, Dorpat, 1888.

46. Stoker, M., and MacPherson, I. *Virology* **14,** 359 (1961).

47. Stoker, M. G., and Rubin, H. *Nature* **215,** 171 (1967).

48. Sumner, J. B., and Howell, S. F. *J. Bacteriol.* **32,** 227 (1936).

49. Svennerholm, L. *J. Neurochem.* **10,** 613 (1963).

50. Svoboda, J. *Int. Rev. Exp. Pathol.* **5,** 22 (1966).

51. Todaro, G. J., and Green, H. *J. Cell Biol.* **17,** 299 (1963).

52. Tunis, M. *J. Immunol.* **92,** 864 (1964).

53. Wu, H. C., Meezan, E., Black, P. H., and Robbins, P. W. *Biochemistry* **8,** 2509 (1969).

54. Yariv, J., Kalb, A. J., and Levitzki, A. *Biochim. Biophys. Acta* **165,** 303 (1968).

Index

References to figures appear in italic.

A

A₂ group: of phospholipases, 18

Aaronson, S. A., 473, *474*

Abrams, A., 34

Absorbance, 230, 280

Absorption: optical, 279–80

Absorption spectroscopy: measurement units, 230–34

Absorption statistics: in CD and ORD measurements of suspensions, 254–263

Accumulation, against concentration gradient: by microorganisms, 332

Acetic acid: dispersion of membrane proteins by, 37

Acetyl CoA carboxylase: in fatty acid synthesis, 10, 12
  metabolic regulation of, 12, 15

Acetyl coenzyme A (acetyl CoA): in fatty acid synthesis, 8, 12

Acids: fatty. *See* Fatty acids
  organic, dispersion of membrane proteins by, 37

Active transport, 290–91
  accompanying covalent modification, 325–26
  of amino acids, 313–18
  binding proteins in, 313–17, 327
  components of systems for, 326–27
  energy sources for, 325, 327–33
  influences on rate, 355–57
  of ions, in bacteria, 318–20
  in *M. laidlawii* membranes, 464–65, 467
  models of, 337–42
  use of metabolic energy in, 335–42

Acyltransferase mutants: of *E. coli*, 349–50

Adam, N. K., 136

Aden, A. L., 264

Adenosine triphosphate (ATP): as energy source for active transport, 325–31
  required for fatty acid synthesis, 10–11

Adhesiveness: of platelets, 256

Adipose tissue: lipids associated with, 3

Adler, Howard, *375*

Adsorbed proteins: as contaminants in membrane preparations, 89–90

Adsorption: alteration of particle density by, 92–93
  of soluble membrane proteins, 29

Aerobacter aerogenes: PTS system in, 299, 304

Affinity chromatography: in subcellular fractionation, 85

Agglutinability: of cells, 480

Agglutination sites: cryptic, on untransformed cells, 478–80
  exposure of, 480–84

Agglutinins: of plants, 475–84

Aggregation: alteration of effective particle size by, 91
  of proteins, in SDS—PAGE system, 47

Alamecithin: in NMR studies, 176–77

Algae: chlorophyll synthesis in, 419–22
  See also names of individual species

Alkaline hydrolysis: measurement of, in γ-stearolactone monolayers, 143

Alkaline phosphodiesterase I: as marker for plasma membrane vesicles, 110

Allison, David, *375*

Ames, G. F., 314–16

Amino acids: in membrane proteins, 50–56, 466
  optical activity of side chains, 237–39
  transport, 313–18

δ-Amino levulinic acid synthetase: in mitochondria, 405

Amoeba proteus: isolation of surface membrane, 107

Amphipathic lipids, 3

Amphotericin B: formation of channels in bilayers, 160